Combined Methods for Elliptic Equations with Singularities, Interfaces and Infinities

Mathematics and Its Applications

Managing Editor:

M. HAZEWINKEL
Centre for Mathematics and Computer Science, Amsterdam, The Netherlands

Volume 444

Combined Methods for Elliptic Equations with Singularities, Interfaces and Infinities

by

Zi Cai Li

Department of Applied Mathematics,
National Sun Yat-sen University,
Kaohsiung, Taiwan

KLUWER ACADEMIC PUBLISHERS
DORDRECHT / BOSTON / LONDON

A C.I.P. Catalogue record for this book is available from the Library of Congress.

ISBN-13: 978-1-4613-3340-1   e-ISBN-13: 978-1-4613-3338-8
DOI: 978-1-4613-3338-8

Published by Kluwer Academic Publishers,
P.O. Box 17, 3300 AA Dordrecht, The Netherlands.

Sold and distributed in North, Central and South America
by Kluwer Academic Publishers,
101 Philip Drive, Norwell, MA 02061, U.S.A.

In all other countries, sold and distributed
by Kluwer Academic Publishers,
P.O. Box 322, 3300 AH Dordrecht, The Netherlands.

Printed on acid-free paper

To my friends,

John Fraser and Elizabeth MacCallum.

CONTENTS

The whole of science is nothing
more than a refinement
of everyday thinking.
——— *Physics and Reality (1936)* ———

Albert Einstein
(1879–1955)

PREFACE

Many numerical methods exist for solving elliptic boundary value problems, such as the finite element method (FEM), the finite difference method (FDM), the finite volume method (FVM), the collocation method (CM), the boundary element method (BEM), the least squares method (LSM), and the Ritz-Galerkin method (RGM) using singular and analytic solutions, etc. Each method possesses its own advantages and drawbacks. Very often, one method can solve certain elliptic problems efficiently, but not others. Single methods encounter some difficulties in handling the complicated problems, those with singularities in particular. Consequently, employing different treatments and numerical methods simultaneously is inevitable in solving elliptic problems.

This book presents a number of useful combinations, for instance, the combination of various FEMs, the combinations of FEM-FDM, RGM-FEM, RGM-FDM, etc. The combined methods have many advantages over single methods: high accuracy of solutions, less CPU time, less computer storage, easily coupling with singularities as well as the complicated boundary conditions, etc. Since coupling techniques are essential to combinations, various matching strategies among different methods are discussed carefully in this book. We provide the matching rules so that optimal convergence, even superconvergence, and optimal stability can be achieved, and also warn of the matching pitfalls to be avoid.

Moreover, the combined methods are also applied to the unbounded domain problems and studied in advance topics, such as the global convergence rates and the domain decomposition methods. This book covers the important research on combinations in the last two decades, and integrates both theory and application. Anyone who knows the numerical methods such as FEM, FDM, BEM, RGM, etc., can understand and apply the combinational algorithms easily. This book may be chosen as a textbook for graduate courses or as a reference book for research in numerical computation.

This book consists of 16 chapters; each chapter is integrated as an indispensable part of the book. Nevertheless, each chapter is also written to stand as an independent research report on an important topic in combinations. Hence readers with specific interests may directly refer the related chapters without much difficulty.

The majority of the material in this book is adapted from my collaborative research work. I also point out important contributions of others by quoting them briefly. A few methods are described in detail: for example, the conformal transformation method of Whiteman and Papmichael(1972), and Rosser and Papamichael (1975), and the Lagrange multiplier of Babuska(1973a), Liang and He(1992), etc. I apologize for other significant contributions not properly recognized in this book. In fact, the combination of FEM and BEM of Yu (1993) in singularity problems is a very significant method; but only described in Section 2.8. Moreover, the local refinements of FEM are the most popular approach for singularity problems with extensive publications. Also the singular functions implemented to FEM by Fix, Gulati and Wakoff (1973) have wide application in singularity problems. Although the refinements and the singular function method can also be categorized into the generalized combinations, this book focuses on the combinations of the important and popular methods which stand already as individual numerical methods.

The principle of selecting materials is based on the guidance of the series of Mathematics and its Applications:
    "Stimulating rather than definitive" and
    "intriguing rather than encyclopedic",
by M. Hazewinked in Series Editor's Preface for the book of Cordero, Dodson and Leon (1989). Representative and heuristic methods are our main concern. For example, the combinations of RGM-FEM are typical, and can be extended to other kinds of combinations. The combination of RGM-FDM in Chapter 11, with emphasis on superconvergence, is representative of other combined methods. Since the unbounded domain and interface problems may also be treated as application of combined methods, and since they have many common characteristics, we describe only the algorithms for unbounded domain problems in Chapters 12, 13 and 16. For combinations for real material interface problems, refer to my previous book Li(1990). The combinations involving artificial interfaces appear almost everywhere in this book. Certainly, the techniques in this book can be easily applied to interface problems.

In fact, combining strategies is commonplace already. For example, the local refinement elements are combined with the quasiuniform elements in singularity problems. In the domain decomposition methods, an entire PDE problem is split into several subproblems; they are then solved separately and integrated together through iterations. This is, indeed, a kind of combination. It is worth noting that parallel implementation of numerical methods can never reduce the total CPU time; but the combinations in this book may greatly reduce the CPU time. We should also point out that the research on combined methods is in its infancy, compared to the study of domain decomposition methods and superconvergence. I sincerely hope that this book may stimulate further developments in this rich subject.

To satisfy the requirements of both mathematicians and engineers is one of the objectives of this book. Compromise in exposition for different sciences is really challenging. Although this text is mathematical in nature, engineers may still find this book useful. First, the most important ideas and combinations are stated concisely, and the significant results are described clearly. Users interested mainly in computation may glance at the theorems and skip all proofs therein. Second, the techniques of programming and derivation for the coupling equations in combinations are also demonstrated in Section 5.2 and 9.5. Moreover, in most chapters, we provide interesting numerical examples. Third, the most accurate solutions of Motz's problem are given. Motz's problem and the crack–infinity problem serve as testing models for new numerical methods of singularity problems. Lastly, we also include many important references in computation and engineering.

Kaohsiung   Zi Cai Li
January 1998

ACKNOWLEDGEMENTS

I am obliged to Professors G. F. D. Duff, T. D. Bui, R. Mathon, G.P. Liang, P. Sermer, K. C. Chou, K. Feng, I. Babuska, G. Strang, J. R. Whiteman, Q. Lin, Z. C. Shi, N. N. Yan, T. T. Lu, S. Wong, L. K. Steinrauf, and to E. MacCallum for their collaboration or valuable suggestions on the research work related to this book. I also express my great thanks to Professor N. N. Yan for carefully reading this manuscript and providing many valuable suggestions. Moreover, I am indebted to C. S. Chang for the typing, and C. C. Hung and S. Y. Wang for the computer figures of this book.

I am grateful to Mr. J. R. Martindale and others at Kluwer Academic Publishers for their collaboration, and in particular for their great encouragement during the lengthy period required to bring this book to fruition. I also wish to thank to Professors V. W. C. Liu and N. J. Ho of University of Sun Yat-sen for their support during the writing process.

Finally, I wish to express my heartfelt appreciation to my wife Jane Guo and sons James H. Lee and Philip H. Li for their support in the face of the many challenges we encountered prior to and during the preparation of this book.

This book and the related research were supported in part by the National Science Council of Taiwan, the Natural Sciences and Engineering Research Council of Canada, and the Fonds pour la Formation de Chercheurs de l'Aide à la Recherche de Quèbec.

ACKNOWLEDGMENTS

Singularity is almost invariably a clue.
The more featureless and commonplace a crime is,
the more difficult is it to bring it home.
——— *The Adventures of Sherlock Holmes (1891).*
                The Boscombe Valley Mystery ———

Sir Airthur Conan Doyle
(1859–1930)

INTRODUCTION

In this book, we will develop the combined methods and apply them to solving singularity problems of elliptic equations. Two important questions arise:

1. Which numerical methods may be combined together? The FEM, FDM, RGM, FVM, etc., may all be combined simultaneously to solve one elliptic boundary value problem. As a model of combinations we combine the most popular FEM and the most efficient Ritz-Galerkin method (RGM) using particular solutions. The admissible functions used in FEM and RGM are most distinct.

2. How can different numerical methods be matched together? This depends on efficient coupling techniques. The objective of analyzing coupling techniques is to achieve optimal convergence rates and optimal numerical stability.

These two questions form the themes of research on the combined method, and hence are crucial in this book. Let us explain this more.

Two methods that have nothing in common can not be combined with each other. To create a combined method, a linkage between two different methods must first be found so that they can be integrated. In this book, the Galerkin description is chosen as a general framework to link various methods. So we intend to interpret different numerical methods as special cases of Galerkin methods using different admissible functions and different integration rules. Hence with Galerkin methods, many numerical methods stated in Chapter 1 may be combined. Of course, other approaches, such as the residual method or the least squares method can also be chosen as a uniform framework to combine different methods as explained in this book. The renovated combined methods may be applied to hyperbolic equations, which, however, are beyond the scope of this book.

In combinations different numerical methods are used in different subdomains. Since different admissible functions do not conform along the common boundary Γ_0 of the different methods used, the coupling techniques on Γ_0 become the most important issues. Since we may "borrow" the methods and analysis for the interior subdomains from the existing textbooks in numerical analysis, we will concentrate on efficient coupling and analysis on Γ_0.

Matching different methods is a fundamental aspect that will be fully developed in Parts II and III. In Part II we employ the straightforward direct constraints. Since the conforming functions for elliptic equations of second order are continuous along Γ_0, we may simply enforce a solution continuity condition at certain isolated points on Γ_0. However, the solution noncontinuity can not be removed on the rest of Γ_0 completely. Hence the nonconforming constraints, techniques and combinations are simply referred to in this book. Note that this kind of nonconforming is slight as pointed out by Feng(1979), as

"the locus root snaps but its fibres stay joined, apparently separated,
  but actually still connected",   藕斷絲連 *(Chinese proverb).*

Other names, such as the Mortar nonconforming method is also referred by Widlund(1997) and Bernardi, Maday and Patera (1993). It is due to such slight nonconformity that the coupling errors which result do not destroy the optimal convergence rates and optimal numerical stability. It is worth noting that the nonconforming constraints are basic in coupling, although the more effective and sophisticated coupling strategies are developed in Part III.

The optimal goal in combinations implies that each method contributes the maximum and balances others. Since the efficiency and flexibility vary with different methods, we may detach the most suitable methods to attack different block-houses of the problems, such an idea was suggested in Strang and Fix (1973). Take the problems with angular singularity as an example. The RGM using the local singular solutions is most suitable for the subdomains near the singular points; the remainder may also be undertaken by the FEM (or FDM). The RGM and FEM have the exponential and polynomial convergence rates respectively. Therefore, fewer singular functions in RGM can balance many finite elements in FEM for a given accuracy. This is an efficient approach for singular problems.

Let us consider the alternative in coupling, where one method may contribute its best but ignore collaboration with other methods. For example, we may choose singular functions as many as possible near the singular subdomain in RGM, even when the finite elements are not considerably small. Initially, we thought, the overwork of RGM would not damage the global accuracy. To our astonishment, however, we discovered that the global energy errors are $O(L\,h^2)$ as derived in Chapter 4, where L is the number of singular functions and h is the maximal boundary of finite elements. Even from the viewpoint of convergence, the solution errors $O(L\,h^2) \longrightarrow \infty$ if L grows faster than $O(1/h^2)$.

The same surprise happens in coupling linear and k-order Lagrange elements in Chapter 5. The boundary effect leads to the energy errors $O\left(\frac{h^{k+1}}{H^{k+\frac{1}{2}}}\right)$, where h and H are the maximal boundary lengths of quasiuniform linear and k-order elements, respectively. A straightforward matching of two methods is that all the element modes along Γ_0 are continuous by choosing the same element nodes or by enforcing direct constraints. Then $H = O(h)$, which leads, unfortunately, to the reduced convergence rate $O(h^{\frac{1}{2}})$. The k-order FEM hinders the combination by excessive calculational effort! In this book we suggest a harmonious collaboration among different methods in combinations. Both the optimality and counterexamples illustrated in coupling constitute a distinctive feature in the analysis of this work, to provide useful guidance in practical application.

The significant advantage of combinations is that the most accurate numerical solutions can be achieved in BAM using local particular solutions given in Chapter 3, since particular solutions may also be chosen to satisfy most effectively the boundary conditions. The most efficient solutions rely on profound understanding of the problems which allow the choice of the most suitable functions. Any specialization of problems, if fully used, is rewarded by efficiency.

The BAM is developed from the classic Galerkin method using uniform analytical functions or particular solutions. Now the use of local particular solutions is permitted in BAM, RGM–FEM, RGM–FDM, etc. This impacts not only numerical analysis but also applied mathematics, because all the analytical expansions, asymptote behavior of PDE solutions in textbooks and research papers may be imbedded into combinations directly. Obviously, such combinations will stimulate more discoveries of solution behavior. The combinations have, indeed, built a bridge between numerical method and applied mathematics. The RGM using the orthogonal polynomials can also be regarded as a spectral method.

For solving the complicated problems, including singularities, interfaces and infinities, we recommend the combined methods, a harmony of numerical methods, where each method contributes the most through collaboration. The performance of combinations is best if the coupling is suitable.

In fact, the combinations of variants, different modifications, and techniques have been growing in all the numerical methods. Once the methods are developed individually, it is the time to combine them to produce better performance: *The solo is good; but the chorus is more pleasing.*

This book covers four parts including 16 chapters and references:
   Part I: Singularities, Treatments and Combinations
   Part II: Combined Methods
   Part III: Coupling Strategy
   Part IV: Application and Advanced Topics
Parts II and III are the core of this book, to respond to the two questions raised above.

Briefly. Part I presents the general ideas: singularity behavior of Laplace equations, many treatments of singularity problems, various combined algorithms and the related error analysis. Extensive references are quoted; interested readers may refer to them for further study or practical applications.

Part II presents three typical combinations: the boundary approximation method (BAM), the combinations of RGM-FEM, and the combinations of various FEMs, based on the nonconforming matching.

Part III presents many other coupling techniques beyond the nonconforming constraints given in Part II. The new techniques include penalty, hybrid, mixed types of penalty and hybrid, Lagrange multiplier, etc. Since the coupling strategies are essential to combined methods, they are developed in this part in detail. Different coupling techniques are also included to provide the maximum informed choices.

Part IV presents wide applications of basic combinations and coupling techniques: e.g., for unbounded domain problems, as well as advanced topics, e.g., for superconvergence rates and domain decomposition methods.

For references, we mention about 600 published items, related to the combined methods, coupling techniques, and singularity problems.

If you know others and know yourself,
you will not be imperiled in a hundred battles;
if you do not know others but know yourself,
you win one and lose one;
if you do not know others and do not know yourself,
you will be imperiled in every single battle.
────── *The Art of War* ──────

Sun Tzu
(4th century B.C.)

PART I

SINGULARITIES, TREATMENTS
AND COMBINATIONS

Part I consists of two chapters:

Chapter 1: Different Numerical Methods.

Chapter 2: Singularities and Treatments.

Chapter 1 briefly describes the main, popular numerical methods for elliptic boundary value problems with important references. We try to interpret different methods as special Galerkin methods. Chapter 2 introduces the singularities of Laplace's equation, and different treatments. Combining different methods, in particular with the Ritz-Galerkin method using the singular particular solutions, is one of most efficient treatments for singularity problems. Part I is also an introductory part that leads to Parts II, III and IV, embodied by combinations, coupling strategies and applications.

Many references for different treatments on singularity problems are also collected in Part I. Moreover, other kinds of combinations and coupling techniques may be further developed by following the explanation in this book.

There are many numerical methods for solving elliptic boundary value problems; we list here the principal methods.

1. The finite element method(FEM), including the h, p−versions and their combination.

2. The finite difference method(FDM).

3. The boundary element method(BEM).

4. The Ritz-Galerkin method (RGM) using analytical or singular functions, including the spectral method and the Sinc method (Stenger(1993)).

5. The boundary approximation method (BAM) of Li, Mathon and Sermer (1987).

6. The finite volume method (FVM), including the conservative schemes, the box method of Varga(1962), and the cell method.

7. The collocation method (CM), including the residual method.

8. The least squares methods (LSM).

Each of the above methods possesses its own advantages and disadvantages. Quite often, one method can solve certain elliptic problems efficiently, but not others. For example, the finite element method(FEM) and the finite difference method(FDM) are widely applied to solve physics and engineering problems without singularities; however, their original methods fail to deal with singularity problems or infinite domain problems. The Ritz-Galerkin method or the boundary approximation method, which uses analytic or singular basis functions, provides high accuracy of solutions for simple problems, but fails to solve general problems. Consequently, it is advantageous to apply several numerical methods simultaneously for solving general, complicated elliptic problems.

It has to be observed that singularity problems are very interesting in both theoretical research and practical application even for Laplace's equation. Three kinds of isolate singularities often occur in elliptic problems: angular singularity, interface singularity, and the infinity in unbounded solution domains. There are also the boundary or moving layers in the convection-diffusion problem, where the singularity arises along curves (see Roos, Stynes and Tobiska (1996)). The line singularity also occurs in 3D elliptic problems. No examples are more important than singularity problems to show the necessity of combined methods.

In this book, we confine ourselves to the isolated singular problems. Since these three kinds of isolate singularities: angular, interface and infinity singularities, have common properties, the same numerical methods may be adopted for all (see Li (1990)). Therefore, we may focus on treatments for the angular singularity, in particular for Motz's problem with the mixed type of Dirichlet and Neumann boundary conditions. The unbounded domains are regarded as important applications in this book. In fact, the artificial interfaces appear everywhere in this book so that the combined methods can be extended to material interface problems easily, provided that the complete particular solutions can be found near the interface singularity (see Li(1990)).

A number of efficient treatments given in Chapter 2 are, indeed, a kind of combinations using different techniques or variants of FEMs, e.g., combining the local refinements of FEM near singular points and the standard FEM in subdomains of smooth solutions.

The combining of different numerical methods *has already appeared*. For examples, the boundary approximation method, the Trefftz function method and the global element method of Delves (1979) may be regarded as the combined methods. Semi-analytical methods, often employed in engineering, can be fully falled into combined methods.

1

DIFFERENT NUMERICAL METHODS

1.1 DESCRIPTIONS OF ELLIPTIC EQUATIONS

To simplify the exposition of different methods and their combinations, consider the self-adjoint elliptic equation (**PDE Problem**).

PDE Problem: Find $u \in C^2(S)$ such that

$$
\begin{aligned}
-\left(\frac{\partial}{\partial x} p \frac{\partial u}{\partial x} + \frac{\partial}{\partial y} p \frac{\partial u}{\partial y}\right) + cu &= f, \quad \text{in } S, \\
u &= g_1, \quad \text{on } \Gamma_D, \\
p \frac{\partial u}{\partial \nu} &= g_2, \quad \text{on } \Gamma_N, \\
p \frac{\partial u}{\partial \nu} + \alpha u &= g_3, \quad \text{on } \Gamma_M,
\end{aligned}
\tag{1.1}
$$

where S is a bounded domain with a piecewise Lipschitz continuous boundary ∂S, which consists of three disjoint subboundaries, Γ_D, Γ_N and Γ_M, the function $p \geq p_0 > 0$ is piecewise continuous, and the functions $c(\geq 0)$, $\alpha(\geq 0)$ and f are smooth enough. $C^k(S)$ denotes the space of functions having k-order continuous partial derivatives over S.

The three boundary conditions in (1.1) are called the Dirichlet, the Neumann and the Robin (or mixed) boundary conditions respectively. Eqs (1.1) have a unique solution except when $\Gamma_N = \partial S$ and $c = 0$, under which Eqs. (1.1) have infinite solutions with an arbitrary constant if the consistent condition is satisfied

$$
\iint_S f ds = -\oint_{\partial S} g_2 d\ell.
$$

We may describe the solution of (1.1) by the following weak form (**Galerkin Problem**), also see Hall and Porsching (1990).

5

Galerkin Problem: Find $u \in H^1_*(S)$ such that:

$$a(u,v) = f(v), \quad \forall v \in H^1_0(S), \tag{1.2}$$

where the notations are

$$a(u,v) = \iint_S (p \nabla u \nabla v + cuv)\, ds + \int_{\Gamma_M} \alpha uv d\ell, \tag{1.3}$$

$$f(v) = \iint_S fv ds + \int_{\Gamma_N} g_2 v d\ell + \int_{\Gamma_M} g_3 v d\ell, \tag{1.4}$$

$$H^1_0 = \left\{ v | v, \ v_x, \ v_y \in L^2(S), v|_{\Gamma_D} = 0 \right\}, \tag{1.5}$$

$$H^1_* = \left\{ v | v, \ v_x, \ v_y \in L^2(S), v|_{\Gamma_D} = g_1 \right\}, \tag{1.6}$$

and $H^1(S)$ denotes the Sobolev space.

Since the elliptic equation in (1.1) is self-adjoint, we have another weak description (**Varitional Problem**).

Variational Problem : Find $u \in H^1_*(S)$ such that

$$I(u) \le I(v), \quad \forall v \in H^1_*(v), \tag{1.7}$$

where the energy is

$$I(v) = \frac{1}{2} a(v,v) - f(v), \tag{1.8}$$

and $a(v,u)$ and $f(v)$ are defined in (1.3) and (1.4), respectively.

Galerkin problem may be used for general elliptic equations with or without self-adjoint property, and even applied to parabolic and hyperbolic equations. Galerkin and Variational Problems are said to be weak because the assumption $u \in H^1_*(S)$ of the solution is much weaker than that $u \in C^2(S)$.

We can prove the following conclusions (see Hall and Porsching (1990)):

$$
\begin{array}{rcl}
Galerkin \ Problem & \equiv & Varitional \ Problem, \\
PDE \ Problem & \Longrightarrow & Galerkin \ Problem, \\
Galerkin \ Problem \wedge (u \in C^2(S)) & \Longrightarrow & PDE \ Problem.
\end{array}
$$

$$\tag{1.9}$$

Note that different problems describing the same elliptic problem will lead to different discretizations, called different numerical methods. For instance, PDE problems lead to the finite difference method (FDM), Galerkin and Variational problems to the finite element method (FEM) and the Ritz-Galerkin method (RGM), etc.

There are also other themes describing the physical problems, such as the conservative law, the single or double layer potential functions, residuals of the equation and their squares; these descriptions may lead to the conservative scheme, the boundary element method, the collocation method and least squares methods.

In this chapter, we introduce different numerical methods with the following emphases.

(1) Basic algorithms and basic theoretical analysis are cited, because the comprehensive contents can be found in many textbooks and monographs describing single methods.

(2) Linkage of one method to another is explored, in particular the finite element method (FEM) to other methods. This linkage enables us to construct combinations of different methods, and to provide the necessary analysis by means of the rich framework of FEM.

(3) Comparisons are provided on different methods by demonstrating their advantages and drawbacks. Such comparisons display the clues how to combine the methods properly.

1.2   FINITE ELEMENT METHOD

To solve elliptic boundary value problems, the finite element method (FEM) is the most important method among all numerical approaches owing to wide applications and deep theoretical analysis.

Let S be a polygon and be divided into quasiuniform triangular elements $\triangle_i$. $S = \cup_i \triangle_i$. Here quasiuniform means that ratio of the maximal and minimal boundary lengths of all $\triangle_i$ is bounded. Choose the Lagrange elements of order k. Denote by $V_h^* \in H_*^1(S)$ and $V_h^0 \in H_0^1(S)$ the finite-dimension collections of such admissible functions satisfying $v|_{\Gamma_D} = g_1$ and $v|_{\Gamma_D} = 0$ respectively. The k-order Lagrange FEM is just a variant of the Galerkin problem:

To find $u_h \in V_h^*$ such that

$$a(u_h, v) = f(v), \quad \forall v \in V_h^0. \tag{1.10}$$

Similarly, FEM can also be given by the Variational problem:

To find $u_h \in V_h^*$ such that

$$I(u_h) = \min_{\forall v \in V_h^*} I(v). \tag{1.11}$$

Both (1.10) and (1.11) lead to the same system of linear algebraic system

$$\mathbf{A}\mathbf{x} = \mathbf{b}, \tag{1.12}$$

where $\mathbf{x}$ is a unknown vector consisting of the components $v_h^k(Q_i)$, where Q_i are the element nodes of $\triangle_i$, $\mathbf{b}$ is a known vector, and $\mathbf{A}$ is a symmetric, sparse, and positive semi-definite matrix. When $\Gamma_N \neq \partial S$ or $c > 0$, or $\alpha > 0$, the matrix $\mathbf{A}$ is also positive definite.

Since triangular elements may match arbitrary geometric domains, and piecewise low-order polynomials may approach the solutions with certain regularity, the FEM has been applied to a very large scale of PDEs. Also, the algebraic equations (1.12) can be formed by assembling stiff sub-matrices, which can be performed by computer. Therefore, the FEM becomes the most powerful numerical methods for PDEs today.

The advantages and the drawbacks of a method are indeed *a twin*. Take medicine as an example for explanation. There does not exist a medicine effective to all diseases. Some medicines, e.g., Aspirin, are widely used, but not very effective to some specific diseases, which need specific medicines and treatments. Similarly, the FEM is widely used for general elliptic problems; but not very effective to some specific elliptic problems. Below we point out two disadvantages of the standard FEM.

(1) Accuracy of the solutions by FEM is not high under a certain CPU time. In other words, the CPU time is large to achieve high accuracy of solutions. The consumption of CPU time in the matrix assembling process for (1.12) is much more than that in the solution process.

(2) When the solution is less smooth, e.g., its derivatives are unbounded near singular points, accuracy of the FEM solution is poor. However, in engineering problems, i.e., fracture mechanics, the solutions near the fracture points are most significant. A remedy is the local refinements, stated in Section 2.3.

For each problem to be solved, the better we acknowledge the solution behavior, the easier we may find the suitable numerical method. Our doctrine is that no methods can be applied to universe without analysis. This is also the philosophy of Sun Tzu (4th century B.C.) guiding the combined methods.

Since the FEM analysis is very comprehensive and flourished and widely, we may employ it to develop the theoretical framework of our combination analysis. Let us introduce the Sobolev space $H^k(S)$, equipped with the norms and the semi-norms (see Sobolev (1963), Adams (1975), Kufner (1980) and Marti (1986)),

$$\|v\|_{k,S} = \left(\sum_{|\alpha| \le k} \iint_S |D^\alpha v|^2 \, ds \right)^{\frac{1}{2}}, \qquad |v|_{k,S} = \left(\sum_{|\alpha| = k} \iint_S |D^\alpha v|^2 \, ds \right)^{\frac{1}{2}}.$$

We provide a main theorem on error estimates to the solution by FEM, whose proof can be found in Strang and Fix (1973) and Ciarlet (1978).

Theorem 1.1 *Let the following bilinear and uniformly V_h-elliptic inequalities hold.*

$$|a(u,v)| \le C_1 \|u\|_{1,S} \, \|v\|_{1,S}, \quad \forall u \in H^1(S), \quad v \in V_h^0, \qquad (1.13)$$
$$C_0 \|v\|_{1,S}^2 \le a(v,v), \quad \forall \, v \in V_h^0, \qquad (1.14)$$

where C_0 and C_1 are two constants independent of u, v and h, and h is the maximal boundary length of triangular elements. Then

$$\|u - u_h\|_{1,S} \le \left(1 + \frac{C_1}{C_0}\right) \inf_{\forall v \in V_h^*} \|u - v\|_{1,S}, \qquad (1.15)$$

where u and v are the true solutions of (1.2) and the FEM solution of (1.10), respectively.

Let us make a few remarks on Theorem 1.1. Eq. (1.15) implies that evaluation on error bounds of the FEM approximation u_h lie in evaluation on approximation errors of u by the admissible functions v in V_h^*. This greatly simplifies the error estimate of FEM solutions, since the latter evaluation is easier and rather straightforward.

Let $u \in H^{k+1}(S)$, and $\bar{u}_h$ be the k-order Lagrange interpolation function of u on the given triangulation. Then from the Sobolev interpolation theory, we have the optimal error estimates

$$\|u - \bar{u}_h\|_{1,S} \le Ch^k |u|_{k+1,S}, \qquad (1.16)$$

where C is a bounded constant independent of u and v. However, where u is less smooth, i.e., $u \in H^l(S)$, integer $l < k + 1$. Then the approximation only yields

$$\|u - \bar{u}_h\|_{1,S} \le Ch^{l-1} |u|_{l,S}. \qquad (1.17)$$

Eq. (1.17) is also valid for $u \in H^\alpha(S)$, where $\alpha(1 < \alpha < k+1)$ is real, thus leading to

$$\|u - \bar{u}_h\|_{1,S} \le Ch^{\alpha-1}\|u\|_{\alpha,S}, \qquad (1.18)$$

where $\|u\|_{\alpha,S}$ is the Sobolev norm with real value α, defined by

$$|u|_{\alpha,S} = |u|_{\lfloor\alpha\rfloor,S} + \left[\int_S \int_S \frac{\left(u^{\lfloor\alpha\rfloor}(X) - u^{\lfloor\alpha\rfloor}(Y)\right)^2}{|X-Y|^{2+2\alpha}} dXdY\right]^{\frac{1}{2}},$$

where $\lfloor\alpha\rfloor$ is the largest integer of α.

Combining (1.16) and (1.18) we have the following corollary.

Corollary 1.1 *Let all conditions in Theorem 1.1 hold. Assume $u \in H^\alpha(S)$, $\alpha > 1$. Then the k-order Lagrange FEM solutions have the error bounds*

$$\|u - u_h\|_{1,S} \le Ch^p|u|_{\alpha,S},$$

where $p = \min(\alpha - 1, k)$.

The proof of the bilinear inequality (1.13) is easy; but the proof of the uniformly V_h-elliptic inequality (1.14) is more troublesome. Now, we give the assumptions:

(A1) $c > 0$.

(A2) $c = 0$, $Meas\ (\Gamma_M) > 0$ and $\alpha > \alpha_0 > 0$.

(A3) $c = 0$, and $Meas\ (\Gamma_D) > 0$.

We obtain a lemma without proofs.

Lemma 1.1 *Let (A1), (A2) or (A3) hold. Then there exists a positive constant C_0 independent of v and h such that*

$$C_0\|v\|_{1,S}^2 \le a(v,v), \qquad \forall v \in V_h^0.$$

Note that Theorem 1.1 and Corollary 1.1 can also be extended to more general cases by suitable variations and further developments. Also we cite a theorem in numerical stability from Strang and Fix (1973).

Theorem 1.2 *When the triangulation is quasiuniform, the condition number of the associated matrix* $\mathbf{A}$ *in (1.12) of FEM has the following bounds*

$$Cond.(\mathbf{A}) = \frac{\lambda_{max}(\mathbf{A})}{\lambda_{min}(\mathbf{A})} \leq Ch^{-2},$$

where $\lambda_{max}(\mathbf{A})$ *and* $\lambda_{min}(\mathbf{A})$ *are the maximal and minimal eigenvalues of* $\mathbf{A}$ *respectively.*

To close this subsection, we mention the reference of FEM. Since Courant (1943) and Zienkiewicz and Cheung (1967), the FEM has become the most popular, widespread numerical methods, see an introduction of Oden (1991). Here we only mention some important books: Zienkiewicz (1977), Babuška and Aziz (1972), Strang and Fix (1973), Oden and Reddy (1976), Oden and Carey (1983), Ciarlet (1978), Fairweather (1978), Lu, Shih and Liem (1992), Akin (1982), Mikhlin (1964, 1971), Sewell (1988), Hackbusch (1992), Ciarlet (1991), Wahlbin (1991), Roberts and Thomas (1991), Axelsson and Barker (1984), Birkhoff and Lynch (1984), and Ames (1977). There are more reports on FEM: Akin (1979a), Aziz and Kellogg (1981), Babuska and Rheinboldt (1978, 1984), Babuska and Zlamal (1973), Barrett and Elliott (1987), Bath and Wilson (1976), Beagles and Whiteman (1986), Bettess, Emson and Chiam (1984), Bramble and Schatz (1977), Burda (1979, 1993), Carey and Oden (1986), Chen and Huang (1995), Dorr (1986), Eisenstat and Schultz (1972), Eriksson (1985a,b,c, 1986), Gardner, Gardner and Dag (1993), Gratkowski (1986), Gui and Babuska (1986), Gupta (1978), Han and Wu (1985), Jensen and Suri (1992), Lascaux and Lesaint (1975), Li and Reed (1995), Lin and Rheinboldt (1994), Lin and Zhu (1994), Lin and Whiteman (1990), Mergelies (1961), Mitchell and Wait (1977), Oganesyan (1990), Percell and Wheeler (1978), Schatz and Wahlbin (1978, 1995), Schatz (1980), Scott (1976), Sgallari (1985), Ladyzhenskaya (1985), Feng et. al. (1978), Feng and Shi (1981), Shi (1985), Thomas (1992), Veidinger (1978), Wathen and Baines (1985), Whiteman (1987), Whiteman and Akin (1980), Wolf and Song (1996), Zhang (1990), Zhou and Rannacher (1996), and Zienkiewicz and Morgan (1983).

1.3 RITZ-GALERKIN METHOD

The FEM is derived by the Galerkin and Variational problems by choosing piecewise low-order polynomials. If the admissible functions are chosen to be analytical or singular functions on S, the Galerkin and Variational problems lead to the Ritz-Galerkin method, a memento to Ritz (1909) and Galerkin (1915), also called spectral method in some literatures if the trigonometrical and orthogonal functions are chosen as admissible functions (see Gottlieb and Orszag (1977)).

Let $\{\Phi_i\} \in H^1(S)$, $i = 1, 2, \ldots$, be the complete, linearly independent basis functions on S, $\forall u \in H^1(S)$, such that

$$\left\| u - \sum_{i=1}^{L} a_i\varphi_i \right\|_{1,S} \longrightarrow 0 \quad \text{as} \quad L \longrightarrow \infty.$$

Choose the admissible functions

$$u_L = \sum_{i=1}^{L} a_i\varphi_i, \tag{1.19}$$

and denote by V_L^*, the finite-dimensional collection of (1.19) satisfying $u|_{\Gamma_D} = g_1$, and by V_L^0 satisfying $u|_{\Gamma_D} = 0$. The Ritz-Galerkin method (RGM) is defined by

To find $u_L \in V_L^*$ such that

$$a(u_L, v) = f(v), \quad \forall v \in V_L^0, \tag{1.20}$$

where $a(u, v)$ and $f(v)$ are defined in (1.3) and (1.4) respectively.

The choice of basis functions φ_i is critical. In fact, the continuous basis functions φ_i in FEM are like the hill functions, which are constructed by piecewise low-order polynomials. In the RGM, the trigonometrical functions, exponential functions, orthogonal polynomials, etc. can be chosen as the basis functions.

Let the true solution of (1.1) be

$$u = \overline{u}_L + R_L,$$

where $\overline{u}_L = \sum_{i=1}^{L} a_i\varphi_i$, the remainder $R_L = \sum_{i=L+1}^{\infty} a_i\varphi_i$ and a_i are the true expansion coefficients. We can easily prove the following theorem

Theorem 1.3 *Suppose*

$$|a(u, v)| \leq C_0\|u\|_{1,S}\|v\|_{1,S}, \quad \forall u \in H^1(S), \ v \in V_L^0,$$
$$C_0\|v\|_{1,S}^2 \leq a(v, v), \quad \forall \ v \in V_L^0,$$

where C_0 and C_1 are the two bounded constants independent of u, v and L. Then there exist the bounds

$$\|u - u_L\|_{1,S} \leq \left(1 + \frac{C_1}{C_0}\right) \inf_{\forall v \in V_L^*} \|u - v\|_{1,S} \leq \left(1 + \frac{C_1}{C_0}\right) \|R_L\|_{1,S}.$$

Based on Theorem 1.3 the evaluation on residual R_L is the main manipulation for error estimates. If the basis functions φ_i can be chosen to match best the true solutions, the RGM will produce high accuracy of numerical solutions, as compared with the FEM. However, finding such φ_i is not trivial for arbitrary S. Another trouble in RGM is severe instability. Accuracy and instability is also like *a twin*. Usually, a method such as RGM yields often high accuracy but poor stability for numerical solutions, except on special solution domains, i.e., a square or a sector, where the orthogonal bases may achieve both high accuracy and good stability. On the other hand, FEM produces good stability but low accuracy. Since the use of analytical or singular functions is promising, in this book we will concentrate on their combinations with other kinds of numerical methods, e.g., the FEM, FDM, etc.

Reference on RGM can be found in Kantorovich and Krylov (1958), Mikhlin (1964, 1971), Duff and Naylor (1966), Tikhonov and Samarskii (1973), Eisenstat (1974), Rektoys (1975), Stenger (1993), Gottlieb and Orszag (1977), Bourlard, Nicaise and Paquet (1990), Bramble and Schatz (1970), Lamp, Schleicher, et. al. (1984), Fletcher (1984), Liang (1980a), McLean (1986), Schatz and Wang (1996), Wahlbin (1995), Falk (1976), Straughan (1992), and Mathon and Johnston (1977).

The spectral method using spectral functions may fall into RGM. Because of high efficiency, many papers have appeared since the book of Gottlieb and Orszag (1977), for example, Chow, Duninger and Miklavcic (1990), Yoseph and Israeli (1986), Pathria and Karniadakis (1995), Begue, Bernardi, et. al. (1989), Canuto and Funaro (1988), Canuto, Hariharan and Lustman (1985), Canuto, Hussaini, et. al. (1988), Boyd (1989), Mercier (1989), Bernardi and Maday (1997), Coulaud, Funaro and Kavian (1990), Hebeker (1987), Karageorghis (1994), Kariadakis, Maxriplis and Patera (1989), Jensen (1991), Deville and Mund (1990), Bernardi and Karageorghis (1996), Zhang, Wong and Hussaini (1982), Greenstadt (1995). However, the stability may still be a serious issue, see an improvement in Heinrichs (1989). Other special functions as Sinc functions are given in Bowers and Lund (1987), Stenger (1979), McArthur, Bowers and Lund (1990) and a book of Stenger (1993).

1.4   BOUNDARY APPROXIMATION METHOD

For simplicity, we consider the Robin condition of (1.1).

$$Lu = f, \quad \text{in } S,$$

$$p\frac{\partial u}{\partial \nu} + \alpha u = g, \quad \text{on } \partial S,$$

where

$$Lu = -\left(\frac{\partial}{\partial x}p\frac{\partial u}{\partial x} + \frac{\partial}{\partial y}p\frac{\partial u}{\partial y}\right) + cu.$$

Suppose that a particular solution Ψ_0 can be found to satisfy

$$L\Psi_0 = f.$$

By the transformation $w = u - \Psi_0$, we obtain the homogeneous elliptic equation

$$Lw = 0, \quad \text{in } S,$$

$$p\frac{\partial w}{\partial \nu} + \alpha w = \bar{g}, \quad \text{in } \partial S, \tag{1.21}$$

where $\bar{g} = g - \left(p\frac{\partial \Psi_0}{\partial \nu} + \alpha \Psi_0\right)$. Also if the particular solutions Ψ_i of (1.21) can be found explicitly,

$$L\Psi_i = 0, \quad i = 1, 2, \ldots$$

Therefore the solution of (1.1) can be spanned by

$$u = \Psi_0 + \sum_{i=1}^{\infty} a_i \Psi_i.$$

The suitable admissible functions are found as

$$v_L = \Psi_0 + \sum_{i=1}^{L} \tilde{a}_i \Psi_i, \tag{1.22}$$

where the coefficients $\tilde{a}_i$ are to be determined.

Since the admissible functions (1.22) already satisfy the equation $Lu = f$, we obtain by the Green theorem

$$\begin{aligned}
a(v_L, w_L) &= \iint_S (p\,\nabla v_L\,\nabla w_L + cv_L w_L)ds + \int_{\Gamma} \alpha v_L w_L d\ell \\
&= \oint_{\partial S} \left(p\frac{\partial v_L}{\partial \nu} + \alpha v_L\right) w_L d\ell + \iint_S w_L L v_L ds \\
&= \oint_{\partial S} \left(p\frac{\partial v_L}{\partial \nu} + \alpha v_L\right) w_L d\ell + \iint_S f w_L ds,
\end{aligned}$$

where

$$w_L = \sum_{i=1}^{L} \tilde{a}_i \Psi_i. \tag{1.23}$$

Denote by V_L^* and V_L the spaces of the functions of (1.22) and (1.23) respectively. The RGM (1.20) leads to :

To find $v \in V_L^*$ such that

$$\int_{\partial S} \left(p \frac{\partial v}{\partial \nu} + \alpha v \right) w \, d\ell = \oint_{\partial S} g w \, d\ell, \quad \forall w \in V_L. \tag{1.24}$$

Since only the boundary condition is involved in (1.24), the boundary approximation method (BAM) is called for. Below we provide another variational approach. Since the admissible functions have satisfied the equation $Lu = f$ already, then the coefficients $\tilde{a}_i$ in (1.22) are chosen to satisfy the boundary conditions as much as possible.

To find $v_L \in V_L^*$ such that

$$I_B(v_L) = \min_{\forall v \in V_L^*} I_B(v),$$

where

$$I_B(v) = \int_{\partial S} \left(p \frac{\partial v}{\partial \nu} + \alpha v - g \right)^2 d\ell.$$

However, when the Dirichlet condition also occurs, we shall solicit the boundary penalty technique. Consider the elliptic equation

$$\begin{cases} Lu = f, & \text{in } S, \\ u = g_1, & \text{in } \Gamma_D, \\ p \frac{\partial u}{\partial \nu} + \alpha v = g_3, & \text{in } \Gamma_M \cup \Gamma_N, \end{cases} \tag{1.25}$$

where $\alpha \geq 0$. Define

$$I_B^*(v) = \int_{\Gamma_D} (v - g_1)^2 \, d\ell + w^2 \int_{\Gamma_M \cup \Gamma_N} \left(p \frac{\partial u}{\partial \nu} + \alpha v - g_3 \right)^2 d\ell, \tag{1.26}$$

where w is a positive weight to balance different boundary conditions. Since the solution derivatives are usually larger than the solutions themselves, we choose $w < 1$. Of course a better choice of w should depend upon error analysis (see Chapter 4). The BAM for (1.25) is: To seek $u_L \in V_L^*$ such that

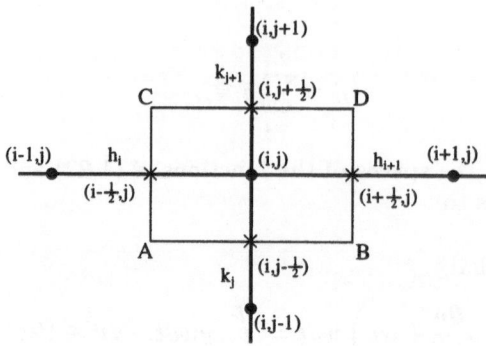

Figure 1.1   The inner node (i, j) for FDM.

$$I_B^*(u_L) = \min_{\forall v \in V_L^*} I_B^*(v). \tag{1.27}$$

Eq (1.27) leads to

$$\mathbf{By} = \mathbf{b},$$

where $\mathbf{B}$ is symmetric and positive definite, but not sparse; $\mathbf{y}$ is the unknown vector consisting of components $\tilde{a}_i$.

A remarkable advantages of BAM is high accuracy owing to exponential convergence rates. BAM is more efficient than the original RGM because the approximation process is reduced to that on ∂S only. The relation between RGM and BAM is analogous to that between FEM and BEM. However, BAM is limited to elliptic equations with constant coefficients where particular solutions can be found. Also, instability is serious when S is arbitrary. In combinations, we may not distinguish RGM and BAM, and call these two methods as the RGM in general.

The references on BAM are given in Delves (1979), Delves and Hall (1979), Delves and Freeman (1981), Li, Mathon and Sermer (1987), Li and Mathon (1990a, b), and Zielinski and Zienkiewicz (1985).

1.5   FINITE DIFFERENCE METHOD

By replacing the partial differentiations in (1.1) by difference quotients at discrete coordinate points, we obtain the finite difference method (FDM). Suppose the solution domain S can be partitioned by the coordinates $x = x_i$ and $y = y_i$ into small rectangles $\square_{ij}$ and small right triangles $\triangle_{ij}$, i.e.,

$$S = \bigcup_{ij} \left(\square_{ij} \cup \triangle_{ij} \right).$$

1.5.1 Interior Equations

Denote u_{ij} (or $u_{i,j}$) $= u(x_i, y_j)$, $h_{i+1} = x_{i+1} - x_i$, and $k_{j+1} = y_{j+1} - y_j$. Then we have the approximation at the interior nodes (i, j) (see Figure 1.1)

$$
\begin{aligned}
\left(\frac{\partial}{\partial x} p \frac{\partial u}{\partial x} \right)_{ij} &\approx \frac{\left(p \frac{\partial u}{\partial x} \right)_{i+\frac{1}{2},j} - \left(p \frac{\partial u}{\partial x} \right)_{i-\frac{1}{2},j}}{\frac{1}{2}(h_{i+1} + h_i)} \\
&\approx \frac{p_{i+\frac{1}{2},j} \frac{u_{i+1,j} - u_{i,j}}{h_{i+1}} - p_{i-\frac{1}{2},j} \frac{u_{i,j} - u_{i-1,j}}{h_i}}{\frac{1}{2}(h_{i+1} + h_i)} \\
&= a_{ij}^x u_{i+1,j} + b_{ij}^x u_{i-1,j} - \left(a_{ij}^x + b_{ij}^x \right) u_{ij},
\end{aligned}
$$

where $p_{i+\frac{1}{2},j} = p(x_{i+\frac{1}{2}}, y_j)$, $x_{i+\frac{1}{2}} = \frac{1}{2}(x_i + x_{i+1})$,

$$
a_{ij}^x = \frac{2 p_{i+\frac{1}{2},j}}{(h_{i+1} + h_i) h_{i+1}}, \qquad b_{ij}^x = \frac{2 p_{i-\frac{1}{2},j}}{(h_{i+1} + h_i) h_i}.
$$

Similarly, we have

$$
\left(\frac{\partial}{\partial y} p \frac{\partial u}{\partial y} \right)_{ij} \approx a_{ij}^y u_{i,j+1} + b_{ij}^y u_{i,j-1} - \left(a_{ij}^y + b_{ij}^y \right) u_{ij},
$$

where

$$
a_{ij}^y = \frac{2 p_{i,j+\frac{1}{2}}}{(k_{j+1} + k_j) k_{j+1}}, \qquad b_{ij}^y = \frac{2 p_{i,j-\frac{1}{2}}}{(k_{j+1} + k_j) k_j}.
$$

Therefore, the discrete equations of (1.1) at interior nodes (i, j) are obtained as

$$
d_{ij} u_{ij} - \left(a_{ij}^x u_{i+1,j} + b_{ij}^x u_{i-1,j} + a_{ij}^y u_{i,j+1} + b_{ij}^y u_{i,j-1} \right) = f_{ij}, \tag{1.28}
$$

where

$$
d_{ij} = \left(a_{ij}^x + b_{ij}^x + a_{ij}^y + b_{ij}^y \right) + c_{ij}.
$$

For the entire Dirichlet conditions $u|_{\partial S} = g$. Let the boundary nodes be located on ∂S, then we may assign

$$
u_{ij} = g_{ij}, \quad \forall (i, j) \in \partial S.
$$

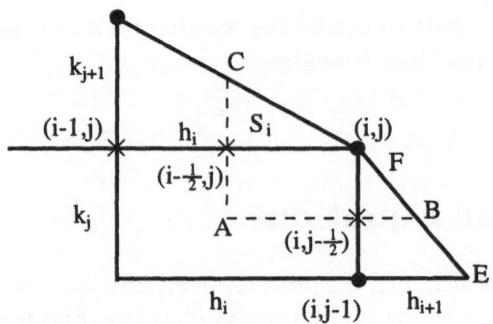

Figure 1.2   The boundary node (i, j) for FDM.

The finite difference equation (1.28) leads to the algebraic equation

$$\mathbf{A}_D \mathbf{x} = \mathbf{b}, \tag{1.29}$$

where the unknown vector $\mathbf{x}$ is composed of interior discrete solutions u_{ij}. The associated matrix $\mathbf{A}_D$ also has good properties: positive definite, symmetric, sparse, diagonally dominant, and with non-positive off-diagonal entries. The finite difference method (FDM) should be chosen if the solution domain S is not so complicated, so that S can be partitioned properly by coordinate lines into quasiuniform rectangles and triangles. Note that formulation of the algebraic equation (1.29) is straightforward, contrasted to that in FEM consuming a lot of CPU time.

1.5.2   Boundary Difference Equations

Other boundary conditions, however, must be treated carefully to retain the good matrix properties stated above. Take as an example the Robin boundary condition on a slant boundary in Figure 1.2. We have

$$-\left[\left(\frac{\partial}{\partial x}p\frac{\partial u}{\partial x}\right)_{ij} + \left(\frac{\partial}{\partial y}p\frac{\partial u}{\partial y}\right)_{ij}\right]$$

$$\approx -\frac{1}{h_i/2}\left[\left(p\frac{\partial u}{\partial x}\right)_{ij} - \left(p\frac{\partial u}{\partial x}\right)_{i-\frac{1}{2},j}\right] - \frac{1}{k_j/2}\left[\left(p\frac{\partial u}{\partial y}\right)_{ij} - \left(p\frac{\partial u}{\partial y}\right)_{i,j-\frac{1}{2}}\right].$$

For the boundary corner point (i, j), we may choose the average of the left and right Robin boundary conditions

$$\left(p\frac{\partial u}{\partial x}\right)_{i,j} \approx \frac{1}{2}\left(p\frac{\partial u}{\partial \nu}\cos(\nu, x)\Big|_{\overline{FC}} + p\frac{\partial u}{\partial \nu}\cos(\nu, x)\Big|_{\overline{BF}}\right)$$

$$\approx \frac{1}{2}\left[\left((g_3)_{ij} - \alpha_{ij}u_{ij}\right)\cos(\nu, x)|_{\overline{FC}} + \left((g_3)_{ij} - \alpha_{ij}u_{ij}\right)\cos(\nu, x)|_{\overline{BF}}\right]$$

$$= \frac{1}{2}\left[(g_3)_{ij} - \alpha_{ij}u_{ij}\right]\left(\frac{k_{j+1}}{\sqrt{h_i^2 + k_{j+1}^2}} + \frac{k_j}{\sqrt{h_{i+1}^2 + k_j^2}}\right).$$

Similarly,

$$\left(p\frac{\partial u}{\partial y}\right)_{i,j} \approx \frac{1}{2}\left[p\frac{\partial u}{\partial \nu}\cos(\nu, y)\bigg|_{\overline{FC}} + p\frac{\partial u}{\partial \nu}\cos(\nu, y)\bigg|_{\overline{BF}}\right]$$

$$\approx \frac{1}{2}\left[\left((g_3)_{ij} - \alpha_{ij}u_{ij}\right)\left(\frac{h_i}{\sqrt{h_i^2 + k_{j+1}^2}} + \frac{h_{i+1}}{\sqrt{h_{i+1}^2 + k_j^2}}\right)\right].$$

Hence, the boundary discrete equations are obtained as

$$d_{ij}u_{ij} - b_{ij}^x u_{i-1,j} - b_{ij}^y u_{i,j-1} = f_{ij} + (g_3)_{ij}e_{ij}, \qquad (1.30)$$

where

$$b_{ij}^x = \frac{2p_{i-\frac{1}{2},j}}{h_i^2}, \quad b_{ij}^y = \frac{2p_{i,j-\frac{1}{2}}}{k_j^2}, \quad d_{ij} = b_{ij}^x + b_{ij}^y + c_{ij} + \alpha_{ij}e_{ij},$$

$$e_{ij} = \left[\frac{1}{\sqrt{h_i^2 + k_{j+1}^2}}\left(\frac{k_{j+1}}{h_i} + \frac{h_i}{k_j}\right) + \frac{1}{\sqrt{h_{i+1}^2 + k_j^2}}\left(\frac{k_j}{h_i} + \frac{h_{i+1}}{k_j}\right)\right].$$

However, when the geometric shape of S is rather arbitrary, quasiuniform difference grids are difficult to construct. Also, the boundary nodes can not be always located just on ∂S. In this case, some additional treatments (see Hall and Porsching (1990)) are necessary to embed three boundary conditions into the algebraic equations. Often here occur violations of the good matrix properties, such as symmetry for the elliptic self-joint equations.

Solutions of FDM are discrete; but solutions of FEM are continuous with discrete support. Also, the classic error analysis of FDM is mainly based on the maximum principle of PDEs and discrete algebraic equations.

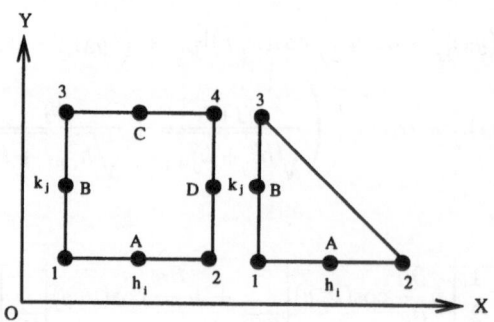

Figure 1.3   The elements $\square_{ij}$ and $\triangle_{ij}$.

1.5.3   View of FDM as FEM

This subsection is intended to establish a relationship between the FDM and FEM. Let us consider

$$-\frac{\partial}{\partial x}\left(p\frac{\partial u}{\partial x}\right) - \frac{\partial}{\partial y}\left(p\frac{\partial u}{\partial y}\right) + cu \ = \ f, \quad (x,y) \in S, \tag{1.31}$$

$$u \ = \ g, \quad (x,y) \in \Gamma_1, \tag{1.32}$$

$$p\frac{\partial u}{\partial \nu} + \alpha u \ = \ q, \quad (x,y) \in \Gamma_2, \tag{1.33}$$

where S is a convex polygon with the boundary $\Gamma = \Gamma_1 \cup \Gamma_2, Meas(\Gamma_1) \neq 0, \nu$ is the outward normal of Γ_2, and the functions p, f, g, q, $c(\geq 0)$ and $\alpha(\geq 0)$ are sufficiently smooth. We suppose that there exists a positive constant p_0 such that $p = p(x,y) \geq p_0 > 0$. The model problem (1.31)-(1.33) is written in a weak form

$$a(u, v) = f(v), \quad \forall v \in H_0^1(S),$$

where the solution $u \in H_*^1(S)$, and the notations are

$$a(u, v) \ = \ \iint_S [p(u_x v_x + u_y v_y) + cuv]ds + \int_{\Gamma_2} \alpha uv d\ell,$$

$$f(v) \ = \ \iint_S fvds + \int_{\Gamma_2} qvd\ell,$$

$$H_0^1(S) \ = \ \left\{ v | v \in H^1(S) \text{ and } v|_{\Gamma_1} = 0 \right\},$$

$$H_*^1(S) \ = \ \left\{ v | v \in H^1(S) \text{ and } v \text{ satisfies (1.32)} \right\}.$$

Piecewise bilinear (or linear) interpolation functions v are chosen to be admissible functions. Such functions v on $\square_{ij}$ or $\triangle_{ij}$ are defined by (see Figure 1.3)

$$
\begin{aligned}
v(x,y) &= \frac{1}{h_i k_j}[(x_2 - x)(y_3 - y)v_1 + (x - x_1)(y_3 - y)v_2 \\
&\quad + (x_2 - x)(y - y_1)v_3 + (x - x_1)(y - y_1)v_4], \\
&\qquad\qquad\qquad\qquad\qquad\qquad \text{for } (x,y) \in \square_{ij}, \qquad (1.34)
\end{aligned}
$$

$$
\begin{aligned}
v(x,y) &= v_1 + \frac{v_2 - v_1}{h_i}(x - x_1) + \frac{v_3 - v_1}{k_j}(y - y_1) \\
&\qquad\qquad\qquad\qquad\qquad\qquad \text{for } (x,y) \in \triangle_{ij}, \qquad (1.35)
\end{aligned}
$$

where v_k denotes the values of v at the corners of $\square_{ij}$ or $\triangle_{ij}$ in Figure 1.3. We note that $v(x,y)$ in (1.34) or (1.35) are linear functions along any straight lines parallel to the axes x and y (i.e., $x = a$ or $y = b$).

Let V_h^* denote the space of such piecewise bilinear (or linear) functions satisfying (1.32), and V_h^0 denote another space of those functions but satisfying $v|_{\Gamma_1} = 0$. Therefore, the bilinear and linear finite element method will yield an approximate solution, $u_h \in V_h^*$, such that

$$
a(u_h, v) = f(v), \quad \forall v \in V_h^0. \qquad (1.36)
$$

If approximations of integrals in (1.36) are necessary, the Gauss integration formulae are used (see Strang and Fix (1973) and Ciarlet (1978)). This is the traditional FEM.

Instead of the popular Gauss integration formulae for (1.36), here we will choose certain approximations of integrals, by which Eq. (1.36) will provide the very finite difference method. Such special integral approximations on $\square_{ij}$ or $\triangle_{ij}$ are given by a number of formulae as follows (see Davis and Rabinowitz (1984)):

$$
\begin{aligned}
\iint_{\square_{ij}} p u_x v_x \, ds &\approx \frac{h_i k_j}{2}\{p u_x v_x(A_{ij}) + p u_x v_x(C_{ij})\} \\
&= \frac{h_i k_j}{2}\left\{p_A \frac{u_2 - u_1}{h_i}\frac{v_2 - v_1}{h_i} + p_C \frac{u_4 - u_3}{h_i}\frac{v_4 - v_3}{h_i}\right\},
\end{aligned}
$$

$$
\begin{aligned}
\iint_{\square_{ij}} p u_y v_y \, ds &\approx \frac{h_i k_j}{2}\{p u_y v_y(B_{ij}) + p u_y v_y(D_{ij})\} \\
&= \frac{h_i k_j}{2}\left\{p_B \frac{u_3 - u_1}{k_j}\frac{v_3 - v_1}{k_j} + p_D \frac{u_4 - u_2}{k_j}\frac{v_4 - v_2}{k_j}\right\},
\end{aligned}
$$

$$
\iint_{\triangle_{ij}} p u_x v_x \, ds \approx \frac{h_i k_j}{2} p u_x v_x(A_{ij}) = \frac{h_i k_j}{2} p_A \frac{u_2 - u_1}{h_i}\frac{v_2 - v_1}{h_i},
$$

$$\iint_{\Delta_{ij}} pu_y v_y \, ds \;\approx\; \frac{h_i k_j}{2} pu_y v_y(B_{ij}) = \frac{h_i k_j}{2} p_B \frac{u_3 - u_1}{k_j} \frac{v_3 - v_1}{k_j}, \qquad (1.37)$$

$$\iint_{\square_{ij}} fv \, ds \;\approx\; \frac{h_i k_j}{4}(f_1 v_1 + f_2 v_2 + f_3 v_3 + f_4 v_4),$$

$$\iint_{\Delta_{ij}} fv \, ds \;\approx\; \frac{h_i k_j}{8}(2 f_1 v_1 + f_2 v_2 + f_3 v_3),$$

$$\int_{\overline{12}} qv \, d\ell \;\approx\; \frac{h_i}{2}(q_1 v_1 + q_2 v_2).$$

Here, the notation A_{ij} denotes the point A on $\square_{ij}$ with respect to the subscripts i and j.

We note that the integration rules used in Eqs (1.37) are quite different from the traditional rules used in the FEM. For example, the integration nodes, A_{ij}, B_{ij}, etc. in (1.37) are the middle points of the boundaries of $\square_{ij}$ and Δ_{ij}, and the integration nodes in the last three rules of (1.37) are the corner points of $\square_{ij}$ and Δ_{ij}.

With the integration formulae (1.37), the bilinear or linear element method (1.36) just leads to the finite difference method:

$$a_h(\tilde{u}_h, v) = f_h(v), \quad \forall v \in V_h^0, \qquad (1.38)$$

where $\tilde{u}_h (\in V_h^*)$ denotes the numerical solution resulting from integration approximation, and the notations $a_h(u, v)$ and $f_h(v)$ are given by

$$
\begin{aligned}
a_h(u, v) \;=\; & \sum_{i,j} \frac{Area(\square_{ij})}{2} \left[pu_x v_x(A_{ij}) + pu_x v_x(C_{ij}) \right. \\
& \left. + pu_y v_y(B_{ij}) + pu_y v_y(D_{ij}) \right] \\
& + \sum_{i,j} Area(\Delta_{ij}) \left[pu_x v_x(A_{ij}) + pu_y v_y(B_{ij}) \right] + \sum_{ij} \frac{Area(\square_{ij})}{4} \sum_{k=1}^{4} cuv(k_{ij}) \\
& + \sum_{ij} \frac{Area(\Delta_{ij})}{4} \sum_{k=1}^{3} \beta_k cuv(k_{ij}) + \sum_{i,j} \frac{Meas(\square_{ij} \cap \Gamma_2)}{2} \sum_{k=1}^{2} \alpha uv(Q_{ij}^{(k)}) \\
& + \sum_{i,j} \frac{Meas(\Delta_{ij} \cap \Gamma_2)}{2} \sum_{k=1}^{2} \alpha uv(Q_{ij}^{(k)}), \qquad (1.39)
\end{aligned}
$$

and

$$f_h(v) \;=\; \sum_{i,j} \frac{Area(\square_{ij})}{4} \sum_{k=1}^{4} fv(k_{ij}) + \sum_{i,j} \frac{Area(\square_{ij})}{4} \sum_{k=1}^{3} \beta_k fv(k_{ij}) \qquad (1.40)$$

$$+ \sum_{i,j} \frac{Meas(\square_{ij} \cap \Gamma_2)}{2} \sum_{k=1}^{2} qv(Q_{ij}^{(k)}) + \sum_{i,j} \frac{Meas(\triangle_{ij} \cap \Gamma_2)}{2} \sum_{k=1}^{2} qv(Q_{ij}^{(k)}).$$

In Eqs. (1.39) and (1.40), $\beta_1 = 2$, $\beta_2 = \beta_3 = 1$, $Q_{ij}^{(k)}(k = 1, 2)$ are the boundary points of the section $\square_{ij} \cap \Gamma_2$ or $\triangle_{ij} \cap \Gamma_2$, and k_{ij} are the corner points k on $\square_{ij}$ and $\triangle_{ij}$.

It is the approach of Eq. (1.38) that results in inner and boundary difference equations as follows (see Figures 1.1 and 1.2):

$$- \frac{k_j + k_{j+1}}{2} \left[p_{i+\frac{1}{2},j} \frac{u_{i+1,j} - u_{i,j}}{h_{i+1}} - p_{i-\frac{1}{2},j} \frac{u_{i,j} - u_{i-1,j}}{h_i} \right]$$

$$- \frac{h_i + h_{i+1}}{2} \left[p_{i,j+\frac{1}{2}} \frac{u_{i,j+1} - u_{i,j}}{k_{j+1}} - p_{i,j-\frac{1}{2}} \frac{u_{i,j} - u_{i,j-1}}{k_j} \right]$$

$$+ \frac{(h_i + h_{i+1})(k_j + k_{j+1})}{4} c_{ij} u_{ij} = \frac{(h_i + h_{i+1})(k_j + k_{j+1})}{4} f_{i,j}, \quad (1.41)$$

and

$$\frac{k_j + k_{j+1}}{2} \left[p_{i-\frac{1}{2},j} \frac{u_{i,j} - u_{i-1,j}}{h_i} \right] + \frac{h_i + h_{i+1}}{2} \left[p_{i,j-\frac{1}{2}} \frac{u_{i,j} - u_{i,j-1}}{k_j} \right]$$

$$+ \frac{1}{2} \left[(h_{i+1}^2 + k_j^2)^{\frac{1}{2}} + (h_i^2 + k_{j+1}^2)^{\frac{1}{2}} \right] (\alpha_{i,j} u_{i,j} - q_{i,j}) + \quad (1.42)$$

$$+ \frac{1}{8} (2h_i k_j + h_i k_{j+1} + h_{i+1} k_j) c_{ij} u_{ij} = \frac{1}{8} (2h_i k_j + h_i k_{j+1} + h_{i+1} k_j) f_{i,j},$$

where the notations are: $h_{i+1} = x_{i+1} - x_i$ with $x_{i+1} > x_i$, $k_{j+1} = y_{j+1} - y_j$ with $y_{j+1} > y_j$, $u_{i,j} = u(x_i, y_j)$, and $u_{i+\frac{1}{2},j} = u(x_{i+\frac{1}{2}}, y_j)$ with $x_{i+\frac{1}{2}} = \frac{1}{2}(x_i + x_{i+1})$.

Dividing the two sides in Eq. (1.41) by the factor, $(h_i + h_{i+1})(k_i + k_{j+1})/4$, we obtain

$$- \left[p_{i+1/2,j} \frac{u_{i+1,j} - u_{i,j}}{h_{i+1}} - p_{i-1/2,j} \frac{u_{i,j} - u_{i-1,j}}{h_i} \right] \Big/ \left(\frac{h_i + h_{i+1}}{2} \right)$$

$$- \left[p_{i,j+1/2} \frac{u_{i,j+1} - u_{i,j}}{k_{j+1}} - p_{i,j-1/2} \frac{u_{i,j} - u_{i,j-1}}{k_j} \right] \Big/ \left(\frac{k_j + k_{j+1}}{2} \right)$$

$$+ c_{ij} u_{ij} = f_{i,j}.$$

This is just (1.28). Therefore, Eq. (1.38), one kind of FEM, also represents the FDM.

Moreover, the boundary discrete equations (1.42) and (1.30) are consistent only when the difference grids are uniform

$$h_i = h, \ \forall i, \ k_j = k, \ \forall j. \tag{1.43}$$

Dividing two sides of (1.42) by $hk/2$ gives

$$\frac{2p_{i-\frac{1}{2},j}}{h^2}(u_{i,j} - u_{i-1,j}) + \frac{2p_{i,j-\frac{1}{2}}}{k^2}(u_{i,j} - u_{i,j-1}) + c_{ij}u_{ij}$$
$$+2\frac{(h^2+k^2)^{\frac{1}{2}}}{hk}(\alpha_{i,j}u_{i,j} - q_{i,j}) = f_{i,j},$$

the same as equation (1.30) under (1.43) by noting $q = q_3$.

On the other hand, Eqs (1.41) and (1.42) can also be derived from box integration approximations (see Varga (1962), Mitchell and Griffiths (1980), and Bellman and Adomian (1985)). See Section 1.6.4.

The main references on FDM are given in Thomas (1995), Mickens (1994), Godunov and Ryabenkii (1987), Hall and Porsching (1990), Thomee (1990), Strikwerda (1989), Ascher, Mattheij and Russell (1988), Marchuk (1982), Forsythe and Wasow (1960), Varga (1962), Samarskii and Andblev (1976), Ames (1977), Mitchell and Griffiths (1980), Birkhoff and Lynch (1984), Ciric and Wong (1986), Dorny (1968), Fornberg (1988), Lakshmikantham and Trigiante (1988), Ladyzhenskaya (1985), Richtmyer and Morton (1967), Bellman and Adomain (1985), Huiskamp (1991), Pereyra, Proskurowski and Widlund (1977), Suli, Jovanovic and Ivanovic (1985), Duran, Muschietti and Rodriguez (1991), Guo (1988) and Ferket and Reusken (1996). The FDM is still developing, for instance, by using Sobolev space in Suli, Jovanovic and Ivanovic (1985), and for high order accuracy in Boisvert (1981), Linde (1974), Pereyra, Proskurowski and Widlund (1977), and for spherical surface in Huiskamp (1991).

1.6   FINITE VOLUME METHOD

1.6.1   Conservative Law

In this section, we describe other kinds of discrete methods, based on conservative law, another variant of description to (1.1). We cite a lemma in calculus.

Lemma 1.2 (The divergence theorem) *Let ∂S be piecewise smooth, and $v = (v_1, v_2) \in [C^1(S)]^2$, then*

$$\iint_S \left(\frac{\partial v_1}{\partial x} + \frac{\partial v_2}{\partial y} \right) ds = \int_{\partial S} v \cdot \nu d\ell,$$

where ν is the unit vector of the exterior normal to ∂S.

Now, we give a theorem.

Theorem 1.4 *Let $u \in C^2(S)$. For any subdomain $S_i \subset S$, the solution of (1.1) also satisfies*

$$- \oint_{\partial S_i} p \frac{\partial u}{\partial \nu} d\ell + \iint_{S_i} cu ds = \iint_{S_i} f ds, \quad \text{if } \partial S_i \cap (\Gamma_N \cup \Gamma_M) = \emptyset; \qquad (1.44)$$

otherwise

$$- \int_{\Gamma_N \cap \partial S_i} g_2 d\ell - \int_{\Gamma_M \cap \partial S_i} (g_3 - \alpha u) \, d\ell - \int_{\Gamma_i^* \cap \partial S_i} p \frac{\partial u}{\partial \nu} d\ell + \iint_{S_i} cu ds$$

$$= \iint_{S_i} f ds, \qquad (1.45)$$

where $\Gamma_i^* = \partial S \backslash (\Gamma_N \cup \Gamma_M)$, the rest of ∂S_i excluding $\Gamma_N \cup \Gamma_M$.

Proof. By integrating on S_i both side of the equation in (1.1), we have

$$\iint_S \left[- \left(\frac{\partial}{\partial x} p \frac{\partial v}{\partial x} + \frac{\partial}{\partial y} p \frac{\partial v}{\partial y} \right) + cu \right] ds = \iint_S f ds.$$

By applying Lemma 1.2,

$$\iint_{S_i} \left(\frac{\partial}{\partial x} p \frac{\partial u}{\partial x} + \frac{\partial}{\partial y} p \frac{\partial u}{\partial y} \right) ds = \oint_{\partial S_i} p \left[\frac{\partial u}{\partial x} \cos(\nu, x) + \frac{\partial u}{\partial y} \cos(\nu, y) \right] d\ell$$

$$= \oint_{\partial S_i} p \frac{\partial u}{\partial \nu} d\ell.$$

The desired results (1.44) and (1.45) are obtained by applying the Neumann and Robin boundary conditions. ∎

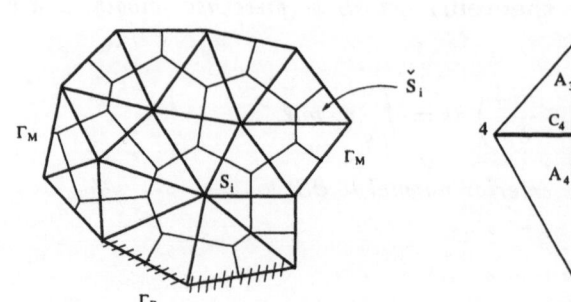

Figure 1.4   A partition of S.

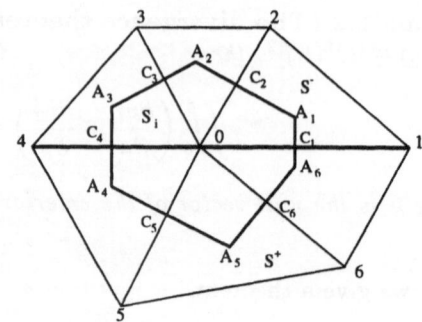

Figure 1.5   The inner node for FVM.

1.6.2   Partition without Obtuse Triangles

Let polygon S be divided into the acute (or right) triangle $\triangle_i$. Then the centers Q_i of circles passing three vertices of $\triangle_i$ are all located in the closed triangle $\overline{\triangle}_i$. Points Q_i are also the intersection points of the perpendicular bisectors ℓ of the triangle boundaries. Therefore, the solution domain S can be divided by ℓ into subdomain S_i, such that $S = \cup_i S_i$ (see Figure 1.4).

Now, let us establish the discrete equations based on Theorem 1.4. We have the approximation from (1.44) (see Figure 1.5)

$$\oint_{\partial S_i} p\frac{\partial u}{\partial \nu}d\ell \;=\; \sum_{i=1}^{n} \int_{\overline{A_i A_{i-1}}} p\frac{\partial u}{\partial \nu}d\ell \approx \sum_{i=1}^{n} p\frac{\partial u}{\partial \nu}\Big|_{C_i} \cdot \overline{A_i A_{i-1}}$$

$$\approx \sum_{i=1}^{n} p(C_i)\frac{u_i - u_0}{|\overline{0i}|}\overline{A_i A_{i-1}}. \tag{1.46}$$

where $A_n = A_0$ and $|\overline{0i}|$ denotes the length of section $\overline{0i}$. Also

$$\iint_{S_i} cuds \approx Area\,(S_i)c_0 u_0, \qquad \iint_{S_i} fds \approx Area\,(S_i)f_0.$$

Therefore, the interior discrete form of (1.44) is given by

$$\sum_{i=0}^{n} a_i u_i = Area\,(S_i)f_0, \tag{1.47}$$

where

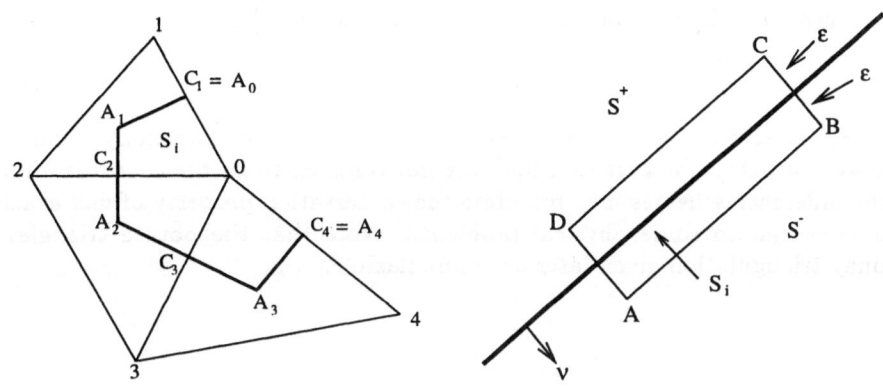

Figure 1.6 The boundary node for FVM.

Figure 1.7 An interface boundary.

$$a_i = -p(C_i)\frac{\overline{A_i A_{i-1}}}{|\overline{0i}|}, \qquad i = 1, 2, \ldots, n,$$

$$a_0 = -\sum_{i=1}^{N} a_i + Area(S_i)c_0.$$

For the Neumann and Robin conditions, we may find from (1.45) the boundary discrete form. Take an example the Robin boundary condition (see Figure 1.6).

$$\int_{\Gamma_i^*} p\frac{\partial u}{\partial \nu} d\ell \approx \sum_{i=1}^{4} p(C_i)\frac{u_i - u_0}{|\overline{0i}|}\overline{A_i A_{i-1}},$$

$$\int_{\Gamma_M \cap S_i} (g_3 - \alpha u) d\ell = ((g_3)_0 - \alpha_0 u_0)\frac{|\overline{01}| + |\overline{04}|}{2}.$$

Hence the boundary difference equation is given by

$$\sum_{i=0}^{4} b_i u_i = Area\,(S_i)f_0 + \frac{1}{2}\left(|\overline{01}| + |\overline{04}|\right)(g_3)_0, \qquad (1.48)$$

where

$$b_i = -p(C_i)\frac{\overline{A_{i-1} A_i}}{|\overline{0i}|}, \qquad i = 1, 2, 3, 4, \qquad (1.49)$$

$$b_0 = -\sum_{i=1}^{4} b_i + Area(S_i)C_0 + \frac{\alpha_0}{2}\left(|\overline{01}| + |\overline{04}|\right).$$

We represent the discrete equations from (1.47) and (1.48) as

$$\mathbf{A}_C \mathbf{x} = \mathbf{b}, \tag{1.50}$$

where $\mathbf{A}_C$ is also positive definite, symmetric sparse, and diagonally dominant. Compared to FDM, the partition lines are not confined to be the coordinate lines, and the difference schemes may maintain the conservation property of flux exactly, which is crucial to some physical problems. Note that the obtuse triangles of Delaunay triangulation given later are more flexible.

1.6.3   Interface Conditions

The conservation scheme may deal well with problems on the mixed media of different materials. Consider

$$-\left(\frac{\partial}{\partial x} p \frac{\partial u}{\partial x} + \frac{\partial}{\partial y} p \frac{\partial u}{\partial y}\right) + cu = f, \qquad \text{in } S^+ \text{ and } S^-, \tag{1.51}$$

$$u^+ = u^-, \quad p^+ \frac{\partial u^+}{\partial \nu} = p^- \frac{\partial u^-}{\partial \nu}, \quad \text{on } \Gamma_0,$$

where Γ_0 is the interior boundary of S^+, ν is the outer normal to ∂S^+, and p may have discontinuity on Γ_0:

$$p^+ \neq p^-, \quad \text{on } \Gamma_0. \tag{1.52}$$

Note that Eqs (1.44) and (1.45) may produce the interface condition automatically

$$p^+ \frac{\partial u^+}{\partial \nu} = p^- \frac{\partial u^-}{\partial \nu}, \quad \text{on } \Gamma_0.$$

Let S_i be a narrow trip with the center line Γ_0 shown in Figure 1.7. The width 2ε is much smaller than the length $L = \overline{AB}$: $2\varepsilon << L$. We have from (1.44)

$$\int_{\overline{CD}} p^+ \frac{\partial u^+}{\partial \nu} d\ell - \int_{\overline{AB}} p^- \frac{\partial u^-}{\partial \nu} d\ell = O(\varepsilon).$$

Based on the mean value theorem of integration, there exist $Q^+ \in \overline{CD}$ and $Q^- \in \overline{AB}$ such that

$$L \left[p^+ \frac{\partial u^+}{\partial \nu} \bigg|_{Q^+} - p^- \frac{\partial u^-}{\partial \nu} \bigg|_{Q^-} \right] = O(\varepsilon).$$

When $\varepsilon \to 0$,

$$p^+ \left.\frac{\partial u^+}{\partial \nu}\right|_{Q^+} = p^- \left.\frac{\partial u^-}{\partial \nu}\right|_{Q^-}, \quad \text{at } Q^\pm \in \Gamma_0.$$

Hence for any point on Γ_0, we may construct such a narrow strip with center Q. If letting $\varepsilon \to 0$, then $L \to 0$ and $Q^+ = Q^- = Q$, we conclude

$$p^+ \frac{\partial u^+}{\partial \nu} = p^- \frac{\partial u^-}{\partial \nu} \quad \text{on } \Gamma_0.$$

The solution continuity $u \in C(S)$ also leads to

$$u^+ = u^-, \quad \text{on } \Gamma_0.$$

Hence, we may easily form the conservative schemes for mixed media consisting of different materials. Let $S = S^+ \cup S^-$, and Γ_0 the boundary between two different materials: $p = p^+$ in S^+ and $p = p^-$ in S^-. Compared to (1.46), we should distinguish the integrals on either S^+ or S^- in Figure 1.5. In fact,

$$
\begin{aligned}
\int_{\overline{A_6 A_1}} p \frac{\partial u}{\partial \nu} d\ell &= \int_{\overline{A_6 C_1}} p^+ \frac{\partial u^+}{\partial \nu} d\ell + \int_{\overline{C_1 A_1}} p^- \frac{\partial u^-}{\partial \nu} d\ell \\
&\approx \left. p^+ \frac{\partial u^+}{\partial \nu}\right|_{C_1} \overline{A_6 C_1} + \left. p^- \frac{\partial u^-}{\partial \nu}\right|_{C_1} \overline{C_1 A_1} \\
&\approx \left(p^+|_{C_1} \overline{A_6 C_1} + p^-|_{C_1} \overline{C_1 A_1} \right) \frac{u_1 - u_0}{|\overline{01}|}.
\end{aligned}
\tag{1.53}
$$

Similarly

$$\int_{\overline{A_3 A_4}} p \frac{\partial u}{\partial \nu} d\ell \approx \left(p^-|_{C_4} \overline{A_3 C_4} + p^+|_{C_4} \overline{C_4 A_4} \right) \frac{u_4 - u_0}{|\overline{04}|}.$$

We obtain the difference equation at node 0 in Figure 1.5

$$\sum_{i=0}^{6} a_i^* u_0 = \text{Area}(S_i) f_0, \tag{1.54}$$

where

$$
\begin{aligned}
a_1^* &= -\left(p^+|_{C_1} \overline{A_6 C_1} + p^-|_{C_1} \overline{C_1 A_1} \right) \frac{1}{|\overline{01}|}, \\
a_4^* &= -\left(p^+|_{C_4} \overline{A_4 C_4} + p^-|_{C_4} \overline{C_4 A_3} \right) \frac{1}{|\overline{04}|}, \\
a_i^* &= a_i, \quad i = 2, 3, 5, 6,
\end{aligned}
$$

and

$$a_0^* = - \sum_{i=1}^{6} a_i^* + Area\ (S_i)c_0.$$

When $p^+ = p^-$, Eq (1.54) leads to (1.47).

Note that the intersection of material interfaces may incur singularity, see Kellogg (1971, 1972, 1975). The study on elliptic problems with interface singularity are also reported in Babuska (1974, 1976a,b, 1977), Canuto and Pietra (1990), Funaro, Quarteroni and Zanolli (1988), Goldstein (1983), Ikebe, Lynn and Timlake (1969), King (1975), MacCamy and Marin (1980), MacKinnon and Carey (1988), Phillips and Davies (1988), Sheldon (1958), Li (1990), and Bramble and King (1997).

1.6.4   View of FVM as FDM

When the subdomains ∂S_i are formed by the coordinate lines as in Section 1.5 we can derive the same interior discrete equations and slightly different boundary discrete equations in FDM. For S_i shown in Figure 1.1, we have

$$\oint_{\partial S_i} p\frac{\partial u}{\partial \nu}d\ell = \int_{\overline{BD}} p\frac{\partial u}{\partial x}d\ell - \int_{\overline{AC}} p\frac{\partial u}{\partial x}d\ell + \int_{\overline{CD}} p\frac{\partial u}{\partial y}d\ell - \int_{\overline{AB}} p\frac{\partial u}{\partial y}d\ell,$$

where

$$\int_{\overline{BD}} p\frac{\partial u}{\partial x}d\ell \approx \left(p\frac{\partial u}{\partial x}\right)_{i+\frac{1}{2},j} \frac{k_j + k_{j+1}}{2} \approx p_{i+\frac{1}{2},j}\frac{u_{i+1,j} - u_{i,j}}{h_{i+1}} \times \frac{k_j + k_{j+1}}{2}.$$

Similarly,

$$\int_{\overline{AC}} p\frac{\partial u}{\partial x}d\ell \approx p_{i-\frac{1}{2},j}\frac{u_{i,j} - u_{i-1,j}}{h_i} \times \frac{k_j + k_{j+1}}{2},$$

$$\int_{\overline{CD}} p\frac{\partial u}{\partial y}d\ell \approx p_{i,j+\frac{1}{2}}\frac{u_{i,j+1} - u_{i,j}}{k_{j+1}} \times \frac{h_i + h_{i+1}}{2},$$

$$\int_{\overline{AB}} p\frac{\partial u}{\partial y}d\ell \approx p_{i,j-\frac{1}{2}}\frac{u_{i,j} - u_{i-1,j}}{k_j} \times \frac{h_i + h_{i+1}}{2}. \tag{1.55}$$

Also, $Area\ (S_i) = \frac{1}{4}(h_i + h_{i+1})(k_j + k_{j+1})$. Then we obtain the five-point difference equation.

$$\overline{a}_{ij}u_{i,j} - \overline{a}_{ij}^x u_{i+1,j} - \overline{b}_{ij}^x u_{i-1,j} - \overline{a}_{ij}^y u_{i,j+1} - \overline{b}_{ij}^y u_{i,j-1} = Area\ (S_i)f_{ij},$$

$$\tag{1.56}$$

where

$$
\begin{aligned}
\bar{a}_{ij}^x &= p_{i+\frac{1}{2},j}\frac{k_j + k_{j+1}}{2}\frac{1}{h_{i+1}}, \quad \bar{b}_{ij}^x = p_{i-\frac{1}{2},j}\frac{k_j + k_{j+1}}{2}\frac{1}{h_i}, \\
\bar{a}_{ij}^y &= p_{i,j+\frac{1}{2}}\frac{h_i + h_{i+1}}{2}\frac{1}{k_{j+1}}, \quad \bar{b}_{ij}^y = p_{i,j-\frac{1}{2}}\frac{h_i + h_{i+1}}{2}\frac{1}{k_j}, \\
\bar{a}_{ij} &= \bar{a}_{ij}^x + \bar{b}_{ij}^x + \bar{a}_{ij}^y + \bar{b}_{ij}^y + Area\ (S_i)c_{ij}.
\end{aligned}
$$

By dividing both sides with the factor $Area\ (S_i)$, it is easy to see that Eq. (1.56) leads to (1.28) by FDM.

Next, we can provide the boundary difference equations at node (i, j) in Figure 1.2. We have

$$
\oint_{\partial S_i} p\frac{\partial u}{\partial \nu}d\ell = -\int_{AB} p\frac{\partial u}{\partial y}d\ell - \int_{AC} p\frac{\partial u}{\partial x}d\ell + \int_{BF\cup FC}(g_3 - \alpha u)d\ell,
$$

where $\int_{AB} p\frac{\partial u}{\partial y}d\ell$ and $\int_{AC} p\frac{\partial u}{\partial x}d\ell$ are given in (1.55), but

$$
\begin{aligned}
\int_{BF\cup FC}(g_3 - \alpha u)d\ell &\approx [(g_3)_0 - \alpha_0 u_0]\left(\overline{BF} + \overline{FC}\right) \\
&= [(g_3)_0 - \alpha_0 u_0]\frac{1}{2}\left(\sqrt{k_{j+1}^2 + h_i^2} + \sqrt{k_j^2 + h_{i+1}^2}\right).
\end{aligned}
$$

Then the boundary discrete equation is given by

$$
\begin{aligned}
\bar{a}_{ij}u_{i,j} &- \bar{b}_{ij}^x u_{i,j-1} - \bar{b}_{ij}^y u_{i,j-1} \\
&= Area\ (S_i)f_{ij} + \frac{(g_3)_{ij}}{2}\left(\sqrt{k_{j+1}^2 + h_i^2} + \sqrt{k_j^2 + h_{i+1}^2}\right).
\end{aligned} \tag{1.57}
$$

where

$$
\bar{a}_{ij} = \bar{b}_{ij}^x + \bar{b}_{ij}^y + c_{ij}Area\ (S_i) + \frac{\alpha_{ij}}{2}\left(\sqrt{k_{j+1}^2 + h_i^2} + \sqrt{k_j^2 + h_{i+1}^2}\right),
$$

and

$$
Area\ (S_i) = \frac{1}{8}\left[2h_i k_j + h_i k_{i+1} + h_{i+1}k_j\right].
$$

Note that Eq. (1.57) leads to (1.42), derived from (1.36).

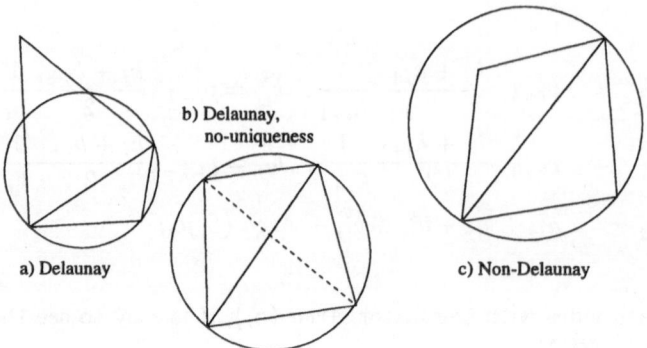

Figure 1.8 Different triangles.

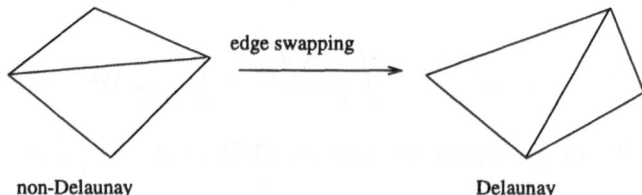

Figure 1.9 Edge swapping.

1.6.5 Partition with Delaunay Triangulation

An auxiliary Voronoi diagram (or Dirichlet tessellation) is formed by the perpendicular bisectors of all triangle edges. The Voronoi polygons of the Delaunay triangulation can be defined by

$$S = S_h = \cup_{i=1}^{N} S_i, \tag{1.58}$$

where the convex polygons

$$S_i = \{x \in S, |x - x_i| < |x - x_j|, x_i \in X_h, j \neq i\}, \tag{1.59}$$

and X_h is the set of all vertices. Eq. (1.59) implies that the *ith* polygon S_i contains all points closest to vertex i.

Since S can be partitioned into small triangles, we may then check whether or not all triangles are of Delaunay triangulation (see Figure 1.8). If not, we may swap the diagonal to reform the non-Delaunay triangles to the Delaunay triangles (see Figure 1.9). Now we provide several properties of Delaunay triangles; their proofs are given in Li and Wang (1997).

Property 1.1 *For a polygonal S, there exists a Delaunay triangulation.*

Property 1.2 *The algebraic length of connected lines between circumcenters of the neighbouring triangles is non-negative in the Delaunay triangulation.*

Property 1.3 *The Voronoi polygons from the Delaunay triangulation have the positive edges and areas.*

With these properties, we may similarly derive the conservative schemes as done in Section 1.6.2, and also obtain the linear algebraic equations

$$\mathbf{Ax} = \mathbf{b}, \tag{1.60}$$

where $\mathbf{A}$ is positive definite, symmetric, sparse and M-matrix-like (see Varga (1962)). Based on Property 1.2 and 1.3 the Delaunay triangulation guarantees stability of the solutions of (1.60).

The straightforward evaluation on entries of $\mathbf{A}$ is very promising, in contrast to the complicated computation in FEM. Since obtuse triangles are allowed, the FVM given in this subsection may also be suited to rather arbitrary shapes of S. Moreover, Eq.(1.60) also reflects the conservative law in physics. This is particularly attractive to many physical problems. Consequently, the FVM may also compete with other numerical methods.

1.6.6   New View of FVM as Galerkin FEM

The finite volume method can be regarded as the Petrov-Galerkin finite method with dual spaces, where the piecewise linear functions are chosen on the Delaunay triangulation, and the piecewise constants on the Voronoi polygons. Since the min-max condition of Ladyzhenskaya-Babuska-Brezzi is not easy to be proven, we will invoke the Galerkin finite method, where both the solution and trial functions are chosen to be the same piecewise linear functions on Delaunay triangulation. Note that here the admissible functions chosen are conforming. For simplicity, let $u|_\Gamma = 0$ in (1.32); Eqs (1.31) and (1.32) can be represented in a weak form:

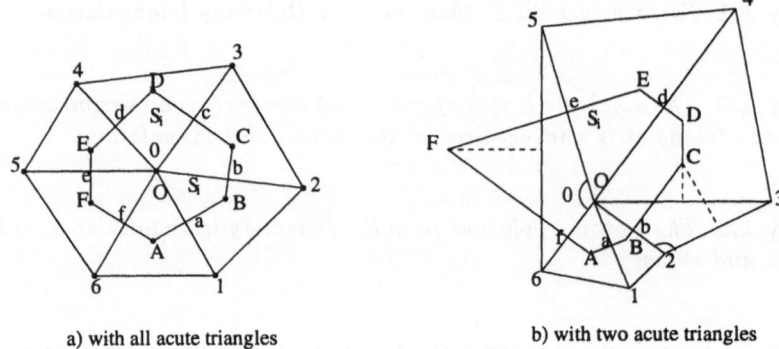

a) with all acute triangles   b) with two acute triangles

Figure 1.10   The inner node O for FVM on Delaunay triangulation.

Find the solution $u \in H_0^1(S)$ such that

$$A(u, v) = f(v), \qquad \forall v \in H_0^1(S),$$

where

$$A(u, v) = \iint_S (p\, \nabla u\, \nabla v + cuv)\, ds, \quad f(v) = \iint_S fv\, ds,$$

where $H_0^1(S)$ $(\subset H^1(S))$ is the Sobolev space such that

$$H_0^1(S) = \left\{v \mid v, v_x, v_y \in L^2(S) \text{ and satisfy } u = 0 \text{ on } \Gamma\right\}.$$

Let $V_h^0 \subset H_0^1(S)$ be the finite dimensional collection of piecewise linear interpolatory functions on the Delaunay triangulation, S_h. We may obtain the FEM with approximation integration:

To seek $\tilde{u}_h \in V_h^0$ such that

$$\hat{A}_h(\tilde{u}_h, v) = \hat{f}(v), \qquad \forall v \in V_h^0, \tag{1.61}$$

where

$$\hat{A}_h(u, v) = \widehat{\iint_S} p\, \nabla u\, \nabla v\, ds + \widehat{\iint_S} cuv\, ds, \quad \hat{f}_h(v) = \widehat{\iint_S} fv\, ds.$$

We denote by $\widehat{\iint}$ and $\hat{\int}$ the approximation of integrals $\iint$ and $\int$. However different rules of integration may be chosen for different integrals. For the Delaunay triangulation in Figure 1.10b,

$$\hat{A}_h(u,v) = \sum_i \left\{ \widehat{\iint_{\triangle_i} p \, \nabla u \, \nabla v ds} + \widehat{\iint_{\triangle_i} cuv ds} \right\}, \qquad (1.62)$$

$$\hat{f}(v) = \sum_i \widehat{\iint_{\triangle_i} fv ds}. \qquad (1.63)$$

Based on Li and Wang (1997),

$$\iint_{\triangle_i} p \, \nabla u \, \nabla v ds = \sum_{k=1}^{3} \iint_{\triangle_i^k} p \left(u_l v_l + u_n v_n \right) ds \approx 2 \sum_{k=1}^{3} \iint_{\triangle_i^k} p u_l v_l ds, \qquad (1.64)$$

where $\triangle_i = \cup_{i=1}^3 \triangle_i^k$, and $\triangle_i^k$ is the sub-triangles of $\triangle_i$ by splitting with the edges between the vertices and circumcenters. In (1.64), u_l and u_n are the derivatives along the outside edges and their outward normals. Hence, Eqs. (1.62) and (1.63) are reduced to

$$\hat{A}_h(u,v) = \sum_i \sum_{k=1}^{3} \left\{ 2 \widehat{\iint_{\triangle_i^k} p u_l v_l ds} + \widehat{\iint_{\triangle_i^k} cuv ds} \right\}, \qquad (1.65)$$

$$\hat{f}_h(v) = \sum_i \sum_{k=1}^{3} \left\{ \widehat{\iint_{\triangle_i^k} fv ds} \right\}. \qquad (1.66)$$

The FVM described in Section 1.6.5 can also derived from (1.61), (1.65) and (1.66) exactly, with the help of special approximation integration, see Li and Wang (1997).

The finite volume method (FVM) may also be called the box integration in Varga (1962) and Hackbusch (1989), conservative scheme, or cell discretizations. The main references on FVM can be found in Varga (1962), McCormick (1989, 1992), Morton (1996), Bank and Rose (1987), Cai(1991), Hackbusch (1992), Hennart, Jaffre and Roberts (1988), Morton (1996), Roos, Stynes and Tobiska (1996), Goldlewski and Raviart (1996), Feistauer, Flcman and Lukacova-Medvidova (1995, 1997), and Miller and Wang (1994).

1.7   BOUNDARY ELEMENT METHOD

1.7.1   Preliminary Lemmas and Theorems

Let us cite the Green formula.

Lemma 1.3 (Green formula) *Let ∂S be piecewise smooth, and u, $v \in C^2(S)$.
Then*

$$\iint_S (vTu - uTv)ds = \oint_{\partial S} \left(vp\frac{\partial u}{\partial \nu} - up\frac{\partial v}{\partial \nu} \right) d\ell,$$

where

$$Tu = \left(\frac{\partial}{\partial x}p\frac{\partial u}{\partial x} + \frac{\partial}{\partial y}p\frac{\partial u}{\partial y} \right).$$

The ideas of boundary element method lie in the concept of using the fundamental or particular solutions, from which the solution of (1.1) within S can be represented by the values of u and $\frac{\partial u}{\partial \nu}$ on the boundary ∂S based on Lemma 1.4 shown later. For each point on ∂S, u and $\frac{\partial u}{\partial \nu}$ can also be solved, purely on the boundary ∂S, based on Lemma 1.5. Hence this approximate method is called the boundary element method (BEM). For simplicity, let us consider the Poisson equation.

$$-\triangle u = f, \quad \text{in } S \tag{1.67}$$

with the boundary conditions in (1.1), where $\triangle = \frac{\partial^2}{\partial x^2} + \frac{\partial^2}{\partial y^2}$.

Choose the fundamental solution

$$w = \frac{1}{2\pi}\ln\frac{1}{r}, \quad r = \frac{1}{\sqrt{(x-x_0)^2 + (y-y_0)^2}}, \quad M_0 = (x_0, y_0) \in S, \tag{1.68}$$

to satisfy $\triangle w = 0$. Now let us provide two important lemmas; proof of which can be derived from Lemma 1.3, or found in the textbooks (Chen and Zhou (1992)).

Lemma 1.4 *Let $\frac{\partial u}{\partial \nu} \in C(\partial S)$, then the solution*

$$u(M_0) = \frac{1}{2\pi}\left\{ \int_{\partial S}\left[\left(\ln\frac{1}{r}\right)\frac{\partial u}{\partial \nu} - u\frac{\partial}{\partial \nu}\left(\ln\frac{1}{r}\right) \right] d\ell + \iint_S \left(\ln\frac{1}{r}\right)f ds \right\}, \forall M_0 \in S.$$

$$\tag{1.69}$$

It can be seen from (1.69) that, if both u and $\frac{\partial u}{\partial \nu}$ on ∂S are known, the interior solutions may be obtained immediately. However, only one of u and $\frac{\partial u}{\partial \nu}$ on ∂S is known from the boundary values in (1.67), the other boundary values can be obtained below.

Lemma 1.5 *Let* $\frac{\partial u}{\partial \nu} \in C(S)$, *then*

$$u(M_0) = \frac{1}{\pi} \left\{ \int_{\partial S} \left[\ln \left(\frac{1}{r} \right) \frac{\partial u}{\partial \nu} - u \frac{\partial}{\partial \nu} \left(\ln \frac{1}{r} \right) \right] d\ell + \right.$$
$$\left. + \iint_S \left(\ln \frac{1}{r} \right) f ds \right\}, \forall M_0 \in \partial S. \tag{1.70}$$

From the boundary conditions in (1.1), we have

$$\int_{\partial S} u \frac{\partial}{\partial \nu} \left(\ln \frac{1}{r} \right) d\ell = \int_{\Gamma_D} g_1 \frac{\partial}{\partial \nu} \ln \left(\frac{1}{r} \right) d\ell + \int_{\Gamma_N \cup \Gamma_M} u \frac{\partial}{\partial \nu} \ln \left(\frac{1}{r} \right) d\ell.$$

$$\int_{\partial S} \ln \left(\frac{1}{r} \right) \frac{\partial u}{\partial \nu} d\ell = \int_{\Gamma_D} \ln \left(\frac{1}{r} \right) \frac{\partial u}{\partial \nu} d\ell + \int_{\Gamma_N} g_2 \ln \left(\frac{1}{r} \right) d\ell + \int_{\Gamma_M} (g_3 - \alpha u) \ln \left(\frac{1}{r} \right) d\ell.$$

Hence we obtain from Lemma 1.5

$$u(M_0) = \frac{1}{\pi} \left\{ \int_{\Gamma_D} \frac{\partial u}{\partial \nu} \ln \left(\frac{1}{r} \right) d\ell - \int_{\Gamma_M} u \left[\frac{\partial}{\partial \nu} \ln \left(\frac{1}{r} \right) + \alpha \ln \left(\frac{1}{r} \right) \right] d\ell + \right.$$
$$- \int_{\Gamma_N} u \frac{\partial}{\partial \nu} \ln \left(\frac{1}{r} \right) d\ell - \int_{\Gamma_D} g_1 \frac{\partial}{\partial \nu} \ln \left(\frac{1}{r} \right) d\ell + \int_{\Gamma_N} g_2 \ln \left(\frac{1}{r} \right) d\ell +$$
$$\left. + \int_{\Gamma_M} g_3 \ln \left(\frac{1}{r} \right) d\ell + \iint_S \ln \left(\frac{1}{r} \right) f ds \right\}, \quad \forall M_0 \in \partial S. \tag{1.71}$$

Hence we may obtain the needed solutions. Below let us describe the discrete form of (1.71).

1.7.2 Discrete Approximation

Let ∂S be divided into quasiuniform subdomains (s_i, s_{i+1}), with the length h_{i+1}, $s_{i+1} = s_i + h_{i+1}$, and $s_0 = s_N$. Also choose piecewise constant and linear functions as the admissible functions of $\frac{\partial v}{\partial \nu}$ and ν respectively,

$$\frac{\partial u}{\partial \nu} = \sum_{i=1}^N (u_\nu)_i \Psi_i(s), \quad u = \sum_{i=1}^N u_i \Phi_i(s),$$

where $\Psi_i(s)$ and $\Phi_i(s)$ are the basis functions shown as

$$\Psi_i(s) = \begin{cases} 1, & s_i - \frac{h_i}{2} < s < s_i + \frac{h_{i+1}}{2}, \\ 0, & \text{otherwise}, \end{cases}$$

and

$$\Phi_i(s) = \begin{cases} \frac{1}{h_i}(s - s_{i-1}), & s_{i-1} \le s \le s_i, \\ 1 - \frac{1}{h_{i+1}}(s - s_i), & s_i \le s \le s_{i+1}, \\ 0, & \text{otherwise}. \end{cases}$$

By integration approximation in (1.71), there are n equations and n unknowns including either u_i or $\left(\frac{\partial u}{\partial \nu}\right)_i$, leading to a linear algebraic system,

$$\mathbf{A}_B \mathbf{x}_B = \mathbf{b}, \tag{1.72}$$

where $\mathbf{A}_B$ is an $n \times n$ matrix, which is neither symmetric nor sparse. Once the solution of (1.72) is obtained, we may seek the solution in S from Lemma 1.4

$$u(M_0) = \frac{1}{2\pi}\left[\sum_{i=1}^{N}(u_\nu)_i \int_{\partial S} \ln\left(\frac{1}{r}\right)\Psi_i(t)d\ell - \sum_{i=1}^{N}u_i \int_{\partial S} \frac{\partial}{\partial \nu}\ln\left(\frac{1}{r}\right)\Phi_i(t)d\ell + \right.$$
$$\left. + \iint_S \ln\left(\frac{1}{r}\right)f ds\right].$$

Let us make a few comments.

BEM in (1.71) is of one-dimensional solver, one dimension lower than that in FEM; the unknown number in BEM is substantially smaller than that in FEM. This is a remarkable advantage. However, when $f \ne 0$, the approximation integration $\iint_S \ln\left(\frac{1}{r}\right)f ds$ may consume much more CPU time than the solution procedure of (1.72). Hence, BEM is strongly be recommended when $f \equiv 0$. It is worth noting that BEM is more beneficial for 3D elliptic problems.

1.7.3   Galerkin Approach

For simplicity, consider the Dirichlet-Laplace equation

$$\Delta u = 0, \text{ in } S, \quad u|_\Gamma = g, \text{ on } \partial S.$$

We rewrite the fundamental functions

$$w = w(X) = \frac{1}{2\pi}\ln\frac{1}{r} = \frac{1}{2\pi}\ln\frac{1}{|X-Y|},$$

where X and Y denote two points on ∂S. Then (1.70) leads to

$$u(Y) = \frac{1}{\pi}\left(\int_{\partial S} \frac{\partial u(X)}{\partial \nu}\ln\frac{1}{|X-Y|}d\ell(X) - \int_{\partial S} g(X)\frac{\partial}{\partial \nu}(\ln\frac{1}{|X-Y|})d\ell(X)\right),$$

where $Y = M_0 \in \partial S$. By multiplying two sides by $v(Y)$ and integrating along ∂S, we obtain

$$\int_{\partial S} u(Y)v(Y)d\ell(Y) = \frac{1}{\pi}\int_{\partial S}\int_{\partial S} \frac{\partial u(X)}{\partial \nu} \ln \frac{1}{|X-Y|} v(Y)d\ell(Y)d\ell(X)$$
$$-\frac{1}{\pi}\int_{\partial S}\int_{\partial S} g(X)\frac{\partial}{\partial \nu}(\ln \frac{1}{|X-Y|})v(Y)d\ell(Y)d\ell(X),$$

where $\frac{\partial u}{\partial \nu} \in H^{-\frac{1}{2}}(\Gamma)$, and $v \in H^{\frac{1}{2}}(\Gamma)$. Hence, find $q(X) \in H^{-\frac{1}{2}}(\Gamma)$ such that

$$Q(q,v) = \int_{\partial S} gvd\ell + \frac{1}{\pi}\int_{\partial S}\int_{\partial S} g(X)\frac{\partial}{\partial \nu}(\ln \frac{1}{|X-Y|})v(Y)d\ell(X)d\ell(Y),$$

where

$$Q(q,v) = \frac{1}{\pi}\int_{\partial S}\int_{\partial S} \ln \frac{1}{|X-Y|} q(X)v(Y)d\ell(X)d\ell(Y).$$

Let V_h be a space of piecewise constant or piecewise linear spaces, and $L_i(\ell)$ be the basis functions. Then $q_h(\ell) = \sum_{j=1}^N q_j L_j(\ell(X))$.

The Galerkin approach is to find $q_h \in S_h$ such that

$$Q(q_h, v_h) = \int_\Gamma gv_h d\ell + \frac{1}{\pi}\int_{\partial S}\int_{\partial S} g(X)\frac{\partial}{\partial \nu}(\ln \frac{1}{|X-Y|})v(Y)d\ell(X)d\ell(Y), \quad \forall v \in V_h.$$

1.7.4 Natural BEM

Consider the pure Neumann-Laplace problem

$$\triangle u = 0 \text{ in } S, \quad \frac{\partial u}{\partial \nu} = g \text{ on } \partial S.$$

The consistent condition

$$\int_{\partial S} \frac{\partial u}{\partial \nu} d\ell = \oint_{\partial S} g d\ell = 0$$

guarantees the solutions with an arbitrary constant. From (1.70), we have

$$u(Y) = \frac{1}{\pi}\left\{\int_{\partial S} \frac{\partial u(X)}{\partial \nu} \ln \frac{1}{|X-Y|}d\ell(X) - \int_{\partial S} u(X)\frac{\partial}{\partial \nu}\ln \frac{1}{|X-Y|}d\ell(X)\right\},$$

to lead to

$$\frac{\partial u(Y)}{\partial \nu_Y} = \frac{1}{\pi}\left\{\int_{\partial S} \frac{\partial u(X)}{\partial \nu}\frac{\partial}{\partial \nu_Y} \ln \frac{1}{|X-Y|}d\ell(X) - \int_{\partial S} u(X)\frac{\partial}{\partial \nu}\frac{\partial}{\partial \nu_Y} \ln \frac{1}{|X-Y|}d\ell(X)\right\}.$$

By applying the Neumann boundary condition,

$$g = \frac{1}{\pi}\left\{\int_{\partial S} g\frac{\partial}{\partial \nu_Y}\ln\frac{1}{|X-Y|}d\ell(Y) - \int_{\partial S} u\frac{\partial}{\partial \nu_Y}\frac{\partial}{\partial \nu_X}\ln\frac{1}{|X-Y|}d\ell(Y)\right\}.$$

Hence, the solutions on ∂S can be sought by the natural FEM:

Find $u \in H^{\frac{1}{2}}(\Gamma)/P_0$ such that

$$D(u, v) = \int_{\partial S} gv d\ell,$$

where P_0 is the space of arbitrary constants, and

$$
\begin{aligned}
D(u, v) = \frac{1}{\pi}\Bigg\{ &\int_{\partial S}\int_{\partial S}\left(g\frac{\partial}{\partial \nu_Y}\ln\frac{1}{|X-Y|}\right)v d\ell(Y)d\ell(X) \\
&- \int_{\partial S}\int_{\partial S}\left(\frac{\partial}{\partial \nu_Y}\frac{\partial}{\partial \nu_X}\ln\frac{1}{|X-Y|}\right)uv d\ell(Y)d\ell(X)\Bigg\}.
\end{aligned}
$$

References on BEM are given in Brebbia (1978), Wendland (1983), Chen and Zhou (1992), Yu (1993), and Brebbia and Dominguez (1989). More study is given in Cruse (1988), Bayliss, Gunzburger and Turkel (1982), Han and Wu (1986), Hsiao (1986, 1988a), Wendland (1983), Yu (1987), and Feng and Yu (1994).

1.8   COLLOCATION METHOD

1.8.1   Algorithms

In FDM, the differentiations are replaced by finite differences of the PDEs at isolated points. In FEM, the admissible functions are chosen as piecewise continuous polynomials, and applied to its weak form of PDEs. Once the continuous admissible functions are differentiable, we may enforce them to satisfy exactly the PDE at certain points. This leads to the collocation method (CM), or the residual method as called in engineering literature, see Percell and Wheeler (1978, 1980). By an equivalence between the Galerkin-FEM and the collocation method, e.g., see Swartz and Wendroff (1974), the algebraic equations can be easily formulated from the CM. However, some difficulties arise in CM:

1. The discrete equations may not be symmetric even for the self-adjoint elliptic equation.

2. The number of unknowns may not be just equal to that of the discrete collocation equations.

Let us rewrite the elliptic problem (1.1) as,

$$\begin{cases} Lu = f, & \text{in } S, \\ Bu = g, & \text{on } \partial S, \end{cases} \tag{1.73}$$

where

$$L = -\frac{\partial}{\partial x} p \frac{\partial}{\partial x} - \frac{\partial}{\partial y} p \frac{\partial}{\partial y} + c,$$

and the operators, $B = 1$, $p\frac{\partial}{\partial \nu}$, $p\frac{\partial}{\partial \nu} + \alpha$, for three kinds of boundary conditions respectively. Choose the smooth basis functions Ψ_i so as to have 2-order differentiation. Then we can span the solution by

$$u = \sum_{i=1}^{N} a_i \Psi_i$$

approximately. Hence (1.73) leads to

$$\begin{cases} \sum_{i=1}^{N} a_i L\Psi_i = f, \\ \sum_{i=1}^{N} a_i B\Psi_i = g. \end{cases} \tag{1.74}$$

Since there are N unknowns, we may choose N collocation points to enforce (1.74) such that

$$\sum_{i=1}^{N} a_i L\Psi_i(Q_i) = f(Q_i), \quad i = 1, 2, \ldots, M, \quad Q_i \in S,$$

$$\sum_{i=1}^{N} a_i B\Psi_i(Q_i) = g(Q_i), \quad i = M+1, \ldots, N, \quad Q_i \in \partial S.$$

Hence we also obtain the discrete equations

$$A_C x_C = b.$$

Suppose that the basis functions are the particular solutions satisfying $Lu = f$ already. Then only the boundary concrete equations

$$\sum a_i B\Psi_i(Q_i) = g(Q_i), \quad Q_i \in \partial S, \quad i = 1, 2, \ldots, N.$$

need to be satisfied. This is the simplified CM, see the boundary solution procedure in Zienkiewicz, Kelley and Bettess (1977), and the Trefftz-complete in Zielinski and Zienkiewicz (1985) and Lefeber(1989).

We may represent (1.73) as the residual form

$$\iint_S R_E w ds = \iint_S (Lu - f) w ds = 0,$$

$$\int_{\partial S} R_B w d\ell = \int_{\partial S} (Bu - g) w d\ell = 0,$$

when R_E and R_B are the residuals of the equation and their boundary conditions, and w is the delta function,

$$w = \delta(P - Q_i) = \begin{cases} \infty, & P = Q_i, \\ 0, & P \neq Q_i, \end{cases}$$

satisfying

$$\iint_S \delta(P - Q_i) f ds = f(Q_i).$$

Then, the CM is also the residual method when w is chosen as the delta function.

1.8.2   Viewpoint of BAM

Choose the middle-point rule of integration

$$\int_a^b f(x)dx \approx \widetilde{\int_a^b} f(x)dx = (b - a)f\left(\frac{a + b}{2}\right).$$

We partition ∂S into subsections with the partition nodes:

$$\Gamma_D : s_0 < s_1 < \ldots < s_M,$$
$$\Gamma_M : s_M < s_{M+1} < \ldots < s_{M+N}.$$

Hence the integrals in (1.26) in Section 1.4 become

$$I_B^*(v) \approx \tilde{I}_B^*(v) = \widetilde{\int_{\Gamma_D}} (v - g)^2 d\ell + \widetilde{\int_{\Gamma_M}} \left(p\frac{\partial u}{\partial \nu} + \alpha u - g_3\right)^2 d\ell$$

$$= \sum_{i=1}^M \left(f_{i-\frac{1}{2}}\right)^2 \delta s_i + w^2 \sum_{i=M+1}^{M+N} \left(f_{i-\frac{1}{2}}^*\right)^2 \delta s_i, \qquad (1.75)$$

where $\delta s_i = s_i - s_{i-1}$, $f = v - g_3$ and $f^* = p\frac{\partial u}{\partial \nu} + \alpha u - g_3$. The BAM involving integration approximation is

$$\widetilde{I}_B(\tilde{u}_L) = \min_{\forall v \in V_L} \widetilde{I}_B(v). \tag{1.76}$$

On the other hand, we may use the CM applied directly to two boundary conditions on the midpoints of $[s_i, s_{i+1}]$

$$\begin{cases} f_{i-\frac{1}{2}} = 0, & i = 1, 2, \ldots, M, \\ f^*_{i-\frac{1}{2}} = 0, & i = M+1, M+2, \ldots, M+N. \end{cases} \tag{1.77}$$

Suppose that $M + N > L$, the number of equations is larger than that of unknown coefficients $\tilde{a}_i$ in (1.23). A natural way is to use the least squares method. We may add some weights to the equations in (1.77) to balance the effects of different boundary conditions and different subsections. We multiply (1.77) by $\sqrt{\delta s_i}$ and $w\sqrt{\delta s_i}$, to lead to

$$\begin{cases} \sqrt{\delta s_i} f_{i-\frac{1}{2}} = 0, & i = 1, 2, \ldots, M, \\ w\sqrt{\delta s_i} f^*_{i-\frac{1}{2}} = 0, & i = M+1, M+2, \ldots, M+N. \end{cases} \tag{1.78}$$

By summing the square of all left sides in (1.78) yields exactly the energy approximation $\widetilde{I}_B^*(v)$ in (1.75). This gives an important linkage of BAM and CM.

In fact, we rewrite (1.78) as the overdetermined system

$$\mathbf{F}\mathbf{x} - \mathbf{b} = 0, \tag{1.79}$$

where $\mathbf{F} \in R^{m \times n}$, $m > n$, $\mathbf{x} \in R^n$ and $\mathbf{b} \in R^m$. Then, from the above equation we obtain

$$\begin{aligned} \widetilde{I}_B^*(v) &= (\mathbf{F}\mathbf{x} - \mathbf{b})^T (\mathbf{F}\mathbf{x} - \mathbf{b}) \\ &= \mathbf{x}^T \mathbf{F}^T \mathbf{F}\mathbf{x} - (\mathbf{b}^T \mathbf{F}\mathbf{x} + \mathbf{x}^T \mathbf{F}^T \mathbf{b}) + \mathbf{b}^T \mathbf{b}. \end{aligned}$$

Minimization of $\widetilde{I}_B^*(v)$ yields the normal equations

$$0 = \frac{\partial}{\partial \mathbf{x}^T} \widetilde{I}_B^*(v) = \mathbf{F}^T \mathbf{F}\mathbf{x} - \mathbf{F}^T \mathbf{b} = 0,$$

i.e.,

$$\mathbf{A}\mathbf{x} = \mathbf{F}^T \mathbf{F}\mathbf{x} = \mathbf{F}^T \mathbf{b}. \tag{1.80}$$

This is also equivalent to the operator multiplying $\mathbf{F}^T$ to the both sides of (1.79). However, the above simple algorithm (1.80) has a large condition number

$$Cond.(\mathbf{A}) = \frac{\lambda_{max}(\mathbf{A})}{\lambda_{min}(\mathbf{A})} = \frac{\lambda^2_{max}(\mathbf{F})}{\lambda^2_{min}(\mathbf{F})},$$

where λ_{max} and λ_{min} are the maximal and minimal singular values. Therefore, a better method is to use the least squares method using the singular value recomposition method in Golub and Loan (1989), to solve (1.79) directly, with much a smaller condition number $Cond.(\mathbf{F}) = \frac{\lambda_{max}(\mathbf{F})}{\lambda_{min}(\mathbf{F})}$.

1.8.3   Viewpoint of FEM

For problems in 2D, the view of equivalence between CM and FEM is nontrivial. In Quarteroni and Valli (1994) the collocation method may be viewed as the spectral FEM without integration approximation.

Consider the homogeneous Dirichlet conditions $u|_{\partial S} = 0$ of the unit solution domain. The orthogonal polynomials are chosen as admissible functions. Denote by V_N the space of algebraic polynomials of degree $\leq N$ with respect to x or y and $V_N^0 \subset V_N$ satisfying $u|_\Gamma = 0$. Hence the Galerkin method is to seek $u_N \in V_N$ such that

$$a(u_N, v) = f(v), \quad \forall v \in V_N^0, \tag{1.81}$$

where

$$a(u, v) = \iint_S p \, \nabla u \, \nabla v \, ds, \quad f(v) = \iint_S f v \, ds.$$

The Legendre Gauss-Lobalt rule in 2D

$$\iint_S f \, ds = \sum_{i,j=0}^{N} w_{ij} f(x_i, y_j) \tag{1.82}$$

is chosen to approximate the integrals involved in (1.81). Their nodes and weights are

$$z_{ij} = (x_i, y_j), \quad w_{ij} = w_i w_j, \quad i, \, j = 0, \ldots, N, \; N \geq 1,$$

where $\{x_i\}$, $i = 0, \ldots, N$, are the real roots of $(1 - x^2)L_N'(x) = 0$,

$$w_j = \frac{2}{N(N+1)} \frac{1}{L_N^2(x_j)}, \quad j = 0, \ldots, N,$$

and $L_N(x)$ are the Legendre polynomials.

Since the integral rule (1.82) is exact for the admissible space, the corresponding collocation equations are

$$\begin{cases} -\nabla(p\,\nabla u_N(x_i, y_j)) = f(x_i, y_j), & (x_i, y_j) \in S, \\ u_N(x_i, y_j) = 0, & \forall (x_i, y_j) \in \partial S. \end{cases}$$

When the Neumann and mixed boundary conditions are involved in, the boundary collocation equations should be modified carefully, see Quarteroni and Zampieri (1992), and Quarteroni and Valli (1994).

The references on CM are given in Karpilovskaja (1963), Kantorovich and Akilov (1964), Lucas and Reddien (1972), Russell and Shampine (1972), Wheeler (1977), Swartz and Wendroff (1974), Oliveriral (1980), Russell (1977), Quarteroni and Zampieri (1992), Quarteroni and Valli (1994), Houstis, Vavalis and Rice (1988) and Dyksen, Housties, et. al. (1984).

1.9   LEAST SQUARES METHOD

For solving (1.73), we may consider the objective energy as

$$\begin{aligned} I(v) = & \iint_S (Lu - f)^2 ds + w_1^2 \int_{\Gamma_D} (u - g_1)^2 d\ell + w_2^2 \int_{\Gamma_N} \left(\frac{\partial u}{\partial \nu} - g_2 \right)^2 d\ell \\ & + w_3^2 \int_{\Gamma_M} \left(\frac{\partial u}{\partial \nu} + \alpha u - g_3 \right)^2 d\ell. \end{aligned}$$

The least squares method (LSM) is described as:

To seek the solution

$$I(u) = \min_{\forall v \in V_h} I(v),$$

where space $V_h (\subset H^1(S))$ is the finite collection of admissible functions without satisfying any boundary conditions. The least squares method can also be viewed as the squares residual method.

The advantages of LSM are that the associated matrix A is always positive definite (or semi-definite) and symmetric, and that rather complicated constraints of the solutions can be easily embedded into the numerical method. However, the disadvantage of LSM is that the condition number will increase quadratically, compared with FEM. Moreover, both essential and natural boundary conditions must be expressed in the energy explicitly.

In fact, the BAM is a special case of LSM. Therefore, we may also adopt the singular value decomposition method to reduce the condition number. If the integration nodes are chosen as the collocation nodes, the least squares method leads to the collocation method.

Considering the continuity of solutions and flux, combinations of different methods may be obtained as follows. Let

$$u = \begin{cases} v^+ & \text{in } S^+, \\ v^- & \text{in } S^-. \end{cases}$$

Then the object energy

$$\widehat{I}(v) = \iint_{S_1} (Lv - f)^2 ds + \iint_{S_2} (Lv - f)^2 ds + D(v),$$

where

$$
\begin{aligned}
D(v) \;=\;& w_4^2 \int_{\Gamma_0} (v^+ - v^-)^2 d\ell + w_5^2 \int_{\Gamma_0} \left(p\frac{\partial v^+}{\partial \nu} - p\frac{\partial v^-}{\partial \nu} \right)^2 d\ell + \\
+\;& w_1^2 \int_{\Gamma_D} (v - g_1)^2 d\ell + w_2^2 \int_{\Gamma_N} \left(p\frac{\partial v}{\partial \nu} - g_2 \right)^2 d\ell + \\
+\;& w_3^2 \int_{\Gamma_M} \left(\frac{\partial v}{\partial \nu} + \alpha v - g_3 \right)^2 d\ell.
\end{aligned}
$$

Least squares method may deal with complicated problems easily, as Navier-Stokes and even hyperbolic equations, see Aziz, Kellogg and Stephens (1985), Bramble and Schatz (1970, 1971, 1977), Bramble and Nitsche (1973), Zhou and Feng (1993), Aziz, Kellogg and Stephen (1985), Bohmer and Locker (1988), Huffel and Vandewalle (1991), Bjorck (1990) and Bochev and Gunzburger (1994). Note that these algorithms require high smoothness of admissible functions. Recently, reduction from PDEs to first-order systems of equations makes LSM reactive, see Carey and Shen (1989).

1.10 COMPARISONS

To close this chapter let us compare the methods again, and point out their relations Usually, the following criteria for different numerical methods are most desirable.

(1) CPU time and computer storage for a certain accuracy, or equivalently, the solution accuracy for a certain CPU time.

(2) Stability, measured by the condition number of the associated matrix obtained.

(3) Maintaince of the corresponding good properties. The corresponding properties between the differential operator L of PDEs and the associated matrix are:

$$self - adjoint \quad \sim \quad symmetric, \tag{1.83}$$

$$L > 0 \quad \sim \quad positive \ definite, \tag{1.84}$$

$$L \geq 0 \quad \sim \quad positive \ or \ positive \ semi\text{-}definite. \tag{1.85}$$

Also the conservative law of physical amount, i.e., the total of flux or mass, is important.

(4) Flexibility of applications to different geometric shapes, and different elliptic equations.

(5) Validity to singularity problems, which is the emphasis of this book.

(6) Simplicity of computer programming.

(7) Ease in theoretical analysis.

Evidently, FEM gains the most popularity owing to high flexibility, in particular for arbitrary geometric shape of S, variable coefficients, and different elliptic equations. It is due to the triangular partition and piecewise low order polynomials (e.g., Lagrange or Hermite elements) that FEM may fit into very wide scale of elliptic problems. Therefore, FEM has been developing in both various algorithms and deep theoretical analysis. In FEM, the mesh generation and entry evaluation of the associated matrix consume a lot of CPU time, although they can be done by computer.

However, for certain type of simple elliptic equations, other numerical methods, i.e., BEM, BAM, etc. may be more efficient. BEM and BAM are confined to certain linear and constant coefficient elliptic equations, where the particular solutions can be found in textbooks or by analytical work. The high efficiency of BAM is not surprising,

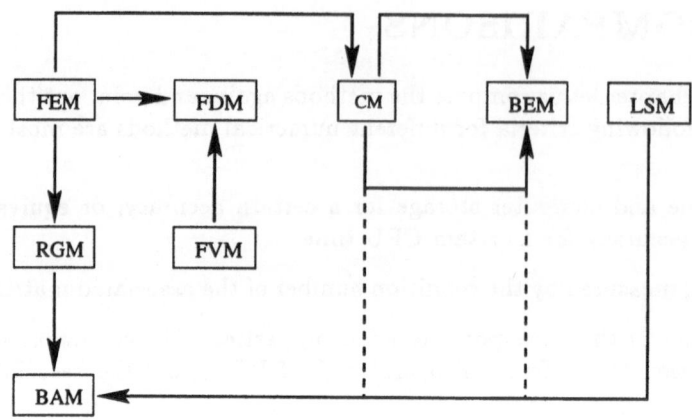

Figure 1.11   A linkage of the important numerical methods.

because the particular solutions are best to approximate the true solutions. Moreover, the high accuracy with the exponential convergence rates can be achieved by BAM. However, such particular solutions may not be found for rather arbitrary S.

From the above analysis of merits and drawbacks, FEM and BAM may compensate to each other. In fact, from the history of FEM, the drawbacks of RGM and BAM are just the motivation of FEM. Unfortunately, FEM also lost the efficiency of RGM and BAM. In this book, we integrate them into a uniform combination, where each method can contribute its best merits, and avoid its disadvantages.

BEM is another efficiency method using the fundamental solutions, whose explicit functions exist only for simple equations, i.e., Laplace's, Possion's equations, etc. Note that the standard BEM is also invalid for singularity problems.

FDM can be applied to any elliptic equation but with rather regular domains; in this case FDM may be more efficient than FEM. To close this chapter the linkage of the important numerical methods is illustrated in Figure 1.11.

We can conclude that *no method is perfect*. Each method has its advantages and drawbacks. The importance in application is to analyze the specific problem, and to choose the most suitable numerical method. However, when the elliptic equations are complicated, in particular with singularities, suitable combinations of different methods will have better efficiency if suitable coupling techniques are chosen.

2

SINGULARITIES AND TREATMENTS

In this chapter, we begin with singularity of Laplace's equation near corners. To deal with such singularities many treatments will be explored in the sequential sections. The conformal transformation method will be first introduced in Section 2.2 with details; other algorithms are stated in Sections 2.3 – 2.9 briefly. Since our emphasis of this book is the combined methods, key aspects on combinations are addressed in the last section, as a linkage to the main body of this book.

Now, let us speak of singularity treatments. The conforming transformation method (CTM) provides the most accurate leading coefficients of singular solutions for Motz's problem, a benchmark of singularity problems. Other treatments of singularly problems in Sections 2.3 – 2.9 may be categorized into three groups, based on the requirements of knowledge on the singularity properties:

I. Local refinements, singular elements and infinite elements, based on the following assumption.

A1: The known location of the singular point, with the known asymptotic behavior

$$u = O(r^\alpha),\ 0 < \alpha < 1\ \text{as}\ r \longrightarrow 0,$$

where the origin is the singular point. Usually, A1 may be satisfied by analyzing the solutions of polygonal domains.

II. The singular function methods of Fix, Gulati and Wakoff (1972) and Wigley (1968), and the dual singular function of Blum and Dobrowolski (1982), which couple the singular solutions and the standard FEM. These methods are based on

A2: The known leading singular solutions:

$$u_{sing.} \approx \sum_{i=1}^{N} \alpha_i \phi_i(r, \theta),\ \text{as}\ r \leq r_0,$$

where α_i are the expansion coefficients which are also called the stress intensity factors. The singular functions for Laplace's equation at the concave points

49

will be explored in Section 2.1 as $\phi_1(r,\theta) = r^\alpha \cos \alpha\theta$, or $r^\alpha \sin \alpha\theta$, $0 < \alpha < 1$. Usually, just one or two singular functions are used enough to get rid of the singular effects.

III. Combined methods, based on local singular and analytical solutions. These methods are based on

A3: The known, complete expansions in the singular subdomain, say, S_2,

$$u = u_{sing.} + u_{anal.} = \sum_{i=1}^{\infty} \alpha_i \phi_i(r,\theta) + \sum_{i=1}^{\infty} \beta_i \psi_i(r,\theta), \quad r \le r_0, \qquad (2.1)$$

where $u_{sing.}$ and $u_{anal.}$ are the singular and analytic parts of u respectively, ψ_i are analytic functions and β_i are coefficients. The combined methods may cover the Trefftz procedures of Trefftz (1926), the global and regional method of Delves (1979), and semi-analytic methods of Cao and Cheung (1992).

The finite form of (2.1) is chosen as the admissible function in S_2

$$v_N = \sum_{i=1}^{N} \alpha_i \phi_i(r,\theta) + \sum_{i=1}^{N} \beta_i \psi_i(r,\theta), \quad \text{in } S_2;$$

other methods may still be used in the outside region of S_2. This is a combined method. When $S = S_2$, the combined methods are simplified as the original Galerkin method, or the spectral methods. Since the singular functions can be fully imbedded into numerical approximation, the semi-analytic functions are found in many literatures (see Li, Shi, et. al. (1981), Liang (1980a), and Cao and Cheung (1992)). Because of the fewer requirements **A1** on the solution behavior, the local refinements are widely used in many engineering problems; their analysis is made extensively. The local refinements should be chosen whenever there is a lack of more singularity knowledge as **A2** and **A3**. The combined methods may achieve higher efficiency than the other two groups, provided that the greater information **A3** can be found. Since the combined methods, however, are the main theme in this book, they will be thoroughly embodied in Parts II–IV. The treatments in Groups I and II are briefly introduced in this chapter, by outlining the basic ideas of the treatments, and by supplying important references.

It has to be observed that singularity problems are very interesting in both theoretical research and practical application. Three kinds of singularities often occur in elliptic problems: angular singularity, interface singularity, and the infinity in unbounded solution domains. There have been many reports on singularity problems; the following reference are concerned with angular singularities and interfaces:

Strang and Fix (1973), Fix, Gulati and Wakeoff (1973), Babuska (1970, 1974, 1977), Babuska and Rosenzweig (1972), Babuska and Rheinboldt (1978, 1984), Babuska and Guo (1988), Babuska and Miller (1984), Birkhoff (1972), Eisenstat (1974), Eisenstat and Schultz (1972), Fox (1971, 1979), Fox and Snaker (1969), Fox, Henrici and Moler (1967), Grisvard (1985), Grisvard, Wendland and Whiteman (1985), Kellog (1971, 1972, 1975), Thatcher (1975, 1976), Whiteman (1982), Whiteman and Akin (1980), Papamichael (1990), Whiteman and Papamichael (1972), Akin (1979a, b), Ames (1977), Barnhill and Whiteman (1975), Bettess (1977), Blum and Dobrowolski (1982), Collatz (1981), Axelsson and Barker (1984), Delves (1979), Delves and Hall (1979), Delves and Freeman (1981), Han (1982), Ying (1978), Wait (1978), Schatz and Wahlbin (1978, 1979), Mitchell and Wait (1977), Lehman (1959), Wigley (1964, 1969, 1987, 1988), Zienkiewicz (1977, 1984), Zienkiewicz, Kelley and Bettess (1977), Zielinski and Zienkiewicz (1985), Kermode, Mckerrell and Delves (1985), Tsamasphyros and Giannakopoulos (1985), Atkinson and Hoog (1984), Yserentant (1986), Piltner (1985), Rapley (1985), Gregory, Fishelov, et. al. (1978), Fares and Li (1989) and Dorr (1986). The study on interface problems are reported in Babuska (1974, 1976a,b, 1977), Kellogg (1971, 1972, 1975), Canuto and Pietra (1990), Funaro, Quarteroni and Zanolli (1988), Goldstein (1983), Ikebe, Lynn and Timlake (1969), King (1975), MacCamy and Marin (1980), MacKinnon and Carey (1988), Phillips and Davies (1988) and Sheldon (1958).

There are more reports: Babuska and Rosenzweig (1972), Barnhill and Whiteman (1974, 1975), Barsoum (1977), Blum (1987, 1988a,b), Blum and Rannacher (1990), Collatz (1981), Cox and Fix (1984), Crank and Furzeland (1977), Dixon, Morgan and Harrison (1979), Dobrowolski (1981, 1985), Donnelly (1969), El Misiery and Ortiz (1986), Eriksson and Thomee (1984), Forsythe (1958), Fares and Li (1989), Fox (1971), Franke (1978), Fujii and Yamaguti (1980), Givoli Rivkin and Keller (1992), Goldstein (1974), Goodrich and Stenger (1970), Greenspan and Warten (1962), Grisvard (1992), Grisvard, Wendland and Whiteman (1985), Guenther (1975), Hoog and Weiss (1976), Laasonen (1958), Li (1987), Lin (1991a,b), McLean (1984), Morley (1973/1974), Morris and Wait (1979), Papamichael (1985), Rapley (1985), Rude (1989), Ruotsalainen (1987), Schatz and Wahlbin (1976), Solecki and Swedlow (1984), Steger (1990), Stephan (1979, 1986, 1988), Stephan and Wendland (1983), Stephan and Whiteman (1988), Stephan and Costabel (1986), Stephan and Suri (1989), Veidinger (1968), Wendland, Schmitz and Bumb (1988), Wendland and Stephan (1990), Atkinson (1989), Atkinson and de Hoog (1984), Hendry and Delves (1979), Hendry, Delves and Phillips (1979), Ingraffea and Manu (1980), Jespersen (1978), Stevenson (1993), Papamichael (1990), Phillips and Delves (1980), Maita (1992), Driscoll (1997), Abdel-Messieh and Thatcher (1990), Schiff, Fishelov and Whiteman (1979), Smith, Bjorstad and Gropp (1996), Whiteman and Goodsell (1988), Whiteman (1968), Whiteman and Barnhill (1973), Whiteman and Schleicher (1984), Whiteman (1970, 1974, 1978, 1981, 1982, 1984, 1986), and Wu and Han (1997).

Recently more and more publications on unbounded domain problems have appeared: Thatcher (1978), Bettess (1977), Bettess, Emson and Chiam (1984), Goldstein (1981a–c, 1982), Johnson and Nedelec (1980), Aziz and Kellogg (1981), Aziz, Dorr and Kellogg (1982), Lin (1985), Canuto, Hariharam and Lustman (1985), Masmoudi (1987), McLean (1986), Medina and Taylor (1983), Kreb and Spassov (1983), Cheung, Guo, et. al. (1986), Stephan and Suri (1989), Orikasa, Honma, et. al. (1983), Amini and Kirkup (1995), Dasgupta (1984), Gerdes and Demkowicz (1996), Grosch and Orszag (1997), Harari and Hughes (1991, 1992), Seager and Carey (1990), Guirguis (1987a, b), Lentini and Keller (1980), Moriya (1985), Schiff (1988), Silverster and Hsieh (1971), Symm (1967), Yu (1985, 1996), Zienkiewicz (1984), Zienkiewicz, Bando, et. al. (1985), and Zienkiewicz, Emson and Bettess (1983).

2.1   SINGULARITY OF LAPLACE EQUATION ON POLYGONS

2.1.1   Neumann-Dirichlet Conditions

Consider Laplace's equation with the Neumann-Dirichlet conditions (see Figure 2.1).

$$\Delta u = \frac{\partial^2 u}{\partial x^2} + \frac{\partial^2 u}{\partial y^2} = 0, \qquad \text{in } S, \qquad (2.2)$$

$$u = g_1 \text{ on } \Gamma_D, \quad \frac{\partial u}{\partial \nu} = g_2 \qquad \text{on } \Gamma_N.$$

Below we describe a way how to obtain the asymptotic solutions near the corners and how to analyze the singularity. Such a singularity damages significantly the efficiency of FEM, FDM, BEM, etc., so that the special remedy must be explored for practical application. This leads to many treatments for singularity problems.

Let us first consider the asymptotic solutions near the angle point O,

$$\left. \frac{\partial u}{\partial \nu} \right|_{\theta=0} = - \left. \frac{\partial u}{r \partial \theta} \right|_{\theta=0} = A,$$

$$u|_{\theta=\Theta} = B, \qquad (2.3)$$

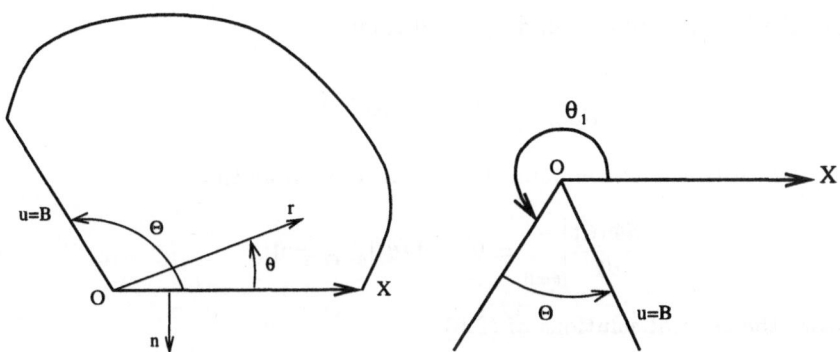

Figure 2.1 The local angular sub-domain.

Figure 2.2 Rotation of the local subdomain.

where A and B are two constants. For different positions of an angular domain, we may use transformation $\theta^* = \theta - \theta_1$ to shift Figure 2.2 to Figure 2.1. By the method of separation variables, assume $u = R(r)\Psi(\theta)$, to lead to

$$0 = \triangle u = \left[\frac{1}{r}\frac{\partial}{\partial r}r\frac{\partial R(r)}{\partial r}\right]\Phi(\theta) + \frac{R(r)}{r^2}\frac{\partial \Phi^2(\theta)}{\partial \theta^2}.$$

Then

$$\frac{r\frac{d}{dr}r\frac{dR(r)}{dr}}{R(r)} = -\frac{\frac{d\Phi^2(\theta)}{d\theta^2}}{\Phi(\theta)} = \alpha^2. \tag{2.4}$$

Since both r and θ are different, independent variables, the equality (2.4) holds if and only if the ratio α^2 is constant. We have

$$\frac{d\Phi^2(\theta)}{d\theta^2} + \alpha^2\Phi(\theta) = 0. \tag{2.5}$$

Then the general solutions are obtained as follows.

(I) When $\alpha \neq 0$, (let $\alpha > 0$),

$$\Phi(\theta) = a\cos\alpha\theta + b\sin\alpha\theta, \quad R(r) = cr^\alpha + dr^{-\alpha}, \tag{2.6}$$

where a, b, c and d are constants. The factor α can be determined by angle Θ and the given boundary conditions, shown below.

(II) When $\alpha = 0$,

$$\Phi(\theta) = a\theta + b, \quad R(r) = c\ln r + d. \tag{2.7}$$

Consider the homogeneous boundary conditions

$$\frac{\partial u}{\partial \nu}\bigg|_{\theta=0} = 0, \quad u|_{\theta=\Theta} = 0. \tag{2.8}$$

For $u = R(r)\Phi(\theta)$, the above boundary conditions imply that

$$\frac{d\Phi(\theta)}{d\theta}\bigg|_{\theta=0} = 0, \quad \Phi(\theta)|_{\theta=\Theta} = 0. \tag{2.9}$$

Then from the general solutions of (2.6)

$$\frac{d\Phi(\theta)}{d\theta}\bigg|_{\theta=0} = (-a\alpha \sin \alpha\theta + b\alpha \cos \alpha\theta)\bigg|_{\theta=0} = b\alpha = 0.$$

Then $b = 0$ if $\alpha \neq 0$. Also another condition in (2.9) gives

$$\Phi(\theta)\bigg|_{\theta=\Theta} = a \cos \alpha\Theta = 0.$$

This leads to

$$\alpha = \alpha_k = \frac{\pi}{\Theta}\left(k + \frac{1}{2}\right).$$

We then obtain the functions $\Phi(\theta) = a \cos \alpha_k \theta$ satisfying (2.8). Therefore, the particular solutions are obtained as

$$u = \sum_{k=0}^{\infty} a_k r^{\alpha_k} \cos \alpha_k \theta. \tag{2.10}$$

Note that the functions $r^{-\alpha_k}$ and (2.7) as $\alpha = 0$ are expelled because of boundness of the solutions.

In fact, the real and the imaginary parts of any analytical complex function are harmonic. Then the functions $r^{\alpha} \cos \alpha\theta$ and $r^{\alpha} \sin \alpha\theta$, $\alpha \in R$ are harmonic from z^{α}, where $z = (x + iy)$. Also, the complex functions, $z^k \ln z$, $k \geq 1$, are analytic, we then obtain more harmonic functions

$$r^k (\ln r \cos k\theta - \theta \sin k\theta), \tag{2.11}$$
$$r^k (\ln r \sin k\theta + \theta \cos k\theta). \tag{2.12}$$

For the nonhomogeneous boundary conditions, suppose that a particular solution w satisfies (2.3). We use a transformation $v = u - w$, to lead to the homogeneous boundary conditions, under which the particular solutions are given in (2.10). Below we find such a particular solution as

$$w = ar \cos \theta + br \sin \theta + c. \tag{2.13}$$

For satisfying the boundary conditions (2.3), we conclude that the coefficients in (2.13) are

$$a = \frac{A \sin \Theta}{\cos \Theta}, \quad \text{when } \Theta \neq \frac{\pi}{2}, \frac{3\pi}{2},$$
$$b = -A, \quad c = B.$$

Then, we obtain the particular solution

$$w = B - Ar \sin \theta + \frac{A \sin \Theta}{\cos \Theta} r \cos \theta.$$

However, when $\Theta = \frac{\pi}{2}, \frac{3\pi}{2}$, we choose other harmonic functions from (2.11) with $k = 1$,

$$w = ar \left(\ln r \cos \theta - \theta \sin \theta \right) + br \sin \theta + c.$$

The boundary conditions (2.3) give

$$-\frac{\partial w}{r \partial \theta}\bigg|_{\theta=0} = \left[a \left(\ln r \sin \theta + \sin \theta + \theta \cos \theta \right) - b \cos \theta \right]\bigg|_{\theta=0} = A,$$

to lead to $b = -A$, then $a = -\frac{A}{\Theta}$, and $c = B$ from the other condition in (2.3). Therefore, we obtain the particular solutions when $\Theta = \frac{\pi}{2}, \frac{3\pi}{2}$

$$w = B - \frac{Ar}{\Theta} \left(\ln r \cos \theta - \theta \sin \theta \right) - Ar \sin \theta.$$

Finally we summarize the asymptotic harmonic solutions of (2.3).

I. When $\Theta \neq \frac{\pi}{2}, \frac{3\pi}{2}$,

$$u = B - Ar \sin \theta + \frac{A \sin \Theta}{\cos \Theta} r \cos \theta + \sum_{k=0}^{\infty} a_k r^{\alpha_k} \cos \alpha_k \theta, \tag{2.14}$$

where $\alpha_k = \frac{\pi}{\Theta} \left(k + \frac{1}{2} \right)$.

II. When $\Theta = \frac{\pi}{2}, \frac{3\pi}{2}$

$$u = B - \frac{Ar}{\Theta}(\ln r \cos \theta - \theta \sin \theta) - Ar \sin \theta + \sum_{k=0}^{\infty} a_k r^{\alpha_k} \cos \alpha_k \theta, \qquad (2.15)$$

where $\alpha_k = 2k + 1$ as $\Theta = \frac{\pi}{2}$; $\alpha_k = \frac{2k+1}{3}$ as $\Theta = \frac{3\pi}{2}$.

Remark 2.1 *The above approaches can be used to seek the non-constant boundary conditions*

$$\left. \frac{\partial u}{\partial \nu} \right|_{\theta=0} = \sum_{i=0}^{\infty} A_i r^i, \quad u \Big|_{\theta=\Theta} = \sum_{i=0}^{\infty} B_i r^i,$$

instead of the conditions in (2.3). We have the following particular solutions for $\Theta \neq \frac{\pi}{2}, \frac{3\pi}{2}$,

$$u = \sum_{k=0}^{\infty} \frac{B_k r^k \cos k\theta}{\cos k\Theta} + \sum_{k=1}^{\infty} \frac{A_{k-1} r^k}{k}\left[-\sin k\theta + \frac{\sin k\Theta}{\cos k\Theta}\cos k\theta\right] + \sum_{k=0}^{\infty} a_k r^{\alpha_k} \cos \alpha_k \theta,$$

where $\alpha_k = \frac{\pi}{\Theta}\left(k + \frac{1}{2}\right)$. When $\Theta = \frac{\pi}{2}, \frac{3\pi}{2}$, the particular solutions can be obtained by means of (2.11).

For the Dirichlet condition $u|_{\theta=0} = 0$ we may need the particular solutions (2.12). The method of separate variables can be adopted to find the asymptotic expansions of Poisson's equation, see Strang and Fix (1973), and Lehman (1959).

Remark 2.2 *When the corner edges are curved, the Laplacian solutions have the general expressions (see Babuska and Suri (1990)):*

$$u(r, \theta) = \sum_{j=1}^{\infty} \sum_{m=0}^{\infty} c_{jm} \psi_{jm}(\theta) r^{\alpha_j + m} + \sum_{j=1}^{\infty} \sum_{m=0}^{\infty} c_{jm}^* \psi_{jm}^*(\theta) r^{\alpha_j + m}(\log r),$$

where the functions $\psi_{jm}(\theta)$ and $\psi_{jm}^*(\theta)$ are smooth and c_{jm} and c_{jm}^* are coefficients.

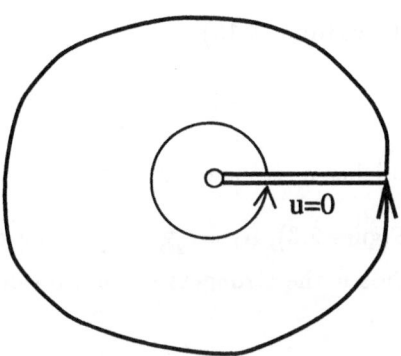

Figure 2.3   A crack problem.

2.1.2   Analysis on Singularity

I. When $\theta = \frac{\pi}{2}$ and $A = 0$, the solutions (2.15) are analytic, infinitely differentiable. This can be viewed as the Dirichlet conditions $u = B$ on the axis y of ∂S. However, when $A \neq 0$, the solution (2.15) involves the function $r \ln r \cos \theta$, to lead to

$$\frac{\partial u}{\partial r} = O\left(|\ln r|\right) \longrightarrow \infty \quad \text{as} \quad r \to 0.$$

II. When $\Theta < \frac{\pi}{2}$, the solutions (2.14) yields

$$u = O\left(r^{\alpha_1}\right), \quad \alpha_1 = \frac{\pi}{2\Theta} > 1.$$

Their derivatives are also bounded

$$\frac{\partial u}{\partial r} = O\left(r^{\alpha_1 - 1}\right) \longrightarrow 0 \quad \text{as} \quad r \to 0.$$

When $\frac{\pi}{4} < \Theta < \frac{\pi}{2}$, $1 < \alpha_1 < 2$. Although the derivatives are bounded, the second order derivatives are unbound:

$$\frac{\partial^2 u}{\partial r^2} = O\left(r^{\alpha_1 - 2}\right) \longrightarrow \infty \quad \text{as} \quad r \to 0.$$

Hence, to guarantee $u \in C^m$ near the corner, the polygon angle must be as small as $\Theta < \frac{\pi}{2m}$.

III. When $\frac{\pi}{2} < \Theta \leq 2\pi$, $\frac{1}{4} \leq \alpha_1 < 1$. The derivatives are unbounded:

$$\frac{\partial u}{\partial r} = O(r^{\alpha_1 - 1}) \longrightarrow \infty \quad \text{as} \quad r \longrightarrow 0.$$

IV. When $\Theta = \frac{3\pi}{2}$, it follows from (2.15)

$$u = O\left(r^{\frac{1}{3}}\right) + O\left(r \ln r\right).$$

The first term $O\left(r^{\frac{1}{3}}\right)$ has a stronger singularity.

V. When $\Theta = 2\pi$ (see Figure 2.3), $\alpha_1 = \frac{\pi}{2\Theta} = \frac{1}{4}$ in the crack problem, the solution $u = O\left(r^{\frac{1}{4}}\right)$. This is the strongest singularity of the Dirichlet-Neumann boundary conditions.

We summarize a few conclusions for angular Dirichlet-Neumann conditions:

- The solutions are analytical only if $\Theta = \frac{\pi}{2}$ and $A = 0$.

- If $\Theta < \frac{\pi}{2}$, then $u \in C^1$.

- If $\Theta > \frac{\pi}{2}$, or $\Theta = \frac{\pi}{2}$ and $A \neq 0$, then $u \notin C^1$.

- If $\Theta = 2\pi$, then $u = O\left(r^{\frac{1}{4}}\right)$, a strong singularity.

2.1.3   Other Particular Solutions

Below we provide other asymptotic solutions; their proofs are similar.

I. For another Dirichlet-Neumann condition, $u|_{\theta=0} = A$ and $\frac{\partial u}{\partial \nu}\big|_{\theta=\Theta} = B$, then $\alpha_k = \frac{\left(k+\frac{1}{2}\right)\pi}{\Theta}$.

(1) When $\Theta \neq \frac{\pi}{2}, \frac{3\pi}{2}$,

$$u = A + \frac{B}{\cos \Theta} r \sin \theta + \sum_{k=0}^{\infty} a_k r^{\alpha_k} \sin \alpha_k \theta.$$

(2) When $\Theta = \frac{\pi}{2}, \frac{3\pi}{2}$,

$$u = A + (-1)^{l+1} \frac{Br}{\Theta} \left(\ln r \sin \theta + \theta \cos \theta\right) + \sum_{k=0}^{\infty} a_k r^{\alpha_k} \sin \alpha_k \theta,$$

where $l = 0$ as $\Theta = \frac{\pi}{2}$, and $l = 1$ as $\Theta = \frac{3\pi}{2}$.

II. For the Neumann-Neumann conditions, $\frac{\partial u}{\partial \nu}\big|_{\theta=0} = A$ and $\frac{\partial u}{\partial \nu}\big|_{\theta=\Theta} = B$, then $\alpha_k = \frac{k\pi}{\Theta}$.

(1) When $\Theta \neq \pi,\ 2\pi$,

$$u = -Ar\sin\theta - \frac{B + A\cos\Theta}{\sin\Theta} r\cos\theta + \sum_{k=0}^{\infty} a_k r^{\alpha_k} \cos\alpha_k\theta.$$

(2) When $\Theta = \pi,\ 2\pi$,

$$u = -Ar\sin\theta + \frac{(-1)^l B - A}{\Theta} \times r\left(\ln r\cos\theta - \theta\sin\theta\right) + \sum_{k=0}^{\infty} a_k r^{\alpha_k} \cos\alpha_k\theta,$$

where $l = 0,\ 1$ for $\Theta = \pi,\ 2\pi$, respectively.

III. For the Dirichlet-Dirichlet conditions, $u|_{\theta=0} = A$ and $u|_{\theta=\Theta} = B$, then $\alpha_k = \frac{k\pi}{\Theta}$,

$$u = A + (B - A)\frac{\theta}{\Theta} + \sum_{k=0}^{\infty} a_k r^{\alpha_k} \sin\alpha_k\theta.$$

The above expansions are still valid for $\theta = \frac{\pi}{2},\ \frac{3\pi}{2}$. We display the singularity when $A \neq B$. In this case, the particular solution

$$w = A + (B - A)\frac{\theta}{\Theta},$$

and

$$\frac{1}{r}\frac{\partial w}{\partial \theta} = \frac{1}{r}\frac{(B - A)}{\Theta} = O\left(\frac{1}{r}\right) \longrightarrow \infty \quad \text{as } r \to 0$$

if $A \neq B$. The singularity at the origin is strongest since the boundary condition given is already discontinuous. When $A = B$ and $\Theta = \frac{\pi}{2}$, the Dirichlet solutions are also analytic.

In this book, the analytic solutions are said if $u \in C^{\infty}(S)$; the singularity exists at some points if the solution derivatives there are unbounded, often being as $u = O(r^{\alpha}),\ 0 < \alpha < 1$ and

$$\frac{\partial u}{\partial r} = O\left(r^{\alpha-1}\right) \longrightarrow \infty \quad \text{as } r \to 0.$$

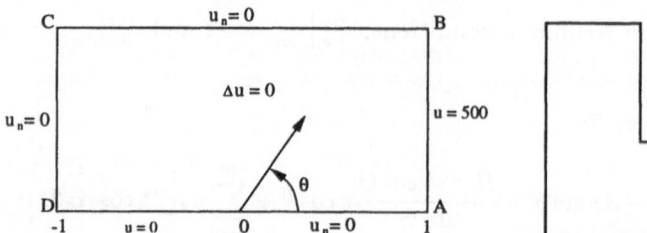

Figure 2.4 Motz's problem.   **Figure 2.5** The L-shape domain.

For the Dirichlet-Dirichlet conditions with $A \neq B$, it is better to use a transformation by passing $\frac{\partial u}{\partial r} = O\left(\frac{1}{r}\right)$, and to consider the homogeneous Dirichlet conditions. Moreover the weak singularity is said if only the second-order derivatives are unbounded. For instant, the solution $u = O\left(r^2 |\ln r|\right)$ has a weak singularity

$$\frac{\partial^2 u}{\partial r^2} = O\left(|\ln r|\right) \longrightarrow \infty \quad \text{as} \quad r \to 0.$$

The regular solutions are referred by $u \in C^2(S)$ based on the PDE problem (1.1). The solutions of (2.2) are regular only in a few ideal cases, and singular usually, e.g., for problems on the concave polygons and unbounded domains, or problems of mixed media consisting of different materials and existing intersection points. Evidently, the combining approaches are inevitable for the complicated elliptic equations, in particular those with singularities.

2.1.4   Motz Problem

Motz's problem was first discussed by Motz (1947) for the relaxation method, then studied in Whiteman and Papamichael (1972), Thatcher (1976), Delves (1979), Fox (1979), Wigley (1988), Li (1983a), and Olson, Georgion and Schultz (1991). Since the singularity $O(r^{\frac{1}{2}})$ of the solution at the origin is stronger than that in the Dirichlet-Laplace problem on the L-shape domain, Motz's problem becomes a benchmark of singularity problems. Therefore, Motz's is often chosen as a typical problem of singularity problems, see Figures 2.4 and 2.5. For instance, in Li, Mathon and Sermer (1987), the accurate solutions of Motz's problem with the relative errors $O(10^{-9})$ are found by BAM with double precision; in Babuska and Oh (1990), Lefeber (1989), and Lu and Horng (1996), Motz's is also selected as a sample of singularity problems for testing effectiveness of numerical methods.

Motz's problem is to seek a Laplacian solution satisfying the mixed Dirichlet-Neumann conditions in a rectangular domain (Figure 2.4).

$$\triangle u = 0, \quad \text{in } S,$$

$$u|_{x=1} = 500, \quad u|_{y=0 \cap -1 < x < 0} = 0,$$

$$\frac{\partial u}{\partial y}\bigg|_{y=1} = \frac{\partial u}{\partial y}\bigg|_{y=0 \cap 0 < x < 1} = 0, \quad \frac{\partial u}{\partial x}\bigg|_{x=-1} = 0.$$

Based on (2.14), we obtain the expansion form:

$$u(r, \theta) = \sum_{i=0}^{\infty} D_i r^{i + \frac{1}{2}} \cos\left(\left(i + \frac{1}{2}\right)\theta\right). \tag{2.16}$$

The coefficients D_i can be determined by different methods. Hence, there exists a singularity point at $(0, 0)$.

The Dirichlet L-shape problem of Laplace's equation is another sample of singularity problems. The treatments of the Dirichlet condition is simple; but the solution with $u = O\left(r^{\frac{2}{3}}\right)$ from Section 2.1.4 has a weaker singularity than that in Motz's problem. Hence we will often choose Motz's problem as the sample for numerical comparisons.

2.1.5 Reduced Convergence Rates Caused by Singularities

Since $u = O(r^\alpha)$, then

$$u \in H^{1+\sigma}(S), \quad \sigma < \alpha < 1. \tag{2.17}$$

From Corollary 1.1 in Section 1.2, we have

$$\|u - u_h\|_{1,S} \le Ch^\sigma \|u\|_{1+\sigma,S}.$$

The solution errors by FEM are of $O(h^\sigma)$ only, worse than $O(h)$ by the linear FEM for regular solutions. A further analysis on the reduced convergence rates is given in many papers, e.g., Bramble, Hubbard and Zlamal (1968) for FDM, Babuska (1970) for FEM, and Schatz and Wahlbin (1979). We cite only the estimates in Blum and Dobrowolski (1982) as

$$C_0 h^{(2-\ell)\alpha} \le \|u - u_h\|_{\ell,S} \le C_1 h^{(2-\ell)\alpha - \delta}, \quad \ell = 0, 1,$$

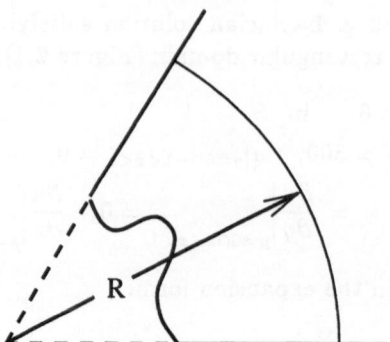

Figure 2.6   An unbounded domain.

where $0 < \delta << 1$, and C_0 and C_1 are two constants independent of h and α.

A remedy is, naturally, to employ different admissible functions, which have the singularity property $O(r^\alpha)$ already. The particular solutions in Sections 2.1.2 – 2.1.4 can be chosen as those admissible functions.

2.1.6   Particular Solutions near Infinity

Consider the unbounded domain

$$\triangle u = 0, \quad \text{in } S,$$
$$u|_{\theta=0} = A, \quad u|_{\theta=\Theta} = B, \quad |u|_{r\to\infty} < C,$$

where S is the unbounded domain (see Figure 2.6). The particular solutions are

$$u = A + (B - A)\frac{\theta}{\Theta} + \sum_{k=1}^{\infty} a_k r^{-\alpha_k} \sin \alpha_k \theta, \quad \alpha_k = \frac{k\pi}{\Theta}.$$

Let us consider other kinds of boundary conditions. Because of the boundness of solutions, the Neumann conditions should be homogeneous. We may easily obtain other particular solutions from Sections 2.1.2 and 2.1.4:

I. Dirichlet-Neumann conditions: $u|_{\theta=0} = A$ and $\frac{\partial u}{\partial \nu}\big|_{\theta=\Theta} = 0$. Then

$$u = A + \sum_{k=0}^{\infty} a_k r^{-\alpha_k} \sin \alpha_k \theta, \quad \alpha_k = \frac{\pi}{\Theta}\left(k + \frac{1}{2}\right).$$

II. Neumann-Dirichlet conditions: $\frac{\partial u}{\partial \nu}\big|_{\theta=0} = 0$ and $u\big|_{\theta=\Theta} = B$. Then

$$u = B + \sum_{k=0}^{\infty} a_k r^{-\alpha_k} \cos \alpha_k \theta, \quad \alpha_k = \frac{\pi}{\Theta}\left(k + \frac{1}{2}\right).$$

III. Neumann-Neumann conditions: $\frac{\partial u}{\partial \nu}\big|_{\theta=0} = \frac{\partial u}{\partial \nu}\big|_{\theta=\Theta} = 0$. Then

$$u = \sum_{k=0}^{\infty} a_k r^{-\alpha_k} \cos \alpha_k \theta, \quad \alpha_k = \frac{k\pi}{\Theta}.$$

The behavior of singularity solutions is analyzed in Kondrat'ev (1967), Fox and Snakar (1969), Lehman (1959), Wigley (1964), Grisvard (1985), Grisvard, Wendland and Whiteman (1985), Nazarov and Plamenevsky (1994), Atkinson (1985), Petersdorff and Stephan (1990), and El-Namoury (1992), as well as Reitich (1991) for plane elasticity, Schmitz, Volk and Wendland (1993) for 3D elasticity. The reduced effects on FEM were given first in Bramble, Hubbard and Zlamal (1968) for FDM, and Babuska (1972) for FEM. An entire chapter in Strang and Fix (1973) is devoted to singularity problems and their treatments. Since then many reports on singularity problems have appeared, see Babuska and Kellogg (1975), Babuska, Kellogg and Pitkaranta (1979), Gregory, Fishelov, et. al. (1978), Babuska and Osborn (1980), Phillips (1986), Wahlbin (1984), E, Huang and Han (1987), Babuska, Stroulis, et. al. (1997), Vorozhtsov and Yanenko (1990), Bank and Scott (1989), Demkowicz, Gropp and Kayes (1992), and Devloo and Oden (1985).

2.2 CONFORMAL TRANSFORMATION METHOD

2.2.1 Basic Methods

It is known from Ahlfords (1979) that the real and imaginary parts of regular complex functions are harmonic, i.e., the Laplacian solutions, and that the Rieman mapping theorem guarantees existence of the transformation from an arbitrary simply-connected domain S to the upper half-plane. In theory, if S can be transformed to a simple domain S^* such that the solutions are explicitly obtained, the harmonic functions on S can also be explicitly obtained. In practice, the concrete transformations must be formulated explicitly, or by fairly easy computation. For polygonal regions and some simple boundary conditions, the explicit transformations using special functions can be found. In this section, the main contents are

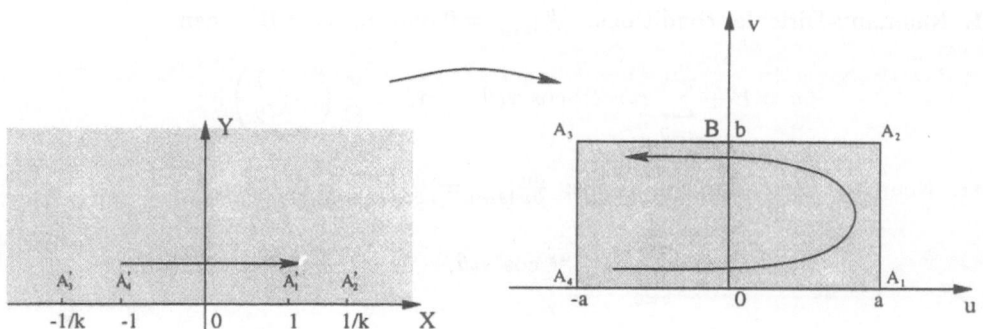

Figure 2.7 A specific Schwarz-Schristoffel transformation.

written, based on Whiteman and Papamichael (1972), Rosser and Papamichael (1975), and Li, Lu, et. al. (1997).

Let n distinct points a_k be in the real axis with the order

$$-\infty < a_1 < a_2 < \cdots < a_n < +\infty,$$

and S be a polygon with n vertices A_k and with the corner angles $\alpha_k \pi$ $(0 < \alpha_k < 2)$. Then the Schwarz-Christoffel transformation in complex variables of Ahlfords (1979) is

$$\omega = f(z) = c \int_{z_0}^{z} \prod_{k=1}^{n} (z - a_k)^{\alpha_k - 1} dz + c_1, \qquad (2.18)$$

where z_0, c and c_1 are three constants. The z-upper half plane is then transformed to the open polygon S; the real axis to the boundary ∂S of S; and the n real points a_i to the n vertices A_i.

A simple case of polygons is the rectangle S with length $2a$ and height b in Figure 2.7, where $n = 4$ and $k < 1$. Then the Schwarz-Christoffel functions (2.18) become

$$
\begin{aligned}
\omega &= F(z,k) = c' \int_0^z \frac{dz}{\sqrt{(z^2 - 1)(z^2 - \frac{1}{k^2})}} + c_1 \\
&= c \int_0^z \frac{dz}{\sqrt{(1 - z^2)(1 - k^2 z^2)}} + c_1,
\end{aligned}
\qquad (2.19)
$$

where $c = kc'$, and $c_1 = 0$ since the origin keeps the same. Also since $A_1'(1, 0) \longrightarrow A_1(a, 0)$ and $A_2'(\frac{1}{k}, 0) \longrightarrow A_2(a, b)$, we have from (2.19)

$$a = c \int_0^1 \frac{dt}{\sqrt{(1 - t^2)(1 - k^2 t^2)}}, \qquad (2.20)$$

and

$$a + ib = c \int_0^{\frac{1}{k}} \frac{dt}{\sqrt{(1-t^2)(1-k^2t^2)}} = a + ic \int_1^{\frac{1}{k}} \frac{dt}{\sqrt{(t^2-1)(1-k^2t^2)}}.$$

Then

$$b = c \int_1^{\frac{1}{k}} \frac{dt}{\sqrt{(t^2-1)(1-k^2t^2)}}. \qquad (2.21)$$

Moreover, by the transformation, $t = \frac{1}{\sqrt{1-(k')^2x^2}}$, where $(k')^2 + k^2 = 1$, Eq (2.21) leads to

$$b = c \int_0^1 \frac{dx}{\sqrt{(1-x^2)(1-(k')^2x^2)}}. \qquad (2.22)$$

The complete elliptic functions are defined by $K(k) = F(1,k)$, in Abramowitz and Stegun (1972), also see Kober (1957),

$$K(k) = f(1,k) = \int_0^1 \frac{dz}{\sqrt{(1-z^2)(1-k^2z^2)}}. \qquad (2.23)$$

We may simply denote the constants from (2.20) and (2.22),

$$c = \frac{a}{K(k)}, \quad b = cK\left(\sqrt{1-k^2}\right).$$

So the ratio of two constants is

$$\frac{a}{b} = \frac{K(k)}{K\left(\sqrt{1-k^2}\right)}, \qquad (2.24)$$

and constant c may be regarded as an enlargement factor. Eq. (2.24) implies the transformation with $a/b \longrightarrow k$ and $k \longrightarrow a/b$. The transformation (2.19) from the upper half plane to the rectangle S is simplified by

$$\omega(z) = c \int_0^z \frac{dz}{\sqrt{(1-z^2)(1-k^2z^2)}}, \qquad (2.25)$$

where k and c are given by (2.24) and (2.23).

The special functions in (2.25)

$$f(z,k) = \int_0^z \frac{dz}{\sqrt{(1-z^2)(1-k^2z^2)}} \qquad (2.26)$$

are called the elliptic functions of first kind in Abramowitz and Stegun (1972). When k is fixed, $f(z,k)$ is of one-to-one correspondence from $z \in S_1(z = x+iy, \, y \geq 0)$ onto $\omega \in S(\omega = u+iv, \, -a < u < a, \, 0 \leq v \leq b)$.

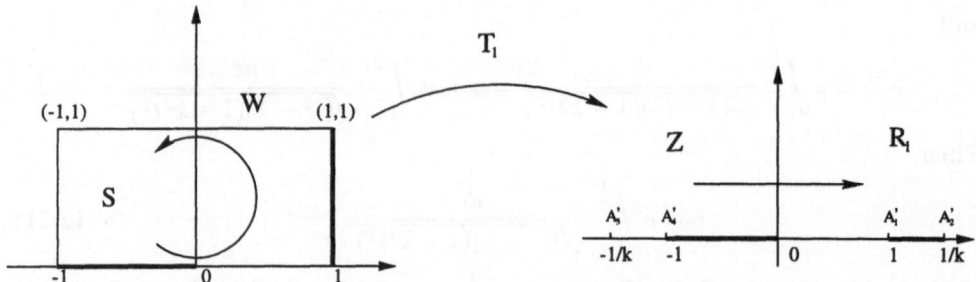

Figure 2.8 Transformation T_1.

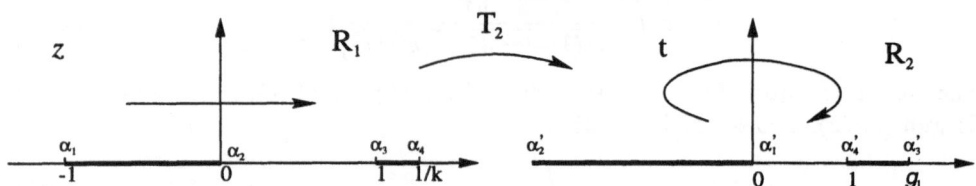

Figure 2.9 Transformation T_2.

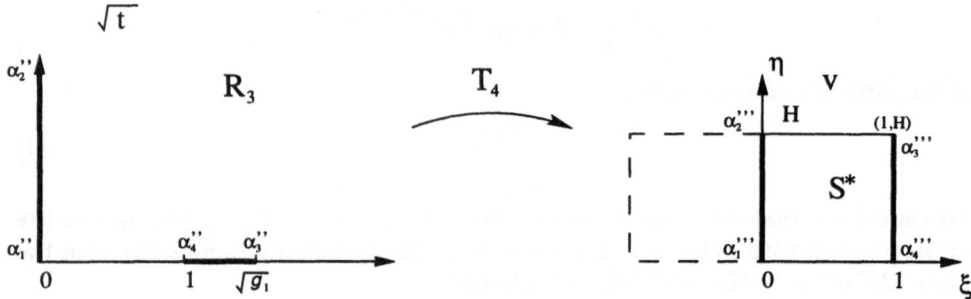

Figure 2.10 Transformation T_4.

On the other hand, the inverse function of $f^{-1}(\omega, k)$ defines a functions from $\omega \in S$ onto $z \in S_1$,

$$z = f^{-1}(\omega, k) = S_n(\omega, k), \tag{2.27}$$

which is called the Jacobian elliptic sinc. Since the Jacobian elliptic functions and their inverse functions (2.27) have been provided by the tables in Abramowitz and Stegun (1972) and the computational subroutines, the conformal transformation method (CTM) is applicable in practical problems.

Now let us apply CTM to Motz's problem in Section 2.1.4.

Consider Motz's problem which solves the Laplace equation on a rectangle $S(-1 < x < 1,\ 0 < y < 1)$

$$\Delta u = \frac{\partial^2 u}{\partial x^2} + \frac{\partial^2 u}{\partial y^2} = 0 \quad \text{in } S, \tag{2.28}$$

with the mixed Neumann-Dirichlet boundary conditions:

$$u \mid_{x<0 \cap y=0} \;=\; 0,\; u \mid_{x=1} = 500, \tag{2.29}$$

$$\frac{\partial u}{\partial y}\bigg|_{y=1} \;=\; \frac{\partial u}{\partial y}\bigg|_{y=0 \cap x>0} = \frac{\partial u}{\partial x}\bigg|_{x=-1} = 0. \tag{2.30}$$

The CTM consists of four steps given below.

Step 1: The transformation T_1 from S onto R shown in Figure 2.8. We obtain from (2.23), (2.25), (2.26) and (2.23) that $c = 1/K(k)$ and

$$\omega(z) = \frac{1}{K(k)} \int_0^z \frac{dz}{\sqrt{(1-z^2)(1-k^2 z^2)}} = \frac{1}{K(k)} f(z,k), \tag{2.31}$$

where the values of k can be determined by (2.24), i.e., $K(k) = K\left(\sqrt{1-k^2}\right)$ by noting $a = b = 1$. The inverse transformation of (2.31) is just the transformation T_1 from S to the upper-half plane R_1 in Figure 2.8. By the Jacobian elliptic sinc in (2.27), we obtain the transformation $T_1 : \omega \longrightarrow z$ in Step 1, $z = S_n\left(K(k)\omega, k\right)$.

Step 2: The rational linear functions T_2,

$$t = \left(\frac{\alpha_4 - \alpha_2}{\alpha_4 - \alpha_1}\right)\left(\frac{z - \alpha_1}{z - \alpha_2}\right), \tag{2.32}$$

transform the upper-half plane R_1 onto the upper-half plane R_2 (see Figure 2.9). Under the transformation T_2 with the point conversion, $\alpha_i \longrightarrow \alpha_i'$ where their coordinates under T_2 are:

$$(0, \alpha_1) \longrightarrow (0,0), \qquad (0, \alpha_2) \longrightarrow (-\infty, 0),$$
$$(0, \alpha_3) \longrightarrow (g, 0), \qquad (0, \alpha_4) \longrightarrow (1, 0),$$

with $\alpha_2 = 0$ and $g_1 = \frac{\alpha_4}{\alpha_4 - \alpha_1} \cdot \frac{\alpha_3 - \alpha_1}{\alpha_3}$.

Step 3: The root function T_3, $\sqrt{t} = (\phi e^{i\varphi})^{\frac{1}{2}} = \sqrt{\phi} e^{\frac{i\varphi}{2}}$ with $\alpha_i' \longrightarrow \alpha_i''$. By T_3 the upper-half plane is converted to the first quadrant, i.e., from the right side of Figure 2.9 to the left side of Figure 2.10.

Step 4: Use the transformation T_4. From (2.31) we obtain $T_4 : \sqrt{t} \longrightarrow v$ with

$$v = \frac{1}{K(m)} \int_0^{\sqrt{t}} \frac{dz}{\sqrt{(1-z^2)(1-m^2 z^2)}} = \frac{1}{K(m)} S_n^{-1} \left(\sqrt{t}, m \right),$$

where $m = 1/\sqrt{g_1}$. The rectangle S^* in Figure 2.9 is with the unit length and the height being given by

$$b = H = \frac{K\left(\sqrt{1-m^2}\right)}{K(m)},$$

based on (2.24).

Finally, under the composite transformations

$$T = T_4 \circ T_3 \circ T_2 \circ T_1$$

defined in Steps 1-4, Motz's problem is reduced to

$$\Delta v(\xi, \eta) = 0, \ 0 < \xi < 1, \ 0 < \eta < H, \tag{2.33}$$

$$\left.\frac{\partial v}{\partial n}\right|_{\xi=0} = \left.\frac{\partial v}{\partial n}\right|_{\xi=1} = 0,$$

$$u(0, \eta) = u_0 = 0, \quad u(1, \eta) = u_1 = 500. \tag{2.34}$$

By using the symmetry, the solutions of (2.33) and (2.34) are found explicitly by

$$v(\xi, \eta) = (u - u_0) \left(\xi + \frac{u_0}{u_1 - u_0} \right) = u(\xi). \tag{2.35}$$

Hence finally we obtain the solutions of Motz's problem, $u(p) = v(p')$, where $p \xrightarrow{T} p'$, $p \in S$ and $p' \in S^*$.

The CTM can also be applied to the L-shape with mixed boundary conditions for multiple singularity. However, a necessary condition for applying CTM is that the polygonal ∂S is split by the Dirichlet and Neumann conditions, separately in two-two fashion (see Papamichael (1990), Papamichael and Whiteman (1973)). This limitation even excludes the L-shaped domain with the pure Dirichlet condition.

2.2.2 Algorithms for Expansion Solutions

Although the transformation T in Section 2.2.1 is explicit in the special functions, the solution approximation of Motz's problem is inevitable. A numerical technique of CTM is developed in Rosser and Papamichael (1975), to evaluate coefficients D_i in (2.16). The 19 leading coefficients are computed with double precision; a comparison with BAM defers in Section 3.6.2. Below we outline their basic techniques with a slight modification, and derive the explicit computational formulas for $D_0 \sim D_9$ with near 1000 significant digits.

Let $\overline{u} = u/500$, we simplify (2.28)-(2.30) to Motz's problem with unit constant on the right boundary $\overline{u}\big|_{x=1} = 1$ at $x = 1$. The expansion solutions

$$\overline{u} = \overline{u}_N(r, \theta) = \sum_{i=0}^{N} \overline{D}_i r^{i+\frac{1}{2}} \cos\left(i + \frac{1}{2}\right)\theta, \qquad (2.36)$$

where $D_i = 500\overline{D}_i$. For easier manipulation in series expansions, the slightly different steps from Rosser and Papamichael (1975) are given as follows.

Step 1: $z_1 = S_n(K(k)z, k)$, where z denotes the complex plane of (x, y).

Step 2: $z_2 = 2 - \frac{2}{1+z_1}$, which is different from T_2 in (2.32).

Step 3: $z_3 = \sqrt{z_2}$,

Step 4: $z_4 = \frac{1}{K(\overline{m})} S_n^{-1}(z_3, \overline{m})$, where $\overline{m} = \frac{2+\sqrt{2}}{4}$.

By noting (2.35), $z_4 = \overline{u}$. Then we can see from Steps 1-4 above,

$$z_3 = \left(2 - \frac{2}{1 + S_n(K(k)z, k)}\right)^{\frac{1}{2}} \qquad (2.37)$$

and

$$z_3 = S_n(K(\overline{m})\overline{u}, \overline{m}). \qquad (2.38)$$

In Rosser and Papamichael (1975), the explicit series of (2.37) are found as

$$\begin{aligned} z_3 &= 2\sqrt{v}\left\{1 - v + v^2 - v^3 + \frac{21}{20}v^4 - \frac{23}{20}v^5 \right. \\ &\quad \left. + \frac{25}{20}v^6 - \frac{27}{20}v^7 + \frac{3487}{2400}v^8 - \frac{1253}{800}v^9 + \ldots\right\}, \end{aligned} \qquad (2.39)$$

where $v = \frac{K(k)Z}{2}$. The transformation can also be expressed by series, based on the formulas of the Jacobian sinc functions in Abramowitz and Stegun (1972). In fact,

$$S_n(v, m) = \frac{2\pi}{\sqrt{m}K(m)} \sum_{n=0}^{\infty} \frac{q^{n+\frac{1}{2}}}{1 - q^{2n+1}} \sin(2n + 1)v,$$

where $v = \frac{\pi u}{2K(m)}$, and $q = e^{-\pi}$. Also the complete elliptic functions of the first kind in Abramowitz and Stegun (1972) are

$$K = K(m) = \frac{\pi}{2} + 2\pi \sum_{n=1}^{\infty} \frac{q^n}{1 + q^{2n}}.$$

By using $\sin x = \sum_{n=0}^{\infty}(-1)^n \frac{x^{2n+1}}{(2n+1)!}$, we have the power series,

$$S_n(u, m) = S_n(v, m) = \frac{2\pi}{\sqrt{m}K(m)} \sum_{r=0}^{\infty} A_r(q) \left(\frac{\pi u}{2K}\right)^{2r+1}, \tag{2.40}$$

where

$$A_r(q) = \frac{(-1)^r}{(2r + 1)!} \sum_{n=0}^{\infty} \frac{q^{n+\frac{1}{2}} (2n + 1)^{2r+1}}{1 - q^{2n+1}}. \tag{2.41}$$

Now we derive the computable formulas for $\overline{D}_0 \sim \overline{D}_9$ in (2.36). Solutions of (2.36) can be expressed in the complex variables

$$\overline{u} = \sum_{n=0}^{N} \overline{D}_n z^{n+\frac{1}{2}}. \tag{2.42}$$

Substituting $\overline{u}$ of (2.42) into (2.38) leads to another power series with $z^{n+\frac{1}{2}}$, by means of (2.40). Since the series solutions z_3 involve the coefficients, $\overline{D}_n$, and since the known power series (2.39) represents the same z_3, we are able to obtain the coefficients by comparing the expansion constants in front of $z^{n+\frac{1}{2}}$.

Let $N = 9$ in (2.36), the solutions (2.38) are expressed as

$$z_3 = \sum_{r=0}^{9} E_r \overline{u}^{2r+1}, \tag{2.43}$$

where E_r are known as

$$E_r = R(\overline{q})A_r(\overline{q}) \left(\frac{\pi}{2}\right)^{2r+1}, \tag{2.44}$$

$$A_r(\bar{q}) = \frac{(-1)^r}{(2r+1)!} \sum_{n=0}^{\infty} \frac{\bar{q}^{n+\frac{1}{2}}(2n+1)^{2r+1}}{1-\bar{q}^{2n+1}}, \tag{2.45}$$

$$\bar{q} = e^{-\pi \frac{\overline{K}'}{\overline{k}}}, \quad R(\bar{q}) = \frac{2\pi}{\bar{k}\sqrt{\overline{m}}}, \quad \bar{k} = K(\overline{m}), \quad \overline{m} = \frac{2+\sqrt{2}}{4}.$$

By substituting (2.42) as $N = 9$ to (2.43) we have

$$z_3 = \sum_{r=0}^{9} E_r \left(\sum_{n=0}^{9} \overline{D}_n z^{n+\frac{1}{2}} \right)^{2r+1}. \tag{2.46}$$

Also we rewrite (2.39) as

$$z_3 = \sum_{i=0}^{9} d_i z^{i+\frac{1}{2}}, \tag{2.47}$$

where the known coefficients are given explicitly

$$d_0 = \sqrt{2k}, \quad d_1 = -2\left(\frac{k}{2}\right)^{3/2}, \quad d_2 = 2\left(\frac{k}{2}\right)^{5/2}, \ldots$$

Combining (2.46) and (2.47) yields

$$\sum_{r=0}^{9} d_r z^{r+\frac{1}{2}} + \text{high order terms} = \sum_{r=0}^{9} E_r \left(\sum_{n=0}^{9} \overline{D}_n z^{n+\frac{1}{2}} \right)^{2r+1}. \tag{2.48}$$

Since N is small, we may expand directly the right side of (2.48), and obtain the following equations by some manipulation in comparison

$$d_0 = E_0 \overline{D}_0, \quad d_1 = E_0 \overline{D}_1 + E_1 \overline{D}_0^3, \tag{2.49}$$
$$d_2 = E_0 \overline{D}_2 + 3E_1 \overline{D}_0^2 \overline{D}_1 + E_2 \overline{D}_0^5, \ldots$$

The leading coefficients $\overline{D}_0, \overline{D}_1, \ldots$, are obtained from (2.49)

$$\overline{D}_0 = \frac{d_0}{E_0}, \quad \overline{D}_1 = \frac{d_1 - E_1 \overline{D}_0^3}{E_0}, \quad \overline{D}_2 = \frac{d_2 - \left(3E_1 \overline{D}_0^2 \overline{D}_1 + E_2 \overline{D}_0^5\right)}{E_0}, \ldots$$

The direct comparison is here developed only for a few leading coefficients (see Li, Lu, et. al. (1997)); a recursion procedure is provided in Rosser and Papamichael (1975) for large N.

2.2.3   First Ten Leading Coefficients

The CTM is the most accurate algorithm for the leading coefficients of solutions of Motz's problem. Based on the algorithms in Section 2.2.2, the coefficients $D_0 \sim D_9$ listed in Table 2.1 are obtained by Mathematica using 1000 significant digits. All digits appearing in the table are significant, based on comparisons with the coefficients by Mathematica using 1200 significant digits in Li, Lu, et. al. (1997), as well as based on error analysis. Note that D_0 has 999 significant digits, with some additional guard digits in computation. Also D_9 has 988 significant digits. These extremely accurate coefficients may provide the test coefficients for other numerical methods. Error analysis on the algorithms is important to achieve very high accuracy of Motz's solutions. We must evaluate carefully, in each computable step of Section 2.2.2, the errors resulting from truncation errors in series calculation and rounding errors because the significant digits used in Mathematica are still limited. Strict analysis is reported in Li, Lu, et. al. (1997).

2.3   LOCAL REFINEMENTS

Let **A1** hold, and origin O be the singular point. Also let the triangles $\widehat{\triangle}_i$ be located with $r_i \leq r \leq r_{i+1}$. We may divide the solution domain S into S_2 and S_1 such that $S = S_2 \cup S_1$, and $S_1 \cap S_2 = \emptyset$. The singular domain S_2 including O may be partitioned into non-quasiuniform triangles $\widehat{\triangle}_i$. From Theorem 1.1, the errors of FEM solutions are, indeed, the interpolation errors of u. Since there exists the solution norm as $r \longrightarrow 0$, $|u|_{1+\alpha-\delta,\triangle_i} < C$, where $\alpha = \min_i \alpha_i$. Then for the linear interpolant u_h of u on S^h, we have

$$\|u - \overline{u}_h\|_{1,S}^2 \leq C \left\{ \sum_{\widehat{\triangle}_i \text{ at } O} h_i^{2(\alpha-\delta)} |u|_{1+\alpha-\delta,\widehat{\triangle}_i}^2 + \sum_{\widehat{\triangle}_i \cap (r>0)} h_i^2 |u|_{2,\widehat{\triangle}_i}^2 + h^2 |u|_{2,S_1}^2 \right\},$$

where h_i is the maximal boundary of $\widehat{\triangle}_i$, h is the maximal boundary of all $\triangle_i$ in S_1, and $0 < \delta << 1$.

To balance the errors in S_1, the finite elements $\widehat{\triangle}_i$ in S_2 should be smaller near point O, since values of $|u|_{2,\widehat{\triangle}_i}$ at $\widehat{\triangle}_i \cap (r \approx 0)$ are large. The smallest elements located at the singular point O should be chosen as $h_i^{\alpha-\delta} = O(h)$. Take $\alpha = \frac{1}{2}$ as an example. Suppose $h = 0.1$ of regular elements. The size of the ending elements should be as about ten times as small as the regular elements.

Table 2.1 The leading coefficients D_i obtained from the CTM using Mathematics with 1000 siginficent digits.

```
D0 = 401.16 24537 45234 42044 31378 39458 16089 26992 51065 85986
       65518 26306 01725 23161 30188 12431 27889 69253 90702 08219
       26757 21171 78423 66904 13922 92491 56495 49727 47287 98159
       24672 04842 37467 00355 56529 56501 76619 38359 05333 08595
       91220 51325 48787 11063 75736 10697 80080 70807 46912 48943
       92102 01268 55190 37637 78030 64921 64176 38629 22141 70165
       49377 84505 19887 53704 42948 60782 36382 13384 44607 60386
       53302 50780 03069 16870 07157 01329 59317 84587 21184 50597
       19358 00882 96788 58901 76440 91206 66222 13144 63834 93065
       60833 97065 50657 80010 58053 15662 15685 07406 00573 24535
       73359 84518 76414 76000 50441 52271 43334 88188 05432 39706
       19272 78851 81600 36058 21534 53066 42304 80152 60901 28907
       53534 90891 78876 15728 44602 53018 34266 69544 12434 49650
       96964 28614 24278 75835 94043 78542 74222 99080 21700 00965
       70113 66475 29262 91380 24584 53682 84720 07714 40279 94422
       32292 62873 66372 31150 03507 93730 73953 80659 44678 86582
       02039 51528 19605 92110 92306 39786 33809 11913 15408 49906
       97851 33656 07159 08119 18927 65537 42447 26764 58584 00556
       71880 11512 57204 05806 53825 12713 49263 91771 94854 16362
       07039 15404 40955 44461 02824 93147 52255 64393 83554 5132

D1 = 87.655 92019 50879 16964 78412 14332 05792 19968 44301 15411
       01582 38050 35569 85547 67842 90603 69409 43880 43016 93304
       17719 21652 06083 99484 40485 94952 44529 68912 67064 71570
       69140 78423 75795 92554 85635 80734 53435 91630 11852 08155
       38613 55022 10610 48577 09893 87861 05910 91838 49741 09594
       39110 89268 41697 95883 74761 63446 58160 00050 66207 38955
       74617 29328 72704 19316 50288 84594 02140 87882 60377 93828
       11467 81620 13234 99078 89419 44073 75397 12719 87480 70983
       55215 35171 53817 18030 24184 49294 41322 03211 90439 76147
       90647 34155 06778 22236 28782 35227 66231 78920 36476 63986
       09495 30930 49573 98885 00141 77369 10085 24004 39895 30402
       88376 63910 91196 10959 05240 55919 92623 02323 38701 87544
       31268 54300 25950 30463 05146 64351 06066 39836 91625 08752
       64051 58228 32188 18859 55163 91759 51692 42946 85613 24375
       66004 72163 03448 12807 33819 50780 48872 63639 08483 02367
       46774 42249 15952 76547 22066 68278 72788 81754 79094 03141
       26594 42987 67105 60657 08978 05048 61402 14802 44267 11007
       88060 48591 09222 14932 56221 55283 21816 69444 07707 19432
       26824 31452 49626 79496 29974 75716 82967 84682 34455 01063
       15553 16941 64559 39224 74681 75652 93294 18237 29493 80
```

```
D2 = 17.237  91507  94468  09287  38603  58611  43932  06015  00491  63038
       74772  78674  59834  22120  18474  43616  27807  31446  12280  00046
       29555  36315  96292  25440  07092  23528  53274  70564  40943  54679
       73778  04531  96200  20166  14305  08786  11111  98441  59095  37484
       49787  48029  90142  32668  01867  03540  67943  27368  39359  66618
       91078  97283  77953  05762  69231  97078  24305  54176  65282  06477
       00223  04237  15620  40116  78971  88169  51416  42451  01074  24022
       99006  69542  17674  99857  67075  38691  32767  01659  90594  85322
       05258  55933  48712  47183  65502  74647  99985  55006  57278  62772
       38083  22915  03735  69857  87024  79120  20790  00497  47006  26938
       15680  96692  89811  23218  94617  31861  88312  17633  16771  45663
       16027  64532  68431  95500  63359  77640  35391  35760  93778  52231
       22041  43014  44381  01329  71297  16725  16120  96006  70080  00628
       36935  97673  78200  04516  62775  59463  31455  80251  40443  37064
       10944  27067  22984  33285  59704  13462  71751  91662  27798  31908
       88848  77510  55594  82652  86535  98882  29062  90658  38813  34213
       53811  89282  12306  44622  39492  54985  99304  44416  54389  99992
       75001  49540  15729  84454  11370  29135  19631  23006  34847  62388
       38048  80482  96928  91938  65903  66593  51289  58997  49348  25875
       53645  80168  75636  12836  38152  25491  54602  83533  87271  88

D3 =-8.0712  15259  69813  44498  22306  15251  29971  94144  54273  59398
       24436  83760  06764  82600  16376  33733  18346  29163  33865  18773
       25777  60587  83964  46374  68559  29572  32794  09470  67810  29377
       47609  09836  75415  68561  19944  39783  01779  68808  95187  84360
       18977  91769  30483  93367  52010  13281  94024  47985  22427  70198
       34792  20888  38716  46561  92860  38403  15707  51291  95432  55432
       21504  97172  00178  68208  79900  32402  68230  61622  77955  17903
       48985  94393  19365  33009  66832  99385  16613  09091  19490  30872
       11474  88514  72031  69155  14915  53171  70105  43742  63861  39542
       99786  60540  92561  44500  80462  56615  42937  40153  54962  03620
       51701  32990  95628  12180  37995  01400  46708  47187  59736  18230
       96316  21754  27449  88109  85764  50753  20647  52815  43982  24569
       28093  35933  59020  72286  39750  41713  86071  18088  46828  16208
       40438  27965  74138  35720  63448  96885  20739  22296  13873  85091
       11358  14698  81806  23682  01417  53541  21923  23188  06749  23040
       07696  96350  78506  25899  93137  40139  69521  93490  40780  40840
       76469  97135  37904  12735  70194  49003  81058  62787  40024  31077
       40272  91287  41718  52437  58591  96977  06709  29140  68433  81287
       58217  86156  31235  76799  02868  91294  09577  70884  20601  18149
       56782  80967  35576  65550  18787  40124  31846  86412  89183

D4 = 1.4402  72717  02285  69808  63415  89244  41287  23699  83453  19793
       30190  31626  48913  31831  84851  76302  67871  05417  05480  23555
       29559  17338  72577  19118  96338  30719  54082  71558  32124  71588
       40392  80725  21601  69072  17049  72356  25549  19505  55313  03518
       73853  00492  06471  36370  89482  33950  70285  39258  67984  54620
       69231  14116  74524  51998  82564  01565  32321  25072  37051  02653
       95373  38633  16605  87772  92146  39164  74271  11061  44008  40218
       53622  22714  85379  61000  96507  54336  72974  26827  40693  82922
       11075  53940  32250  91066  27491  34041  56421  59613  98401  52839
       53497  82033  99969  24460  03310  77072  50204  75908  01593  10981
       86327  42380  17155  38141  87045  11136  30710  38236  79146  56400
       39744  68118  89619  95080  90376  78717  79242  40464  27941  48574
       78068  83643  53255  53834  75756  80106  35228  73843  09630  23486
       48903  69900  68807  87279  72409  78198  44297  06970  92725  25344
       35527  70088  43279  78104  59341  26762  27916  00504  01208  82168
       72794  54788  58366  14012  39725  15914  31467  73798  91531  17975
       24541  22850  17372  20611  25676  32722  67847  24946  35731  26159
       54518  28260  72026  79530  84318  91417  65376  34978  30154  73289
       69504  02479  31841  86822  34524  18900  35165  10181  72542  04903
       38454  65514  08371  22005  95048  51251  47562  94397  7170
```

D5 = 0.3310 54885 92073 60575 39355 97222 15690 59674 72734 66373
08382 52961 06175 64611 82462 92851 05106 12438 24598 70317
80416 11839 64522 53799 50078 28471 66304 75936 97678 67345
13989 54529 28847 99698 02998 45178 86706 18017 63322 88809
34649 99973 12878 20966 22039 50123 14992 54160 97706 74305
10257 58958 09449 18626 17764 36235 25243 93723 78473 94053
27472 10029 93519 60664 08075 41510 58673 55750 47206 36549
01616 91852 01177 06025 21994 94912 67643 27141 46423 51103
32803 04149 51972 01271 49051 40830 71416 02942 51306 08108
68856 08551 05730 96060 34748 15185 83789 83617 90688 27514
70811 43529 01470 16470 35307 03884 72457 39415 14411 33341
56610 00607 88840 01155 70082 51547 63091 07404 32407 31585
45670 80241 01730 63072 39544 87941 99438 39961 18666 81433
62882 29955 80138 61181 79251 16452 40420 00614 55630 10834
81824 94543 17770 81319 52970 22854 14702 95062 64328 74506
84272 49297 75312 27079 61659 10511 92291 64129 03053 98856
64189 89895 15957 00707 64593 21881 04786 78013 57154 08067
57394 50606 90306 26137 02154 46805 09644 43122 01781 77234
47532 90153 28084 02417 53298 53733 49591 49062 88994 81847
69871 02001 38406 84424 35805 51968 09354 12033 719

D6 = 0.2754 37344 50918 08000 91497 68396 20101 30442 71374 97142
64600 71742 26588 18328 56686 90273 97704 05794 25299 27248
14062 95168 32332 04357 22938 45277 28743 90055 02318 78701
65433 81695 66392 13312 47471 47604 48289 37554 23654 12177
10645 11987 75135 72471 94077 28356 32580 87204 83986 30020
00011 16365 67879 28387 25702 75588 27337 36371 25661 71532
58733 56344 77500 08180 20994 51197 62832 06204 53627 58402
13457 19541 78233 09723 54490 76601 94357 40954 22411 41782
49925 85433 47867 29000 08296 20499 87226 47161 46844 57343
80934 16810 98383 55494 38230 76806 84989 55077 85605 62583
79075 34502 11066 58022 44695 72970 69118 16473 74088 20903
24509 51635 49987 63598 96343 45476 03816 87919 26075 57172
98911 33230 31907 50401 96802 05999 84336 52804 88885 07888
03819 29758 88460 26248 58229 51861 62301 05391 98376 08292
41587 23971 87395 65470 27769 32494 41800 61843 27690 01671
57510 19072 92875 87014 53409 50110 88446 19110 04042 09824
76425 06859 85830 17906 60950 52685 11505 61438 18189 31729
98035 07598 28282 98574 17067 72689 60484 50210 17419 42347
95953 87649 08321 57829 88747 26312 54791 67532 73622 14033
85538 84783 59898 28257 24527 26869 03274 52626 05

D7 =-0.0869 32994 52553 66886 42224 70541 85982 42973 21276 65959
62309 33521 82043 84275 27013 93617 84622 77676 70461 03856
06443 52162 49997 56065 97521 74360 26581 52653 85896 80585
66219 68241 18223 05580 55952 03164 00389 66017 93084 46028
55401 74470 35949 90149 26672 17721 74433 97874 32910 65896
15543 51612 56033 93813 52225 96970 89106 70416 01965 53648
32523 91967 57117 94471 12984 46938 29375 82301 46404 53352
14304 88682 36520 45833 21859 27764 79006 77745 44135 98738
36692 54670 91477 05878 40407 48039 68069 30242 38288 74984
18397 19987 38698 28501 88373 26842 38628 55169 62055 50596
18766 42031 36953 65443 40559 78784 53908 78980 78505 00629
47883 43390 79955 60878 59253 08068 26004 70079 53785 53965
08257 06795 35610 25322 88736 54302 76883 95847 03388 07301
79818 32935 25952 45196 88251 24893 83191 35748 04757 70380
68947 05098 00218 76708 52711 92977 53760 62501 94777 38939
23505 79863 79208 64417 08882 97853 24020 09502 20010 10232
74931 86531 68974 53279 26128 34057 74469 54834 91228 68006
97261 72689 24327 48101 27929 90128 01920 01021 72900 24783
32937 45592 23605 44751 66647 88810 59844 97969 96771 09904
80666 59246 91263 74163 85220 22593 30628 32774 91

```
D8 = 0.0336 04878 42653 69511 27652 65990 94305 56060 44858 80036
        91875 24040 59780 69018 49304 87046 42150 38009 43327 43068
        24224 86727 64074 64792 67102 82778 22809 62168 58857 59469
        30620 15951 28266 14580 13414 45289 01836 14368 44734 32831
        46438 05885 96285 64530 59716 06571 24894 30441 89796 48933
        08781 13642 65110 45990 84386 25320 52686 14785 07044 18303
        28233 15316 77696 50676 32594 31072 86273 06767 41362 65116
        82160 59507 31822 36529 24406 08815 63492 27720 57251 92334
        08043 24737 17618 57829 29255 52721 57474 31777 90352 23975
        82234 19496 34657 66227 39947 63671 24373 67684 82821 35166
        55499 82754 40182 62211 24296 51000 67832 66903 31154 13582
        88810 81344 31320 59118 48125 52101 27109 20034 57169 15694
        66995 15257 99741 07368 57126 37657 91888 71567 83717 73992
        41365 55634 21710 33877 50365 44652 80216 41960 05451 50785
        48150 71480 67521 42484 99370 02355 90515 66880 42568 96043
        85835 73381 11474 65700 22168 56055 75478 44860 76038 46049
        49982 01316 42760 86521 78738 13088 13326 42879 37106 85205
        04968 64136 84807 60339 17127 99392 92883 54938 35249 58692
        09911 44038 09709 99742 92834 17294 61140 10431 21264 09795
        72651 61346 61182 25688 88461 85220 78054 01606 3

D9 = 0.0153 84374 48256 94020 24433 56713 40205 89478 25053 06866
        87172 15587 43954 68107 46159 63996 28935 62349 34987 87620
        11994 51438 69223 15921 41971 68112 40526 88848 14975 93441
        99943 29097 69737 71361 64773 29756 60281 81908 21208 64624
        56770 23945 11711 17749 49852 97311 38583 10348 46561 98140
        16778 24574 33541 57680 64074 13437 40617 39906 54028 28793
        24628 81341 22552 37529 94103 84932 25424 59952 83956 11867
        15659 85191 85939 04254 81374 66369 89154 76083 95680 52304
        37552 53525 63073 50530 82872 56935 45664 09417 12747 77935
        96248 41957 30601 66644 61956 98797 99094 73289 22568 65728
        47403 54064 71595 16370 92645 87375 58079 23935 86146 65259
        66461 23788 93153 21968 66015 12227 59068 14217 54823 07875
        81656 70378 23071 24225 18448 05049 64023 07680 14897 10114
        41861 09335 89968 01829 50915 70233 02200 96118 20390 65333
        79857 00269 21304 98587 61420 44855 54725 57091 15287 07549
        07482 52893 95964 12606 14065 22784 31218 29606 19309 95531
        83007 54376 56512 18289 82646 50371 98978 80894 08827 43229
        50905 19962 27566 82701 52809 35955 47366 18018 84342 87817
        58083 15993 72109 22687 47316 92422 05940 90876 17755 90935
        86762 69133 88944 14048 37718 98441 40768 29506
```

An efficient transition techniques choosing h_i is given in Wahlbin (1991), p. 429, based on maximal error estimates in Schatz and Wahlbin (1977, 1978, 1979). The conforming non-quasiuniform elements can be formed easily, also see Gregory, Fishelov, et. al. (1978). The transferring from large elements to small elements may also be nonconforming; but the resulted errors should be estimated so not to reduce the convergence rates.

Some advanced topics involving local refinements are also developed, see Goldstein (1981a,b) for the exterior Helmholtz problem, Eriksson (1985) for high order convergence rates; Eriksson and Johnson (1988) for adaptive FEM; Babuska and Miller (1984a, 1984b) for post processing, Zenger and Gietl (1978) for extension to FDM, Blum and Rannacher (1988) for extrapolation techniques, and Blum and Dobrowolski (1982) and Vasilopoulos (1988) for stress analysis near corners. Moreover, interesting engineering applications are provided in Zienkiewicz (1990) and Zienkiewicz, Gallagher and Lewis (1994).

2.4 SINGULAR ELEMENTS

Let **A1** hold. Since the solution derivatives near the singular point have the asymptotic behavior

$$\frac{\partial u}{\partial r} = O(r^{\alpha-1}), \quad \text{as } r \longrightarrow 0, \tag{2.50}$$

we may design the ad hoc elements to model the singular properties (2.50), called singular elements developed from the standard elements. Usually only the elements near the singular point are chosen as the singular elements. So a good strategy is to combine the singular elements with the refinements in Section 2.3. Based on such a mixed treatment, the numerical results obtained in Wait (1977) are satisfactory.

Assume that the solutions near the origin are

$$u(r, \theta) = C_0 r^\alpha \sin(\beta\theta) + \ldots, \tag{2.51}$$

with $C_0 \neq 0$. Then

$$\min_{v \in V_h} \|u - v\|_{1,S} \geq C_0 h^\alpha. \tag{2.52}$$

We cite a lemma from Wahlbin (1991, p.427).

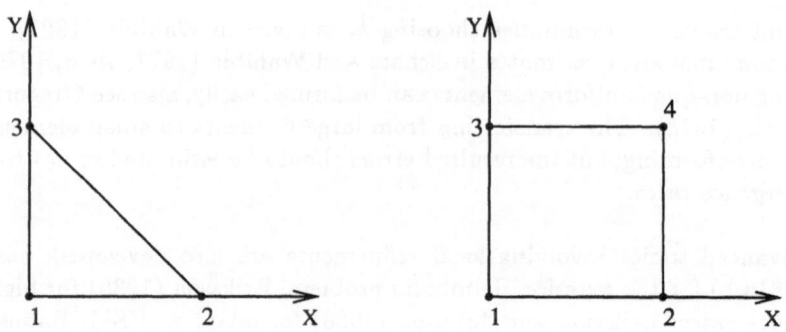

Figure 2.11   The standard elements.

Lemma 2.1 *Let (2.51) and (2.52) be given, also*

$$A_d = \left\{ x : d \le \sqrt{x^2 + y^2} \le 2d \right\} \cap S.$$

Then for any δ, $\exists$ C_0 and d_0 such that for $h^{1-\delta} \le d \le d_0$, then

$$\|u - u_h\|_{0,\infty,A_d} \ge C_0 h^{2\alpha} \cdot d^{-\alpha}.$$

Lemma 2.1 implies that the defect errors may pollute the FEM solutions outside the singular elements if no other special treatments are undertaken. As a consequence, using only the singular finite elements may mitigate but never eradicate the imperfection of singularity. However, the refinements with the ending, singular elements are still beneficial. Also, the exploition of singular elements in 3D elasticity problems, or complicated problems is also worthwhile, see Thompson and Whiteman (1985), Michavila, Gavete and Diez (1988), Ingraffea and Manu (1980) and Hughes and Akin (1980). We may adopt a moderate singular element using the k-order Lagrange polynomials with respect to abscissa $x = r^\beta$. Similar approaches are used in the infinite elements of Bettess (1977) and Cheung, Guo, et. al. (1986).

There are two kinds of conforming singular elements. Akin's elements, see Akin (1976) and Michavila, Gavete and Diez (1988), and quarter points elements for the solution $O(r^{\frac{1}{2}})$, which may also be extended to $O(r^\alpha)$, see Wait (1977), Henshell and Shaw (1975), and Barsoum (1976). A justification on the quarter point elements is given in Ying (1982).

Below, we give only a brief description; the comprehensive exposition is given in Thompson and Whiteman (1985), and Michavila, Gavete and Diez (1988). Some special finite elements for (2.50) and (2.51), which match the FEM conformally, can be modified from standard elements as follows.

2.4.1 Modifying Shape Functions

For the triangle $((x,\ y),\ 0 \le x+y \le 1)$ in Figure 2.11, the shape functions are

$$u(x,y) = \sum_{i=1}^{3} H_i(x,y)u_i,$$

where $u_1 = u(0,0)$, $u_2 = u(1,0)$, $u_3 = u(0,1)$, and the modified basis functions are

$$H_1(x,y) = 1 - (x+y)^\alpha,$$
$$H_3(x,y) = \frac{y}{(x+y)^{1-\alpha}}, \quad H_2(x,y) = \frac{x}{(x+y)^{1-\alpha}}.$$

Next, for the unit square element $\{(x,y),\ 0 \le x,\ y \le 1\}$ in Figure 2.11, the shape functions are

$$u(x,y) = \sum_{i=1}^{4} H_i(x,y)u_i,$$

$u_1 = u(0,0)$, $u_2 = u(1,0)$, $u_3 = u(0,1)$, $u_4 = u(1,1)$, and the basis functions

$$H_1(x,y) = 1 - (x+y-xy)^\alpha, \quad H_2(x,y) = \frac{x(1-y)}{(x+y-xy)^{1-\alpha}},$$

$$H_3(x,y) = \frac{y(1-x)}{(x+y-xy)^{1-\alpha}}, \quad h_4(x,y) = \frac{xy}{(x+y-xy)^{1-\alpha}}.$$

More useful shape functions are given in Akin (1976).

2.4.2 Modifying Reference Nodes

For $O(\alpha^{1/2})$, the quarter point elements are provided in Wait (1977), Henshell and Shaw (1975), and Barsoum (1976). Consider the transformation $T{:}(\xi,\eta) \longrightarrow (x,y)$. We have the following transformation

$$x = \xi(\xi+\eta), \quad y = \eta(\xi+\eta), \tag{2.53}$$

for the linear element on Figure 2.12, and

$$x = \xi\eta + \xi^2(1-\eta), \quad y = \xi\eta + \eta^2(1-\xi). \tag{2.54}$$

for the bilinear element on Figure 2.13. Note that only the midpoints on the x and y axes are changed to quarter points. Based on analysis in Wait (1977), the shape functions (2.53) and (2.54) may well describe $O(r^{\frac{1}{2}})$ behavior along any radius direction from the singular points, and the shape functions for $O(r^\alpha)$ are also established by k-order elements.

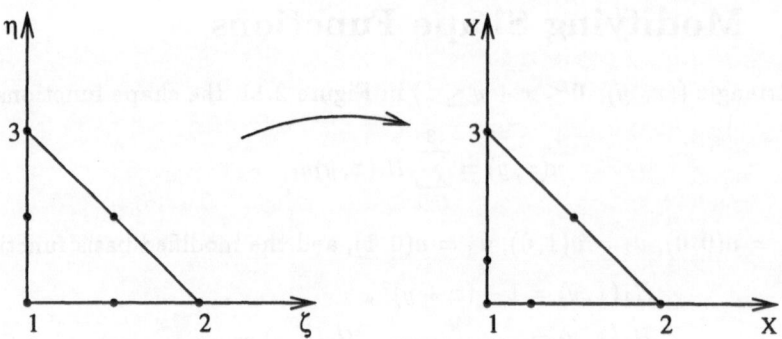

Figure 2.12   The triangular element with quarter points.

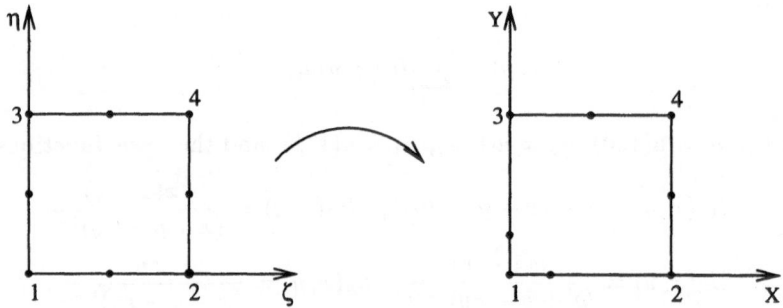

Figure 2.13   The rectangular element with quarter points.

2.5   INFINITY ELEMENT METHOD

Consider the Laplace equation. If $h_{i+1} > h_i$ and $k_{j+1} > k_j$ in such a way

$$h_{i+1} = (1+\alpha)h_i, \quad k_{j+1} = (1+\alpha)k_j, \quad \alpha > 0.$$

Let $h_i = \beta k_j$, the interior discrete equations (1.28) in Section 1.5.1 of Chapter 1 of the interior difference equations are the same

$$\frac{1}{\beta^2}\left[\frac{1}{1+\alpha}u_{i+1,j} + u_{i-1,j}\right] + \left[\frac{1}{1+\alpha}u_{i,j+1} + u_{i,j-1}\right] - \left[\frac{2+\alpha}{1+\alpha}\left(1+\frac{1}{\beta^2}\right)\right]u_{i,j} = 0.$$

Hence, the solution of the infinite algebraic equations as

$$
\begin{bmatrix}
A_0 & -B_0 & & & \\
-B_0^T & A & -B & & \\
& -B^T & A & -B & \\
& & \cdot & \cdot & \cdot \\
& & & \cdot & \cdot & \cdot
\end{bmatrix}
\begin{bmatrix}
v_0 \\
v_1 \\
v_2 \\
\cdot \\
\cdot \\
\cdot
\end{bmatrix}
=
\begin{bmatrix}
b_0 \\
0 \\
0 \\
\cdot \\
\cdot \\
\cdot
\end{bmatrix}
$$

leads to the recursive equations

$$ -B^T v_{p-1} + A v_p - B v_{p+1} = b_p, \quad p = 2, \ 3, \ldots $$

which may be solved by an eigenvalue problem in Thatcher (1976) or by an iteration method in Han and Ying (1979), and Ying (1978) for plane elastic problems. Other reports are given in Thatcher (1978) for unbounded problems, Han (1982b) for interface problems, and Ying (1986) for axial symmetric Stokes flow. The infinite element method is, indeed, an infinite process of local refinements. Hence it may fall into the group of refinements. The application is, however, confined to the homogeneous equations $f = 0$ and without the term cu.

The conception of infinity element methods is interesting, since the refinements can go over and over without an end. However, from the viewpoint of application, the local refinements do not need to go to the limitation, for the accuracy required in application is not very small, though. Note that the examples shown in Thatcher (1976, 1978) are a kind of combinations, the combinations of FEM and the infinitesmall elements methods.

2.6   SINGULAR FUNCTION METHODS

Let **A2** hold. From $r^\alpha \sin \alpha\theta$, we may form an additional conforming basis function $\phi_i(r, \theta) = \rho(r) r^{\alpha_i} \sin \alpha_i \theta$, where $\rho(r)$ is a sufficiently smooth cut-off function defined by

$$ \rho(r) = \begin{cases} 1, & 0 \le r \le r_0, \\ 0, & r > r_1, \end{cases} \tag{2.55} $$

where $r_0 < r_1$. The function (2.55) are augmented to FFE, to lead to the singular function method, reported in Fix, Gulati and Wakoff (1973). The computational results are encouraging, to show optimal convergence rates. A similar method is also studied in Wait and Mitchell (1971). Of course, more than one singular function can be added to FEM. An error analysis of this method is given in Bourlard, Dauge and Nicaise (1990), and analysis on the leading coefficients in Blum (1982), Blum and Dobrowolski (1982) and Bourlard, Dauge, et. al. (1992).

Since the singular function is different from the polynomials FEM, the stiffness matrices are different from those in the standard FEM. The singular function method is also another kind of combinations, the combinations using two different admissible functions overlapped in the same singular subdomain. The method of Fix amends FEM by implementing the leading singular functions. On the other hand, the method of Wigley removes the singular function from the FEM solutions, see Wigley (1968, 1987, 1988).

Let the solution near the singularity point O be

$$u(r,\theta) = \sum_{i=1}^{\infty} D_i \phi_i = \sum_{i=1}^{\infty} D_i r^{\alpha_i} \sin \alpha_i \theta,$$

where the coefficients

$$D_i = \frac{2}{\Theta} r_0^{-\alpha_i} \int_0^{\Theta} u(r_0, \theta) \sin \alpha_i \theta d\theta. \tag{2.56}$$

When the approximate solution u_h of linear FEM is obtained, we may compute the leading approximate coefficient $\tilde{D}_i$ from (2.56). Let the solution be subtracted from the leading singular functions:

$$w = u - \sum_{i=1}^{L} \tilde{D}_i r^{\alpha_i} \sin \alpha_i \theta = u - \sum_{i=1}^{L} \tilde{D}_i \phi_i.$$

Then the corresponding boundary condition is changed to

$$w|_{\Gamma_D} = u|_{\Gamma_D} - \sum_{i=1}^{L} \tilde{D}_i \phi_i \bigg|_{\Gamma_D}, \qquad \frac{\partial w}{\partial n}\bigg|_{\Gamma_N} = \frac{\partial u}{\partial n}\bigg|_{\Gamma_N} - \sum_{i=1}^{L} \tilde{D}_i \frac{\partial \phi_i}{\partial n}\bigg|_{\Gamma_N}.$$

The FEM solution is sought again under the above modified boundary conditions.

For the k-order FEM used, the errors of the leading coefficients for this method are given in Wigley (1987) with $|D_i - \tilde{D}_i| = O(h^k)$, $k \geq 2$, $1 \leq i \leq L$. Since the singularity effect is expelled, the solution errors near the singular point can be computed by

$$|u(r,\theta) - \tilde{u}_L(r,\theta)| = \left| \sum_{i=1}^{L} (D_i - \tilde{D}_i) r^{\alpha_i} \sin \alpha_i \theta \right| + |R_L|$$

$$\leq Ch^k r^{\alpha_i} + O(r^{\alpha_{N+1}}),$$

where $\tilde{u}_L(r,\theta) = \sum_{i=1}^{L} \tilde{D}_i r^{\alpha_i} \sin \alpha_i \theta$. When $r \leq Ch$, then $|u(r,\theta) - \tilde{u}_L(r,\theta)| = O(h^{k+\beta})$ with $\beta = \min \alpha_i$.

The singular part of solutions is removed partially, due to the approximation of $\tilde{D}_0$. Therefore, the complete removal of $\sum_{i=1}^{L} D_i \phi_i$ should be repeated until their defect is insignificant. The numerical results in Wigley (1988) show that a few iterations are good enough for practical applications.

The iteration process of Wigley is similar to the Schwarz alternative method, see Chapter 16. Wigley method is also a combination of FEM with the singular functions by subtracting operations.

Blum and Dobrowolski (1982) propose the dual singular function method (DSFM), which is developed from both Fix's and Wigley's methods. The DSFM is extended to the biharmonic equation in Blum (1982), and a more analysis of SFM and DSFM is reported in Bourlard, Dauge and Nicaise (1990), and Bourlard, Dauge, et. al. (1992). In the DSFM, the stress intensity factors, i.e., the coefficients D_i, can be computed explicitly in numerical methods; this is particularly favourable to engineering problems. The determination of singular fracture intensity factors with high order are discussed in Vasilopoulos (1988).

An improvement of SFM is also given in Liang, Kong and He (1996), by setting the nodes as variables

$$v = \sum_{i=1}^{M}(a_i + b_j \phi_1)\psi_i + \sum_{i=M+1}^{N} a_i \psi_i,$$

where ψ_i are the basis functions of FEM, and ϕ_1 is the leading singular function. As an example, $\phi_1 = r^{\frac{1}{2}} \cos(\frac{\theta}{2})$ for Motz's problem, and a_i are the node values in FEM of approximation solutions. In the singular domain, the leading singular function ϕ_1 is imbedded by implementing the terms $b_j \phi_1 \psi_i$ with new variables b_j. The advantage of this method is to treat the stiffness in a uniform way; however, more unknowns b_i are required.

2.7 COMBINATIONS OF H- AND P-VERSIONS

The p-version FEM is developed by Babuska and his colleagues. To raise the efficiency of the standard h-version FEM, the high order p of piecewise polynomials can be employed so that the elements need not be small, also to reach the same accuracy of numerical solutions by the h-version FEM. The optimal convergences of k-order Lagrange elements and p-version FEM are

$$\begin{aligned} \|u - u_h\|_E &= O(h^k), \quad \text{as } h \longrightarrow 0. \\ \|u - u_p\|_E &= O(e^{-C_1\sqrt{\rho}}), \quad \text{as } p \longrightarrow \infty, \end{aligned} \tag{2.57}$$

respectively in the energy norms, which $C_1(> 0)$ is a bounded constant independent of integer p. For the singularity problems, the errors from the p-version are given in Babuska and Suri (1987),

$$\|u - u_p\|_E \leq Cp^{-2\alpha+\delta}, \quad 0 < \delta << 1, \quad \alpha = \min_i \alpha_i,$$

twice the rates of the h-version with k order elements, $k \geq 1$. See Babuska and Suri (1987), Gui and Babuska (1986), Babuska and Guo (1988, 1989), Guo and Babuska (1986), and Babuska, Guo and Osborn (1989). When the smoothness of solutions is not uniform on the entire solution domain, the coupling of h- and p-version is necessary. The using of $h-p$ versions is natural by both increasing p and decreasing h. Note that the combinations of $h - p$ versions are conforming; their errors for singularity problems are

$$\|u - u_{h,p}\|_E \leq Ch^{Min(p,k)}p^{-\mu}\|u\|_{\mu,S},$$

where $\mu = \alpha - \delta$, and $C(> 0)$ is a constant independent of p and h.

Still the singularity damages the efficiency of p and $h - p$ versions. By applying the local refinements or the singular functions methods near the corners, the exponential convergence rates (2.57) of p-version can be recovered, see Babuska and Oh (1990), Oh and Babuska (1992), and Guo and Oh (1994).

The p- and h-p versions were first addressed by Babuska, Szabo and Katz (1981) and Babuska and Dorr (1981), then many contributions have been made. A overview on the p and h-p versions is given in Babuska and Suri (1990).

2.8   COMBINATIONS OF FEM AND BEM

Since FEM is more flexible, but BEM is more efficiency owing to the variables with less degree needed, much attention has been paid on their coupling, see Zienkiewicz, Kelley and Bettess (1977), Givoli (1992) and Yu (1993). In fact, the coupling of FEM and BEM is of the typical combined methods called in this book, since both

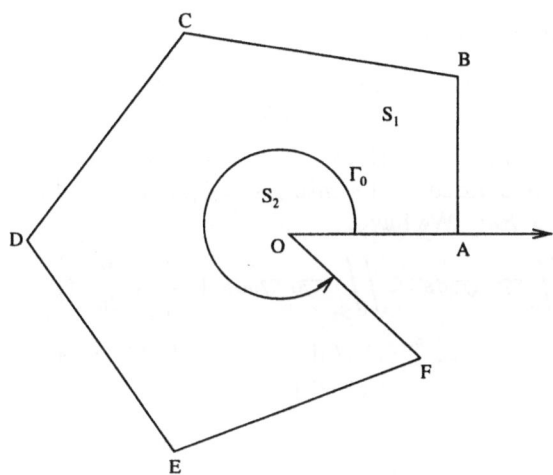

Figure 2.14 A concave domain.

BEM and FEM stand as independent numerical methods. The BEM can be certainly imbedded into the combined family, since the techniques matching RGM and FEM in Parts II and III later can also be used for BEM with other numerical methods, e.g., RGM, FDM, etc.

2.8.1 Combination for Angular Singularity

Consider the Laplace equation with the Neumann conditions on the concave polygon as shown in Figure 2.14

$$\begin{cases} \triangle u = 0, & \text{in } S, \\ \frac{\partial u}{\partial n} = 0, & \text{on } \overline{OA} \cup \overline{OF}, \\ \frac{\partial u}{\partial n} = g, & \text{on } A\widehat{BCDE}F. \end{cases} \tag{2.58}$$

The existence of solutions to (2.58) is granted by the consistent condition

$$\int_{A\widehat{BCDE}F} g \, d\ell = 0.$$

The solutions of (2.58) are unique in the quotient space $H^1(S)/P_0$, i.e., the uniqueness with an arbitrary constant. Hence the Galerkin problem of (2.58) is expressed as follows.

To seek $u \in H^1(S)/P_0$ such that

$$\iint_S \nabla u \, \nabla v \, ds = \oint_{\widehat{ABCDEF}} g v \, ds.$$

Choose the circular arc, $\Gamma_0 = \{(r, \theta), \ r = R_0, \ 0 \le \theta \le \Theta\}$, as an artificial boundary. Then S is divided by Γ_0 into $S_2 = \{(r, \theta), \ r \le R_0, \ 0 \le \theta \le \Theta\}$ and $S_1 = S \cap \{(r, \theta), \ r > R_0\}$. We have

$$\iint_S \nabla u \, \nabla v \, ds = \iint_{S_1} \nabla u \, \nabla v \, ds + \int_{\Gamma_0} v \frac{\partial u}{\partial n} d\ell,$$

where the normal derivatives, $\frac{\partial u}{\partial n}(R_0, \theta)|_{\Gamma_0} = \frac{\partial}{\partial r} u(R_0, \theta)$, satisfy the natural integral equation from Section 1.7 (see Yu (1993))

$$\frac{\partial u}{\partial n}(R_0, \theta) = -\frac{\pi}{4\Theta^2 R} \int_0^\Theta \left(\frac{1}{\sin^2 \frac{\theta - \theta'}{2\Theta} \pi} + \frac{1}{\sin^2 \frac{\theta + \theta'}{2\Theta} \pi} \right) u(R_0, \theta') d\theta', \quad 0 < \theta \le \Theta.$$

Hence we obtain the combining form:

To seek $u \in H^1(S)$ such that

$$\iint_{S_1} \nabla u \, \nabla v \, ds + D(u, v) = \int_{\widehat{ABCDEF}} g v \, d\ell, \quad \forall v \in H^1(S),$$

where

$$D(u, v) = \frac{-\pi}{4\Theta^2} \int_0^\Theta \int_0^\Theta \left(\frac{1}{\sin^2 \frac{\theta - \theta'}{2\Theta} \pi} + \frac{1}{\sin^2 \frac{\theta + \theta'}{2\Theta} \pi} \right) u(R_0, \theta') v(R_0, \theta) d\theta' d\theta.$$

We may use FEM in S_1 and the discrete approximation on $D(u, v)$, to lead to the combinations of FEM and BEM. Note that the double integral $D(u, v)$ with strong singularity is defined on Γ_0; only the bounded subdomain S_1 is involved in computation. The numerical integrations on $D(u, v)$ are given in Yu (1993). The algebraic equations

$$\mathbf{Ax} = \mathbf{b} \tag{2.59}$$

can be obtained using the approximation of FEM in S_1 and BEM in S_2 (see Sections 1.2 and 1.7), where $\mathbf{A}$ is the positive semi-definite and symmetric. With an arbitrary constant permitted, the solutions of (2.59) are unique under a consistent condition.

When the k-order Lagrange element is chosen in S_1, the optimal errors for the solutions u_h^c by the combination of FEM-BEM are given in Yu (1993)

$$\|u - u_h^c\|_{\ell, S_1} + \|u - u_h^c\|_{\frac{1}{2}, \Gamma_0} \le C h^{k+1-\ell} \|u\|_{k+1, S}, \quad \ell = 0, \ 1.$$

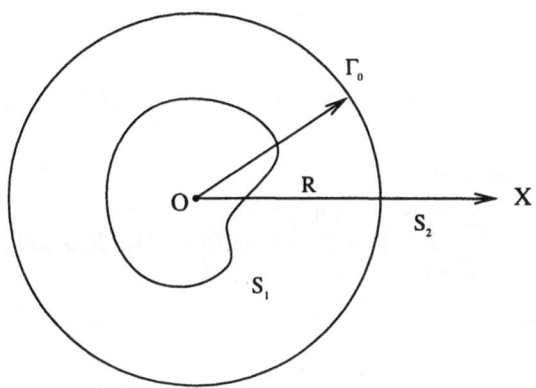

Figure 2.15 A unbounded domain.

2.8.2 Combination for Unbounded Domain Problems

Consider the equation

$$-\left(\frac{\partial}{\partial x}p\frac{\partial u}{\partial x} + \frac{\partial}{\partial y}p\frac{\partial u}{\partial y}\right) = f, \qquad \text{in } S,$$

$$u = g \text{ on } \partial S, \qquad u \longrightarrow 0 \text{ as } r \longrightarrow \infty, \tag{2.60}$$

under the conditions

$$p = 1, \quad f = 0, \quad \text{when } r \geq R. \tag{2.61}$$

We divide the domain S by the circle $(r = R)$ into S_1 and S_2, where $S_2(r > R)$ is unbounded (see Figure 2.15). From (2.61), we restate (2.60) as

$$-\frac{\partial}{\partial x}\left(p\frac{\partial u}{\partial x}\right) - \frac{\partial}{\partial y}\left(p\frac{\partial u}{\partial y}\right) = f, \qquad \text{in } S_1, \tag{2.62}$$

$$\triangle u = 0, \qquad \text{in } S_2,$$

$$u = g, \qquad \text{on } \partial S, \tag{2.63}$$

$$u \longrightarrow 0, \qquad \text{as } r \longrightarrow \infty. \tag{2.64}$$

The explicit solutions for the Laplace equations in S_2 with (2.64) are obtained for Section 2.2.1

$$u(r, \theta) = \sum_{n=1}^{\infty} r^{-n}\left(a_n \cos n\theta + b_n \sin n\theta\right), \tag{2.65}$$

where the coefficients

$$a_n = \frac{R^n}{\pi} \int_0^{2\pi} u(R,\theta) \cos n\theta d\theta, \quad b_n = \frac{R^n}{\pi} \int_0^{2\pi} u(R,\theta) \sin n\theta d\theta.$$

Hence we obtain the solutions

$$u(r,\theta) = \sum_{n=1}^{\infty} \int_0^{2\pi} \left(\frac{R}{r}\right)^n \frac{1}{\pi} \cos n(\theta - \theta') u(R,\theta') d\theta',$$

and their derivatives

$$\left. \frac{\partial u}{\partial r} \right|_{r=R} = -\sum_{n=1}^{\infty} \int_0^{2\pi} \frac{n}{\pi R} \cos n(\theta - \theta') u(R,\theta') d\theta'$$

$$= -\int_0^{2\pi} M_n(\theta,\theta') u(R,\theta') d\theta', \tag{2.66}$$

where

$$M_n(\theta,\theta') = \sum_{n=1}^{\infty} \frac{n}{\pi R} \cos n(\theta - \theta'). \tag{2.67}$$

We may reduce the unbounded problem (2.62)–(2.64) to the bounded problem with (2.62), (2.63) and (2.65). Note that (2.65) is exact on the artificial boundary Γ_0 and that there is also a singularity in (2.66), because of the divergent series (2.67).

Based on an equality in the generalized functions

$$\frac{1}{\pi} \sum_{n=1}^{\infty} n \cos n\theta = -\frac{1}{4\pi \sin^2 \frac{\theta}{2}}, \tag{2.68}$$

we have the simple expression but with a strong singularity

$$M_n(\theta,\theta') = -\frac{1}{4\pi R \sin^2 \frac{(\theta-\theta')}{2}},$$

to express (2.66) as

$$\left. \frac{\partial u}{\partial r} \right|_{r=R} = \int_0^{2\pi} \frac{u(R,\theta')}{4\pi R \sin^2 \left(\frac{\theta-\theta'}{2}\right)} d\theta' = \oint_{\Gamma_0} \frac{1}{4\pi R^2} \frac{u(R,\theta')}{\sin^2 \left(\frac{\theta-\theta'}{2}\right)} d\ell.$$

Note the equality (2.68) is valid in the sense of generalized functions, although the series on the left diverges (see Yu (1993)).

Now, consider the Galerkin problem:

$$\iint_{S_1} p \,\nabla u \,\nabla v ds + \iint_{S_2} \nabla u \,\nabla v ds = \iint_{S_1} f v ds,$$

where

$$\iint_{S_2} \nabla u \,\nabla v ds = \oint_{\Gamma_0} \frac{\partial u}{\partial n} v d\ell = - \oint_{\Gamma_0} \frac{\partial u}{\partial r} v d\ell.$$

Hence we obtain

$$\iint_{S_1} p \,\nabla u \,\nabla v ds + \oint_{\Gamma_0} \oint_{\Gamma_0} \frac{1}{4\pi R} \frac{u(R,\theta')}{\sin^2\left(\frac{\theta-\theta'}{2}\right)} v(R,\theta) d\theta d\theta' = \iint_{S_1} f v ds. \qquad (2.69)$$

If the FEM is used in S_1, the combination of BEM-FEM is formulated easily for solving the unbounded problems.

2.8.3 Comparisons with Combined Methods of RGM and FEM

From the above unbounded domain problems, we may find some linkage between the BEM and the BAM using particular solutions. In (2.65), the infinite terms are chosen, and the coefficients are replaced by the integrals of the true solution u. Hence, only the solution $u(r,\theta)$ appear in the final equation (2.69). However, since the leading coefficients a_k and b_k in (2.65) are so important, we had better retain the expansion of (2.65) but with finite terms

$$u \approx u_L(r,\theta) = \sum_{n=1}^{L} r^{-n}(a_n \cos n\theta + b_n \sin n\theta).$$

Two different kinds of admissible functions are chosen as

$$v = \begin{cases} v^- = v_k, & \text{in } S_1, \\ v^+ = v_L(r,\theta), & \text{in } S_2, \end{cases}$$

where v_k is k-order Lagrangian FEM, but

$$v_L^+(r,\theta) = \sum_{n=1}^{L} r^{-n} \left(\tilde{a}_n \cos n\theta + \tilde{b}_n \sin n\theta \right).$$

Now we face a new difficulty how to link v^+ and v^-, which will be fully discussed in Parts II and III. Let us take as an example the penalty techniques using the additional integral on Γ_0

$$\frac{P_c}{h^\sigma} \int_{\Gamma_0} (v^+ - v^-)^2 d\ell,$$

where P_c is a bounded constant, h is the maximal boundary length of $\triangle_i$ in S_1, and $\sigma(> 0)$ is a power constant to be determined later. Then the unbounded domain problems may be solved approximately by the combined method of RGM-FEM, as called in this book.

$$\iint_{S_1} p \nabla u \nabla v ds + \iint_{S_2} p \nabla u \nabla v ds + \frac{P_c}{h^\sigma} \int_{\Gamma_0} (v^+ - v^-)^2 d\ell = \iint_{S_1} f v ds, \quad (2.70)$$

where

$$\iint_{S_2} \nabla u \nabla v ds = \oint_{\Gamma_0} \frac{\partial u}{\partial n} v d\ell.$$

We may rewrite (2.70) as

$$\iint_{S_1} p \nabla u \nabla v ds + \oint_{\Gamma_0} \frac{\partial u}{\partial n} v d\ell + \frac{P_c}{h^\sigma} \int_{\Gamma_0} (v^+ - v^-)^2 d\ell = \iint_{S_1} f v ds.$$

This also represents the problem just on the bounded domain S_1, by expelling away the unbounded subdomain S_2. The advantages of the combinations of RGM-FEM over the combination of BEM-FEM are that the leading coefficients may also be solved explicitly and that the singular integration in Section 2.8.2 may also be bypassed.

In Zielinski and Zienkiewicz (1985), similar combinations, called the T-complete boundary solution functions are proposed but with different coupling techniques, which are similar to hybrid or weight residual techniques. Also see Zielinski (1988), and Cheung, Jin and Zienkiewicz (1989).

The study on coupling FEM and BEM is extensive. The early justification on coupling BEM and FEM was given by Johnson and Nedelec (1980), then by Gatica and Hsiao (1989), Stephan and Wendland (1984a,b), Stephan (1992), Ervin and Stephan (1990), Bossavit (1991), Hsiao and Porter (1988), Carstensen (1996), Bielak and MacCamy (1991), Meddahi, Valdes, et. al. (1996), Theocaris, Tsamasphyros and Theotokoglou (1984), He and Li (1990), Li and He (1989), Jiang and Li (1987), Belytschko and Lu (1994), and Gatica and Hsiao (1992).

Many reports on coupling techniques have appeared: Berger, Warnecke and Wendland (1994), Bossavit (1988), Bremi (1977), Brezzi and Johnson (1979), Brezzi,

Johnson and Nedelec (1978), Brichau and Deconinck (1992), Brink and Stephan (1996), Cai, Lee and Oh (1993), Carstensen and Stephan (1995), Cecot and Ork-isz (1987), Colin and Floquet (1986), Costabel (1987, 1989), Costabel, Ervin and Stephan (1990), Costabel and Stephan (1990), Dorr (1989, 1991), Grannell (1987, 1988), Guirguis (1987a), Hsiao (1988b, 1990), Li (1980, 1989a–c, 1992a,b, 1996a, 1997a–c), Li, Lin and Yan (1997), Li and Bui (1990b), Margulies (1981), Rangogni (1986), Wendland (1988, 1990), Zienkiewicz (1984) and Meddahi, Valdes, et. al. (1997).

It is noted that other methods or treatments may also be adopted in the boundary element method, see Stephan and Suri (1989), Babuska, Guo and Stephan (1990), and Rank (1989) for using p, h-p versions, and Babuska, Guo and Stephan (1990) for local refinements.

2.9   COMBINED METHODS

2.9.1   Introduction

The complete particular solutions, $u = \sum_{i=1}^{\infty} a_i \phi_i$ are recommended in Zielinski and Zienkiewicz (1985), in order to compete with BEM, because the difficulty of singular integration can be avoided. They also referred this kind of methods to Trefftz (1926). Suppose that the admissible functions, $u \approx u_L = \sum_{i=1}^{L} \tilde{a}_i \phi_i$, are chosen, the coefficients $\tilde{a}_i$ can be computed directly. Note that the leading coefficient a_1 of ϕ_1 is significant, as the fracture stress factor. The semi-analytical methods are also called when the finite expressions are combined with FEM in the neighbouring subdomains, see Li, Shi, et. al. (1981), Liang and Gu (1978), Cheung, Jin and Zienkiewicz (1991), and Matveyenko and Borzenkov (1996). The semi-analytic methods are favourable to engineers because a few leading particular solutions may be found in many practical problems, and because using those singular functions may provide satisfactory and easy analysis in engineering, see Cao and Cheung (1992). Other reports on Trafftz functions are given in Piltner (1985) for internal cracks, Jirousek and Guex (1986) for the hybrid-trefftz method, and Cheung, Jin and Zienkiewicz (1989) for the Helmholtz equation. Also Cao and Cheung (1992) summrizes the semi-analytical method with extensive literatures; however, there is no error analysis stated therein. Besides, the singular functions have been fully used in Ogen and Schiff (1983), to be matched with FEM.

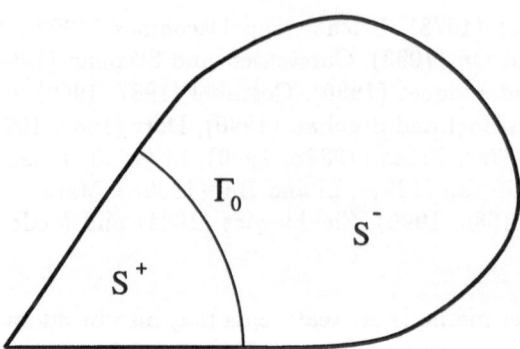

Figure 2.16   An angular subdomain.

The computational results by the combined methods in engineering are very promising . In this book, both the important combinations and various coupling techniques are provided, error analysis and optimal matching are derived carefully. Useful guidance in application can also be found in this book. Combining different particular solutions is reported in Delves (1979), Hendry and Delves (1979), Hendry, Delves and Phillips (1979), Phillips and Delves (1980) and Delves and Hall (1979), by means of the coupling techniques similar to penalty and hybrid couplings, given in Chapters 6–8. Developments of global element method, which has a common feature of BAM, are found in Delves and Freeman (1981), Hendry and Delves (1979) and Kermode, McKerrell and Delves (1985). In 1977 by Li, the author, and Chou (1977), the asymptotic solutions were used near the infinity, and combined with FEM, to explore an enzyme reaction problem on the unbounded domain. Since the computed results using combining techniques are very attractive, also see Li, Shi, et. al. (1981), the author has been making a great effort to study systematically the combined methods, their coupling strategies and advanced topics. Besides, error and stability analysis are given in Liang and Gu (1978) for semi-analytic method. Liang has many important contributions on the combined methods, for instance, the combination of different finite elements in Liang (1979) and Li and Liang (1980), combination of RGM-FEM for singularity problems in Liang and Gu (1978), Liang (1980a,b), and Li and Liang (1980, 1983), and a new finite element space in Liang, Kong and He (1996).

2.9.2   General Description

For simplicity, we consider the combination of two methods. Let S in Figure 2.16 be partitioned into S^+ and S^- with or without overlaps. The admissible functions are chosen as

$$v = \begin{cases} v^+, & \text{in } S^+, \\ v^-, & \text{in } S^-. \end{cases}$$

Since v^+ and v^- are different admissible functions, they may not satisfy the continuity conditions along the common boundary Γ_0 of S^+ and S^-:

$$v^+ \neq v^-, \qquad \text{on } \Gamma_0. \tag{2.71}$$

Therefore, the combinations are nonconforming, usually.

The key study is how to match two methods along Γ_0. For (2.71), the straightforward treatments are the direct constraints of v^+ and v^- on the element nodes $Q_i \in \Gamma_0$, i.e.,

$$v^+(Q_i) = v^-(Q_i), \qquad \text{on } \Gamma_0. \tag{2.72}$$

This is analogous to the case of connecting two pieces of clothes with buttons. Such methods are called the nonconforming combinations in this book. In part II, we will provide different numerical methods matched with (2.72) in details.

Since the direct constraints in (2.72) involve some programming of elimination, we may use an additional integral on Γ_0 instead. The simplest one is the penalty integral.

$$I_p(v) = w^2 \int_{\Gamma_0} \left(v^+ - v^- \right)^2 d\ell,$$

where w is a suitable weight function. Also the solution derivatives, called hybrid techniques, may also be employed into the additional integrals as well, resulting in different coupling techniques. We describe them in Part III.

Let us consider a model problem:

$$Lu = 0 \text{ in } S, \quad u|_{\partial S} = g, \tag{2.73}$$

where $L = -\nabla p \nabla + c$. For the solution singularity existing in S_2, we choose the singular particular solutions ϕ_i such that $L\phi_i = 0$ in S_2. The particular solutions ϕ_i can be found in Section 2.1 and in PDE textbooks; otherwise, exposition of particular solutions or asymptotic expansions should be done.

The singular solutions ϕ_i are very well suited to approaching the solution near the singular points; but not to that in other parts of S with arbitrary shapes. A better idea is to combine the singular solutions ϕ_i for the singular domain, say S^+, and other admissible functions in S^- (see Figure 2.16). Then we express them as

$$v_{L-h} = \begin{cases} v_L, & \text{in } S^+, \\ v_h, & \text{in } S^-, \end{cases} \tag{2.74}$$

where $v^+ = v_L = \sum_{i=1}^{L} c_i \phi_i$, and v_h, for instance, are piecewise k-order Lagrange polynomials in FEM. Denote by V_h^* and V_h^0 the space of functions (2.74) satisfying (2.72), and $u|_{\partial S} = g$ and $u|_{\partial S} = 0$, respectively. We obtain the nonconforming combinations of RGM-FEM.

Find $u_{L-h} \in V_h^*$ such that

$$a_c(u_{L-h}, v) = f(v), \quad \forall v \in V_h^0,$$

where

$$a_c(u, v) = \iint_{S+} p \, \nabla u \, \nabla v \, ds + \iint_{S-} p \, \nabla u \, \nabla v \, ds + \iint_{S} cuv \, ds, \tag{2.75}$$

$$f(v) = \iint_{S} fv \, ds.$$

In the original RGM, the uniform functions are used in the entire region of S. This confines its application only to the simple problems, such as those on a sectorial or a rectangular domain. The combinations in this book may adopt local particular solutions, to widen greatly their applications.

2.9.3   Aspects in Analysis

When different methods are combined by the chosen coupling techniques, i.e., the nonconforming constraints, penalty techniques, etc., analysis on solution error and stability is important. It is required that the coupling strategies do not reduce the optimal convergence rates. Define a norm

$$\|v\|_1 = \left(\|v\|_{1,S+}^2 + \|v\|_{1,S-}^2 \right)^{\frac{1}{2}}.$$

Obviously, the optimal coupling strategies should be chosen to balance two error bounds

$$\|u - u_L^+\|_{1,S+} = O\left(\|u - u_h^-\|_{1,S-} \right).$$

Each method contributes itself just to balance its opponent. However, if one method does its best to overwork against its opponent, reduced convergence rates may occur owing to the errors resulting from the matching boundary Γ_0, see the counterexamples in Part II.

This implies that the collaborating is important in combinations. The main analysis in this book is devoted to estimate the coupling error bounds along the common boundary Γ_0 of different methods. Our objects are to discover the couplings that lead to optimal (or reduced) convergence rates and optimal (or worse) stability. This is also the distinct feature in analysis of this book from other books.

Many literatures cited in this book are benificial to my research on combined methods; among them, two monographs, Strang and Fix (1973) and Ciarlet (1978), are most prominent. The ideas of combined methods and the samples of singularity problems in Strang and Fix (1973) have stimulated my research motivation, and Strang's lemmas of nonconforming elements and Ciarlet's techniques of the FEM analysis are the main tools for analysis of the combined methods given in this book.

In the beginning, we said that a perfect method does not exist. Then what drawbacks do the combined methods have? A drawback is that more efforts paid to acknowledge the solution behavior, to choose coupling techniques, and to analyze their effects. Another drawback is that more programming is needed in dealing with the coupling along Γ_0. This is, however, a fair tradeoff of combinations, to gain effectiveness in solving the singularity problems.

To close this chapter, an overview of singularity treatments is illustrated in Figures 2.17 and 2.18. The ideal numerical methods in the 21 century should be like the combined methods, where all methods can be employed together, and integrated in a very harmonious way such that to ulitize fully their merits and also to avoid their shortcomings. This book is dedicated to such a new trend of numerical methods for solving elliptic boundary value problems.

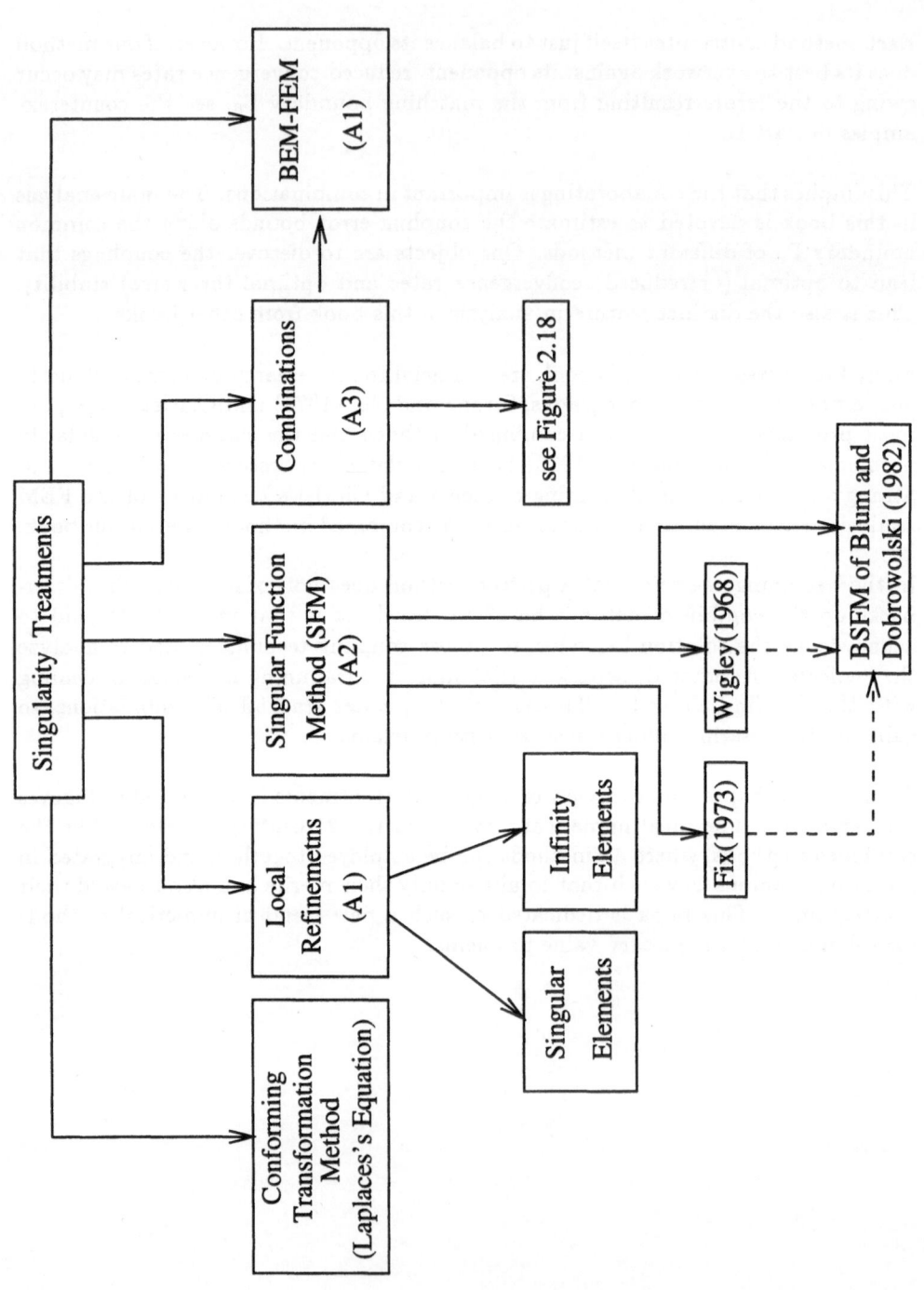

Figure 2.17 Overview of singularity treatments.

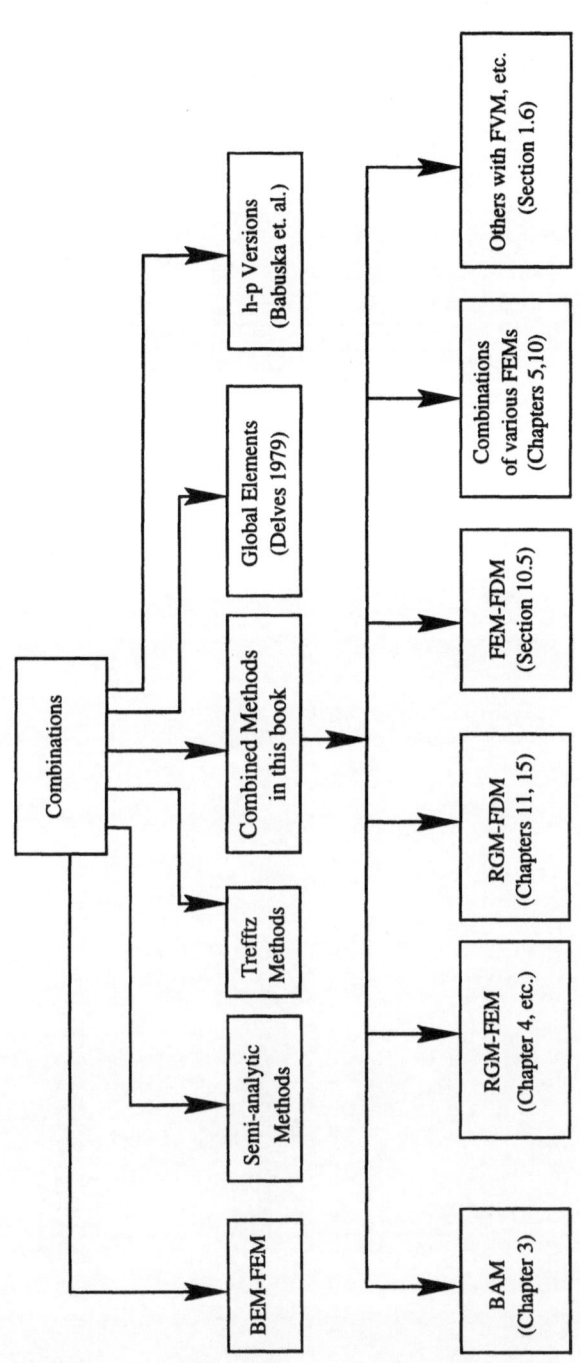

Figure 2.18 Overview of combinations.

The worst and best are both inclined
To snap like vixens at the truth;
But, O, beware the middle mind
That purrs and never shows a tooth!
—— *Nonsense Rhyme* ——

Elinor Hoyt Wylie
(1885–1928)

PART II

COMBINED METHODS

This part presents three typical kinds of combinations coupling different admissible functions:

Chapter 3: Boundary Approximation Methods, coupling different singular and analytical functions.

Chapter 4: Combinations of RGM and FEM, coupling singular (or analytical) functions and piecewise polynomials.

Chapter 5: Combinations of Various FEMs, coupling different piecewise polynomials.

The contents of this part are adapted mainly from Li (1983b, 1986, 1990, 1991), Li and Mathon (1990a, b), Li, Mathon and Sermer (1987), and Li and Liang (1981).

Under the generalized Galerkin frame, different numerical methods may be viewed as the same variational form, but with different admissible functions. In the combinations described in this book, different numerical methods, i.e., different admissible functions, are chosen in different subdomains. When two different methods are matched, the entire admissible functions consist of two different admissible functions,

$$v = \left\{ \begin{array}{ll} v_1, & \text{in } S_1, \\ v_2, & \text{in } S_2. \end{array} \right.$$

Obviously, the functions v are, usually, not continuous along the common boundary Γ_0. Aimed at the requirements, $v \in C(S)$, for conforming functions, we may enforce v to satisfy the constraint conditions on discrete nodes Q_i on Γ_0: $v_1(Q_i) = v_2(Q_i)$. Since the remedied functions v are still *nonconforming*, the nonconforming coupling and the nonconforming combination are called, and described in Chapters 4 and 5. In Chapter 3, the two interior continuity conditions

$$v^+ = v^-, \quad \frac{\partial v^+}{\partial n} = \frac{\partial v^-}{\partial n} \text{ on } \Gamma_0$$

are also enforced on the admissible functions v.

By the above straightforward constraints, the algorithms of the stated three combinations are developed. However, analysis must be carried out, to estimate the error bounds resulting from the coupling conditions. Our goals in analysis are to achieve the optimal convergence rates and the optimal numerical stability, and to discover the worse situation that should be avoided in use. In this part, the combination algorithms, error analysis, stability analysis and numerical experiments are provided.

The algorithms and analysis in combinations may be considered simply as those on the interior boundary Γ_0; they are different from those in the nonconforming FEMs (see Ciarlet (1978)), because nonconformity in combinations in this book occurs only on Γ_0 not on all element edges. Note that the theoretical and numerical results are surprising, and are stated later in details.

3

BOUNDARY APPROXIMATION METHODS

For solving homogeneous elliptic equations, boundary approximation methods use particular solutions to approximate the boundary conditions as accurately as possible, usually in a least squares sense. In the interior of a given region, such an approximate solution satisfies the differential equation exactly. The advantage of boundary approximation methods over a standard finite difference or finite element method is that it can cope easily with complicated boundaries and boundary conditions, as well as with singularities and infinite domains.

Approximation by particular solutions using harmonic polynomials was first applied by Kantorovich and Krylov (1958) to solve Laplace's equation. Fox, Henrici and Moler (1967) used particular solutions to find eigenvalues of the Laplace operator. Bergman's (Bergman (1969), Bergman and John (1961, 1965) and Vekua's (Vekua (1967)) integral representations of solutions yielded particular solution expansions for a large class of elliptic equations. The error analysis for the Bergman-Vekua method was worked out by Eisenstat (1974). The book of Lefeber (1989) described the BAM without interfaces. However, the approaches given in this chapter will provide better results due to using the least squares method in Golub and Loan (1989), which may greatly reduce the condition numbers, and then raise accuracy of the numerical solutions in double precisions.

From a computational point of view, boundary approximation methods are easy to use, and beneficial from the reduced complexity of a boundary approximation. In addition, it is often possible to control the errors of the approximate solutions by the computable errors on the boundary, even for elliptic problems which do not possess the maximum principle (see Mathon and Sermer (1982)).

However, difficulties may arise in the situations where a large number of particular solutions is needed to achieve a satisfactory approximation, mainly due to ill-conditioning of the associated least squares matrices. For the problems with material interfaces, singularities and unbounded domains, it may not be possible to find a single, uniform expansion into particular solutions which is valid in the entire domain. It is therefore necessary to subdivide a domain into several subdomains and to use different expansions in each of them. A solution is then obtained by approximating both the exterior boundary conditions and the interior continuity

103

conditions across the various interfaces. We note that for a good subdivision, only a few terms are needed in each subdomain to achieve a highly accurate approximation, and to mitigate the numerical instability. Moreover, the numerical stability is greatly improved and the least squares matrices have many zero entries.

We will study the boundary approximation methods for solving homogeneous self-adjoint elliptic equations that combine several expansions of particular solutions from different parts of the region. In this chapter, we will carry out the analysis first for the Laplace equation, and then for the Debye–Huckel equation $- \triangle u + u = 0$ in the case of different subdomains. Error bounds are obtained for the approximate solutions, and guidelines are given for the choice of weights in the least squares function. Finally, we illustrate the methods on the Motz problem and a crack-infinity problem, which can be regarded as two testing models of singularity problems. The contents in this chapter are adapted from Li (1990), Li, Mathon and Sermer (1987), and Li and Mathon (1990a, 1990b).

3.1   NOTATIONS AND PRELIMINARIES

Let S be a bounded polygon in the plane R^2 with a piecewise smooth boundary ∂S. We will consider the problem

$$\triangle u \quad = \quad \frac{\partial^2 u}{\partial x^2} + \frac{\partial^2 u}{\partial y^2} = 0 \ \text{ in } S, \tag{3.1}$$

$$u \quad = \quad f \ \text{ on } \Gamma_D, \ \ u_\nu = g \ \text{ on } \Gamma_N, \tag{3.2}$$

where f and g are sufficiently smooth functions, Γ_D and Γ_N are compact subsets of ∂D, $\Gamma_D \neq \emptyset$, $\Gamma_D \cup \Gamma_N = \partial S$, and u_ν is the normal derivative.

Let Γ_0 be an interface which divides S into two subdomains, S^+ and S^-, and let u^+ and u^- be defined on S^+ and S^- respectively. If u^+ and u^- satisfy (3.1) and (3.2) on the corresponding subdomains and boundaries, and

$$u^+ = u^-, \quad u_\nu^+ = u_\nu^- \ \text{ on } \Gamma_0, \tag{3.3}$$

then $u^+ = u$ in S^+ and $u^- = u$ in S^-.

For an integer $k \geq 0$, $H^k(S)$ denote the Sobolev spaces of order k of real-valued functions with the Sobolev norms $\| \cdot \|_{k,S}$ and seminorms $| \cdot |_{k,S}$:

$$\|v\|_{k,S} = \left\{ \sum_{|\alpha| \leq k} \iint_S |D^\alpha v|^2 ds \right\}^{\frac{1}{2}}, \quad |v|_{k,S} = \left\{ \sum_{|\alpha| = k} \iint_S |D^\alpha v|^2 ds \right\}^{\frac{1}{2}}.$$

If S is divided by an interface Γ_0 into S^+ and S^-, we will use a new space H:

$$H = \{v \in L_2(S) | v \in H^1(S^+), \; v \in H^1(S^-), \text{ and } \triangle v = 0 \text{ in } S^+ \text{ and } S^-\},$$

where $L_2(S)$ is the usual space of square integrable functions in S with the inner product $(u, v) = \iint_S uv ds$ and the norm $\|u\|_0 = (u, u)^{\frac{1}{2}}$. The space H is equipped with the norm and seminorm:

$$\|v\|_1 = \left\{\|v\|^2_{1,S+} + \|v\|^2_{1,S-}\right\}^{\frac{1}{2}}, \quad |v|_1 = \left\{|v|^2_{1,S+} + |v|^2_{1,S-}\right\}^{\frac{1}{2}}.$$

If γ is a curve, then $|\cdot|_{k,\gamma}$ denotes the norms in the Sobolev spaces $H^k(\gamma)$ of functions defined on γ.

For $v \in H$, $f \in L_2(\Gamma_D)$ and $g \in L_2(\Gamma_N)$ given by (3.2), we define a functional

$$
\begin{aligned}
I(v) \; = \; & \int_{\Gamma_D} (v - f)^2 d\ell + w^2 \int_{\Gamma_N} (v_\nu - g)^2 d\ell + \int_{\Gamma_0} (v^+ - v^-)^2 d\ell + \\
& + w^2 \int_{\Gamma_0} (v_\nu^+ - v_\nu^-)^2 d\ell,
\end{aligned}
$$

where w is some positive weight. On $H \times H$, we will consider a bilinear form $[\cdot, \cdot]$ given by

$$
\begin{aligned}
[u, v] \; = \; & \int_{\Gamma_D} uv d\ell + w^2 \int_{\Gamma_N} u_\nu v_\nu d\ell + \int_{\Gamma_0} (u^+ - u^-)(v^+ - v^-) d\ell \\
& + w^2 \int_{\Gamma_0} (u_\nu^+ - u_\nu^-)(v_\nu^+ - v_\nu^-) d\ell.
\end{aligned}
$$

Clearly, $[\cdot, \cdot]$ is an inner product which induces the norm on Γ_0

$$|v|_B^2 = [v, v]. \tag{3.4}$$

We need the following lemma.

Lemma 3.1 *If $v \in H$, then there exists a positive constant C independent of v such that*

$$\|v\|_1 \leq C\left\{|v|_1 + |v|_{0,\Gamma_D} + |v^+ - v^-|_{0,\Gamma_0}\right\}.$$

Proof. For a curve $\gamma \in \partial S$ and $v \in H^1(S)$, we have (Sobolev (1963), p.68)

$$\|v\|_{1,S} \leq C\{|v|_{1,S} + |v|_{0,\gamma}\}. \tag{3.5}$$

Applying this inequality to S^+ and S^-, we obtain

$$\|v\|_1 \leq C\{|v|_1 + |v|_{0,\Gamma_D \cap S^-} + |v^+|_{0,\Gamma_0}\}$$

$$\leq C\{|v|_1 + |v|_{0,\Gamma_D} + |v^-|_{0,\Gamma_0} + |v^+ - v^-|_{0,\Gamma_0}\}, \qquad (3.6)$$

where we assume, without loss of generality, that $\Gamma_D \cap S^- \neq \emptyset$. Using the Sobolev imbedding theorem (Sobolev (1963), p. 84) and (3.5), we obtain

$$|v^-|_{0,\Gamma_0} \leq C\|v\|_{1,S^-} \leq C\{|v|_{1,S^-} + |v|_{0,\Gamma_D \cap S^-}\}$$

$$\leq C\{|v|_1 + |v|_{0,\Gamma_D}\}. \qquad (3.7)$$

The desired result follows from (3.6) and (3.7). ∎

3.2 APPROXIMATION PROBLEMS

Let $\{\psi_i^+\}$ and $\{\psi_i^-\}$ ($i = 1,2,...$) be complete sets of particular solutions of (3.1) on S^+ and S^- respectively. Consider finite-dimensional subspaces $S_{m,n} \subseteq H$ defined by

$$S_{m,n} = \left\{ v \,\middle|\, v = v^+ = \sum_{i=1}^m c_i \psi_i^+ \text{ on } S^+, \text{ and } v = v^- = \sum_{i=1}^n d_i \psi_i^- \text{ on } S^- \right\},$$

where c_i and d_i are real coefficients.

A function $u_{m,n} \in S_{m,n}$ will be called a boundary approximation to the solution u of (3.1) – (3.3) if it minimizes I over $S_{m,n}$, i.e.,

$$I(u_{m,n}) \leq I(v), \ \forall \ v \in S_{m,n}. \qquad (3.8)$$

The space $S_{m,n}$ is assumed to satisfy the following two properties:

Inverse Property: For any $v \in S_{m,n}$, we have

$$|v_\nu|_{0,\Gamma_D} \leq k_{m,n}\|v\|_1, \quad |v_\nu^+|_{0,\Gamma_0} \leq k_{m,n}\|v\|_1, \qquad (3.9)$$

where $k_{m,n}$ is unbounded as $m, n \longrightarrow \infty$.

Approximatability Property: For any $u \in H$, there exists a function $v \in S_{m,n}$ such that

$$\|u^+ - v^+\|_{l,\partial S^+} \leq \alpha_{n,k,l}^+ |u^+|_{k,\partial S^+}, \ 0 \leq l \leq k, \qquad (3.10)$$

and

$$\|u^- - v^-\|_{l,\partial S-} \le \alpha^-_{m,t,l} |u^-|_{t,\partial S-}, \quad 0 \le l \le t, \tag{3.11}$$

where $\alpha^+_{n,k,l}$, $\alpha^-_{m,t,l} \to 0$ as m, $n \to \infty$. The bounds of the constants, $\alpha^+_{n,k,l}$ and $\alpha^-_{m,t,l}$, can be obtained from Cheney (1966) and Eisenstat (1974). In (3.10) and (3.11), the Sobolev norms on the boundary ∂S are defined by

$$\|v\|_{l,\partial S} = \left\{ \sum_{|\alpha| \le l} \int_{\partial S} \left(\frac{\partial^\alpha v}{\partial s^\alpha} \right)^2 d\ell \right\}^{\frac{1}{2}}, \quad |v|_{l,\partial S} = \left\{ \sum_{|\alpha| = l} \int_{\partial S} \left(\frac{\partial^\alpha v}{\partial s^\alpha} \right)^2 d\ell \right\}^{\frac{1}{2}}.$$

We require the following estimates.

Lemma 3.2 *If $v \in S_{m,n}$ and the inverse property holds, then*

$$\|v\|_1 \le C(k_{m,n} + w^{-1})|v|_B, \tag{3.12}$$

where C is a positive constant independent of m and n.

Proof. By using Green's theorem and that $\triangle v = 0$ in S^+ and S^-, we have

$$0 = \iint_{S_1} v \triangle v ds + \iint_{S_2} v \triangle v ds = -|v|_1^2 + \int_{\partial S+} v v_\nu \, d\ell + \int_{\partial S-} v v_\nu \, d\ell.$$

Hence, from Holder's inequality it follows that

$$\begin{aligned}
|v|_1^2 &= \int_{\Gamma_D} v v_\nu d\ell + \int_{\Gamma_N} v v_\nu d\ell + \int_{\Gamma_0} [(v^+ - v^-)v_\nu^+ + v^-(v_\nu^+ - v_\nu^-)]d\ell \\
&\le |v|_{0,\Gamma_D}|v_\nu|_{0,\Gamma_D} + |v|_{0,\Gamma_N}|v_\nu|_{0,\Gamma_N} \\
&\quad + |v^+ - v^-|_{0,\Gamma_0}|v_\nu^+|_{0,\Gamma_0} + |v^-|_{0,\Gamma_0}|v_\nu^+ - v_\nu^-|_{0,\Gamma_0}.
\end{aligned} \tag{3.13}$$

By applying the Sobolev imbedding theorem and the inverse property, we obtain $|v|_1^2 \le Cq\|v\|_1$, where

$$q = k_{m,n}(|v|_{0,\Gamma_D} + |v^+ - v^-|_{0,\Gamma_0}) + |v_\nu|_{0,\Gamma_N} + |v_\nu^+ - v_\nu^-|_{0,\Gamma_0}.$$

From the imbedding theorem, Lemma 3.1 and that $k_{m,n} \ge 1$ we have

$$\begin{aligned}
\|v\|_1^2 &\le C\{|v|_1^2 + |v|_{0,\Gamma_D}^2 + |v^+ - v^-|_{0,\Gamma_0}(|v^+|_{0,\Gamma_0} + |v^-|_{0,\Gamma_0})\} \\
&\le C\{|v|_1^2 + (|v|_{0,\Gamma_D} + |v^+ - v^-|_{0,\Gamma_0})\|v\|_1\} \\
&\le Cq\|v\|_1.
\end{aligned}$$

Dividing both sides by $\|v\|_1$ and noting that $q \leq (k_{m,n} + w^{-1})|v|_B$, we obtain the desired result. ∎

Concerning the existence, uniqueness and stability of $u_{m,n}$ defined by (3.8), we have the following lemma.

Lemma 3.3 *Suppose that u is the solution of (3.1) and (3.2). Then for any $w > 0$, there exists a unique function $u_{m,n} \in S_{m,n}$ which satisfies*

$$[u_{m,n}, v] = \int_{\Gamma_D} fv\,d\ell + w^2 \int_{\Gamma_N} gv_\nu\,d\ell, \ \forall \ v \in S_{m,n}, \tag{3.14}$$

$$[u - u_{m,n}, v] = 0, \ \forall \ v \in S_{m,n}, \tag{3.15}$$

and

$$\|u_{m,n}\|_1 \leq C(k_{m,n} + w^{-1})\{|f|_{0,\Gamma_D} + w|g|_{0,\Gamma_N}\}. \tag{3.16}$$

Also, $u_{m,n}$ minimizes $I(v)$ over $S_{m,n}$ if and only if it satisfies (3.14).

Proof. Since $f = u$ on Γ_D, $g = u_\nu$ on Γ_N, and u and u_ν are continuous across Γ_0, the right-hand side expression in (3.14) equals $[u, v]$, and so (3.15) easily follows from (3.14). In (3.14), let $v = u_{m,n}$ and apply the Schwarz inequality to obtain

$$|u_{m,n}|_B^2 = [u_{m,n}, u_{m,n}] = [u, u_{m,n}] \leq |u|_B|u_{m,n}|_B.$$

Consequently,

$$|u_{m,n}|_B \leq |u|_B = \{|f|_{0,\Gamma_D}^2 + w^2|g|_{0,\Gamma_N}^2\}^{\frac{1}{2}},$$

and therefore there exists a unique solution of (3.14) since $S_{m,n}$ is finite-dimensional. By applying the last inequality to (3.12), we obtain (3.16) which expresses stability of the approximate solutions with respect to perturbations of the boundary data f and g.

To prove the last part of the lemma, assume that $\eta \in S_{m,n}$. Then,

$$I(u_{m,n} + \eta) = I(u_{m,n}) + 2[u_{m,n} - u, \eta] + |\eta|_B^2.$$

If $u_{m,n}$ satisfies (3.15), then

$$I(u_{m,n} + \eta) = I(u_{m,n}) + |\eta|_B^2 \geq I(u_{m,n}),$$

and hence it minimizes $I(v)$ over $S_{m,n}$. On the other hand, if $u_{m,n}$ minimizes I, then for any $v \in S_{m,n}$ and scalar α:

$$\frac{\partial I(u_{m,n} + \alpha v)}{\partial \alpha} = 2[u_{m,n} - u, v] + 2\alpha |v|_B^2.$$

Setting $\alpha = 0$ shows that $u_{m,n}$ satisfies (3.15). ∎

3.3   ERROR ESTIMATES

In this section, error estimates are derived for $u - u_{m,n}$ in the space H. Since for any $v \in H$, $\|v\|_0 \leq \|v\|_1$, this will automatically imply convergence of the approximation in the L_2-norm.

Theorem 3.1 *Let $u \in H^1(S)$ be the solution of (3.1) and (3.2), and let $u_{m,n} \in S_{m,n}$ be the boundary approximation satisfying (3.14). If the inverse property holds, then for any $w > 0$, there exists a positive constant C independent of m, n and u such that*

$$\|u - u_{m,n}\|_1 \leq \inf_{\forall v \in S_{m,n}} \left\{ \|u - v\|_1 + C(k_{m,n} + w^{-1})|u - v|_B \right\}, \tag{3.17}$$

where $k_{m,n}$ is defined in (3.9).

Proof. For $v \in S_{m,n}$, let $\eta = v - u_{m,n}$. Then since $\eta \in S_{m,n}$, Lemma 3.2 implies that

$$
\begin{aligned}
\|u - u_{m,n}\|_1 &\leq \|u - v\|_1 + \|\eta\|_1 \\
&\leq \|u - v\|_1 + C(k_{m,n} + w^{-1})|\eta|_B.
\end{aligned}
\tag{3.18}
$$

Applying the orthogonality property (3.15), we obtain

$$|\eta|_B^2 = [\eta, \eta] = [v - u, \eta] \leq |u - v|_B |\eta|_B.$$

Hence, $|\eta|_B \leq |u - v|_B$ and (3.17) follows directly from (3.18). ∎

Next we estimate bounds of the error in H, in terms of errors on the boundary and the interface.

Theorem 3.2 *Let u and $u_{m,n}$ be the same as in Theorem 3.1 and let $v \in S_{m,n}$ be an approximation to u. If the inverse property holds and $w = 1/k_{m,n}$, then there exists a positive constant C independent of m, n, u and v such that*

$$\|u - u_{m,n}\|_1 \leq C\left\{(|r|_{0,\Gamma_D}|r_\nu|_{0,\Gamma_D})^{\frac{1}{2}} + (|r|_{0,\Gamma_N}|r_\nu|_{0,\Gamma_N})^{\frac{1}{2}} \right.$$
$$+ k_{m,n}(|r|_{0,\Gamma_D} + |r^+|_{0,\Gamma_o} + |r^-|_{0,\Gamma_o})$$
$$\left. + |r_\nu|_{0,\Gamma_N} + |r_\nu^+|_{0,\Gamma_o} + |r_\nu^-|_{0,\Gamma_o}\right\}, \qquad (3.19)$$

where $r = u - v$.

Proof. From (3.17) we obtain the following bounds:

$$\|u - u_{m,n}\|_1 \leq \|r\|_1 + Ck_{m,n}|r|_B. \qquad (3.20)$$

To estimate $\|r\|_1$, we have from Lemma 3.1 and the triangular inequality

$$\|r\|_1 \leq C\left\{|r|_1 + |r|_{0,\Gamma_D} + |r^+|_{0,\Gamma_o} + |r^-|_{0,\Gamma_o}\right\}.$$

Since $\triangle r = 0$ on S^+ and S^-, the inequality (3.13) is also valid for r and so

$$|r|_1^2 \leq |r|_{0,\Gamma_D}|r_\nu|_{0,\Gamma_D} + |r|_{0,\Gamma_N}|r_\nu|_{0,\Gamma_N}$$
$$+ (|r^+|_{0,\Gamma_o} + |r^-|_{0,\Gamma_o})(|r_\nu^+|_{0,\Gamma_o} + |r_\nu^-|_{0,\Gamma_o}).$$

Using the Minkowski inequality and the inequality $\sqrt{ab} \leq \frac{a+b}{2}$, we obtain

$$\|r\|_1 \leq C\left\{(|r|_{0,\Gamma_D}|r_\nu|_{0,\Gamma_D})^{\frac{1}{2}} + (|r|_{0,\Gamma_N}|r_\nu|_{0,\Gamma_N})^{\frac{1}{2}} \right.$$
$$\left. + |r|_{0,\Gamma_D} + |r^+|_{0,\Gamma_o} + |r^-|_{0,\Gamma_o} + |r_\nu^+|_{0,\Gamma_o} + |r_\nu^-|_{0,\Gamma_o}\right\}, \qquad (3.21)$$
$$|r|_B \leq |r|_{0,\Gamma_D} + |r^+|_{0,\Gamma_o} + |r^-|_{0,\Gamma_o}$$
$$+ w\left\{|r_\nu|_{0,\Gamma_N} + |r_\nu^+|_{0,\Gamma_o} + |r_\nu^-|_{0,\Gamma_o}\right\}. \qquad (3.22)$$

Substitution of (3.21) and (3.22) into (3.20) and the fact that $k_{m,n} \geq 1$ lead to the desired result (3.19). ∎

Assuming that (3.9) – (3.11) are valid in $S_{m,n}$, we obtain the following theorem.

Theorem 3.3 *Let u and $u_{m,n}$ be the same as in Theorem 3.1. If both the inverse and approximatability properties hold and $w = k_{m,n}^{-1}$, then there exists a positive constant C independent of m, n and u such that*

$$\|u - u_{m,n}\|_1 \leq C\left\{\left(k_{m,n}\alpha_{n,k,0}^+ + \alpha_{n,k,1}^+\right)|u^+|_{k,\partial S+} \right.$$
$$\left. + \left(k_{m,n}\alpha_{m,\ell,0}^- + \alpha_{m,\ell,1}^-\right)|u^-|_{\ell,\partial S-}\right\}, \qquad (3.23)$$

where k, $\ell > 1$.

Proof. Note that both square roots on the right-hand side of (3.19) are bounded from above by

$$\left\{ \left(|r^+|_{0,\partial S^+} + |r^-|_{0,\partial S^-} \right) \left(|r^+_\nu|_{0,\partial S^+} + |r^-_\nu|_{0,\partial S^-} \right) \right\}^{\frac{1}{2}}$$
$$\leq\ C\left\{ k_{m,n} \left(|r^+|_{0,\partial S^+} + |r^-|_{0,\partial S^-} \right) + |r^+_\nu|_{0,\partial S^+} + |r^-_\nu|_{0,\partial S^-} \right\},$$

here again we use the inequality $\sqrt{ab} \leq \frac{a+b}{2}$. Consequently, from (3.19) it follows that

$$\|u - u_{m,n}\|_1 \leq C\left\{ k_{m,n} \left(|r^+|_{0,\partial S^+} + |r^-|_{0,\partial S^-} \right) + |r^+_\nu|_{0,\partial S^+} + |r^-_\nu|_{0,\partial S^-} \right\}. \quad (3.24)$$

Finally, the equation (3.23) is obtained from (3.10), (3.11) and (3.24). ∎

Note that the approximation $u_{m,n}$ may have a different number of terms in S^+ from that in S^-, depending on the smoothness of u on ∂S^+ and ∂S^-. Hence, k and t in (3.10) and (3.11) may have different values.

Let us now investigate two special cases in which the error estimates (3.19) and (3.23) apply.

First, assume that the admissible functions $v \in S_{m,n}$ satisfy also the continuity conditions (3.3) on Γ_0. Then we can modify (3.19) as follows.

Corollary 3.1 *Suppose that in addition to the conditions of Theorem 3.2, the functions in $S_{m,n}$ satisfy (3.3). Then there exists a positive constant C such that*

$$\|u - u_{m,n}\|_{1,S} \leq C\left\{ \left(|r|_{0,\Gamma_D}|r_\nu|_{0,\Gamma_D}\right)^{\frac{1}{2}} + \left(|r|_{0,\Gamma_N}|r_\nu|_{0,\Gamma_N}\right)^{\frac{1}{2}} + k_{m,n}|r|_{0,\Gamma_D} + |r_\nu|_{0,\Gamma_N} \right\}.$$

Second, assume that the admissible functions $v \in S_{m,n}$ also satisfy the boundary conditions (3.2) on ∂S. Then we can write (3.23) as follows.

Corollary 3.2 *Suppose that u is a solution of (3.1) and (3.2). Let $u_{m,n}$ be the boundary approximation satisfying (3.14) in $S_{m,n}$ and (3.2) on ∂S. If the inverse property is satisfied, $w = k_{m,n}^{-1}$, and for any $v \in S_{m,n}$ and $0 \leq l \leq k$,*

$$\|u - v^+\|_{l,\Gamma_0} \leq \beta^+_{n,k,l}|u|_{k,\Gamma_0}, \quad (3.25)$$

and

$$\|u - v^-\|_{l,\Gamma_0} \leq \beta^-_{m,k,l}|u|_{k,\Gamma_0}. \quad (3.26)$$

Then there exists a positive constant C such that for $k \geq 1$,

$$\|u - u_{m,n}\|_1 \leq C \left\{ k_{m,n} \left(\beta_{n,k,0}^+ + \beta_{m,k,0}^- \right) + \beta_{n,k,1}^+ + \beta_{m,k,1}^- \right\} |u|_{k,\Gamma_0}.$$

Note that the bounds (3.25) and (3.26) can be conveniently applied to problems with singularities on the boundary $\partial S = \Gamma_D \cup \Gamma_N$, because the differentiability of u is required only on the interior boundary Γ_0.

We will close this section with an estimate of the factor $k_{m,n}$ in (3.9) as a function of m and n. Consider the special case when Γ_0 is a circular arc.

From Section 2.1 of Chapter 2 a harmonic polynomial is

$$v = \sum_{i=1}^{n} \rho^{\mu_i} \left(a_i \cos \mu_i \theta + b_i \sin \mu_i \theta \right), \quad (\rho, \ \theta) \in S_0, \tag{3.27}$$

where S_0 is a simply connected region, a_i and b_i are coefficients, and the powers μ_i (not necessarily integers) are arranged in ascending order.

Let Γ_0 be a circular arc ($\rho = R_0$, $0 \leq \theta \leq \Theta \leq 2\pi$), and denote by S_0 the corresponding sector ($0 \leq \rho \leq R_0$, $0 \leq \theta \leq \Theta$). Suppose that the trigonometric functions $\cos \mu_i \theta$ and $\sin \mu_i \theta$ form an orthogonal system on $[0, \Theta]$. For a function v of the form (3.27), the orthogonality and Sobolev's imbedding theorem imply that

$$
\begin{aligned}
R_0^2 |v_\nu|_{0,\Gamma_0}^2 &= \Theta \sum_{i=1}^{n} \mu_i^2 R_0^{2\mu_i + 1} \frac{(a_i^2 + b_i^2)}{2} \leq \mu_n^2 \Theta \sum_{i=1}^{n} R_0^{2\mu_i + 1} \frac{(a_i^2 + b_i^2)}{2} \\
&= \mu_n^2 |v|_{0,\Gamma_0}^2 \leq C \mu_n^2 \|v\|_{1,S_0}^2.
\end{aligned}
$$

Consequently, we have

$$|v_\nu|_{0,\Gamma_0} \leq C \mu_n \|v\|_{1,S_0}.$$

Hence, if S^+ is a sector with circular boundary Γ_0 and the admissible functions v^+ are of the form (3.27), then

$$k_{m,n} = C \mu_n. \tag{3.28}$$

We will use (3.28) for the weight selections in the numerical experiments in Section 3.6.

3.4 DEBYE–HUCKEL EQUATION

Consider an elliptic boundary value problem on a domain divided into several subdomains by artificial or material interfaces. If the admissible functions consist of particular solutions of the underlying elliptic equation on the subdomains, an approximate solution can be obtained by satisfying the exterior boundary conditions and the continuity conditions on the interior boundary as much as possible in a least squares sense. Since the approximation is performed only on the interior and exterior boundaries, we call this method the *boundary approximation method* (BAM).

The advantages of boundary approximate methods (BAM) are again summarized as follows:

1. It is easy to solve the problems with corners and interface singularities as well as with unbounded domains, with which the standard finite element and finite difference methods have difficulties coping.

2. The solution procedure is simple to carry out because only the interior and exterior boundary conditions are taken into account in the solution process.

3. A very accurate solution can be obtained by using relatively few expansion terms of particular solutions (approximation in one lower dimension), thus great saving on CPU time and storage space.

4. It is possible to estimate errors of the approximate solutions, although the exact solution of the physical problem is unknown. In this section, a useful relation for the error behavior will be established:

$$\|\varepsilon\|_1 = O(M^\alpha |\varepsilon|_B), \tag{3.29}$$

where $\alpha = 1$ or $\frac{1}{2}$, and M is the total number of unknown coefficients in the piecewise expansions used. The formula (3.29) is significant in practical calculation because we can evaluate error $\|\varepsilon\|_1$ in the domain in terms of the errors on the boundary, $|\varepsilon|_B$, which is naturally obtained from the BAM. Once the errors of solutions are known, we can easily control the calculation procedure.

However, the following two difficulties arise in using BAM.

1. Piecewise particular solutions of elliptic equations have to be known. For the most important elliptic equations in application, we may find useful particular solutions in textbooks of partial differential equations, e.g., Tikhonov and

Samarskii (1973). Nevertheless quite often, an analysis is essential to yield asymptotic expansions near singular points and infinity, once the behavior of the solution near them is unknown or unclear.

2. Stability of numerical solutions is also important. In fact, stability will rely substantially on both the choice of piecewise particular solutions and the partition of the solution domain. Our intention here is to use stability analysis to guide us in choosing partitions so that better solutions can be obtained via BAM.

It is worth pointing out that the boundary approximation methods may fall into the class of general, weighted least squares methods for elliptic systems of Aziz, Kellogg and Stephen (1985), which are applied both within small elements (viewed here as subdomains) and on their boundaries (viewed as interfaces). Since only particular solutions are chosen to be admissible functions, the number of unknown coefficients decreases drastically. A good boundary approximation can be obtained even if several subdomains are used (see the numerical examples in Section 3.6). It would also be interesting to develop the least squares methods in which the number of trial functions (as in this part) and the number of subdomains (as in Aziz, Kellogg and Stephen (1985)) are both changing, as done in the h-p version of FEM.

In this and the next sections, we will present an analysis of errors and in particular stability for the Debye–Huckel equation

$$- \Delta u + u = 0. \tag{3.30}$$

It should be noted that the infinite elements of Han (1982b) and Thatcher (1976, 1978) cannot be applied to (3.30).

Kellogg's singular functions for interface problems (Kellogg (1971, 1972, 1975)) are not complete when the intersection angles of interfaces are $\Theta = \pi/n, \quad n = 2, 3, \ldots$ The additional analytic functions found in Li (1988) together with Kellogg's function form a complete set for interface problems. Of course, a complete system of particular solutions is essential not only to theoretical research, but also to numerical methods, such as the BAM and the combined methods discussed in this book.

Moreover, we will establish an error norm relation (3.29), analyze the stability of boundary approximations and investigate different shapes of subdomains divided by circular arcs and straight lines.

Let the bounded domain S be divided by Γ_0 into two subdomains S^+ and S^-, i.e., $S = S^+ \cup S^-$. In the rest of this chapter, we will consider the piecewise equations

$$- \triangle u + u = 0 \text{ in } S^+ \text{ and } S^-, \tag{3.31}$$

with the interior and exterior boundary conditions

$$u^+ = u^-, \quad \frac{\partial u^+}{\partial \nu} = \frac{\partial u^-}{\partial \nu}, \text{ on } \Gamma_0, \tag{3.32}$$

$$u = f \text{ on } \Gamma_D, \quad u_\nu = g \text{ on } \Gamma_N. \tag{3.33}$$

For (3.31)–(3.33), the boundary approximation method is exactly the same as (3.14).

For the space $S_{m,n}$, we assume that the inverse property (3.9), and the approximatability property (3.10) and (3.11) still hold. Approximatability properties may be found in Cheney (1966) for some spaces, or in Eisenstat (1974) for the Bergman-Vekua space. For the equation $- \triangle u + u = 0$, the approximatability properties of $S_{m,n}$ can be obtained only when the given subdomains are embedded into sectors of a circle, which are within the solution domains. In other cases, further study of the approximating spaces needs to be done (refer to the density study in Aziz, Dorr and Kellogg (1982) and Browder (1962)).

To provide more precise estimates, we assume that $S_{m,n}$ also satisfies the following inverse property.

Second inverse property. For any $v \in S_{m,n}$,

$$|v|_{0,\Gamma_N} \leq q_{m,n} \|v\|_1, \quad |v|_{0,\Gamma_0} \leq q_{m,n} \|v\|_1, \tag{3.34}$$

where $q_{m,n}$ is constant.

Obviously, the constant $q_{m,n}$ is bounded because of the Sobolev imbedding theorem (Sobolev (1963)); but in some cases, there may exist better estimates:

$$q_{m,n} = o(1) \text{ as } m, n \longrightarrow \infty,$$

for the errors $v = u - u_{m,n}$.

Under these assumptions, we will provide error estimates and establish the relation (3.29). First, we give two lemmas.

Lemma 3.4 *Let* $v \in S_{m,n}$ *and suppose that the inverse property (3.9) and the second inverse property hold. Then when* $w > 0$ *, there exist the norm bounds*

$$\|v\|_1 \le (k_{m,n} + q_{m,n}/w)|v|_B.$$

Proof. By using Green's theorem and the fact

$$-\Delta v + v = 0 \quad \text{in } S^+ \text{ and } S^-,$$

we obtain

$$
\begin{aligned}
0 &= \iint_{S^+} (\Delta v^+ - v^+) v^+ \, ds + \iint_{S^-} (\Delta v^- - v^-) v^- \, ds \\
&= -\iint_{S^+} [(v_x^+)^2 + (v_y^+)^2 + (v^+)^2] ds - \iint_{S^-} [(v_x^-)^2 + (v_y^-)^2 + (v^-)^2] ds \\
&\quad + \int_{\partial S^+} (v_\nu^+) v^+ \, ds + \int_{\partial S^-} (v_\nu^-) v^- \, ds.
\end{aligned}
$$

Then we have for $v = v^\pm$ in $S^\pm$

$$
\begin{aligned}
\|v\|_1^2 &\le \left| \int_{\partial S^+} (v_\nu^+) v^+ \, d\ell + \int_{\partial S^-} (v_\nu^-) \, v^- \, d\ell \right| \\
&\le \left\{ \left| \int_{\Gamma_D} (v_\nu) v d\ell \right| + \left| \int_{\Gamma_N} (v_\nu) v d\ell \right| + \left| \int_{\Gamma_0} [(v^+ - v^-)(v_\nu^+) + v^-(v_\nu^+ - v_\nu^-)] d\ell \right| \right\} \\
&\le \{ |v|_{0,\Gamma_D} |v_\nu|_{0,\Gamma_D} + |v|_{0,\Gamma_N} |v_\nu|_{0,\Gamma_N} \\
&\quad + |v^+ - v^-|_{0,\Gamma_0} |v_\nu^+|_{0,\Gamma_0} + |v^-|_{0,\Gamma_0} |v_\nu^+ - v_\nu^-|_{0,\Gamma_0} \}.
\end{aligned}
$$

Therefore, from the inverse property (3.9) and the second inverse property (3.34), we can obtain

$$
\begin{aligned}
\|v\|_1^2 &\le \{ k_{m,n} |v|_{0,\Gamma_D} + q_{m,n} |v_\nu|_{0,\Gamma_N} + k_{m,n} |v^+ - v^-|_{0,\Gamma_0} \\
&\quad + q_{m,n} |v_\nu^+ - v_\nu^-|_{0,\Gamma_0} \} \|v\|_1.
\end{aligned}
$$

The desired result is obtained by dividing both sides by $\|v\|_1$ and by noting the definition $|v|_B$ in (3.4). ∎

We can prove the following lemma by using the same arguments as in Lemma 3.3.

Lemma 3.5 *Let* u *be the solution of (3.31)-(3.33). Then for any* $w > 0$ *, there exists a unique function,* $u_{m,n} \in S_{m,n}$ *, such that*

$$[u_{m,n}, v] = \int_{\Gamma_D} f v d\ell + w^2 \int_{\Gamma_N} g v_\nu d\ell, \quad \forall \, v \in S_{m,n}, \qquad (3.35)$$

$$[u - u_{m,n}, v] = 0, \quad \forall \, v \in S_{m,n}, \qquad (3.36)$$

and

$$\|u_{m,n}\|_1 \leq (k_{m,n} + q_{m,n}/w)\{|f|_{0,\Gamma_D} + w|g|_{0,\Gamma_N}\}.$$

Also, $u_{m,n}$ minimizes I over $S_{m,n}$ if and only if (3.35) holds.

Now, we have an error estimate theorem.

Theorem 3.4 *Let $u \in H^1(S)$ be the solution of (3.31)–(3.33), and let $u_{m,n} \in S_{m,n}$ be the boundary approximation satisfying (3.35). If the inverse property (3.9) and the second inverse property hold, then for any $w > 0$ there exists a constant C independent of m, n and u such that*

$$\|u - u_{m,n}\|_1 \leq \inf_{v \in S_{m,n}} \{\|u - v\|_1 + (k_{m,n} + q_{m,n}/w)|u - v|_B\}, \qquad (3.37)$$

where $k_{m,n}$ is defined in (3.9), and $q_{m,n}$ in (3.34).

Proof. Let $v \in S_{m,n}$ and $\eta = v - u_{m,n}$. We have from Lemma 3.4

$$\begin{aligned}\|u - u_{m,n}\|_1 &\leq &\|u - v\|_1 + \|\eta\|_1 \\ &\leq &\|u - v\|_1 + (k_{m,n} + q_{m,n}/w)|\eta|_B. \qquad (3.38)\end{aligned}$$

Applying the orthogonality property (3.36), we obtain

$$|\eta|_B^2 = [\eta, \ \eta] = [v - u, \ \eta] \leq |u - v|_B|\eta|_B.$$

Hence $|\eta|_B \leq |u - v|_B$, and the desired inequality (3.37) follows from (3.38). ∎

Obviously, Theorem 3.1 is a special case of Theorem 3.4 because the constant $q_{m,n} \leq C$.

Similar error bounds as Theorems 3.2 and 3.3, and Corollaries 3.1 and 3.2 can be easily derived. However, here we only investigate the relation between $\| u - u_{m,n}\|_1$ and $|u - u_{m,n}|_B$.

Let u be the solution of (3.31)–(3.33), then the error norms satisfy

$$\begin{aligned}|\varepsilon|_B^2 &= &|u - u_{m,n}|_B^2 = \int_{\Gamma_D} (u_{m,n} - f)^2 d\ell \\ &&+ w^2 \int_{\Gamma_N} \left(\frac{\partial u_{m,n}}{\partial \nu} - g\right)^2 d\ell + \int_{\Gamma_0} (u_{m,n}^+ - u_{m,n}^-)^2 d\ell \\ &&+ w^2 \int_{\Gamma_0} \left(\frac{\partial u_{m,n}^+}{\partial \nu} - \frac{\partial u_{m,n}^-}{\partial \nu}\right)^2 d\ell. \qquad (3.39)\end{aligned}$$

We note that the explicit, true solution u disappears in $|\varepsilon|_B$ (see (3.39)) and the values of $|\varepsilon|_B$ are easily computed in the least squares procedure employed in the BAM. We are then interested in evaluating $\|\varepsilon\|_1$ in terms of $|\varepsilon|_B$. Such a relation of these two norms is given in the following theorem:

Theorem 3.5 *Let $u \in H^1(S)$ be the solution of (3.31)-(3.33) and $u_{m,n} \in S_{m,n}$ be the boundary approximation (3.35). If the inverse property (3.9) and the second inverse property hold for the difference $u - u_{m,n}$, then for any $w > 0$ there exists a constant C independent of m, n and u such that*

$$\|u - u_{m,n}\|_1 \le (k_{m,n} + q_{m,n}/w)|u - u_{m,n}|_B. \tag{3.40}$$

Proof. Letting $v = u_{m,n} \in S_{m,n}$, we have from (3.32)

$$
\begin{aligned}
\|u - v\|_1^2 &\le \left\{ \left| \int_{\partial S^+} (u_\nu^+ - v_\nu^+)(u^+ - v^+)d\ell + \int_{\partial S^-} (u_\nu^- - v_\nu^-)(u^- - v^-)d\ell \right| \right\} \\
&\le \left\{ \left| \int_{\Gamma_D} (u_\nu - v_\nu)(f - v)d\ell \right| + \left| \int_{\Gamma_N} (g - v_\nu)(u - v)d\ell \right| \right. \\
&\qquad \left. + \left| \int_{\Gamma_0} \left[(v^+ - v^-)(u_\nu^+ - v_\nu^+) + (u^- - v^-)(v_\nu^+ - v_\nu^-) \right] d\ell \right| \right\} \\
&\le \left\{ |u_\nu - v_\nu|_{0,\Gamma_D}|f - v|_{0,\Gamma_D} + |g - v_\nu|_{0,\Gamma_N}|u - v|_{0,\Gamma_N} \right. \\
&\qquad \left. + |v^+ - v^-|_{0,\Gamma_0}|u_\nu^+ - v_\nu^+|_{0,\Gamma_0} + |u^- - v^-|_{0,\Gamma_0}|v_\nu^+ - v_\nu^-|_{0,\Gamma_0} \right\} \\
&\le \left\{ k_{m,n}|f - v|_{0,\Gamma_D} + q_{m,n}|g - v_\nu|_{0,\Gamma_N} \right. \\
&\qquad \left. + k_{m,n}|v^+ - v^-|_{0,\Gamma_0} + q_{m,n}|v_\nu^+ - v_\nu^-|_{0,\Gamma_0} \right\} \|u - v\|_1.
\end{aligned}
$$

Dividing both sides by $\| u - v\|_1$ and using the definition (3.4), we obtain

$$\|u - v\|_1 \le (k_{m,n} + q_{m,n}/w)|u - v|_B.$$

The inequality (3.40) follows from the substitution $v = u_{m,n}$. ∎

Based on Theorem 3.5, we can easily obtain the following two corollaries as the desired result (3.29). By noting $q_{min} \le C$ Theorem 3.5 leads to the following corollary.

Corollary 3.3 *Let the weight $w = 1/M$, and all the conditions in Theorem 3.5 hold. Also, suppose that the constant $k_{m,n}$ in the inverse property satisfies*

$$k_{m,n} \le CM, \tag{3.41}$$

where C is a bounded constant independent of m, n and u. Then

$$\|u - u_{m,n}\|_1 = O(M|u - u_{m,n}|_B).$$ (3.42)

Corollary 3.4 *Let the weight $w = \frac{1}{M}$, where $M = Max(m, n)$ and $m = O(n)$, and all the conditions in Theorem 3.5 hold. Also, suppose that for $u - u_{m,n}$ the constants $k_{m,n}$ and $q_{m,n}$ in the inverse property and the second inverse property satisfy*

$$k_{m,n} \leq C\sqrt{M}, \quad q_{m,n} \leq C/\sqrt{M}.$$ (3.43)

Then,

$$\|u - u_{m,n}\|_1 = O(M^{\frac{1}{2}}|u - u_{m,n}|_B).$$ (3.44)

3.5 STABILITY ANALYSIS

The bounds of the constants $k_{m,n}$ and the integral approximation involving in BAMs may refer Li (1990). In this section, we will present stability analysis for BAM based on domain decomposition, and discuss the choice of geometric shapes for the subdomains used.

In order to discuss stability of the solution $u_{m,n}$, we need to estimate the values of condition numbers using the least squares method (see Golub and Loan (1989))

$$Cond. = \left(\frac{\lambda_{Max}(\mathbf{B})}{\lambda_{Min}(\mathbf{B})} \right)^{\frac{1}{2}}.$$

Here the associated coefficient matrix $\mathbf{B}$ is defined by

$$\mathbf{x}^T \mathbf{B} \mathbf{x} = |v|_B^2, \quad v \in S_{m,n},$$

where the vector $\mathbf{x}$ is composed of the unknown coefficients.

Let $\underline{S}^{\pm}$ and $\overline{S}^{\pm}$ be the bounded domains such that

$$\underline{S}^+ \subseteq S^+ \subseteq \overline{S}^+, \quad \underline{S}^- \subseteq S^- \subseteq \overline{S}^-.$$ (3.45)

We define two matrices as follows:

$$\mathbf{F}_{S^+} = (f_{i,j}^+), \quad \mathbf{F}_{S^-} = (f_{i,j}^-),$$

where the matrix elements $f_{i,j}^{\pm}$ are

$$f_{i,j}^+ = (\psi_i^+, \psi_j^+)_{S+} + \left(\frac{\partial \psi_i^+}{\partial x}, \frac{\partial \psi_j^+}{\partial x}\right)_{S+} + \left(\frac{\partial \psi_i^+}{\partial y}, \frac{\partial \psi_j^+}{\partial y}\right)_{S+}$$

and

$$f_{i,j}^- = (\psi_i^-, \psi_j^-)_{S-} + \left(\frac{\partial \psi_i^-}{\partial x}, \frac{\partial \psi_j^-}{\partial x}\right)_{S-} + \left(\frac{\partial \psi_i^-}{\partial y}, \frac{\partial \psi_j^-}{\partial y}\right)_{S-}.$$

Then we have

$$(\mathbf{x}^+)^T \mathbf{F}_{S+} \mathbf{x}^+ = \|v^+\|_{1,S+}^2, \quad (\mathbf{x}^-)^T \mathbf{F}_{S-} \mathbf{x}^- = \|v^-\|_{1,S-}^2,$$

where the notations are

$$v^+ = \sum_{i=1}^m c_i \psi_i^+, \qquad v^- = \sum_{i=1}^n d_i \psi_i^-,$$

$$\mathbf{x}^+ = (c_1, c_2, ..., c_m)^T, \quad \mathbf{x}^- = (d_1, d_2, ..., d_n)^T.$$

We now prove a main theorem.

Theorem 3.6 *Suppose that for any $v \in S_{m,n}$, the following bounds are satisfied:*

$$|v_\nu|_{0,\Gamma_N} \text{ and } |v_\nu^\pm|_{0,\Gamma_0} \leq k_{m,n}\|v\|_1, \tag{3.46}$$

$$|v|_{0,\Gamma_D} \text{ and } |v^\pm|_{0,\Gamma_0} \leq q_{m,n}\|v\|_1, \tag{3.47}$$

with the positive constants, $k_{m,n}$ and $q_{m,n}$. Then for any $w > 0$, there exists a bounded constant C independent of m, n and u such that

$$Cond. \leq w \left[k_{m,n} + q_{m,n}/w\right]^2 \times \left\{\frac{Max[\lambda_{Max}(\mathbf{F}_{\overline{S}+}), \lambda_{Max}(\mathbf{F}_{\overline{S}-})]}{Min[\lambda_{Min}(\mathbf{F}_{\underline{S}+}), \lambda_{Min}(\mathbf{F}_{\underline{S}-})]}\right\}^{\frac{1}{2}}, \tag{3.48}$$

Proof. We have from Lemma 3.4:

$$|v|_B^2 \geq \frac{\|v\|_1^2}{(k_{m,n} + q_{m,n}/w)^2} \geq \frac{\|v^+\|_{1,\underline{S}+}^2 + \|v^-\|_{1,\underline{S}-}^2}{(k_{m,n} + q_{m,n}/w)^2}. \tag{3.49}$$

Then,

$$\lambda_{Min}(\mathbf{B}) = \min \frac{|v|_B^2}{\mathbf{x}^T\mathbf{x}} \geq \frac{1}{(k_{m,n} + q_{m,n}/w)^2} \min \frac{\|v^+\|_{1,\underline{S}^+}^2 + \|v^-\|_{1,\underline{S}^-}^2}{\mathbf{x}^T\mathbf{x}}.$$

Let the matrix $\mathbf{T}$ be denoted by

$$\mathbf{T} = \begin{bmatrix} \mathbf{F}_{\underline{S}^+} & 0 \\ 0 & \mathbf{F}_{\underline{S}^-} \end{bmatrix},$$

then we obtain a relation for the smallest eigenvalues of the matrices $\mathbf{T}$, $\mathbf{F}_{\underline{S}^+}$ and $\mathbf{F}_{\underline{S}^-}$:

$$\lambda_{Min}(\mathbf{T}) = \min \frac{\|v^+\|_{1,\underline{S}^+}^2 + \|v^-\|_{1,\underline{S}^-}^2}{\mathbf{x}^T\mathbf{x}} = \min[\lambda_{Min}(\mathbf{F}_{\underline{S}^+}), \lambda_{Min}(\mathbf{F}_{\underline{S}^-})].$$

Obviously,

$$\lambda_{Min}(\mathbf{B}) \geq \frac{1}{(k_{m,n} + q_{m,n}/w)^2} \min[\lambda_{Min}(\mathbf{F}_{\underline{S}^+}), \lambda_{Min}(\mathbf{F}_{\underline{S}^-})]. \tag{3.50}$$

Similarly, from the assumption (3.46), (3.47) and the Sobolev imbedding theorem we can see that

$$|v|_B^2 \leq (q_{m,n} + wk_{m,n})^2 \|v\|_1^2 \leq (q_{m,n} + wk_{m,n})^2 \left[\|v^+\|_{1,\overline{S}^+}^2 + \|v^-\|_{1,\overline{S}^-}^2 \right].$$

Therefore,

$$\lambda_{Max}(\mathbf{B}) \leq (q_{m,n} + wk_{m,n})^2 \max[\lambda_{Max}(\mathbf{F}_{\overline{S}^+}), \lambda_{Max}(\mathbf{F}_{\overline{S}^-})]. \tag{3.51}$$

It follows by computing (3.50) and (3.51) that

$$\frac{\lambda_{Max}(\mathbf{B})}{\lambda_{Min}(\mathbf{B})} \leq w^2 [k_{m,n} + q_{m,n}/w]^4 \times \frac{\max[\lambda_{Max}(\mathbf{F}_{\overline{S}^+}), \lambda_{Max}(\mathbf{F}_{\overline{S}^-})]}{\min[\lambda_{Min}(\mathbf{F}_{\underline{S}^+}), \lambda_{Min}(\mathbf{F}_{\underline{S}^-})]}.$$

Now Eq. (3.48) is obtained. ■

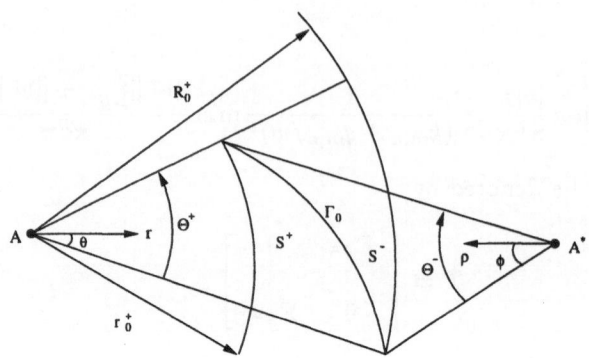

Figure 3.1   A division with $S = S^+ \cup S^-$.

Below, we will apply Theorem 3.6 to the problem using the division displayed in Figure 3.1 and the admissible functions:

$$v^+ = \sum_{i=1}^{m} c_i \frac{I_{\mu_i^+}(r)}{I_{\mu_i^+}(R_0^+)} \sin \mu_i^+ \theta, \ (r,\theta) \in S^+ \tag{3.52}$$

and

$$v^- = \sum_{i=1}^{n} d_i \frac{I_{\mu_i^-}(\rho)}{I_{\mu_i^-}(R_0^-)} \sin \mu_i^- \phi, \ (\rho,\phi) \in S^-, \tag{3.53}$$

where c_i and d_i are unknown coefficients, $\mu_i^{\pm} = i\pi/\Theta^{\pm}$, $\Theta^{\pm}$ are the intersection angles; (r, θ) and (ρ, ϕ) are the polar coordinates with the origins at A and A^*, respectively, and the radii $R_0^{\pm}$ are defined by the following formulae (3.55) below. In (3.52) and (3.53). The Bessel functions $I_\mu(r)$ for a purely imaginary argument can be expressed by (Abramowitz and Stegun (1980))

$$I_\mu(r) = \frac{\left(\frac{r}{2}\right)^\mu}{\Gamma\left(\frac{1}{2}\right)\Gamma\left(\mu+\frac{1}{2}\right)} \int_{-1}^{1} e^{\pm rt}(1-t^2)^{\mu-\frac{1}{2}} dt. \tag{3.54}$$

Let the sectors $\overline{S}^{\pm}$ and $\underline{S}^{\pm}$ satisfy (3.45) such that

$$\begin{aligned}
\overline{S}^- &= \left\{ 0 < \rho < R_0^- \text{ and } 0 < \phi < \Theta^- \right\}, \\
\underline{S}^- &= \left\{ 0 < \rho < r_0^- \text{ and } 0 < \phi < \Theta^- \right\}, \\
\overline{S}^+ &= \left\{ 0 < r < R_0^+ \text{ and } 0 < \theta < \Theta^+ \right\}, \\
\underline{S}^+ &= \left\{ 0 < r < r_0^+ \text{ and } 0 < \theta < \Theta^+ \right\}.
\end{aligned} \tag{3.55}$$

Then we have the following corollary.

Corollary 3.5 *Let $v^{\pm}$ be admissible functions given by (3.52) and (3.53) on the division in Figure 3.1. If $m = O(n)$ and the conditions in Theorem 3.6 are satisfied, then there exists a bounded constant C independent of m, n and u such that*

$$\text{Cond.} \leq C w \left[k_{m,n} + q_{m,n}/w \right]^2 \max \left\{ \left[\frac{R_0^+}{r_0^+} \right]^{\mu_m^+}, \left[\frac{R_0^-}{r_0^-} \right]^{\mu_n^-} \right\}, \tag{3.56}$$

where the radii, $R_0^{\pm}$ and $r_0^{\pm}$, are defined in (3.55).

Proof. Using the orthogonality of $\sin \mu_i^+ \theta$, we can see that

$$\|v^+\|_{1,\overline{S}^+}^2 = \frac{\Theta^+}{2} \sum_{i=0}^{m} c_i^2 \int_0^{R_0^+} r G_i^+(r) dr$$

and

$$\|v^+\|_{1,S^+}^2 = \frac{\Theta^+}{2} \sum_{i=0}^{m} c_i^2 \int_0^{r_0^+} r G_i^+(r) dr,$$

where

$$G_i^+(r) = \frac{\left[[I_{\mu_i^+}'(r)]^2 + \frac{(\mu_i^+)^2}{r^2} I_{\mu_i^+}^2(r) + I_{\mu_i^+}^2(r) \right]}{I_{\mu_i^+}^2(R_0^+)}.$$

On the other hand, there exist the bounds in terms of the definition $I_\mu(r)$ in (3.54):

$$\beta_\mu r^\mu e^{-Max\ r} \leq I_\mu(r) \leq \beta_\mu r^\mu e^{Max\ r},$$

where the constants are

$$\beta_\mu = \frac{(\frac{1}{2})^\mu}{\Gamma(\frac{1}{2})\Gamma(\mu + \frac{1}{2})} \int_{-1}^{1} (1 - t^2)^{\mu - \frac{1}{2}} dt,$$

with $\beta_{\mu+1} \leq \beta_\mu$. Moreover, by noting the formula (see Gradshteyn and Ryzhik (1980)):

$$I_\mu'(r) = \frac{\mu}{r} I_\mu(r) + I_{\mu+1}(r),$$

we can obtain the following bounds:

$$\int_0^{R_0^+} r G_i^+(r) dr \leq C \mu_i^+ e^{4R_0^+}, \quad \int_0^{r_0^+} r G_i^+(r) dr \geq \delta_0 \mu_i^+ e^{-4R_0^+} \left[\frac{r_0^+}{R_0^+} \right]^{2\mu_i^+},$$

with constants $0 < \delta_0 < C < \infty$. This yields

$$\lambda_{Max}(\mathbf{F}_{\overline{S}+}) \le C\mu_m^+, \quad \lambda_{Min}(\mathbf{F}_{\underline{S}+}) \ge \delta_0 \min\left[1, \mu_m^+\left(\frac{r_0^+}{R_0^+}\right)^{2\mu_m^+}\right].$$

Similarly, we obtain

$$\lambda_{Max}(\mathbf{F}_{\overline{S}-}) \le C\mu_n^-, \quad \lambda_{Min}(\mathbf{F}_{\underline{S}-}) \ge \delta_0 \min\left[1, \mu_n^-\left(\frac{r_0^-}{R_0^-}\right)^{2\mu_n^-}\right].$$

At least, one of the following two inequalities:

$$r_0^+ < R_0^+ \quad \text{and} \quad r_0^- < R_0^-$$

must hold. Therefore,

$$\min[\lambda_{Min}(\mathbf{F}_{\underline{S}+}), \lambda_{Min}(\mathbf{F}_{\underline{S}-})]$$

$$= \min\left(1, \mu_m^+\left[\frac{r_0^+}{R_0^+}\right]^{2\mu_m^+}, \mu_n^-\left[\frac{r_0^-}{R_0^-}\right]^{2\mu_n^-}\right)$$

$$= \min\left(\mu_m^+\left[\frac{r_0^+}{R_0^+}\right]^{2\mu_m^+}, \mu_n^-\left[\frac{r_0^-}{R_0^-}\right]^{2\mu_n^-}\right),$$

for some large numbers μ_m^+ and μ_n^-. Consequently, Theorem 3.6 yields

$$Cond. \le Cw(k_{min} + q_{m,n}/w)^2$$

$$\times \left[\frac{\max[\mu_m^+, \mu_n^-]}{\min[\mu_m^+(r_0^+/R_0^+)^{2\mu_m^+}, \mu_n^-(r_0^-/R_0^-)^{2\mu_n^-}]}\right]^{\frac{1}{2}}.$$

The desired inequality (3.56) is obtained from the fact

$$\mu_m^+ = O(\mu_n^-),$$

which results from $\mu_i^\pm = i\pi/\Theta^\pm$ and the assumption $m = O(n)$. ∎

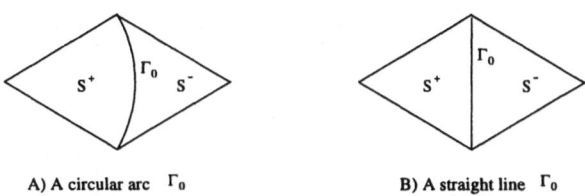

A) A circular arc Γ_0 B) A straight line Γ_0

Figure 3.2   Division of a rhomboid.

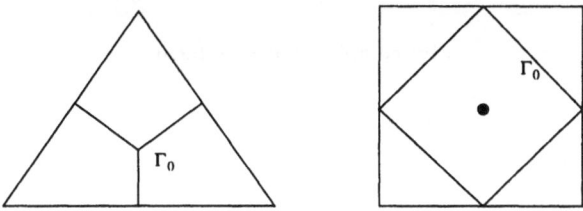

Figure 3.3   Good divisions for boundary approximation methods.

As a result of Corollary 3.5, the following formula holds for $w = 1/M$:

$$Cond. = O\left\{ M \times \max\left[\left(\frac{R_0^+}{r_0^+}\right)^{\mu_m^+}, \left(\frac{R_0^-}{r_0^-}\right)^{\mu_n^-} \right] \right\} \qquad (3.57)$$

provided that the bounds (3.41) are satisfied. To end this section, let us consider the geometric shapes of solution domains while using the BAM. For the solution domain as a rhomboid (Figure 3.2), the values of *Cond.* in the case of B with a straight line Γ_0 are smaller than those in the case of A with a circular arc Γ_0, based on (3.56) or (3.57). Also the divisions in Figure 3.3 will yield good stability of numerical solutions by BAMs.

3.6   TWO MODELS OF SINGULARITY PROBLEMS

In the last section, the BAM is applied to two models of singularity problems, accompanied with the numerical solutions in double decision, which may be treated as testing samples for other numerical methods.

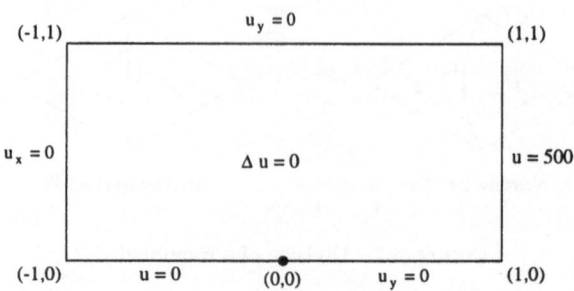

Figure 3.4   Motz's problem.

3.6.1   Motz Problem

Consider Motz's problem stated in Section 2.1.4 in Chapter 2, also see Figure 3.4. First, we approximate u in the entire domain S by the particular solutions

$$v = \sum_{l=0}^{L} D_l r^{l+\frac{1}{2}} \cos\left(l + \frac{1}{2}\right)\theta. \tag{3.58}$$

Since v satisfies the boundary conditions on the side $y = 0$, it suffices to approximate the rest of the boundary conditions only on $y = 1$ and $x = \pm 1$. A summary of results is given in Table 3.1. Here $w = (L+1)^{-1}$ is the weight in the boundary approximation functional

$$|v|_B = \left\{ \int_{\Gamma_D} v^2 d\ell + w^2 \int_{\Gamma_N} v_\nu^2 d\ell \right\}^{\frac{1}{2}},$$

and $Cond.$ is the condition number associated with the least squares problem

$$Cond. = (\lambda_{Max}/\lambda_{Min})^{\frac{1}{2}}, \tag{3.59}$$

where λ_{Max} and λ_{Min} are the largest and smallest eigenvalues of the normal matrix respectively.

As we can see from Table 3.1, the boundary approximation method exhibits an exponential rate of convergence:

$$\begin{aligned}
|u - v|_B &\longrightarrow 1.3 \times 0.56^L, \\
|u - v|_{\infty,\, x=1} &\longrightarrow 2.1 \times 0.56^L, \\
Cond. &\longrightarrow 0.5 \times 1.68^L.
\end{aligned}$$

L	w	$\lvert u - v \rvert_B$	$\lvert u - v \rvert_{\infty, x=1}$	*Cond.*
10	1/11	1.06×10^{-2}	8.56×10^{-3}	1.07×10^2
18	1/19	5.66×10^{-5}	6.84×10^{-5}	2.58×10^3
26	1/27	4.21×10^{-7}	5.28×10^{-7}	9.94×10^4
34	1/35	3.64×10^{-9}	5.47×10^{-9}	3.97×10^7

Table 3.1 The error norms and condition numbers by BAM for Motz's problem.

l	D_l	l	D_l
0	$+.4011624537450 \times 10^3$	18	$+.1153487716826 \times 10^{-4}$
1	$+.8765592019410 \times 10^2$	19	$-.5293093500170 \times 10^{-5}$
2	$+.1723791508189 \times 10^2$	20	$+.2289544439905 \times 10^{-5}$
3	$-.8071215260659 \times 10^1$	21	$+.1062433367985 \times 10^{-5}$
4	$+.1440272716286 \times 10^1$	22	$+.5306818163622 \times 10^{-6}$
5	$+.3310548866399 \times 10^0$	23	$-.2449487241451 \times 10^{-6}$
6	$+.2754373447498 \times 10^0$	24	$+.1085273915317 \times 10^{-6}$
7	$-.8693299479662 \times 10^{-1}$	25	$+.5104036199208 \times 10^{-7}$
8	$+.3360487813999 \times 10^{-1}$	26	$+.2536655670930 \times 10^{-7}$
9	$+.1538437470706 \times 10^{-1}$	27	$-.1100379627543 \times 10^{-7}$
10	$+.7302301925366 \times 10^{-2}$	28	$+.4902479844290 \times 10^{-8}$
11	$-.3184113785070 \times 10^{-2}$	29	$+.2327568240758 \times 10^{-8}$
12	$+.1220645640712 \times 10^{-2}$	30	$+.1140160534364 \times 10^{-8}$
13	$+.5309654900256 \times 10^{-3}$	31	$-.3405273585694 \times 10^{-9}$
14	$+.2715121637438 \times 10^{-3}$	32	$+.1498656324429 \times 10^{-9}$
15	$-.1200450364895 \times 10^{-3}$	33	$+.7203673588029 \times 10^{-10}$
16	$+.5053869450987 \times 10^{-4}$	34	$+.3402718132400 \times 10^{-10}$
17	$+.2316636224195 \times 10^{-4}$		

Table 3.2 The coefficients by BAM with double precision for Motz's problem when $L = 34$ and $w = 1/35$.

The coefficients $D_0, ..., D_{34}$ are listed in Table 3.2. Note that the stability of the least squares problem quickly deteriorates as L grows large. In fact, the approximation with $L = 34$ is computed in double precision, and it is the most accurate global solutions over the entire domain S. As to the leading coefficients D_i, the CTM provides the most accurate values; see the next subsection and Section 2.2.3 of Chapter 2. Next, we consider three subdivisions, A, B and C, of the domain S

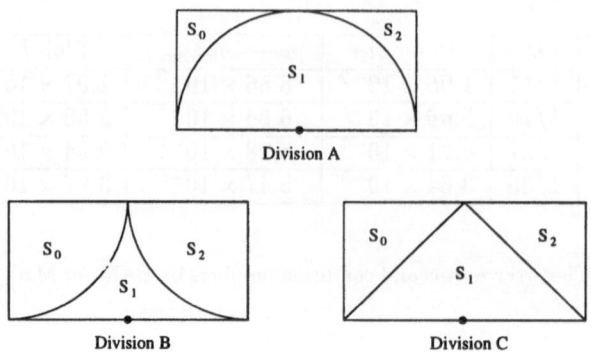

Figure 3.5   Three subdivisions of the solution domain.

into three subdomains, S_0, S_1 and S_2 (see Figure 3.5). In three cases, we choose the following particular solutions:

$$v|_{S_0} = \sum_{l=0}^{N} a_l \xi^{2l} \cos 2l\phi, \quad v|_{S_1} = \sum_{l=0}^{L} D_l r^{l+\frac{1}{2}} \cos\left(l + \frac{1}{2}\right)\theta$$

and

$$v|_{S_2} = 500 + \sum_{l=0}^{M} c_l \eta^{2l+1} \sin(2l+1)\psi,$$

where (ξ, ϕ), (r, θ) and (η, ψ) are polar coordinates with the origins at $(-1, 1)$, $(0, 0)$ and $(1, 1)$ respectively. More numerical experiments are reported in Li, Mathon and Sermer(1987) and Lu and Horng(1996). The Decision C using straight line Γ_0 results in the solution with the best stability.

Estimation of the leading coefficients, i.e., the stress intensity factors, is significant for practical problems. Below we give a lemma.

Lemma 3.6 *Assume*

$$u = \sum_{l=0}^{\infty} d_l r^{l+\frac{1}{2}} \cos\left(l + \frac{1}{2}\right)\theta, \quad u_L = \sum_{l=0}^{L} D_l r^{l+\frac{1}{2}} \cos\left(l + \frac{1}{2}\right)\theta,$$

where d_l and D_l are the true and approximate coefficients by the BAM. Also let $S_R = \{(r, \theta), 0 \le r \le R, 0 \le \theta \le \pi\}$. Then the errors have the bounds

$$|u - u_L|_{1,S_R}^2 = \frac{\pi}{2}\left[\sum_{l=0}^{L}(D_l - d_l)^2\left(l + \frac{1}{2}\right)R^{2l+1}\right.$$

$$+ \sum_{l=L+1}^{\infty} d_l^2 \left(l + \frac{1}{2} \right) R^{2l+1} \right]. \tag{3.60}$$

Proof : We have

$$|v|_{1,S_R}^2 = \iint_{S_R} \left[v_r^2 + \left(\frac{1}{r} \frac{\partial v}{\partial \theta} \right)^2 \right] ds.$$

By using the orthogonality of trigonometric functions, we obtain by some manipulation for $v = u - u_L$,

$$\iint_{S_R} v_r^2 ds = \int_0^R \int_0^{\pi} \left[\sum_{l=0}^{L} (D_l - d_l) \left(l + \frac{1}{2} \right) r^{l-\frac{1}{2}} \cos \left(l + \frac{1}{2} \right) \theta \right.$$

$$+ \sum_{l=L+1}^{\infty} d_l \left(l + \frac{1}{2} \right) r^{l-\frac{1}{2}} \cos \left(l + \frac{1}{2} \right) \theta \right]^2 r dr d\theta \tag{3.61}$$

$$= \frac{\pi}{4} \left[\sum_{l=0}^{L} (D_l - d_l)^2 \left(l + \frac{1}{2} \right) R^{2l+1} + \sum_{l=L+1}^{\infty} d_l^2 \left(l + \frac{1}{2} \right) R^{2l+1} \right].$$

Similarly we obtain

$$\iint_{S_R} \left(\frac{1}{r} \frac{\partial v}{\partial \theta} \right)^2 ds \tag{3.62}$$

$$= \frac{\pi}{4} \left[\sum_{l=0}^{L} (D_l - d_l)^2 \left(l + \frac{1}{2} \right) R^{2l+1} + \sum_{l=L+1}^{\infty} d_l^2 \left(l + \frac{1}{2} \right) R^{2l+1} \right].$$

Combining (3.61) and (3.62) leads to (3.60). ∎

Theorem 3.7 *Let all conditions in Lemma 3.6 and Corollaries 3.3 and 3.4 hold. Then the coefficient errors satisfy*

$$|\triangle D_\ell| = |D_\ell - d_\ell| = O \left(\frac{L^\alpha \|u - u_L\|_B}{\sqrt{\ell + \frac{1}{2}}} \right), \tag{3.63}$$

$$\left[\sum_{\ell=0}^{L} (D_\ell - d_\ell)^2 \right]^{\frac{1}{2}} = O(L^\alpha \|u - u_L\|_B), \tag{3.64}$$

where $\alpha = 1$ and $1/2$.

Proof :   Letting $v = u - u_L$, we have

$$|v|_{1,S_R} \le \|v\|_{1,S_R} \le \|v\|_{1,S}$$

and from Lemma 3.6

$$\frac{\pi}{2}(D_l - d_l)^2 \left(l + \frac{1}{2}\right) R^{2l+1} \le |v|_{1,S_R}^2.$$

Hence we obtain

$$\Delta D_l = |D_l - d_l| \le \sqrt{\frac{2}{\pi}} \frac{|v|_{1,S_R}}{R^{l+\frac{1}{2}}\sqrt{l+\frac{1}{2}}}.$$

Also by Corollaries 3.3 and 3.4

$$|D_l - d_l| \le \sqrt{\frac{2}{\pi}} \frac{\|v\|_{1,S}}{R^{l+\frac{1}{2}}\sqrt{l+\frac{1}{2}}} = O\left(\frac{L^\alpha \|u - u_L\|_B}{R^{l+\frac{1}{2}}\sqrt{l+\frac{1}{2}}}\right). \tag{3.65}$$

Eq. (3.63) follows by letting $R = 1$.

Next, we also have from (3.60) as $R = 1$

$$\sum_{l=0}^{L}(D_l - d_l)^2 \le \frac{4}{\pi}\left[\frac{\pi}{2}\sum_{l=0}^{L}(D_l - d_l)^2\left(l + \frac{1}{2}\right)R^{2l+1}\right] \le \frac{4}{\pi}|u - u_L|_{1,S_R}^2. \tag{3.66}$$

Eq. (3.64) follows by the same arguments in (3.65). ∎

First note in (3.63) that when l increases, the absolute errors ΔD_l diminish as $O(\frac{1}{\sqrt{l}})$. This result is significantly different from Rosser and Papamichael (1975) where ΔD_l deteriorates substantially as l increases. Form Table 3.1, $\|u - u_L\|_B = 3.6 \times 10^{-9}$ as $L = 34$. It then follows from (3.63) that

$$|\Delta D_l| = |D_l - d_l| = O\left(\frac{\sqrt{34} \times 3.6 \times 10^{-9}}{\sqrt{l+\frac{1}{2}}}\right) = O\left(\frac{10^{-8}}{\sqrt{l+\frac{1}{2}}}\right). \tag{3.67}$$

In fact, if compared with the more accurate solutions in Table 3.3, Table 2.1 of Chapter 2 and Lu and Horng (1996), we find that

$$\Delta D_0 = 10^{-9}, \quad \frac{\Delta D_0}{D_0} = 10^{-12}; \tag{3.68}$$

$$\Delta D_9 = 10^{-9}, \quad \frac{\Delta D_9}{D_9} = 10^{-8}; \qquad \Delta D_{19} = 10^{-9}, \quad \frac{\Delta D_{19}}{D_{19}} = 10^{-4};$$

$$\Delta D_{29} = 10^{-9}, \quad \frac{\Delta D_{29}}{D_{29}} = 10^{-1}; \qquad \Delta D_{34} = 3 \times 10^{-10}, \quad \frac{\Delta D_{34}}{D_{34}} = \frac{1}{2}.$$

ℓ	D_ℓ	ℓ	D_ℓ
0	401.16245374523442024	10	0.007302301674
1	87.65592019508791703	11	-0.003184113917
2	17.2379150794468094	12	0.00122064611
3	-8.0712152596981344	13	0.00053096548
4	1.440272717022858	14	0.0002715122
5	0.331054885920737	15	-0.0001200464
6	0.27543734450918	16	0.000050540
7	-0.08693299452554	17	0.000023167
8	0.033604878427	18	0.00001154
9	0.015384374483	19	-0.00000530

Table 3.3 The coefficients by CTM in Section 2.2.2 with double precision.

A few interesting conclusions can be made from these data. First, the theoretical estimates (3.63) coincide well with the practical errors. Second, the absolute errors $\triangle D_i$ do not decrease as $i \longrightarrow 34$. Third, the last coefficient D_{34} has the absolute error 3×10^{-10}; its relative errors are large, since the true coefficients are very small, $D_{34} = 6 \times 10^{-10}$. Although D_{34} has poor accuracy, it is indispensable as a part of the global optimal solutions. Usually, the last coefficient D_L will modify itself to minimize the global errors. The shapes of the solutions u, u_x and u_y by the BAM are shown in Figure 3.6 cited from Lu and Horng (1996), to display the singular behavior at the origin.

3.6.2   Comparisons with CTM

The BAM and the CTM in Section 2.2 are the most accurate algorithms for solving Motz's problem. We cite in Table 3.3 the leading coefficients D_ℓ in (3.58) from Rosser and Papamichael (1975) by multiplying the factor 500.

Under double precision, the leading coefficient D_0 obtained by CTM is extremely accurately, with the relative error $O(10^{-17})$. However, since the sequential coefficients D_i deteriorate very quickly, one significant digit is lost from D_i to D_{i+2}. Since the absolute values of D_i also diminish quickly, by one digit decreasing from D_i to D_{i+3}, the computed coefficients D_i with larger i are very poor (worse than those by BAM). Coefficients D_9 and D_{19} in Table 3.3 have ten and three significant digits, respectively. This drawback leads to less efficiency of the global solutions.

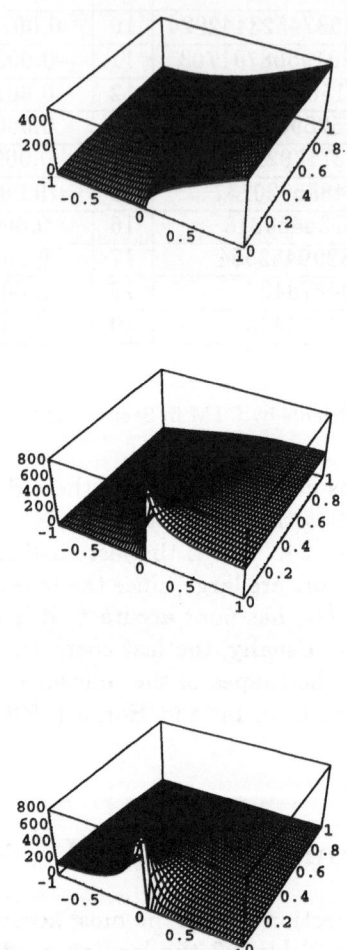

Figure 3.6 The profiles of solutions for Motz's problem: the top of picture for the solution u; the middle for $\frac{\partial u}{\partial x}$; the bottom for $\frac{\partial u}{\partial y}$.

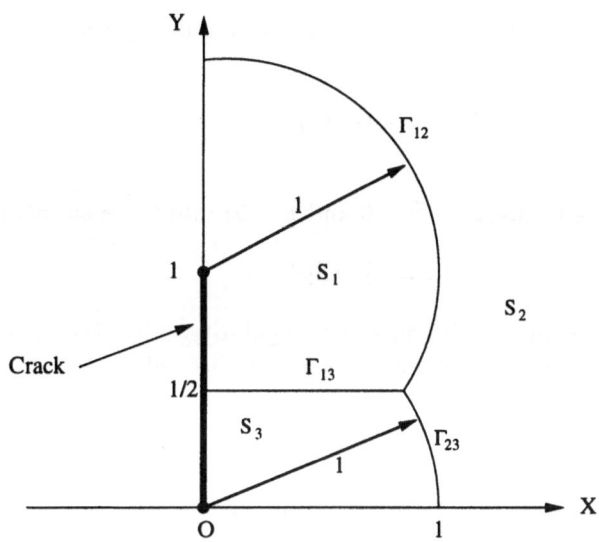

Figure 3.7 A good division of S for the crack-infinity problem.

In summary, CTM is the most accurate method for the leading coefficients; BAM may produce the best global solutions although the coefficients in the tail end are still poor.

3.6.3 Crack-Infinity Problem

Let S^* be the upper semiplane, excluding the section $(x = 0) \cap (0 \le y \le 1)$. Consider the Debye-Huckel equation in S^*:

$$-\Delta u + u = 0, \text{ in } S^*, \tag{3.69}$$

and the Dirichlet conditions (see Figure 3.7):

$$u = 1 \text{ for } x = 0 \text{ and } 0 < y < 1,$$

and

$$u = 1 \text{ for } y = 0.$$

In two dimensions (see Tikhonov and Samarskii (1973)), the condition of solutions at the infinity is: $u|_{\rho \to \infty} < C$, where $\rho = (x^2 + y^2)^{\frac{1}{2}}$. The uniform particular solutions cannot be found for the crack-infinity problem, because there exist more than two singularities: a crack singularity, a corner singularity and the mild singularity of the solution as

$$u = O(1) + O(\rho^2 \ln \rho) \text{ as } \rho \rightarrow 0.$$

Therefore, we have to divide $S(x > 0 \text{ and } y > 0)$ into three subdomains

$$S = S_1 \cup S_2 \cup S_3,$$

such that each of them includes only one singularity. After trying several divisions, we have found a good division shown in Figure 3.7, where the subdomains S_i are described as follows:

$$S_1 : (r < 1) \cap (y > \frac{1}{2}); \quad S_3 : (\rho < 1) \cap (y < \frac{1}{2}); \tag{3.70}$$

$$S_2 : \text{the rest of the solution domain } S, \quad i.e., \quad S_2 = S \backslash (S_1 \cup S_3).$$

On the basis of the particular solutions given in Li and Mathon (1990b), we choose the following piecewise expansions as admissible functions in BAM:

$$v = v_L^{(1)} = cosh(r \sin \theta) + \sum_{l=0}^{L} \tilde{a}_l \frac{I_{l+\frac{1}{2}}(r)}{I_{l+\frac{1}{2}}(1)} \sin \left(l + \frac{1}{2}\right) \theta, \text{ for } (r, \ \theta) \in S_1, \tag{3.71}$$

$$v = v_N^{(2)} = e^{(-\rho \sin \phi)} + \sum_{l=0}^{N} \tilde{c}_l \frac{K_{2l+1}(\rho)}{K_{2l+1}(1)} \sin(2l + 1)\phi, \text{ for } (\rho, \ \phi) \in S_2, \tag{3.72}$$

and

$$v = v^{(3)} \ = \ 1 + \sum_{l=0}^{N_E} w_{2l+1}^*(\rho) \sin 2(2l + 1)\phi$$

$$+ \sum_{l=0}^{K} \tilde{d}_l \frac{I_{2l}(\rho)}{I_{2l}(1)} \sin 2l\phi \text{ for } (\rho, \ \phi) \in S_3, \tag{3.73}$$

where the explicit functions $w_{2l+1}^*(\rho)$ are given in Li and Mathon (1990b), $\tilde{a}_l$, $\tilde{c}_l$ and $\tilde{d}_l$ are the unknown coefficients to be solved, and the integer $N_E \gg 1$. In (3.71)–(3.73), the functions $I_{l+\frac{1}{2}}(1)$, etc., serve as scale factors. In (3.72) $K_n(\rho)$ is the Hankel function for a purely imaginary argument defined by

$$K_n(\rho) = \frac{1}{2} \int_{-\infty}^{\infty} e^{-\rho \cosh \eta - n\eta} d\eta.$$

Note that the admissible functions (3.71), (3.72) or (3.73) satisfy (3.69) in S_1, S_2 or S_3, and all the following exterior boundary conditions

$$u = 1 \text{ for } x = 0 \text{ and } 0 < y < 1, \tag{3.74}$$

$$\frac{\partial u}{\partial x} = 0 \text{ for } x = 0 \text{ and } y > 1, \tag{3.75}$$

$$u = 1 \text{ for } y = 0 \text{ and } x > 0. \tag{3.76}$$

Then BAMs are carried out for seeking the coefficients $\tilde{a}_l$, $\tilde{c}_l$ and $\tilde{d}_l$ to satisfy the continuity conditions on the interior boundary Γ_{ij} only, i.e.,

$$v^{(i)} = v^{(j)} \text{ and } \frac{\partial v^{(i)}}{\partial \nu} = \frac{\partial v^{(j)}}{\partial \nu}, \qquad (x, \, y) \in \Gamma_{ij}, \, i < j,$$

where $\Gamma_{ij} = S_i \cap S_j$, and ν is the normal of Γ_{ij}.

Define a space H by

$$H = \left\{ v \in L_2(S) | -\Delta v + v = 0 \text{ in } S_i, \text{ and } v \in H^1(S_i), \text{ for } i = 1, 2, 3 \right\},$$

and a norm $|v|_B$ over H such that

$$|v|_B^2 = [v, \, v] = \sum_{\substack{i,j=1 \\ i<j}}^{3} \int_{\Gamma_{ij}} \left\{ (v^{(i)} - v^{(j)})^2 + w^2 \left(\frac{\partial v^{(i)}}{\partial \nu} - \frac{\partial v^{(j)}}{\partial \nu} \right)^2 \right\} d\ell, \tag{3.77}$$

where w is a weight constant. Hence the coefficients $\tilde{a}_l$, $\tilde{c}_l$ and $\tilde{d}_l$ are chosen to minimize the norm $|u - v|_B$, i.e.,

$$|u - \tilde{v}|_B = \min_{v \in H} |u - v|_B, \tag{3.78}$$

where $\tilde{v}$ is called a boundary approximation.

In application, the total number N_0 of integration nodes (i.e., the control points) must be larger than the total number M of the unknown coefficients used in (3.71)-(3.73). In our calculation $M = 3 \sim 60$, we have chosen $N_0 = 50 \sim 200$ so that $N_0 >> M$. Letting the weight constant

$$w = \frac{1}{\max(L, 2K, 2N+1)}, \tag{3.79}$$

we solve the problem (3.69), (3.74)–(3.76) by the BAM, (3.78). Table 3.4 gives the error norms $|v|_B$ and the condition numbers *Cond.* of the numerical solutions, from which we can see the following asymptotic relations:

$$|\varepsilon|_B = |u - \tilde{v}|_B = O(0.755^M), \quad Cond. = O(1.21^M),$$

L	N	K	N_E	$\|u - v\|_B$	$Cond$
2	0	1	299	8.475×10^{-2}	3.3
5	2	3	299	3.873×10^{-3}	13.3
8	4	5	299	4.682×10^{-4}	34.8
11	6	7	299	7.053×10^{-5}	118.9
14	8	9	299	1.193×10^{-5}	458.0
17	10	11	299	2.166×10^{-6}	1832
20	12	13	999	4.163×10^{-7}	7504
23	14	15	999	8.378×10^{-8}	31426
25	16	17	999	1.750×10^{-8}	134795

Table 3.4   Error norms and condition numbers for the crack-infinity problem.

l	a_l	l	a_l
0	$-.1134321797585 \times 10^1$	13	$+.3130281737391 \times 10^{-3}$
1	$-.2035664113452 \times 10^0$	14	$+.2391633360670 \times 10^{-3}$
2	$+.1802645475372 \times 10^0$	15	$+.1815954074885 \times 10^{-3}$
3	$+.4985166606623 \times 10^{-1}$	16	$+.1355865655678 \times 10^{-3}$
4	$+.1448626726660 \times 10^{-1}$	17	$+.9845758791387 \times 10^{-4}$
5	$+.7317028845157 \times 10^{-2}$	18	$+.6874549027162 \times 10^{-4}$
6	$+.4021067893764 \times 10^{-2}$	19	$+.4560095585229 \times 10^{-4}$
7	$+.2397388765168 \times 10^{-2}$	20	$+.2835875926942 \times 10^{-4}$
8	$+.1542368986040 \times 10^{-2}$	21	$+.1627821904107 \times 10^{-4}$
9	$+.1050063183825 \times 10^{-2}$	22	$+.8453706771470 \times 10^{-5}$
10	$+.7462738867242 \times 10^{-3}$	23	$+.3861170957769 \times 10^{-5}$
11	$+.5477558835601 \times 10^{-3}$	24	$+.1482617318115 \times 10^{-5}$
12	$+.4112992391382 \times 10^{-3}$	25	$+.4392926896278 \times 10^{-6}$

Table 3.5   Coefficients a_l for the crack-infinity problem when $L = 25$, $N = 16$, $K = 17$, $N_E = 999$ and $w = 1/34$.

where M denotes the total number of coefficients defined by $M = L + N + K + 2$.

When $L = 25$, $N = 16$, $K = 17$, $N_E = 999$ and $w = 1/34$, the approximate coefficients are given in Tables 3.5 and 3.6 with high accuracy and good stability:

$$|\varepsilon|_B = O(10^{-8}), \quad Cond. = O(10^5), \tag{3.80}$$

by using only 60 unknown coefficients. The notation $O(10^{-8})$ means a quantity: $\alpha \times 10^{-8}$ with $1 \leq \alpha < 10$. We notice that the error norm (3.80) is very small. Such a solution can then be regarded as a "true solution" of (3.69), (3.74)–(3.76) for both theoretical and practical purposes. More details in computations are given in Li and Mathon (1990b).

l	c_l	l	d_l
0	$+.2769005233385 \times 10^0$	1	$-.4015781884664 \times 10^0$
1	$-.9066552957337 \times 10^{-1}$	2	$+.5890728222103 \times 10^{-1}$
2	$+.4561996989200 \times 10^{-1}$	3	$-.3532773398943 \times 10^{-1}$
3	$-.2828278554347 \times 10^{-1}$	4	$+.2354001734508 \times 10^{-1}$
4	$+.1964155709458 \times 10^{-1}$	5	$-.1702868100792 \times 10^{-1}$
5	$-.1462958139045 \times 10^{-1}$	6	$+.1301850746566 \times 10^{-1}$
6	$+.1140500948182 \times 10^{-1}$	7	$-.1033923809290 \times 10^{-1}$
7	$-.9131571726473 \times 10^{-2}$	8	$+.8397076489702 \times 10^{-2}$
8	$+.7346837942624 \times 10^{-2}$	9	$-.6830460487149 \times 10^{-2}$
9	$-.5765893519271 \times 10^{-2}$	10	$+.5400913572576 \times 10^{-2}$
10	$+.4249769307289 \times 10^{-2}$	11	$-.3994984244826 \times 10^{-2}$
11	$-.2816615820485 \times 10^{-2}$	12	$+.2647170401599 \times 10^{-2}$
12	$+.1601231751265 \times 10^{-2}$	13	$-.1500088526740 \times 10^{-2}$
13	$-.7398900405725 \times 10^{-3}$	14	$+.6896932422165 \times 10^{-3}$
14	$+.2587534935069 \times 10^{-3}$	15	$-.2398708862801 \times 10^{-3}$
15	$-.6080346016446 \times 10^{-4}$	16	$+.5609893058375 \times 10^{-4}$
16	$+.7228909240972 \times 10^{-5}$	17	$-.6651837194904 \times 10^{-5}$

Table 3.6 Coefficients c_l and d_l for the crack-infinity problem when $L = 25$, $N = 16$, $K = 17$, $N_E = 999$ and $w = 1/34$.

4

COMBINATIONS OF RGM AND FEM

We begin with the first typical combination, the combination of RGM and FEM, where the local particular solutions v^+ and piecewise polynomials v^- are chosen in the singular and regular subdomains, S^+ and S^-, respectively. The nonconforming constraints are used to match two admissible functions along their common boundary Γ_0:

$$v^+(Q_i) = v^-(Q_i), \quad Q_i \in \Gamma_0,$$

where Q_i are the nodes of finite elements, and

$$v = \begin{cases} v^+ = \sum_{i=0}^{L} D_i\phi_i, & \text{in } S^+, \\ v^- = v_k, & \text{in } S^-. \end{cases} \tag{4.1}$$

In (4.1), $\{\phi_i\}$ are the local particular solutions, D_i are unknown coefficients, and v_k are k-order Lagrange elements. The important results are given in Theorem 4.2 for Laplace's equation by using linear elements in S^-:

$$\|u - u_h^*\|_1 \le C\left\{ h + \left(\frac{R_2}{R}\right)^{\tau(L+3/2)} + [\tau(L + \frac{1}{2})]^{3/2}h^2 \right\}, \tag{4.2}$$

where $\tau > 0$ and $R_2 < R$, and in Theorem 4.5 for elliptic equations by using k-order elements in S^-

$$\|u - u_h^*\|_1 \le C\left\{ h^k|u|_{k+1,S_1} + \|R_L\|_{1,S_2} + h^{k+\frac{1}{2}}|u|_{k+1,\Gamma_0} + \right.$$

$$\left. + \frac{1}{\sqrt{h}}\|R_L\|_{0,\Gamma_0} + h^{k+\frac{1}{2}}|R_L|_{k+1,\Gamma_0} + h^{k+1}L^{2(k+1)}\left|\frac{\partial u}{\partial n}\right|_{0,\Gamma_0} \right\}, \tag{4.3}$$

where R_L is the remainder, and h is the maximal element boundary length. Note that the term, $[\tau(L + \frac{1}{2})]^{3/2}h^2$ or $h^{k+1}L^{2(k+1)}$ results from the coupling condition on Γ_0. Significant consequences are induced from (4.2) and (4.3).

The first result is that the optimal convergence rates are obtained if

$$[\tau(L + \frac{1}{2})]^{3/2}h \le C, \text{ or } hL^{2(k+1)} \le C.$$

139

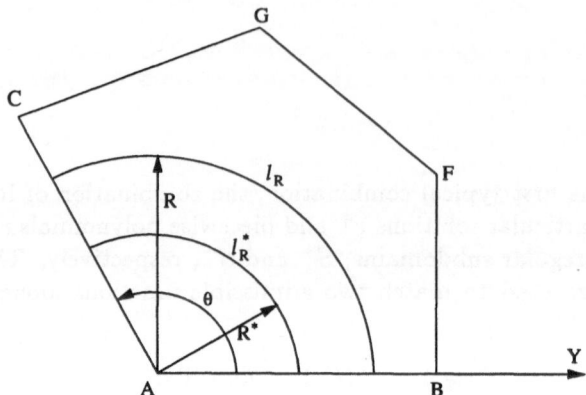

Figure 4.1 An angular singularity problem of Laplace's equation.

Otherwise, the reduced rates incur.

The second important result of (4.2) is that the best choice of term L is achieved when the two other error terms are balanced:

$$\left(\frac{R_2}{R}\right)^{\tau(L+\frac{3}{2})} = O(h),$$

thus leading to $L = O(|\ln h|)$. The numerical experiments in Section 4.5 show that four or five terms of particular solutions are good enough for practical application.

The third result is that good stability is obtained if the term L is chosen to be small, as so to balance the errors of FEM. The materials in this chapter are adopted from Li (1986, 1991), and Li and Liang (1981).

4.1   NONCONFORMING APPROACH FOR LAPLACE EQUATION

In this section, we introduce the nonconforming approach for combining the Ritz-Galerkin method (RGM) and the finite element method (FEM).

For simplicity, let the solution domain S be a polygon shown in Figure 4.1, where $\Gamma(= ABFGCA)$ is the domain boundary. We consider the Laplace equation in S

with the Dirichlet-Neumann type of boundary conditions:

$$\Delta u = \frac{\partial^2 u}{\partial x^2} + \frac{\partial^2 u}{\partial y^2} = 0, \quad (x, y) \in S, \tag{4.4}$$

$$u = 0, \quad (x, y) \in \overline{AC}, \tag{4.5}$$

$$\frac{\partial u}{\partial \nu} = 0, \quad (x, y) \in \overline{AB}, \tag{4.6}$$

and

$$\frac{\partial u}{\partial \nu} = g, \quad (x, y) \in \overline{BF} \cup \overline{FG} \cup \overline{GC}, \tag{4.7}$$

where ν is the outward normal of Γ, and g is a sufficiently smooth function. As is well known, the Eqs. (4.4)–(4.7) can also be written in a weak form:

$$I(u, v) = g(v), \quad \forall v \in H_0^1(S), \ u \in H_0^1(S),$$

where the bilinear form is

$$I(u, v) = \iint_S (u_x v_x + u_y v_y) ds,$$

and the notations are

$$g(v) = \int_{\overline{BF} \cup \overline{FG} \cup \overline{GC}} g v d\ell$$

and

$$H_0^1(S) = \left\{ v | v, v_x, v_y \in L^2(S), \text{ and } v|_{\overline{AC}} = 0 \right\}.$$

Denote $\Theta = \angle CAB$, then $0 < \Theta \le 2\pi$. The solutions near the angle point A, satisfying the Laplace equation (4.4) and the Dirichlet-Neumann conditions, (4.5) and (4.6), are given in Section 2.1 of Chapter 2 as

$$u(r, \theta) = \sum_{n=0}^{\infty} D_n \left(\frac{r}{R} \right)^{\tau(n + \frac{1}{2})} \cos(\tau(n + \frac{1}{2})\theta), (r \le R), \tag{4.8}$$

where the notation $\tau = \pi/\Theta$, r and θ are the polar coordinates with the original A, $(r/R)^{\tau(n+\frac{1}{2})} \cos \tau(n + \frac{1}{2})\theta$ are basis functions, and D_n are expansion coefficients, which can be defined by

$$D_n = \frac{2}{\Theta} \int_0^{\Theta} u(R, \theta) \cos(\tau(n + \frac{1}{2})\theta) d\theta. \tag{4.9}$$

In order to guarantee a positive radius R in (4.8), we assume both $Meas(\overline{AB}) \neq 0$ and $Meas(\overline{AC}) \neq 0$. Also since $Meas(\overline{AC}) \neq 0$, there exists a unique solution for (4.4)-(4.7).

Of course, let $u \in L^2(S)$, then the coefficients D_n in (4.9) are bounded. We therefore learn from (4.8) the following characteristics of solutions, also see Section 2.1 of Chapter 2:

1. The solution is analytic if $\Theta = \pi/2k$, and k is integer (i.e., $\tau = \pi/\Theta = 2k$).

2. The solution is singular when $\Theta > \pi/2$. (i.e., $\tau < 2$), because the partial derivative $\partial u/\partial r$ at A is unbounded. In this case the angular point A is called *the singularity point*.

In this chapter, a radius R^* is chosen such that

$$R^* < R \tag{4.10}$$

and the solution domain S is divided into two subdomains: $S_1(r > R^*)$ and $S_2(r < R^*)$, by a circular arc $l_{R^*}(r = R^*, 0 \leq \theta \leq \Theta)$, which is called *the common boundary* of both methods.

For simplicity, we discuss only one singularity: The angle point A with $\Theta > \pi/2$ in the subdomain S_2, called *the singular domain*. We naturally assume that the solution u in the remaining domain S_1 is smooth enough such that

$$u \in H^2(S_1). \tag{4.11}$$

First, we still use the linear finite element method in S_1 because of the assumption (4.11). Thus we need to divide S_1 again into many small triangular elements, $\triangle_k$ (see Figure 4.2), and choose (R^*, θ_j) to be the element nodes of triangulation on l_{R^*}. Evidently, the circular arc l_{R^*} is approximated by a small piecewise straight line $\hat{l}_{R^*}$, and the subdomain S_1 by $\hat{S}_1^h(= \underset{k}{\cup}\triangle_k)$. As FEM we choose the piecewise linear interpolation functions v_h on the triangulation in $\hat{S}_1^h$ to be the admissible functions.

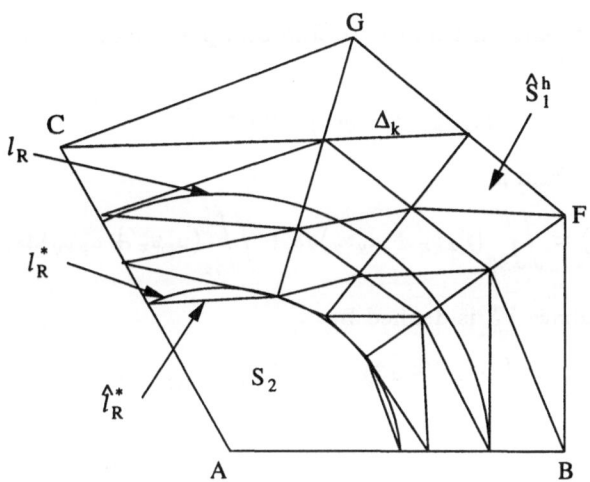

Figure 4.2   The division of the solution domain for the combined method.

Second, we use the RGM in the singular domain $S_2(r < R^*)$. By noting the solution expansions (4.8), we choose

$$f_L(\tilde{D}_l) = \sum_{l=0}^{L} \tilde{D}_l \left(\frac{r}{R}\right)^{\tau(l+\frac{1}{2})} \cos(\tau(l+\frac{1}{2})\theta), \quad r \leq R^*, \qquad (4.12)$$

to be the admissible functions in S_2, where $\tilde{D}_l$ are approximate coefficients to be sought, $\left(\frac{r}{R}\right)^{\tau(\frac{l+1}{2})} \cos(\tau(\frac{l+1}{2})\theta)$ are singular basis functions, and $(L+1)$ is the total number of singular functions chosen. So, the admissible functions v_h in the entire solution domain S are represented in the form:

$$v_h = \begin{cases} v_h, & (x,y) \in \hat{S}_1^h, \\ f_L(\tilde{D}_l), & (x,y) \in S_2. \end{cases} \qquad (4.13)$$

Lastly, we combine both methods, written as the RGM-FEM, by adding the following continuity conditions of v_h only on the element nodes (R^*, θ_j):

$$v_h(R^*, \theta_j) = \sum_{l=0}^{L} \tilde{D}_l \left(\frac{R^*}{R}\right)^{\tau(l+\frac{1}{2})} \cos(\tau(l+\frac{1}{2})\theta_j), \quad \forall \theta_j. \qquad (4.14)$$

This method is called *the nonconforming method* because the admissible functions v_h are not continuous on the entire common boundary l_{R^*}. The aim in using this nonconforming combined method is to find an approximate solution, $u_h^* \in V_h^0$, of (4.4)-(4.7) such that

$$I_h(u_h^*, v) = g(v), \quad \forall v \in V_h^0, \tag{4.15}$$

where the bilinear form is

$$I_h(u, v) = \iint_{\tilde{S}_1^h} (u_x v_x + u_y v_y)ds + \iint_{S_2} (u_x v_x + u_y v_y)ds, \tag{4.16}$$

and the function space V_h^0 is defined by

$$V_h^0 = \left\{ v_h \text{ in (4.13) satisfying } v_h|_{\overline{AC}} = 0 \right\}.$$

After elimination of the unknown $V_h(R^*, \theta_j)$ in (4.15) by the constraints (4.14), the dimensionality of V_h^0 is $L+1+K-K^*$, where K is the total number of finite element nodes on $\hat{S}_1^h$, and K^* is the total number of element nodes on l_{R^*}. Therefore, we obtain a linear system of algebraic equations

$$\mathbf{Tx = b}, \tag{4.17}$$

where $\mathbf{x}$ is the unknown vector which consists of the components, $\tilde{D}_l$ and $v_h(r_i, \theta_j)$ with $r_i > R^*$, $\mathbf{b}$ is a known vector, and $\mathbf{T}$ is the associated matrix of linear algebraic equations. The unknown vector $\mathbf{x}$ (i.e. the numerical solution u_h^*) can then be easily obtained, since the coefficient matrix $\mathbf{T}$ is positive definite and symmetric, as well as sparse in the part of $\mathbf{T}$ coming from the FEM.

Below, we will show why the matrix $\mathbf{T}$ in (4.17) is positive definite and symmetric. In the nonconforming combinations, a general technique for the incorporation of constraints into the stiffness matrix is required, which we describe below.

Let $\psi_i(x, y), i = 1, 2, \ldots, N$, be a linearly independent set of basis functions. A general function in Span ψ_i has the form

$$w = \sum_{i=1}^{N} a_i \psi_i(x, y), \tag{4.18}$$

where a_i are the coefficients. Consider imposing a set of m linear constraints on the coefficients of a_i in the form

$$\mathbf{Da = 0}, \tag{4.19}$$

where $\mathbf{D}$ is an $N \times m (m < N)$ matrix with rank m, and $\mathbf{a}$ is an m dimensional vector. Then the resulting functions, which are determined by (4.18) and (4.19), can be denoted by v and they form an $N - m$ dimensional subspace of Span $\{\psi_i\}$. We denote this constrained space by Span $\{\psi_i, \mathbf{D}\}$. The form of a finite element functional $E(w)$ for a function w in Span ψ_i is

$$E(w) = \frac{1}{2} I_h(w, w) - (g, w) = \frac{1}{2} \sum_i \sum_j a_i a_j I_h(\psi_i, \psi_j) - \sum_i a_i(g, \psi_i), \quad (4.20)$$

where $I_h(w, w)$ and (g, w) may be defined as (4.16) and

$$(g, w) = \int_{\overline{BF \cup FG \cup GC}} gw d\ell.$$

Clearly, the stiffness matrix $\mathbf{T}_h = (\mathbf{T}_h(\psi_i, \psi_j))$ is positive definite and symmetric.

The form of $F(v)$ for a function, $v \in \text{Span}\{\psi_i, \mathbf{D}\}$, can be obtained by solving (4.19) for m of the coefficients because of the matrix $\mathbf{D}$ with rank m. For notational convenience, assume that we can find for $a_{N-m+1}, a_{N-m+2}, ..., a_N$, i.e.,

$$a_{N-m+i} = \sum_{k=1}^{N-m} b_{N-m+i,k} a_k + d_i, \quad i = 1, 2, ..., m, \quad (4.21)$$

where $B = (b_{N-m+i,k})$ is an $m \times (N - m)$ dimensional constant matrix, and $\mathbf{d} = (d_1, d_2, ..., d_m)^T$ is a constant vector.

Substituting (4.21) into (4.20) for a_{N-m+i}, $i = 1, 2, ..., m$, we obtain a new quadratic form

$$F(v) = \frac{1}{2} \sum_{i=1}^{N-m} \sum_{j=1}^{N-m} T_{ij} v_i v_j - \sum_{i=1}^{N-m} b_i v_i.$$

The minimum of this form leads to

$$\mathbf{Tx} = \mathbf{b},$$

i.e., Eq. (4.17), where $\mathbf{x} = (v_1, v_2, ..., v_{N-m})^T$ is a unknown vector. The $(N-m) \times (N - m)$ dimensional matrix $\mathbf{T} = (T_{ij})$ and the vector $\mathbf{b} = (b_1, b_2, ..., b_{N-m})^T$ are computed from $I_h(\psi_i, \psi_j)$, (g, ψ_i) and $b_{N-m+i,k}$ appropriately.

Clearly, the matrix $\mathbf{T}$ is symmetric. By noting the matrix theory of Courant and Hilbert (1953), the minimal eigenvalue of the matrix $\mathbf{T}$ is larger than that of the matrix $\mathbf{T}_h$ of the unconstrained problem (also see (4.52) in Section 4.4 below):

$$\lambda_{min}(\mathbf{T}) \geq \lambda_{min}(\mathbf{T}_h) > 0,$$

where the matrix $\mathbf{T}_h = (I_h(\psi_i, \psi_j))$. Consequently, the matrix $\mathbf{T}$ is also positive definite.

4.2 ERROR ESTIMATES

Now we denote the space

$$H_1 = \left\{ v \mid \text{both } v \in H^1(\hat{S}_1^h) \text{ and } v \in H^1(S_2) \right\},$$

and define a new norm over H_1:

$$\|v\|_1 = \left\{ \|v\|_{1,\hat{S}_1^h}^2 + \|v\|_{1,S_2}^2 \right\}^{\frac{1}{2}}. \tag{4.22}$$

Since this combined method is nonconforming, the space V_h^0 of the admissible functions v_h will satisfy $v_h \in V_h^0 \not\subset H^1(S)$; but $V_h^0 \subset H_1$. Consequently, we will give the error estimates in the new norm $\| \cdot \|_1$. Then we have the following theorem:

Theorem 4.1 *Suppose the bilinear form I_h in (4.16) is uniformly V_h^0-elliptic, i.e., there exists a positive constant α independent of h, L and v_h such that*

$$\alpha \|v_h\|_1^2 \leq I_h(v_h, v_h), \quad \forall v_h \in V_h^0. \tag{4.23}$$

Then the numerical solution u_h^* of the nonconforming combined method (4.15) has the error bounds:

$$
\begin{aligned}
\|u - u_h^*\|_1 \leq C \Bigg\{ & \inf_{v_h \in V_h^0} \|u - v_h\|_1 \\
+ & \sup_{w_h \in V_h^0} \left| \int_{\hat{l}_{R^*}} \frac{\partial u}{\partial \nu} w_h^- \, d\ell - \int_{l_{R^*}} \frac{\partial u}{\partial r} w_h^- \, d\ell \right| \bigg/ \|w_h\|_1 \\
+ & \sup_{w_h \in V_h^0} \left(\int_{l_{R^*}} \left(\frac{\partial u}{\partial r} \right)^2 d\ell \right)^{\frac{1}{2}} \left(\int_{l_{R^*}} (w_h^+ - w_h^-)^2 d\ell \right)^{\frac{1}{2}} \bigg/ \|w_h\|_1 \Bigg\},
\end{aligned}
\tag{4.24}
$$

where ν (or r) is the outward normal on $\hat{l}_{R^*}$ (or l_{R^*}) of S_2; C is a constant independent of h and L; h is the maximal boundary length of triangular elements; and the notations are: $w_h^+ = w_h|_{S_2} = f_L(D_l)$ and $w_h^- = w_h|_{\hat{S}_1^h} = v_h$.

Proof. When (4.23) holds, we see from Strang and Fix (1973) and Ciarlet (1978):

$$\|u - u_h^*\|_1 \leq C \left\{ \inf_{v_h \in V_h^0} \|u - v\|_1 + \sup_{w_h \in V_h^0} \frac{|g(w_h) - I_h(u, w_h)|}{\|w_h\|_1} \right\},$$

where C is a constant independent of h and L. By applying Green's theorem and (4.4), we have

$$|g(w_h) - I_h(u, w_h)| = \left| - \int_{l_{R^*}} \frac{\partial u}{\partial r} w_h^+ d\ell + \int_{\hat{l}_{R^*}} \frac{\partial u}{\partial \nu} w_h^- d\ell \right|$$

$$\leq \left| \int_{\hat{l}_{R^*}} \frac{\partial u}{\partial \nu} w_h^- d\ell - \int_{l_{R^*}} \frac{\partial u}{\partial r} w_h^- d\ell \right| + \left| \int_{l_{R^*}} \frac{\partial u}{\partial r} (w_h^- - w_h^+) d\ell \right|.$$

Also by applying Schwarz's inequality:

$$\left| \int_{l_{R^*}} \frac{\partial u}{\partial r} (w_h^- - w_h^+) d\ell \right| \leq \left[\int_{l_{R^*}} \left(\frac{\partial u}{\partial r} \right)^2 d\ell \right]^{\frac{1}{2}} \left[\int_{l_{R^*}} (w_h^+ - w_h^-)^2 d\ell \right]^{\frac{1}{2}},$$

we obtain (4.24) immediately. ■

We notice that in (4.24), the second term in the braces is an error resulting from the inconsistency between $\hat{l}_{R^*}$ and l_{R^*}, and the third term is an error resulting from the noncontinuity of v_h on the common boundary l_{R^*}.

Lemma 4.1 *Let $Meas(\overline{AC} \cap \hat{S}_1^h) \neq 0$, then the uniformly V_h^0- elliptic inequality (4.23) holds.*

Proof. Since $Meas(\overline{AC} \cap \hat{S}_1^h) \neq 0$, we have from the Poincare–Friedrichs inequality (Ciarlet (1978))

$$\|v_h\|_{1,\hat{S}_1^h} \leq C|v_h|_{1,\hat{S}_1^h}, \quad \forall v_h \in V_h^0, \tag{4.25}$$

where $C(> 0)$ is a bounded constant independent of h and L.

Also the assumption $Meas(\overline{AC}) \neq 0$ implies that $Meas(\overline{AC} \cap S_2) \neq 0$. We then obtain the following estimates by using the Poincare–Friedrichs inequality again:

$$\|v_h\|_{1,S_2} \leq C|v_h|_{1,S_2}, \quad \forall v_h \in v_h^0. \tag{4.26}$$

Hence, the uniformly V_h^0-elliptic inequality (4.23) follows from (4.16), (4.22), (4.25) and (4.26). ■

Let an analytic cut-off function be $\phi(r) \in C^\infty[R^*, R]$ such that

$$\begin{cases} \phi(r) = 0, & \text{for } r \leq R_1, \\ \phi(r) = 1, & \text{for } r \geq R_2, \end{cases} \tag{4.27}$$

where $R^* < R_1 < R_2 < R$.

We define an auxiliary function v:

$$v = \begin{cases} u, & \text{for } r \geq R, \\ f_L(D_l) + \phi(v) \displaystyle\sum_{l=L+1}^{\infty} D_l \, (r/R)^{\tau(l+\frac{1}{2})} \cos(\tau(l+\tfrac{1}{2})\theta), & \text{for } R^* \leq r \leq R, \\ f_L(D_l), & \text{for } r \leq R^* (\text{ i.e., } S_2), \end{cases} \qquad (4.28)$$

where the functions are

$$f_L(D_l) = \sum_{l=0}^{L} D_l \, (r/R)^{\tau(l+\frac{1}{2})} \cos(\tau(l+\frac{1}{2})\theta)$$

and the expansion coefficients D_l are given by (4.9).

On the basis of the auxiliary function v, we construct a particular function $\overline{w}_h \in V_h^0$ such that

$$\overline{w}_h = \begin{cases} \overline{v}_h, & (x,y) \in \hat{S}_1^h, \\ f_L(D_l), & (x,y) \in S_2, \end{cases}$$

where $\overline{v}_h$ are piecewise linear interpolation functions of v on the triangulation in $\hat{S}_1^h$. Then we need the following lemma.

Lemma 4.2 *Let (4.10) and (4.11) hold; and suppose that the family of triangular elements in $\hat{S}_1^h$ is quasiuniform. There then exists a constant C independent of h and L such that*

$$\|u - \overline{w}_h\|_1 \leq C \left[h + \left(\frac{R_2}{R} \right)^{\tau(L+\frac{3}{2})} \right], R^* < R_2 < R. \qquad (4.29)$$

Proof. There exists a triangular inequality,

$$\|u - \overline{w}_h\|_1 \leq \|u - v\|_1 + \|v - \overline{w}_h\|_1, \qquad (4.30)$$

where the function v is chosen as the auxiliary function (4.28). On the basis of Eqs (4.5), (4.6) and (4.11), the coefficients, D_n in (4.9), have the bounds

$$\begin{aligned} |D_n| &= \frac{2}{\tau(n+\frac{1}{2})\Theta} \left| \int_0^\Theta \frac{\partial u}{\partial \theta}(R,\theta) \sin(\tau(\frac{n+1}{2})\theta) d\theta \right| \\ &\leq \frac{2}{\tau(n+\frac{1}{2})\sqrt{\Theta R}} |u|_{1,l_R} \leq \frac{C}{\tau(n+\frac{1}{2})} \|u\|_{2,S_1} \\ &= O\left(\frac{1}{(n+\frac{1}{2})} \right). \end{aligned}$$

We then obtain from (4.10), (4.22), (4.27) and (4.28)

$$
\begin{aligned}
\|u - v\|_1 &\leq \|u - v\|_{1,\hat{S}_1^h} + \|u - v\|_{1,S_2} \\
&= \|u - v\|_{1,\hat{S}_1^h \cap (r \leq R_2)} + \|u - v\|_{1,S_2} \\
&\leq C \left(\frac{R_2}{R} \right)^{\tau(L+\frac{3}{2})}.
\end{aligned}
$$

Next, since $u \in H^2(S_1)$, then $v \in H^2(S_1)$. So, for the piecewise linear interpolation functions $\bar{v}_h$ on the quasiuniform triangulation, there exist the bounds

$$
\|v - \overline{w}_h\|_1 = \|v - \bar{v}_h\|_{1,\hat{S}_1^h} \leq Ch. \tag{4.31}
$$

The inequality (4.29) is therefore obtained. ∎

Lemma 4.3 *There exist the bounds*

$$
\left[\int_{l_{R^*}} (w_h^+ - w_h^-)^2 d\ell \right]^{\frac{1}{2}} \leq C \left(h^{3/2} + \left[\tau \left(L + \frac{1}{2} \right) \right]^{3/2} h^2 \right) \|w\|_1. \tag{4.32}
$$

Proof: The admissible functions w_h are continuous at the element nodes (R^*, θ_j), i.e.

$$
w_h(R^*, \theta_j) = w_h^+(R^*, \theta_j) = w_h^-(R^*, \theta_j),
$$

where $\theta_0 < \theta_1 < \dots < \theta_j < \theta_{j+1} < \dots < \theta_n$. We can thus construct a piecewise linear function $p(\theta)$ with respect to θ such that

$$
p(\theta) = w_h(R^*, \theta_j) + \frac{w_h(R^*, \theta_{j+1}) - w_h(R^*, \theta_j)}{(\theta_{j+1} - \theta_j)}(\theta - \theta_j), \text{ for } \theta_j \leq \theta \leq \theta_{j+1}.
$$

Clearly, $p(\theta)$ is the piecewise interpolation function of both w_h^+ and of w_h^-. Here, we have the triangular inequality:

$$
\left[\int_{l_{R^*}} (w_h^+ - w_h^-)^2 d\ell \right]^{\frac{1}{2}} \leq \left[\int_{l_{R^*}} (w_h^+ - p(\theta))^2 d\ell \right]^{\frac{1}{2}} + \left[\int_{l_{R^*}} (p(\theta) - w_h^-)^2 d\ell \right]^{\frac{1}{2}}. \tag{4.33}
$$

Also from Ciarlet (1978):

$$
\int_{l_{R^*}} (w_h^+ - p(\theta))^2 d\ell \leq Ch^4 \int_{l_{R^*}} \left(\frac{\partial^2 w_h^+}{\partial \theta^2} \right)^2 d\ell, \tag{4.34}
$$

where

$$w_h^+ = \sum_{l=0}^{L} \tilde{D}_l \left(\frac{r}{R}\right)^{\tau(l+\frac{1}{2})} \cos(\tau(l+\frac{1}{2})\theta).$$

Through integration calculations, we can obtain two equalities:

$$\int_{l_{R^*}} \left(\frac{\partial^2 w_h^+}{\partial \theta^2}\right)^2 d\ell = \frac{\Theta R^*}{2} \sum_{l=0}^{L} [\tau(l+\frac{1}{2})]^4 \tilde{D}_l^2 \left(\frac{R^*}{R}\right)^{\tau(2l+1)}$$

and

$$\iint_{S_2} \left(\frac{1}{r}\frac{\partial w_h^+}{\partial \theta}\right)^2 ds = \frac{\Theta}{4} \sum_{l=0}^{L} \tau(l+\frac{1}{2}) \tilde{D}_l^2 \left(\frac{R^*}{R}\right)^{\tau(2l+1)}. \tag{4.35}$$

Therefore, by noting (4.34), (4.35) and the following inequality:

$$\iint_{S_2} \left(\frac{1}{r}\frac{\partial w_h^+}{\partial \theta}\right)^2 ds \leq |w_h^+|_{1,S_2}^2 \leq \|w_h\|_1^2,$$

we obtain the bounds for one term in (4.33):

$$\int_{l_{R^*}} (w_h^+ - p(\theta))^2 d\ell \leq C[\tau(L+\frac{1}{2})]^3 h^4 \iint_{S_2} \left(\frac{1}{r}\frac{\partial w_h^+}{\partial \theta}\right)^2 ds$$

$$\leq C[\tau(L+\frac{1}{2})]^3 h^4 \|w_h\|_1^2.$$

On the other hand, since

$$w_h^- = a_i + b_i x + c_i y = a_i + b_i r \cos\theta + c_i r \sin\theta, \quad (x,y) \in \Delta_i,$$

where a_i b_i and c_i are constants, we find that

$$\int_{\Gamma_0} (p(\theta) - w_h^-(\tilde{\theta}))^2 \leq C \sum_i h_i^4 \int_{\theta_i}^{\theta_{i+1}} \left(\frac{\partial^2 w_h^-}{\partial \theta^2}\right)^2 R^* d\theta$$

$$= CR^* \sum_i h_i^4 \int_{\theta_i}^{\theta_{i+1}} (b_i \cos\theta + c_i \sin\theta)^2 R d\theta$$

$$\leq 2CR^* \sum_i h_i^4 R(\theta_{i+1} - \theta_i)(b_i^2 + c_i^2)$$

$$= 2CR^* \sum_i h_i^4 \frac{R(\theta_{i+1} - \theta_i)}{Meas(\Delta_i)} \int_{\Delta_i} \left[\left(\frac{\partial w_h^-}{\partial x}\right)^2 + \left(\frac{\partial w_h^-}{\partial y}\right)^2\right]$$

$$\leq Ch^3 \|w_h\|_1^2.$$

In the last step of the above equations we have used the Sobolev imbedding theorem and the inequality

$$R^*(\theta_{i+1} - \theta_i)/Meas(\triangle_i) \leq \frac{C_1}{h},$$

for quasiuniform triangular elements $\triangle_i$. Then we obtain

$$\int_{l_{R^*}} (p(\theta) - w_h^-(\theta))^2 d\ell \leq Ch^3 \|w_h\|_1^2.$$

Consequently, the inequality (4.33) leads to the bounds (4.32). ∎

Theorem 4.2 *Suppose that $Meas(\overline{AC} \cap \hat{S}_1^h) \neq 0$ and that the assumptions in Lemma 4.2 hold. Then the numerical solutions u_h^* by the nonconforming combined method of the RGM-FEM have the error bounds*

$$\|u - u_h^*\|_1 \leq C \left\{ h + \left(\frac{R_2}{R}\right)^{\tau(L+\frac{3}{2})} + [\tau(L + \frac{1}{2})]^{3/2} h^2 \right\}, \tag{4.36}$$

where $R^* < R_2 < R, \tau = \pi/\Theta$, and C is a bounded constant independent of h and L.

Proof. Since $Meas(\overline{AC} \cap \hat{S}_1^h) \neq 0$, the bilinear from I_h is uniformly V_h^0-elliptic based on Lemma 4.1. Hence the inequality (4.24) in Theorem 4.1 holds, in which each term will be evaluated afterwards.

Let $v_h = \overline{w}_h (\in V_h^0)$; then we have from Lemma 4.2

$$\inf_{v_h \in V_h^0} \|u - v_h\|_1 \leq \|u - \overline{w}_h\|_1 \leq C \left\{ h + \left(\frac{R_2}{R}\right)^{\tau(L+\frac{3}{2})} \right\}. \tag{4.37}$$

Also, we see from Li (1983a)

$$\left| \int_{l_{R^*}} \frac{\partial u}{\partial \nu} w_h^- d\ell - \int_{l_{R^*}} \frac{\partial u}{\partial r} w_h^- d\ell \right| \leq Ch^{3/2} \|w_h\|_1. \tag{4.38}$$

Finally, from Lemma 4.3 and

$$\int_{l_{R^*}} \left(\frac{\partial u}{\partial r}\right)^2 d\ell \leq C \|u\|_{2,S_2}^2 = O(1),$$

we obtain the inequality (4.36) from (4.24), (4.37) and (4.38). ∎

It has to be noted that in Theorem 4.2, the first and second terms in the braces of (4.36) are the error bounds resulting from the finite element method in S_1 and the RGM in S_2, respectively; and the third term is from the nonconforming coupling on the common boundary l_{R^*}.

Based on Theorem 4.2, it is easy to obtain error estimates for approximate coefficients as follows.

Corollary 4.1 *If the assumptions in Theorem 4.2 hold, the coefficients $D_l (1 \leq l \leq L)$ by the nonconforming combined method have the error bounds*

$$|D_l - \tilde{D}_l| \leq \frac{C\,(R/R^*)^{\tau(l+\frac{1}{2})}}{[\tau(l+\frac{1}{2})]^{\frac{1}{2}}} \left\{ h + \left(\frac{R_2}{R}\right)^{\tau(L+\frac{3}{2})} + [\tau(L+\frac{3}{2})]^{3/2} h^2 \right\}, \quad (4.39)$$

where $R^* < R_2 < R$.

Proof. We have from (4.8), (4.12) and (4.13)

$$u - u_h^* = \sum_{l=0}^{L} (D_l - \tilde{D}_l) \left(\frac{r}{R}\right)^{\tau(l+\frac{1}{2})} \cos(\tau(l+\frac{1}{2})\theta)$$

$$+ \sum_{l=L+1}^{\infty} D_l \left(\frac{r}{R}\right)^{\tau(l+\frac{1}{2})} \cos(\tau(l+\frac{1}{2})\theta), (r, \theta) \in S_2.$$

Then it follows

$$\|u - u_h^*\|_1^2 \geq \iint_{S_2} \left[\frac{1}{r}\frac{\partial(u - u_h^*)}{\partial \theta}\right]^2 ds \geq \frac{\Theta}{4}(D_l - \tilde{D}_l)^2 \tau(l+\frac{1}{2}) \left(\frac{R^*}{R}\right)^{\tau(2l+1)},$$

where $0 \leq l \leq L$. The desired inequality (4.39) is obtained from Theorem 4.2. ∎

4.3   COUPLING STRATEGY

The error estimates of Theorem 4.2 may suggest a good coupling relation between $(L+1)$ and h.

We see from Theorem 4.2 evidently that when $h = \mathit{constant}$ but $L+1 \longrightarrow \infty$, the errors of the numerical solutions u_h^* will be unbounded:

$$\|u - u_h^*\|_1 = O((L + \frac{1}{2})^{3/2}h^2) \to \infty.$$

This means that while the total number $(L + 1)$ of singular basic functions is too large, the errors from the nonconforming coupling could cause divergence of the numerical solutions.

By contrast, the total $(L + 1)$ should be small; but what is the best choice for $(L + 1)$? In order to answer this question, we recall the error bounds in the linear FEM:

$$\|u - \tilde{u}_h\|_{1,S} \leq Ch, \quad \forall h,$$

where $\tilde{u}_h$ is the numerical solution of the linear FEM. If we let the error estimates (4.36) also satisfy the similar bounds $O(h)$, i.e.,

$$\|u - u_h^*\|_1 \leq Ch, \quad \forall h, \tag{4.40}$$

the following two inequalities must be fulfilled:

$$[\tau(L + \frac{1}{2})]^{3/2}h \leq C, \quad \forall h \tag{4.41}$$

and

$$\left(\frac{R_2}{R}\right)^{\tau(L+\frac{3}{2})} \leq Ch, \quad \forall h. \tag{4.42}$$

Since inequality (4.42) results in (4.41), we have to choose $L + 1$ such that

$$\left(\frac{R_2}{R}\right)^{\tau(L+\frac{3}{2})} = Ch, \quad \forall h. \tag{4.43}$$

This then gives us the new coupling relation between $(L + 1)$ and h:

$$L + 1 = \frac{\ln h + \ln C}{\tau \ln (R_2/R)} - \frac{1}{2}, \quad \tau = \frac{\pi}{\Theta}, \quad R^* < R_2 < R. \tag{4.44}$$

We present this result as a corollary:

Corollary 4.2 *The error bounds (4.40) are valid if Eq. (4.44) (or (4.43)) and the assumptions in Theorem 4.2 hold.*

When $h \to 0$, Eq. (4.44) leads to an interesting, asymptotic relation:

$$L + 1 = O(|\ln h|). \tag{4.45}$$

As a result of (4.45) or (4.44), the total $(L + 1)$ is still not large even though h is small enough in practical calculations, such as $h = 0.001$. Therefore, the coupling relation, (4.45) or (4.44), will be more important when we demand higher precision of numerical solutions for singularity problems.

Another useful formula

$$L_{h_1} + 1 = (L_h + 1) + \frac{|\ln(h_1/h)|}{\tau |\ln(R_2/R)|} \tag{4.46}$$

is also derived from (4.44), where the notation is

$$L_h + 1 = \frac{\ln h + \ln C}{\tau \ln(R_2/R)} - \frac{1}{2}$$

with respect to a fixed h. Consequently, if a suitable total $L_h + 1$ of singular functions used has been known, we can evaluate directly from (4.46), another suitable total $(L_{h_1} + 1)$ for a new triangulation with a smaller $h_1(< h)$.

Take the numerical model in Section 4.8 later as an example where

$$\Theta = \pi, \quad \tau = \frac{\pi}{\Theta} = 1, \quad \text{and} \quad \frac{R^*}{R} = \frac{1}{2}.$$

By noting $R^* < R_2 < R$ and letting $R_2 \to R^*$, we have from (4.46)

$$L_{h/2} + 1 \approx (L_1 + 1) + 1. \tag{4.47}$$

In other words, almost one more singular function will be required on S_2 when all boundary lengths of triangular elements in S_1 proportionally decrease to their halves (i.e., when the number of element nodes in S_1 increases four times). This fact clearly displays how efficient the nonconforming method using this coupling strategy is!

4.4 STABILITY OF NUMERICAL SOLUTIONS

In order to discuss the stability of numerical solutions, it is sufficient to examine the condition numbers of the coefficient matrix $\mathbf{T}$ in (4.17):

$$Cond.(\mathbf{T}) = \lambda_{Max}(\mathbf{T})\big/\lambda_{Min}(\mathbf{T}),$$

where $\lambda_{Max}(\mathbf{T})$ and $\lambda_{Min}(\mathbf{T})$ are the maximal and minimal eigenvalues of $\mathbf{T}$ respectively.

Theorem 4.3 *Suppose that the inequality (4.10) and $Meas(\overline{AC} \cap \hat{S}_1^h) \neq 0$ hold; and the admissible functions are chosen as (4.13) with the constraint conditions (4.14). Then the condition numbers of $\mathbf{T}$ have the bounds*

$$Cond.(\mathbf{T}) \leq Max \left\{ Ch_{Min}^{-2}, C \left(\frac{R^*}{R} \right)^{-\tau(2L+1)} \bigg/ [\tau(L + \tfrac{1}{2})] \right\}, \tag{4.48}$$

where C is a bounded constant independent of h and L, and h_{Min} is the minimal boundary length of triangular elements in $\hat{S}_1^h$.

Proof. First, if we only use the finite element method in $\hat{S}_1^h$, the maximal eigenvalue and the minimal eigenvalue of the coefficient matrix $\mathbf{F}$ have the following bounds because of the assumption $Meas(\overline{AC} \cap \hat{S}_1^h) \neq 0$ (Strang and Fix (1973)):

$$\lambda_{Max}(\mathbf{F}) \leq C; \quad \lambda_{Min}(\mathbf{F}) \geq C_p h_{Min}^2, \tag{4.49}$$

where the constant C_p has a positive inferior limit independent of h.

On the other hand, if we only use the RGM in S_2, the coefficient matrix $\mathbf{G}$ is composed of the elements $g_{i,j}$, i.e.,

$$\mathbf{G} = (g_{i,j}),$$

where the matrix entries are

$$g_{i,j} = \frac{1}{2} \iint_{S_2} \left[\frac{\partial u_i}{\partial r} \frac{\partial u_j}{\partial r} + \frac{1}{r^2} \frac{\partial u_i}{\partial \theta} \frac{\partial u_j}{\partial \theta} \right] ds,$$

and the basis functions are

$$u_l = \left(\frac{r}{R} \right)^{\tau(l+\frac{1}{2})} \cos(\tau(l + \tfrac{1}{2})\theta).$$

With the aid of the orthogonality of trigeometric functions, we can easily obtain the eigenvalues of G:

$$\lambda_i(G) = \frac{\Theta}{4}\tau(l + \frac{1}{2})\left(\frac{R^*}{R}\right)^{\tau(2l+1)}, \quad l = 0, 1, ..., L.$$

It then follows that the maximal and minimal eigenvalues have the bounds

$$\lambda_{Max}(\mathbf{G}) \leq C, \quad \lambda_{Min}(\mathbf{G}) \geq C_p\tau(L + \frac{1}{2})\left(\frac{R^*}{R}\right)^{\tau(2L+1)}. \tag{4.50}$$

Next, we construct a collective matrix such that

$$\tilde{\mathbf{T}} = \begin{bmatrix} \mathbf{F} & O \\ O & \mathbf{G} \end{bmatrix}.$$

This is just the same coefficient matrix as that obtained from (4.15) but without the constraint conditions (4.14). We then obtain from (4.49) and (4.50)

$$\lambda_{Max}(\tilde{\mathbf{T}}) \leq C,$$
$$\lambda_{Min}(\tilde{\mathbf{T}}) \geq \min\left\{C_p h_{Min}^2, \ C_p\tau(L + \frac{1}{2})\left(\frac{R^*}{R}\right)^{\tau(2L+1)}\right\}. \tag{4.51}$$

As for the nonconforming combined method, the constraint conditions (4.14) have to be performed. Furthermore, we know from the matrix theory (Courant and Hilbert (1953)):

$$\lambda_{Max}(\mathbf{T}) = \max_{\substack{(\tilde{x},\tilde{x})\neq 0, \\ L_j(\tilde{x})=0}} \frac{\tilde{x}^T\tilde{\mathbf{T}}\tilde{x}}{(\tilde{x},\tilde{x})} \leq \max_{(\tilde{x},\tilde{x})\neq 0} \frac{\tilde{x}^T\tilde{\mathbf{T}}\tilde{x}}{(\tilde{x},\tilde{x})} = \lambda_{Max}(\tilde{\mathbf{T}}),$$

$$\lambda_{Min}(\mathbf{T}) = \min_{\substack{(\tilde{x},\tilde{x})\neq 0 \\ L_j(\tilde{x})=0}} \frac{\tilde{x}^T\tilde{\mathbf{T}}\tilde{x}}{(\tilde{x},\tilde{x})} \geq \min_{(\tilde{x},\tilde{x})\neq 0} \frac{\tilde{x}^T\tilde{\mathbf{T}}\tilde{x}}{(\tilde{x},\tilde{x})} = \lambda_{Min}(\tilde{\mathbf{T}}), \tag{4.52}$$

where the notation $L_j(\tilde{x}) = 0$ represents the constraint equation (4.14). Therefore, both the maximal eigenvalue $\lambda_{Max}(\mathbf{T})$ and the minimal eigenvalue $\lambda_{Min}(\mathbf{T})$ also satisfy (4.51). The inequality (4.48) holds. ∎

We can see from Theorem 4.3 that if the total number $(L+1)$ of singular functions is too large, there does occur serious instability of the numerical solutions. For example, when $h = constant$ but $L + 1 \rightarrow \infty$, the bounds (4.48) of condition numbers will be

$$Cond.(\mathbf{T}) = O(a^{L+1}) \rightarrow \infty, \tag{4.53}$$

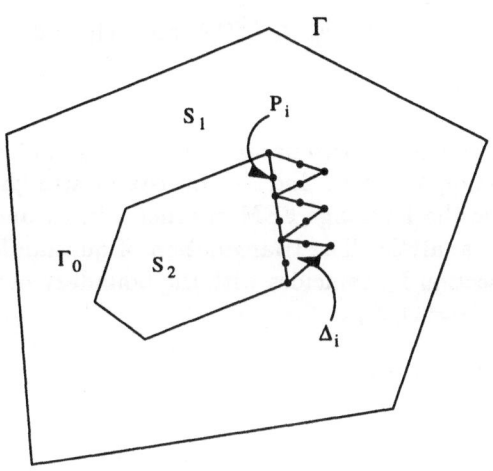

Figure 4.3 Partition of S.

where the constant $a = (R/R^*)^{2\tau} > 1$. However, the total number $(L+1)$ is not large in general cases of application, on the ground of the coupling relation (4.44). The stability of numerical solutions by the nonconforming approach is, in fact, as good as that by the FEM (see numerical examples in the Section 4.8).

4.5   ELLIPTIC EQUATIONS

In this section, we extend the combinations of RGM-FEM to general elliptic equations. Consider a two-dimensional model problem (see Figure 4.3):

$$\begin{cases} -\frac{\partial}{\partial x}\left(\beta\frac{\partial u}{\partial x}\right) - \frac{\partial}{\partial y}\left(\beta\frac{\partial u}{\partial y}\right) = f, & (x,y) \in S, \\ u = 0, & (x,y) \in \Gamma, \end{cases} \tag{4.54}$$

where S is a convex polygonal domain, Γ is the boundary of S, the functions β and f are smooth enough, and there exists a positive constant β_0 such that

$$\beta = \beta(x,y) \geq \beta_0 > 0. \tag{4.55}$$

The elliptic problem (4.54) is equivalent to the weak form

$$a(u,v) = f(v), \quad \forall v \in H_0^1(S), \tag{4.56}$$

where

$$a(u,v) = \iint_S \beta(u_x v_x + u_y v_y)ds, \quad f(v) = \iint_S fvds, \tag{4.57}$$

$$H_0^1(S) = \{v|v, \ v_x, \ v_y \ \in L^2(S) \ \text{and} \ v|_\Gamma = 0\}, \tag{4.58}$$

with $v_x = \frac{\partial v}{\partial x}$ and $v_y = \frac{\partial v}{\partial y}$.

Let us divide the polygon S into two parts: $S = S_1 \cup S_2$, and suppose $Meas(S_1 \cap \Gamma) \neq 0$. The intersection Γ_0 of S_1 and S_2 consists of straight line segments(see Figure 4.3). We choose the Lagrange FEM of order k in S_1 and the RGM in S_2 if the solution u in S_2 is analytic. The triangulation of quasiuniform family in S_1 is chosen, and the intersection Γ_0 coincides with the boundary of triangular elements adjacent to Γ_0. Then $S_1 = \cup_i \triangle_i$.

Take the admissible trial function as

$$v = \left\{ \begin{array}{ll} v^- = v_k, & (x, y) \in S_1, \\ v^+ = f_L(\tilde{D}_l), & (x, y) \in S_2, \end{array} \right. \tag{4.59}$$

where v_k is a piecewise interpolation polynomial of order k, and the analytic functions are

$$f_L = f_L(\tilde{D}_l) = \sum_{l=0}^{L} \tilde{D}_l \varphi_l. \tag{4.60}$$

In (4.60) $\tilde{D}_l$ are the expansion coefficients, and $\{\varphi_l\}$ is a complete set of linearly independent bases, which are the polynomials, in particular, the orthogonal polynomials. The admissible function v_h at the triangular element nodes P_i on Γ_0 satisfies the continuity condition

$$v_k(P_i) = \sum_{l=0}^{L} D_l \varphi_i(P_i), \quad P_i \in \Gamma_0. \tag{4.61}$$

We denote by V_h^0 such admissible functional spaces satisfying the boundary condition $u|_\Gamma = 0$. Hence we can obtain the combination of RGM-FEM:

To seek $u_h^* \in V_h^0$ such that

$$a_h(u_h^*, v) = f(v), \quad \forall v \in V_h^0, \tag{4.62}$$

where

$$a_h(u, v) = \sum_i \iint_{\triangle_i} \beta(u_x v_x + u_y v_y)ds + \iint_{S_2} \beta(u_x v_x + u_y v_y)ds, \tag{4.63}$$

and

$$f(v) = \sum_i \iint_{\triangle_i} f v ds + \iint_{S_2} f v ds, \tag{4.64}$$

where $\triangle_i$ are the triangular elements in S_1. Define the norm

$$\|v\|_1 = \left(\|v\|_{1,S_1}^2 + \|v\|_{1,S_2}^2 \right)^{1/2}.$$

We have the following theorem.

Theorem 4.4 *Suppose the bilinear form in (4.62) is uniformly V_h^0-elliptic, i.e., there exists a positive constant α_p independent of h and L such that*

$$\alpha_p \|v\|_1^2 \leq a_h(v,v), \qquad \forall v \in V_h^0. \tag{4.65}$$

Then the numerical solution u_h^* of (4.62) obtained by the combined method of RGM-FEM has the error estimates

$$\|u - u_h^*\|_1 \leq C \left\{ \inf_{v_h \in V_h^0} \|u - v_h\|_1 \right.$$

$$\left. + \sqrt{\int_{\Gamma_0} \left(\beta \frac{\partial u}{\partial n} \right)^2 d\ell} \sup_{v_h \in V_h^0} \frac{\sqrt{\int_{\Gamma_0} (w^+ - w^-)^2 d\ell}}{\|w_h\|_1} \right\}, \tag{4.66}$$

where n is the outward normal of the boundary Γ_0 of S_2, C the constant independent of h and L, h the maximal boundary length of triangular elements, and

$$w_h^+ = w_h|_{S_2} = f_L(\tilde{D}_l), \qquad w_h^- = w_h|_{S_1} = v_k.$$

Proof. From Strang and Fix (1973) and Ciarlet (1978) we have

$$\|u - u_h^*\|_1 \leq C \left\{ \inf_{v_h \in V_h^0} \|u - v_h\|_1 + \sup_{w_h \in V_h^0} \frac{|f(w_h) - a_h(u, w_h)|}{\|w_h\|_1} \right\}.$$

Applying the Green's theorem, (4.54), (4.62) and (4.63), we obtain

$$|f(w_h) - a_h(u, w_h)| = \left| \int_{\partial S_1} \beta \frac{\partial u}{\partial n} w_h^- d\ell + \int_{\partial S_2} \beta \frac{\partial u}{\partial n} w_h^+ d\ell \right|,$$

where ∂S_1 and ∂S_2 are the boundary of S_1 and S_2 respectively, and n the normal of ∂S_1 (or ∂S_2). Since the normal flux $\beta \frac{\partial u}{\partial n}$ on Γ_0 is continuous because f and β are continuous functions, we obtain from the Schwarz inequality that

$$\left| \int_{\partial S_1} \beta \frac{\partial u}{\partial n} w_h^- d\ell + \int_{\partial S_2} \beta \frac{\partial u}{\partial n} w_h^+ d\ell \right|$$

$$= \left| \int_{\Gamma_0} \beta \frac{\partial u}{\partial n} (w_h^+ - w_h^-) d\ell \right| \leq \sqrt{\int_{\Gamma_0} \left(\beta \frac{\partial u}{\partial n} \right)^2} \sqrt{\int_{\Gamma_0} (w_h^+ - w_h^-)^2 d\ell}.$$

Thus, the inequality (4.66) is obtained.   ■

Suppose that the solution u in S_2 is analytic, and can be uniformly approximated by the linearly independent complete bases $\{\varphi_l\}$, i.e.,

$$u = \sum_{l=0}^{\infty} D_l \varphi_l = f_L(D_l) + R_L, \quad (x,y) \in S_2, \tag{4.67}$$

where D_ℓ are the expansion coefficients, and R_L the remainder as

$$R_L = \sum_{l=L+1}^{\infty} D_l \varphi_l. \tag{4.68}$$

We construct an auxiliary function

$$W_h = \begin{cases} \overline{u}_h, & (x,y) \in S_1, \\ f_L(D_l), & (x,y) \in S_2, \end{cases} \tag{4.69}$$

where $\overline{u}_h$ is the piecewise Lagrange polynomial of order k in S_1, but the values of $\overline{u}_h$ at the nodes P_i of triangular elements $\triangle_i$ are defined by

$$\overline{u}_h(P_i) = \begin{cases} u(P_i), & P_i \notin \Gamma_0, \\ f_L(D_l)(P_i), & P_i \in \Gamma_0. \end{cases}$$

Obviously $W_h \in V_h^0$, and we have the following lemma.

Lemma 4.4 *Let the expansion (4.67) hold. Then for the auxiliary function W_h defined by (4.69) there exist the bounds*

$$\|u - W_h\|_1 \le C\left\{ h^k |u|_{k+1,S_1} + \|R_L\|_{1,S_2} + h^{k+\frac{1}{2}} |u|_{k+1,\Gamma_0} \right.$$
$$\left. \frac{1}{\sqrt{h}} \|R_L\|_{0,\Gamma_0} + h^{k+\frac{1}{2}} |R_L|_{k+1,\Gamma_0} \right\}, \tag{4.70}$$

where h is the maximal boundary length of regular triangular elements in S_1, and R_L is the remainder term (4.68).

Proof. Since

$$\|u - W_h\|_1 \le \|u - \overline{u}_h\|_{1,S_1} + \|R_L\|_{1,S_2}$$
$$\le \|u - u_h\|_{1,S_1} + \|u_h - \overline{u}_h\|_{1,S_1} + \|R_L\|_{1,S_2}, \tag{4.71}$$

where u_h is the piecewise k-order Lagrange interpolation of u in S_1, we have

$$\|u - u_h\|_{1,S_1} \leq C h^k |u|_{k+1,S_1}.$$

From Lemma 5.2, later shown in Section 5.5 in Chapter 5, we obtain

$$\|u_h - \bar{u}_h\|_{1,S_1} \leq C \frac{1}{\sqrt{h}} \|u_h - \bar{u}_h\|_{0,\Gamma_0}.$$

Then it follows from (4.67) that

$$
\begin{aligned}
& \|u_h - \bar{u}_h\|_{0,\Gamma_0} \\
\leq {} & \|u - u_h\|_{0,\Gamma_0} + \|u - f_L(D_l)\|_{0,\Gamma_0} + \|\bar{u}_h - f_L(D_l)\|_{0,\Gamma_0} \\
\leq {} & C \left(h^{k+1} |u|_{k+1,\Gamma_0} + \|R_L\|_{0,\Gamma_0} + h^{k+1} |f_L(D_l)|_{k+1,\Gamma_0} \right) \\
\leq {} & C \left(2 h^{k+1} |u|_{k+1,\Gamma_0} + \|R_L\|_{0,\Gamma_0} + h^{k+1} |R_L|_{k+1,\Gamma_0} \right).
\end{aligned}
\tag{4.72}
$$

Finally, we obtain (4.70) from (4.71)–(4.72). ∎

Lemma 4.5 *Suppose Γ_0 is made up of finite sections of straight lines, and the admissible function v_h in S_2 is an L-order polynomial. Then there exists a constant C independent of L such that*

$$\sup_{w_h \in V_h^0} \frac{|w_h^+|_{k+1,\Gamma_0}}{\|w_h\|_1} \leq C L^{2(k+1)}. \tag{4.73}$$

Proof. First, let us prove that the following inequality for the L-order polynomial $\rho_L(x)$ on the segment $[-1,1]$:

$$\int_{-1}^{1} [\rho_L'(x)]^2 dx \leq C L^4 \int_{-1}^{1} [\rho_L(x)]^2 dx. \tag{4.74}$$

The Legendre polynomials are defined by

$$P_n(x) = \frac{(-1)^n}{2^n n!} \frac{d^n}{dx^n} \left[(1 - x^2)^n \right], \text{ on } [-1,1].$$

The orthogonality of $P_n(x)$ is given by

$$\int_{-1}^{1} P_m(x) P_n(x) dx = \left\{ \begin{array}{ll} 0, & m \neq n, \\ \frac{2}{2n+1}, & m = n. \end{array} \right.$$

In fact, the Legendre polynomials $P_n(x)$ have

$$P_{n+1}'(x) - P_{n-1}'(x) = (2n+1) P_n(x).$$

Since $P_0(x) = 1, P_1(x) = x$, we obtain easily

$$\begin{cases} P'_{2n}(x) = \sum_{i=1}^{n}(4i-1)P_{2i-1}(x), \\ P'_{2n+1}(x) = \sum_{i=0}^{n}(4i+1)P_{2i}(x). \end{cases} \tag{4.75}$$

Any L-order polynomial $\rho_L(x)$ can be expanded by Legendre polynomials,

$$\rho_L(x) = \sum_{i=0}^{L} \alpha_i P_i(x),$$

where α_i are the expansion coefficients. We have

$$[\rho'_L(x)]^2 = \left[\sum_{i=1}^{L}\alpha_i P'_i(x)\right]^2 \le \left(\sum_{i=1}^{L}\frac{2\alpha_i^2}{2i+1}\right)\left(\sum_{i=1}^{L}\frac{2i+1}{2}[P'(x)]^2\right). \tag{4.76}$$

Again applying (4.75), (4.76) and the orthogonality of Legendre polynomials, we obtain

$$\int_{-1}^{1}[\rho'_L(x)]^2dx \le \left(\sum_{i=1}^{L}\frac{2\alpha_i^2}{2i+1}\right)\left(\sum_{i=1}^{L}\frac{2i+1}{2}\int_{-1}^{1}[P'_i(x)]^2dx\right).$$

Also from (4.75) and (4.74)

$$\int_{-1}^{1}(P'_i(x))^2dx \le \sum_{j=0}^{i}(2j+1)^2\int_{-1}^{1}P_j^2(x)dx$$

$$\le \sum_{j=0}^{i}2(2j+1) = 2(i+1)^2. \tag{4.77}$$

Hence we obtain from (4.74):

$$\int_{-1}^{1}[\rho'_L(x)]^2dx \le \left(\int_{-1}^{1}[\rho_L(x)]^2dx\right)\left(\sum_{i=1}^{L}\frac{(2i+1)}{2}2(i+1)^2\right)$$

$$\le CL^4\int_{-1}^{1}[\rho_L(x)]^2dx.$$

With the help of (4.74), it follows that

$$\int_{-1}^{1}[\rho_L^{k+1}(x)]^2dx \le C(L-k)^4\int_{-1}^{1}[\rho_L^{k}(x)]^2dx$$

$$\leq \quad C^{k+1}(L-k)^4(L-k+1)^4 \dots L^4 \int_{-1}^{1}[\rho_L(x)]^2 dx$$

$$\leq \quad C^{k+1}L^{4(k+1)} \int_{-1}^{1}[\rho_L(x)]^2 dx. \tag{4.78}$$

From Sobolev (1963), we have

$$|w_h^+|_{0,\Gamma_0} \leq C|w_h^+|_{1,S_2} \leq C\|w_h\|_1.$$

Since $\Gamma_0 = \cup_{i=1}^{m}\Gamma_i$, where Γ_i is a segment of straight lines, and m is a finite integer, we obtain

$$\frac{|w_h^+|_{k+1,\Gamma_0}}{\|w_h\|_1} \leq C\frac{|w_h^+|_{k+1,\Gamma_0}}{|w_h^+|_{0,\Gamma_0}} \leq C\sum_{i=1}^{m}\frac{|w_h^+|_{k+1,\Gamma_i}}{|w_h^+|_{0,\Gamma_i}}. \tag{4.79}$$

The length of Γ_i is denoted as $2a_i$, where $a_i > 0$. Since $v_h|_{S_2}$ is an L-order polynomial, $w_h^+|_{S_2}(= (u_h^* - v_h)|_{S_2})$ is an L-order polynomial, too. Applying the linear mapping F: $\hat{x} \in [-1,1] \to F(\hat{x}) = a_i\hat{x} + b_i$, where b_i is a constant, we can transfer $w_h^+|_{\Gamma_i}$ into an L-order polynomial $\hat{W}_h^+(\hat{x})$ for $\hat{x} \in [-1,1]$. According to the semi-norm relation we have

$$|w_h^+|_{0,\Gamma_0}^2 = \int_{\Gamma_i}(w_h^+)^2 d\ell = a_i \int_{-1}^{1}(\hat{W}_h^+)^2 d\hat{x},$$

$$|w_h^+|_{k+1,\Gamma_i}^2 = \int_{-1}^{1}\left(\frac{\partial^{k+1}w_h^+}{\partial l^{k+1}}\right)^2 d\ell = \frac{a_i}{a_i^{2(k+1)}}\int_{-1}^{1}\left(\frac{\partial^{k+1}\hat{W}_h^+}{\partial\hat{x}^{k+1}}\right)^2 d\hat{x}.$$

It follows from (4.78) that

$$\frac{|w_h^+|_{k+1,\Gamma_i}^2}{|w_h^+|_{0,\Gamma_i}^2} = \frac{\int_{-1}^{1}\left(\frac{\partial^{k+1}\hat{W}_h^+}{\partial\hat{x}}\right)^2 d\hat{x}}{a_i^{2(k+1)}\int_{-1}^{1}(\hat{W}_h^+)^2 d\hat{x}} \leq \frac{C^{k+1}}{a_i^{2(k+1)}}L^{4(k+1)}. \tag{4.80}$$

Again from (4.79) we obtain (4.73). ∎

Theorem 4.5 *Let* $u \in H^{k+1}(S_1)$, $u \in H^{k+1}(\Gamma_0)$ *and* u *in* S_2 *be analytic with expansion (4.67). Assume that the uniformly* V_h^0-*elliptic inequality (4.65) holds, then the numerical solution* u_h^* *of the combination of RGM-FEM has the error bounds:*

$$\|u - u_h^*\|_1 \leq C\left\{h^k|u|_{k+1,S_1} + \|R_L\|_{1,S_2} + h^{k+\frac{1}{2}}|u|_{k+1,\Gamma_0} + \frac{1}{\sqrt{h}}\|R_L\|_{0,\Gamma_0} + \right.$$

$$\left. + h^{k+\frac{1}{2}}|R_L|_{k+1,\Gamma_0} + h^{k+1}L^{2(k+1)}\left|\frac{\partial u}{\partial n}\right|_{0,\Gamma_0}\right\}. \tag{4.81}$$

Proof. For $W_h \in V_h^0$, we have

$$\inf_{v_h \in V_h^0} \|u - v_h\|_1 \leq \|u - W_h\|_1. \tag{4.82}$$

Since $w_h^-|_{\Gamma_0}$ is a piecewise k-order polynomial interpolation of $w_h^+|_{\Gamma_0}$, we obtain

$$\int_{\Gamma_0} (w_h^+ - w_h^-)^2 d\ell \leq Ch^{2(k+1)} \sum_i \int_{\Gamma_0 \cap \Delta_i} \left(\frac{\partial^{k+1} w_h^+}{\partial l^{k+1}} \right)^2 d\ell$$

$$= Ch^{2(k+1)} |w_h^+|^2_{k+1, \Gamma_0}.$$

By applying Lemma 4.5 it follows that

$$\sup_{w_h \in V_h^0} \frac{\sqrt{\int_{\Gamma_0} (w_h^+ - w_h^-)^2 d\ell}}{\|w_h\|_1} \leq Ch^{k+1} L^{2(k+1)}. \tag{4.83}$$

Obviously

$$\int_{\Gamma_0} \left(\beta \frac{\partial u}{\partial n} \right)^2 d\ell \leq C \left| \frac{\partial u}{\partial n} \right|^2_{0, \Gamma_0}.$$

By using Theorem 4.4, Lemma 4.4, (4.82) and (4.83), we have proved the estimates (4.81). ∎

4.6   CHOICE OF POLYNOMIAL BASES

In this section, we will discuss the choice of bases $\{\varphi_l\}$ and the number L in (4.59).

In (4.59) we choose

$$f_L(\tilde{D}_l) = \sum_{l=0}^{L} \sum_{i+j=l} \tilde{D}_{ij} X_i(x) Y_j(y), \tag{4.84}$$

where $\tilde{D}_{ij}$ are the coefficients, and $\{X_i(x)\}$ $(i = 0, 1, \ldots)$ are the complete i-order polynomials of linear independence, in particular, orthogonal polynomials. The definition of bases $\{Y_j(y)\}$ $(j = 0, 1, \ldots)$ is similar to $\{X_i(x)\}$.

An L-order polynomial can be expressed as (4.84). Then when $f_L(\tilde{D}_l)$ is chosen as Taylor's expansion of $u(x, y)$ at a point in S_2,

$$u(x, y) = f_L(D_l) + R_L,$$

where the remainder term R_L has the norm estimates

$$\|R_L\|_{1,S_2} \leq C\frac{c^L}{L!}M_{L+1}, \quad |R_L|_{k+1,\Gamma_0} \leq C\frac{c^{L-k+1}}{(L-k+1)!}M_{L+1}. \tag{4.85}$$

In (4.85), c is the diameter of S_2, and the notation $M_l = \max_{\substack{(x,y)\in S_2 \\ i+j\leq l}} \left|\frac{\partial^{i+j}u}{\partial x^i \partial y^j}\right|$. Using the Stirling formula (see Courant and Hilbert (1953))

$$N! = \left(\frac{N}{e}\right)^N \sqrt{2\pi N}(1+O(N^{-\frac{1}{5}})),$$

we obtain the error estimate of u_h^* from Theorem 4.5 and (4.85)

$$\begin{aligned}
\|u - u_h^*\|_1 \leq \; & C\left\{h^k|u|_{k+1,S_1} + \frac{1}{\sqrt{L}}\left(\frac{ce}{L}\right)^L M_{L+1} + h^{k+\frac{1}{2}}|u|_{k+1,\Gamma_0}\right. \\
& + \frac{1}{\sqrt{hL}(L+1)}\left(\frac{ce}{L}\right)^L M_{L+1} + h^{k+\frac{1}{2}}\frac{(ce)^{L-k+1}}{(L-k+1)!}M_{L+1} \\
& \left. + h^{k+1}L^{2(k+1)}\left|\frac{\partial u}{\partial n}\right|_{0,\Gamma_0}\right\}.
\end{aligned} \tag{4.86}$$

Suppose that the errors of u_h^* have the same convergence rates of k-order Lagrange FEM, then

$$\|u - u_h^*\|_1 = O(h^k). \tag{4.87}$$

Hence we require from (4.86),

$$L \leq Ch^{-\frac{1}{2(k+1)}}, \tag{4.88}$$

which is easily satisfied as shown below.

Suppose that the l-order partial derivatives M_l of u do not increase very rapidly, satisfying

$$M_L \leq Cd^L, \tag{4.89}$$

where d is a constant independent of L. Obviously, for (4.87) we may balance the error terms in (4.86), if

$$\frac{1}{\sqrt{L}(L+1)}\left(\frac{ce}{L}\right)^L M_{L+1} \leq Ch^{k+\frac{1}{2}},$$

to lead to

$$L = O\left(T^{-1}\left(h^{-\frac{k+\frac{1}{2}}{cde}}\right)\right),\qquad(4.90)$$

where $T^{-1}(x)$ is the inverse function of x^x.

The error term, $\frac{1}{\sqrt{hL(L+1)}}\left(\frac{ce}{L}\right)^L M_{L+1}$, displays the exponential convergence rates of the solutions by the RGM. Then the total term $(L+1)(L+2)/2$ satisfying (4.90) is very small, thus saving a lot of calculation and storage by the combined method.

In the above analysis, we do not specify the specific polynomial bases $\{X_i(x)\}$ and $\{Y_j(y)\}$; we may choose the orthogonal polynomials, i.e., the Legendre polynomial on $[-1, 1]$

$$X_i(x) = P_i(x),\qquad(4.91)$$

under a suitable linear transformation.

4.7   UNIFORM V_H–ELLIPTICITY AND STABILITY

Theorem 4.6 *Let (4.88) be given and the admissible function be $v_h \in V_h^0$ in (4.59) satisfying $u|_\Gamma = 0$ and (4.61). Then when $h \le h_0(h_0 = constant)$, the uniformly V_h^0-elliptic inequality (4.65) holds.*

Proof. From (4.55), we have

$$a_h(v, v) \ge C_0\left(|v|_{1,S_1}^2 + |v|_{1,S_2}^2\right), \qquad \forall v \in V_h^0.\qquad(4.92)$$

The desired result (4.65) follows from the following bounds,

$$|v|_{1,S_1}^2 + |v|_{1,S_2}^2 \ge C_1\left\{\|v\|_{1,S_1}^2 + \|v\|_{1,S_2}^2\right\}, \qquad \forall v \in V_h^0,\qquad(4.93)$$

where C_1 is a positive constant independent of $h(\le h_0)$ and L. Now, let us prove (4.93). Since $Meas(S_1 \cap \Gamma) \ne 0$, we have

$$\|v\|_{1,S_1} \le C|v|_{1,S_1}\qquad(4.94)$$

and

$$|v^-|_{0,\Gamma_0} \le C|v|_{1,S_1}.\qquad(4.95)$$

From (4.79), (4.80) and (4.88), we obtain

$$
\begin{aligned}
|v^+|_{0,\Gamma_0} &\leq |v^+ - v^-|_{0,\Gamma_0} + |v^-|_{0,\Gamma_0} \\
&\leq Ch^{k+1}L^{2(k+1)}|v^+|_{0,\Gamma_0} + |v^-|_{0,\Gamma_0} \\
&\leq C_2 h^k |v^+|_{0,\Gamma_0} + |v^-|_{0,\Gamma_0},
\end{aligned}
$$

where C_2 is a constant independent of h and L. Let $h_0 = \left(\frac{1}{2C_2}\right)^{1/k}$, then when $h \leq h_0$,

$$
|v^+|_{0,\Gamma_0} \leq \frac{|v^-|_{0,\Gamma_0}}{1 - C_2 h^k} \leq 2|v^-|_{0,\Gamma_0}. \tag{4.96}
$$

Moreover, combining (4.95) and (4.96) gives

$$
\|v\|_{1,S_2} \leq C \left(|v|_{1,S_2} + |v^+|_{0,\Gamma_0} \right) \leq C_2 \left(|v|_{1,S_1} + |v|_{2,S_2} \right). \tag{4.97}
$$

Therefore we have proved (4.93) from (4.94) and (4.97). ■

Let us discuss stability of the numerical solutions. To solve (4.62) we obtain a system of linear algebraic equations

$$
\mathbf{A}\mathbf{x} = \mathbf{b}, \tag{4.98}
$$

where matrix $\mathbf{A}$ is symmetric and positive definite, $\mathbf{x}$ is the unknown vector consisting of $V_{i,j}$ and $\tilde{D}_l$, and $\mathbf{b}$ is a known vector. To study stability of the numerical solutions, we estimate the condition number, $Cond.(\mathbf{A}) = (\lambda_{max}(\mathbf{A})/\lambda_{min}(\mathbf{A}))$, of matrix $\mathbf{A}$, where $\lambda_{max}(\mathbf{A})$ (or $\lambda_{min}(\mathbf{A})$) is the maximum (or minimum) eigenvalue of $\mathbf{A}$.

We choose the special basis functions $\varphi_i (i = 0, 1, \ldots, L)$ in S_2, and denote

$$
\Phi_{ij} = (\varphi_i, \varphi_j) = \iint_{S_2} \varphi_i \varphi_j \, ds + \iint_{S_2} \left(\frac{\partial \varphi_i}{\partial x} \frac{\partial \varphi_j}{\partial x} + \frac{\partial \varphi_i}{\partial y} \frac{\partial \varphi_j}{\partial y} \right) ds.
$$

For matrix $\Phi = (\Phi_{ij})$, suppose

$$
0 < \overline{\lambda}_{min} \leq \lambda_i(\Phi) \leq \overline{\lambda}_{max}, \quad i = 0, 1, \ldots, L. \tag{4.99}
$$

Then we have the following theorem.

Theorem 4.7 *Suppose that (4.99) and all conditions of Theorem 4.6 hold. Then the condition number of $\mathbf{A}$ in (4.98) has the bounds*

$$
Cond.(\mathbf{A}) \leq C \frac{max(1, \overline{\lambda}_{max})}{min(h_{min}^2, \overline{\lambda}_{min})}, \tag{4.100}
$$

where h_{min} *is the minimal boundary length of quasiuniform triangular elements in* S_1.

Proof. We have

$$\lambda_{min}(\mathbf{A})\mathbf{x}^T\mathbf{x} \leq \mathbf{x}^T\mathbf{A}\mathbf{x} = a_h(v_h, v_h) \leq \lambda_{max}(\mathbf{A})\mathbf{x}^T\mathbf{x}, \qquad v_h \in V_h^0. \qquad (4.101)$$

First we do not consider the constraint condition (4.61) temporarily. We obtain

$$\iint_{S_1} \beta(v_x^2 + v_y^2)ds = \sum_i \iint_{\Delta_i} \beta(v_x^2 + v_y^2)ds \leq C \sum_{i=1}^{M+M'} v_i^2,$$

where v_{M+1}, v_{M+2}, $\ldots$, $v_{M+M'}$ are the functions at the nodes on Γ_0. Let $v = \sum_{i=0}^{L} \tilde{D}_l \varphi_l$ in S_2, then we obtain from (4.99)

$$\iint_{S_2} \beta(v_x^2 + v_y^2)ds \leq C\|v\|_{1,S_2}^2 \leq C\overline{\lambda}_{max} \sum_{i=0}^{L} \tilde{D}_l^2. \qquad (4.102)$$

Therefore from (4.63), (4.101) and (4.102), we have

$$a_h(v,v) \leq Cmax(1,\overline{\lambda}_{max}) \left\{ \sum_{i=1}^{M+M'} v_i^2 + \sum_{i=0}^{L} \tilde{D}_l^2 \right\}.$$

Based on the matrix theory, when the constraint condition (4.61) has been applied to the unknowns $v_i(i > M)$ on Γ_0 (i.e., $v_h \in V_h^0$), the maximum eigenvalue of the matrix will not increase, i.e.,

$$\lambda_{max}(\mathbf{A}) \leq C\ max(1,\overline{\lambda}_{max}). \qquad (4.103)$$

Similarly

$$\lambda_{min}(\mathbf{A}) \geq C_b \min\left\{h_{min}^2, \overline{\lambda}_{min}\right\}, \qquad (4.104)$$

with the constant $C_b > 0$. Combining (4.103) and (4.104) leads to the desired result (4.100). ∎

From (4.100) we know that if the maximum (or minimum) eigenvalue of Φ is not too large (or too small), the condition number $Cond.(\mathbf{A})$ of A in (4.98) will not be too large. Therefore the numerical solution of the combined method has a good stability. Obviously, it is better to choose orthogonal polynomials as the bases $\{\varphi_l\}$ in Section 4.6.

By letting $v = \sum_{l=0}^{L} \dfrac{\tilde{D}_l}{\sqrt{\overline{\lambda}_{max}}} \varphi_l$ in S_2, we obtain the better bounds than (4.100):

$$Cond.(\mathbf{A}) \le C \frac{1}{min(h_{min}^2, \overline{\lambda}_{min}/\overline{\lambda}_{max})}. \tag{4.105}$$

Therefore, if $\overline{\lambda}_{max}/\overline{\lambda}_{min} \le Ch_{min}^{-2}$, the stability of the numerical solutions obtained by the combined method will not be worse than that of the numerical solutions by FEM.

4.8   NUMERICAL EXPERIMENTS FOR MOTZ PROBLEM

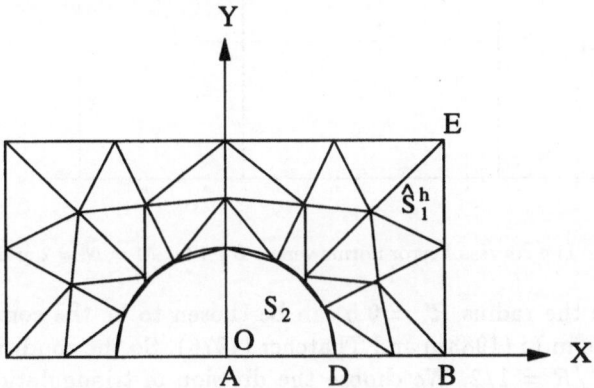

Figure 4.4   The division for Motz's problem with $M = N = 2$ and $R^* = 0.5$.

Consider Motz's problem in Section 2.1.4. In Figure 4.4, $\Theta = \pi$, then $\tau = \pi/\Theta = 1$. Hence, the origin $(0, 0)$ is really a singularity. Note that the radius $R = 1$, and that

$L+1$	$\|\varepsilon^+\|_{\infty, l_{R^*}}$	$\|\varepsilon^+\|_{0, l_{R^*}}$	$\|\varepsilon^+\|_{1, l_{R^*}}$	$\|\varepsilon\|_{\infty, S}$	$\|\varepsilon\|_{0, S}$	$\|\varepsilon\|_1$
2	4.238	2.982	14.59	4.825	2.861	37.24
3	2.087	1.426	4.998	4.233	2.468	35.77
4	1.597	1.314	3.100	4.214	2.422	35.54
5	1.647	1.387	3.210	4.208	2.419	35.53
6	1.693	1.321	3.277	4.209	2.418	35.53
8	1.717	1.322	3.324	4.211	2.417	35.53
12	1.730	1.327	3.338	4.213	2.417	35.53
14	1.987	1.397	14.67	4.299	2.426	35.61
16	21.26	10.84	257.5	21.26	3.546	58.90

Table 4.1 Error norms of the numerical solutions for $M = N = 2$ and $R^* = 0.5$ while increasing $(L+1)$. In the table, $\varepsilon^+ = u - (u_h^*)^+$ and $(u_h^*)^+ = f_L(\tilde{D}_l)$.

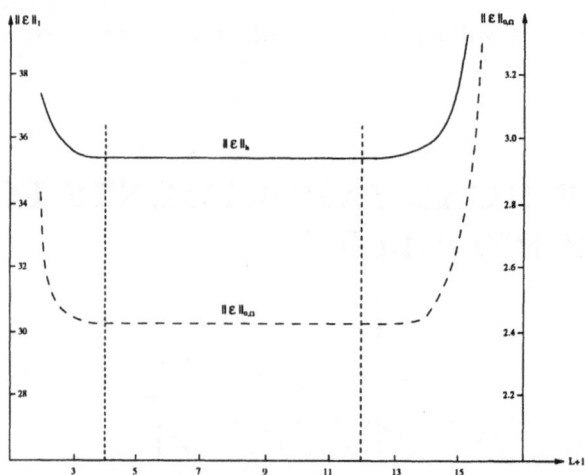

Figure 4.5 The curves of error norms versus $L+1$ for $M = N = 2$ and $R^* = 0.5$.

a semicircle with the radius $R^* = 0.5$ can be chosen to be the common boundary l_{R^*} just as is done in Li (1983a) and Thatcher (1976). So the condition (4.10) holds and the ratio $R^*/R = 1/2$. We choose the division of triangulation S_1 shown in Figure 4.4 which is symmetric to the axis y, and which in the part $(y > 0)$ is also symmetric to the straight line $y = x$. Let M and N denote the element numbers along the domain boundaries $\overline{DB}$ and $\overline{BE}$ in an equidistribution, respectively, e.g., $M = N = 2$ for Figure 4.4. Because we have made comparisons between this method and other methods in Li (1983a), we investigate only the true errors of

$L+1$	D_0	D_1	D_2	D_3	D_7	D_{11}	D_{15}
2	399.459	88.4264					
3	399.175	88.1875	18.6909				
4	399.177	88.2152	18.7371	-11.9038			
5	399.175	88.2163	18.7389	-11.8995			
6	399.174	88.2217	18.7354	-11.9013			
8	399.174	88.2203	18.7416	-11.8971	3.38050		
12	399.173	88.2210	18.7425	-11.8959	2.22900	-52.7236	
14	399.153	88.2061	16.9777	-10.2253	2.05272	-52.8649	
16	390.412	82.7628	16.6300	-10.2842	1.69082	-45.4343	412678
True	401.162	87.6559	17.2379	-8.07122	$-.08693$	$-.00318$	$-.00012$

Table 4.2   The coefficients for $M = N = 2$ and $R^* = 0.5$ while increasing $(L + 1)$.

numerical solutions.

In fact, the expansions of the true solution

$$u(r, \theta) = \sum_{n=0}^{\infty} D_n r^{(n+\frac{1}{2})} \cos((n + \frac{1}{2})\theta)$$

hold not only for $r \leq R$ but also for the whole solution domain S (see Chapter 2). The exact coefficients $D_0 - D_{34}$ have been calculated in Section 3.6 (see Table 3.2 in Chapter 3). With the use of the true solutions obtained, we will analyze the true errors of numerical solutions in three aspects as follows.

1. For the division in Figure 4.4 (i.e. $R^* = 0.5$, $M = N = 2$), the error norms have been calculated while $(L+1)$ increases, and shown in Table 4.1 and Figure 4.5. It is then seen that the error norms are almost constant when $L + 1 = 4 \approx 12$; but they become much larger when $L + 1 > 14$. It is also noteworthy that when $L + 1$ is too large, the calculated value of D_{15} is 412678 in Table 4.2; nevertheless, the true value of D_{15} is only -0.00012. Evidently, all these results coincide with the conclusions in Sections 4.2 – 4.4.

2. Let the division of S_1 in Figure 4.4 be finer by increasing the numbers M and N (i.e. by decreasing h). According to the coupling formulae (4.47), we may increase just one more singular function in S_2 while the numbers M and N of

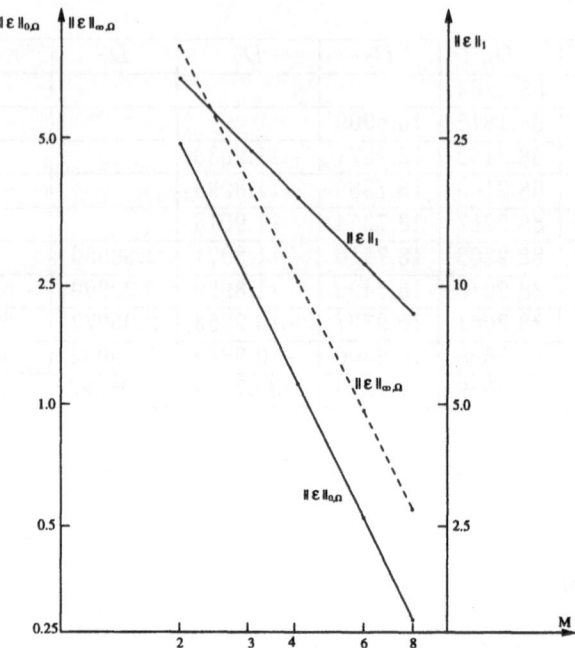

Figure 4.6 The curves of error norms $\|\varepsilon\|_h$, $\|\varepsilon\|_{0,S}$ and $\|\varepsilon\|_{0,\infty}$ versus M with $M = N$.

divisions increase to their doubles. For example, considering that $L + 1 = 4$ is good for $M = N = 2$ because of the minimal calculational work required (see Table 4.1 or Figure 4.5), we simply choose $L + 1 = 5$ for $M = N = 4$, and $L + 1 = 6$ for $M = N = 8$. The error norms of numerical solutions and approximate coefficients have been calculated in Tables 4.3 and 4.4, while the number M increases with $M = N$. The curves of error norms versus $M(= N)$ have been depicted in Figures 4.6.

Since $h = O(1/M)$, we can see from Figure 4.6 that the error norms satisfy the asymptotic formulas

$$\|\varepsilon\|_1 = O(h), \quad \|\varepsilon\|_{0,S} = O(h^2) \tag{4.106}$$

and

$$\|\varepsilon\|_{\infty,S} = \max_S |\varepsilon| = O(h^{2-\delta}), \text{ and } 0 < \delta << 1,$$

where $\varepsilon = u - u_h^*$. We notice that all these asymptotic formulas are the same as those in the FEM. Furthermore, the left asymptotic expansion in (4.106) is consistent with Corollary 4.2 concerning the analyses for the coupling strategy.

$L+1$	$\|\delta^+\|_{\infty,l_{R^*}}$	$\|\delta^+\|_{0,l_{R^*}}$	$\|\varepsilon^+\|_{1,l_{R^*}}$	$\|\varepsilon\|_{\infty,S}$	$\|\varepsilon\|_{0,S}$	$\|\varepsilon\|_1$
$M = N = 2$ $L+1 = 4$	1.597	1.314	3.100	4.214	2.422	35.53
$M = N = 3$ $L+1 = 5$	0.7274	0.5941	1.612	1.942	1.032	23.31
$M = N = 4$ $L+1 = 5$	0.4127	0.3418	1.015	1.103	0.5727	17.40
$M = N = 6$ $L+1 = 6$	0.1856	0.1565	0.5165	0.4856	0.2521	11.56
$M = N = 8$ $L+1 = 6$	0.1065	0.08936	0.3095	0.2746	0.1413	8.665

Table 4.3   Error norms of the numerical solutions for $N = M$ and $R^* = 0.5$ while increasing M.

Additional calculations have been done for the solution errors on the coupling boundary l_{R^*}. It is shown from Table 4.3 that the error norms on l_{R^*} also satisfy

$$\|\varepsilon^+\|_{0,l_{R^*}} = O(h^2), \qquad \|\varepsilon^+\|_{1,l_{R^*}} \approx O(h^\alpha), \qquad \alpha \approx 5/3$$

and

$$\|\delta^+\|_{\infty,l_{R^*}} = \max_{l_{R^*}} |\varepsilon^+| = O(h^{2-\delta}),$$

where the notations are

$$\varepsilon^+ = u - (u_h^*)^+ = u - f_L(\tilde{D}_l)$$

and

$$\|\varepsilon\|_{1,l_{R^*}} = \left(\|\varepsilon\|_{0,l_{R^*}}^2 + \left\| \frac{\partial \varepsilon}{R^* \partial \theta} \right\|_{0,l_{R^*}}^2 \right)^{\frac{1}{2}}.$$

These calculated results imply that the errors of numerical solutions are small on the common boundary l_{R^*}, where the admissible functions are not continuous.

More attention has been paid to error analyses for the approximate coefficients $\tilde{D}_l$; we can again see from Table 4.4 that

$$\delta D_l = |\tilde{D}_l - D_l| = O(h^{\alpha_l}),$$

where $\alpha_l \approx 2$, $l = 0 \sim 3$.

The precision of the first coefficient D_0 is most interesting in theoretical research and engineering application; there always appears to exist an asymptotic formula $\delta D_0 = O(h^2)$. In fact, even though $L + 1 = 6$, a good approximation for D_0 is given by $\tilde{D}_0 = 401.037$, with a small relative error 0.0003.

We should again notice that using six singular functions is good enough for matching the finest division of S_1 with $M = N = 8$. This fact also shows that the combination of RGM-FEM using the new coupling relation yields satisfactory solutions for singularity problems, even with a modest computational effort.

Division	D_0	D_1	D_2	D_3	D_4	D_5
$M = N = 2$ $L + 1 = 4$	399.177	88.2152	18.7371	-11.9038		
$M = N = 3$ $L + 1 = 5$	400.283	87.8632	17.9605	-10.3612	1.39800	
$M = N = 4$ $L + 1 = 5$	400.665	87.7679	17.6683	-9.56311	1.79985	
$M = N = 6$ $L + 1 = 6$	400.940	87.7051	17.4399	-8.82600	1.78304	-0.0940075
$M = N = 8$ $L + 1 = 6$	401.037	87.6837	17.3542	-8.51892	1.68181	0.0445709
True coeffs.	401.162	87.6559	17.2379	-8.07122	1.44027	0.331055

Table 4.4   The coefficients for $N = M$ and $R^* = 0.5$ while increasing M.

5

COMBINATIONS OF VARIOUS FEMS

Usually, the smoothness of solutions in the interior domain is higher, and the smoothness near the boundary is lower. One natural combination is to choose high order Lagrange-FEM in the interior subdomain, and lower order Lagrange-FEM in the boundary layer. This chapter displays the significance of combinations to solve not only the singularity problems but also the general elliptic equations. Also, such a combination of various FEMs is also a representative for many other kinds of combined methods using different admissible functions. The $k-$ order Lagrange FEM can be used on the solution domain S if the solution $u \in H^{k+1}(S)$. However we have to use the linear finite element method on S if $u \in H^2(S)$ only. The smoothness of the solutions of elliptic equations is, generally, different on different subdomains. For example, the solution u satisfies (Figure 5.1):

$$u \in H^2(S), \quad u \notin H^3(S) \quad \text{and} \quad u \in H^{k+1}(S_2) \quad (k \geq 2),$$

where S consists of two subdomains, S_1 and S_2. Since $u \in H^2(S)$ only, the linear FEM has to be used on S_1 with relatively small linear elements. Clearly, the k-order Lagrange FEM ought to be used on S_2 because $u \in H^{k+1}(S_2)$ $(k \geq 2)$. We hope that relatively large, k-order elements are applied in S_2 such that calculation work and storage space can be saved.

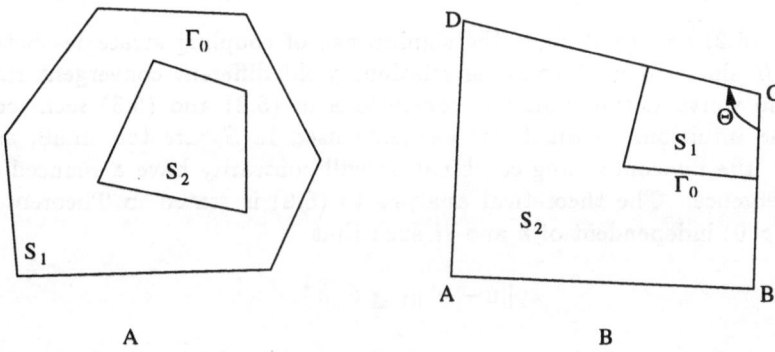

Figure 5.1 A division.

Nonconforming combinations of various finite element methods (FEMs) for solving elliptic boundary value problems are presented in this chapter. Piecewise polynomials with different orders are constrained to be continuous only at the element nodes on the common boundary Γ_0 of S_1 and S_2, on which the $m(\geq 1)-$ and $k(\geq m+1)-$ order Lagrange FEMs are used, respectively. This chapter gives a strict analysis on such combinations. The following important results will be reported.

1. The most important results are from Theorem 5.1: when linear and $k-$order elements are matching by the nonconforming constraints, the errors are

$$\|u - u_h^*\|_1 \leq C \left\{ h + H^k + \frac{h^2}{H^{3/2}} \left\| \frac{\partial u}{\partial n} \right\|_{0,\Gamma_0} + \cdots \right\}, \tag{5.1}$$

where h and H are the maximal boundary lengths of linear and k-order elements. From (5.1), the following interesting conclusions can be drawn, to yield coupling guidance in practical application.

2. For the nonconforming combination of the linear and the quadratic Lagrange FEM, i.e., $k = 2$, the asymptotic convergence rates are derived from (5.1) for $\varepsilon = u - u_h^*$

$$\|\varepsilon\|_1 = O(h), \quad \text{when } H = O(h^{\frac{1}{2}}), \tag{5.2}$$

$$\|\varepsilon\|_1 = O(h^{\frac{1}{2}}), \quad \text{when } H = O(h). \tag{5.3}$$

Eqs. (5.2) and (5.3) show the significance of coupling strategies between h and H since different coupling relations yield different convergent rates. It is also worth noting that the conclusions in (5.2) and (5.3) seem contrary to the intuition. If quadratic elements used in S_2 are too small, as $H = O(h)$, the nonconforming combination will contrarily have a reduced rate of convergence. The theoretical analysis to (5.2) is stated in Theorem 5.3 by $\exists C_0 (> 0)$ independent of h and H such that

$$\|u - u_h^*\|_1 \geq C_0 h^{\frac{1}{2}}. \tag{5.4}$$

3. The programming techniques given in Section 5.3, which are suited not only to combinations of various FEMs, but also to all kinds of nonconforming combinations. By such programming, the nonconforming constraints can be handled easily, and the resulted associated matrix is positive definite and symmetric.

4. Numerical experiments, which are carried out for solving sample problems by the combination of linear and quadratic FEMs, coincide with the conclusions

made. In particular, a computable sample has been found to verify perfectly the reduced convergence rates of order $O(h^{\frac{1}{2}})$. On the whole, this chapter emphasizes the importance of theoretical analysis which can predict numerical results and guide practical applications.

Remark 5.1 *The conclusion of Section 5.6 is based on the asymptotic analysis (i.e., h is an infinitesimal). Certainly, the term* $(h^-)^{\frac{1}{2}}|\partial u/\partial n|_{0,\Gamma_0}$ *might not be very troublesome in the case that the mesh sizes in the calculation are not very small and*

$$\left|\frac{\partial u}{\partial n}\right|_{0,\Gamma_0} << |u|_{k+1,S_2},$$

such that

$$h^{\frac{1}{2}}\left|\frac{\partial u}{\partial n}\right|_{0,\Gamma_0} \leq CH^k|u|_{k+1,S_2}.$$

Then the combined method as (5.2) with the reduced convergence rate $O(h^{1/2})$ *in Section 5.6 might be used.*

The computational results also give the same hint that the disastrous errors may not be so serious if h is not so small. Hence the users had better carry out the combinations with one more refinement, say $h/2$, and to compare the solution behavior. If the reality of numerical solutions is applicable, one may accept it. However, one must bear in mind that a *devil* is still in, but not active at this moment yet. The contents of this chapter are adapted from Li (1983b, 1991).

5.1   DESCRIPTION OF METHODS

Consider a two-dimensional problem:

$$-\frac{\partial}{\partial x}\left(\beta\frac{\partial u}{\partial x}\right) - \frac{\partial}{\partial y}\left(\beta\frac{\partial u}{\partial y}\right) = f, \quad (x,y) \in S, \tag{5.5}$$

$$u = 0, \quad (x,y) \in \Gamma, \tag{5.6}$$

where S is the convex polygon, Γ is the boundary, the functions β and f are sufficiently smooth, $\beta = \beta(x,y) \geq \beta_0 > 0$, and β_0 is a positive constants. The model problem (5.5) and (5.6) can be written in the weak-form:

$$a(u,v) = f(v), \quad \forall v \in H_0^1(S), \tag{5.7}$$

where

$$a(u,v) = \iint_S \beta(u_x v_x + u_y v_y)ds, \quad f(v) = \iint_S fvds, \tag{5.8}$$

$$H_0^1(S) = \{v|v, \ v_x, \ v_y \in L^2(S) \text{ and } v|_\Gamma = 0\}.$$

For simplicity, we first choose the combination of the linear and $k(\geq 2)$-order Lagrange FEMs. The combinations of other finite element methods can be similarly discussed (see Section 5.4).

Let the solution domain S be divided into S_1 and S_2 by a piecewise straight line Γ_0, i.e., $S = S_1 \cup S_2$ (see Figure 5.1). Also let S_1 be the "boundary layer" subdomain always on the boundary Γ such that $Meas(S_1 \cap \Gamma) \neq 0$. The linear finite element method is still used on S_1, but the $k(\geq 2)$-order Lagrange FEM will be used on S_2. Let the triangular elements in S_1 and S_2 be regular, and

$$S_1 = \bigcup_i \triangle_i^-, \quad S_2 = \bigcup_i \triangle_i^+,$$

where $\triangle_i^-$ and $\triangle_i^+$ are the elements in S_1 and S_2, respectively.

The admissible functions are written as

$$v_h = \begin{cases} V_1, & (x,y) \in S_1, \\ V_k, & (x,y) \in S_2, \end{cases} \tag{5.9}$$

where V_1 are piecewise linear interpolation functions, V_k are piecewise $k(\geq 2)$-order Lagrange interpolation polynomials, and v_h also satisfy the boundary condition (5.6).

Figure 5.2 represents a nonconforming combination of linear and three-order Lagrange FEMs, where $\overline{Q}_j^*$ and $\overline{Q}_j^{(k)}$ denote the nodes on Γ_0 of the small linear elements and large k-order elements, respectively. Let all nodes of k-order elements be just located on the nodes of linear elements. Then the admissible functions v_h on all k-order element nodes are, naturally, continuous,

$$V_1\left(\overline{Q}_j^{(k)}\right) = V_k\left(\overline{Q}_j^{(k)}\right), \quad \overline{Q}_j^{(k)} \in \Gamma_0 \text{ and } \overline{Q}_j^{(k)} \in \left\{\overline{Q}_j^*\right\}. \tag{5.10}$$

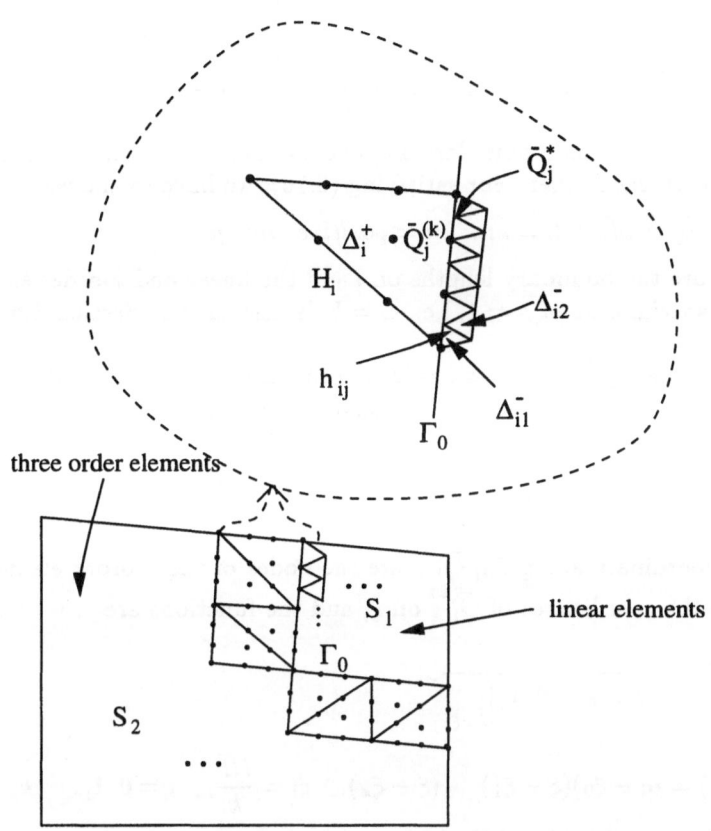

Figure 5.2 A coupling scheme for the nonconforming combination of linear and cubic Lagrange finite element methods.

Moreover, on the remaining nodes $\overline{Q}_j^*$ on Γ_0 of small linear elements, we add the continuity conditions for v_h

$$v_h\left(\overline{Q}_j^*\right) = V_1\left(\overline{Q}_j^*\right) = V_k\left(\overline{Q}_j^*\right), \qquad \overline{Q}_j^{(k)} \in \Gamma_0. \tag{5.11}$$

Let V_h^* denote the space of the admissible function v_h, which are defined by (5.9) and constrained by (5.10). Obviously, V_h^* is a nonconforming space, i.e., $V_h^* \not\subset H_0^1(S)$. Therefore, the nonconforming combined method of the linear and $k(\geq 2)$-order Lagrange FEMs is as follows.

Find an approximate solution $u_h^* \in V_h^*$ such that

$$a_h(u_h^*, v) = f(v), \qquad \forall v \in V_h^*, \tag{5.12}$$

where the bilinear form is

$$a_h(u,v) = \iint_{S_1} \beta(u_x v_x + u_y v_y)ds + \iint_{S_2} \beta(u_x v_x + u_y v_y)ds.$$

Suppose that on the common boundary Γ_0, one k-order element and one L-linear element match with each other. For satisfying (5.10), we have to choose

$$\overline{H} = L\overline{h}, \quad L = kn, \quad n = positive\ integers, \tag{5.13}$$

where $\overline{h}$ and $\overline{H}$ are the boundary lengths on Γ_0 of the linear and k-order elements, respectively. A special case of $n = 1$, i.e., $L = k$, is discussed in Section 5.6.

Now, let us give more explicit expressions for the coupling conditions (5.10). Clearly, V_k on a section, $\Gamma_0 \cap \Delta_i^+$, is a k-order Lagrange interpolation polynomial given by

$$V_k|_{\Gamma_0 \cap \Delta_i^+}(\xi) = \sum_{l=0}^{k} \alpha_l(\xi) V_k(\overline{Q}_l^{(k)}), \quad 0 \le \xi \le \overline{H},$$

where ξ is the coordinate along Γ_0, $\overline{Q}_l^{(k)}$ are the nodes of the k-order elements on $\Gamma_0 \cap \Delta_i^+$, ξ_l are the coordinates of $\overline{Q}_l^{(k)}$ on ξ, and the functions are

$$\alpha_l(\xi) = \frac{w(\xi)}{(\xi - \xi_l) \left.\frac{\partial w(\xi)}{\partial \xi}\right|_{\xi=\xi_l}},$$

$$w(\xi) = (\xi - \xi_0)(\xi - \xi_1)\ldots(\xi - \xi_k), \quad \xi_l = \frac{l\overline{H}}{k}, \quad l = 0, 1, \ldots, k.$$

Then the coupling conditions (5.11) yield

$$v_h\left(\overline{Q}_j^*\right) = V_1\left(\overline{Q}_j^*\right) = V_k\left(\overline{Q}_j^*\right) = \sum_{l=0}^{k} \alpha_l(\xi_j^*) V_k(\overline{Q}_l^{(k)}), \quad j = 0, 1, \ldots, L, \tag{5.14}$$

where $\overline{Q}_j^*$ are all nodes of the linear elements on the section $\Gamma_0 \cap \Delta_i^+$, with the coordinates $\xi_j^* = j\overline{h}$.

For $k = 2$, Eqs. (5.14) lead to

$$V_1(\overline{Q}_j^*) = A_j V_2(\overline{Q}_0^{(2)}) + B_j V_2(\overline{Q}_1^{(2)}) + C_j V_2(\overline{Q}_2^{(2)}), \quad j = 0, 1, \ldots, L, \quad L = 2n,$$

where the constants are

$$A_j = \alpha_0(\xi_j^*) = \frac{2}{\overline{H}^2}(\xi_j^* - \overline{H})\left(\xi_j^* - \frac{\overline{H}}{2}\right) = 2\left(\frac{j}{L} - 1\right)\left(\frac{j}{L} - \frac{1}{2}\right),$$

$$B_j = \alpha_1(\xi_j^*) = 4\frac{j}{L}\left(1 - \frac{j}{L}\right), \quad C_j = \alpha_2(\xi_j^*) = 2\frac{j}{L}\left(\frac{j}{L} - \frac{1}{2}\right).$$

5.2 PROGRAMMING TECHNIQUE

The nonconforming combined method (5.12) can be equivalently described in the variational problem, see Section 1.1:

Find an approximate solution, $u_h^* \in V_h^*$, such that

$$I_h(u_h^*) = \min_{v_h \in V_h^*} I_h(v_h),$$

where a quadratic functional

$$I_h(v_h) = \frac{1}{2} a_h(v_h, v_h) - f(v_h), \qquad v_h \in V_h^*.$$

Note that when the values $V_1\left(\overline{Q}_j^*\right)$ in the functional $I_h(v_h)$ have been substituted by $V_k\left(\overline{Q}_l^{(k)}\right)$, the function $I_h(v_h)$ is still quadratic with respect to $V_k\left(Q_l^{(k)}\right)$ and $V_1\left(Q_j^*\right)(Q_j^* \notin \Gamma_0)$, i.e.,

$$I_h(v_h) = \frac{1}{2}\mathbf{x}^T \mathbf{A} \mathbf{x} - \mathbf{x}^T b,$$

where x is the unknown vector with the components $V_k\left(Q_j^{(k)}\right)$ and $V_1\left(Q_j^*\right)(Q_j^* \notin \Gamma_0)$, $Q_j^{(k)}$ are k-order element nodes on S_2, Q_j^* are linear element nodes on S_1, b is a known vector, and the constant matrix $\mathbf{A}$ is positive definite, symmetric and sparse. Therefore, the system of linear algebraic equations is obtained

$$\mathbf{A}\mathbf{x} = \mathbf{b} \tag{5.15}$$

by minimizing $I_h(v_h)$ under $v_h \in V_h^*$.

In practice, we develop a programming technique for easily constructing the concrete, difference equations of (5.15), which may also be applied to other kinds of nonconforming combinations, such as combinations of RGM-FEM, RGM-FDM, etc.

We may regard

$$I_h(v_h) = I_h \left(V_k(Q_j^{(k)})(Q_j^{(k)} \notin \Gamma_0), V_k \underbrace{\left(\overline{Q}_j^*\right), V_1(\overline{Q}_j^*)}_{through\ (5.14)}, V_1(Q_j^*)(Q_j^* \notin \Gamma_0) \right), \tag{5.16}$$

as a composite function with respect to $V_k(\overline{Q}_j^{(k)})$, where the element nodes $\overline{Q}_j^{(k)} \in \Gamma_0$ and $\overline{Q}_j^* \in \Gamma_0$. Then the programming technique consists of two steps.

Step I. We form the difference equations on element nodes:

$$L_1(Q_j^*) = \frac{\partial I_h(v_h)}{\partial V_1(Q_j^*)}\bigg|_{\text{without (5.14)}} = 0, \quad Q_j^* \in S_1, \tag{5.17}$$

and

$$L_k(Q_j^*) = \frac{\partial I_h(v_h)}{\partial V_k(Q_j^{(k)})}\bigg|_{\text{without (5.14)}} = 0, \quad Q_j^{(k)} \in S_2, \tag{5.18}$$

where Q_j^* and $Q_j^{(k)}$ are all nodes of linear and k- order elements, respectively. In fact, the procedure of (5.17) and (5.18) is exactly the same as in the conventional FEM.

Step II. Having imposed the constraint conditions (5.14) on v_h, we have to modify the above difference equations by the following cases:

Case 1. On the element nodes $\overline{Q_j^*} \in (\Gamma_0 \cap S_1)$, the values $V_1(\overline{Q_j^*})$ are always substituted by (5.14) so that $V_1(\overline{Q_j^*})$ are no longer regarded as unknown variables.

Case 2. On $Q_j^* \notin \Gamma_0$ and $Q_j^{(k)} \notin \Gamma_0$, Eqs. (5.17) and (5.18), in which Case 1 will be applied if there exist the values $V_1(Q_j^*)$, still serve as the difference equations for the combinations.

Case 3. On the quadratic element nodes $\overline{Q_j^{(k)}} \in (\Gamma_0 \cap S_2)$, the difference equations for the combinations are obtained by applying (5.16), (5.14), and the differential chain rule:

$$\begin{aligned}
0 &= \frac{\partial I_h(v_h)}{\partial V_k(\overline{Q_j^{(k)}})}\bigg|_{\text{with (5.14)}} \\
&= \frac{\partial I_h(v_h)}{\partial V_k(\overline{Q_j^{(k)}})}\bigg|_{\text{without (5.14)}} + \sum_l \frac{\partial I_h(v_h)}{\partial V_l(\overline{Q_l^*})}\bigg|_{\text{without (5.14)}} \times \frac{dV_l(\overline{Q_l^*})}{dV_k(\overline{Q_j^{(k)}})} \\
&= L_k(\overline{Q_j^{(k)}}) + \sum_l L_1(\overline{Q_l^*})\alpha_j(\xi_l^*),
\end{aligned}$$

when L_1 and L_k have already been formed in (5.17) and (5.18). Of course, Case 1 will again be applied for $V_1(\overline{Q_j^*})$.

Consequently, all those equations in Cases 2 and 3 serve as the difference equations in (5.15). The matrix structure of A in (5.15) is only a little different from that in the conventional FEMs because the constraint conditions (5.14) are related to the solutions $V_k(\overline{Q}_l^{(k)})$ only at the element nodes $\overline{Q}_l^{(k)}$ on Γ_0. Therefore, the linear algebraic equations (5.15) can be easily solved by the Gaussian elimination method in Golub and Loan (1989).

5.3 ERROR BOUNDS AND COUPLING STRATEGY

We give error bounds of the numerical solutions in the following theorem, whose proof is deferred to the next section.

Theorem 5.1 *Let the bilinear form $a_h(\cdot, \cdot)$ in (5.12) be uniformly V_h^*-elliptic, i.e., there exists a positive constant α independent of h and v such that*

$$\alpha\|v\|_1^2 \le a_h(v, v), \quad \forall v \in V_h^*. \tag{5.19}$$

Moreover, suppose that the following inverse assumptions of triangulation hold:

$$H \Big/ \min_i(H_i) \le C, \quad h \Big/ \min_{i,j}(h_{ij}) \le C, \tag{5.20}$$

where h_{ij} and H_i are the maximal boundary lengths of linear and k-order elements on Γ_0, respectively, h and H are the maximal boundary lengths of the elements on S_1 and S_2, respectively, and C is a constant independent of h, H and u. With these assumptions, when $u \in H^2(S_1)$, $u \in H^{k+1}(S_2)$, and $u \in H^{k+1}(\Gamma_0)$, $k \ge 2$, then

$$\|u - u_h^*\|_1 \le C \left\{ h|u|_{2,S_1} + H^k|u|_{k+1,S_2} + \frac{h^2}{H^{\frac{3}{2}}} \left|\frac{\partial u}{\partial n}\right|_{0,\Gamma_0} \right.$$
$$\left. + h^{\frac{3}{2}}|u|_{2,\Gamma_0} + \left(\frac{1}{\sqrt{h}}H^{k+1} + h^{\frac{3}{2}}H^{k-1}\right)|u|_{k+1,\Gamma_0} \right\}. \tag{5.21}$$

In the right braces of (5.21), the first and second terms are the errors resulting from the linear and $k(\ge 2)$-order Lagrange FEMs respectively, and the other terms are the errors resulting from the nonconforming coupling on Γ_0.

From Theorem 5.1, we have the following corollary.

Corollary 5.1 *Suppose that the conditions in Theorem 5.1 hold and $H = O(h)$. Then the numerical solutions by the nonconforming combination of the linear and $k(\geq 2)$-order Lagrange FEMs have the error bounds:*

$$\|u - u_h^*\|_1 \leq Ch^{\frac{1}{2}}.$$

Corollary 5.1 coincides with the analysis in Section 5.6 later where $H = kh$. Now, we try to find more efficient combinations.

If we choose the boundary length H as

$$C_0 h^{\frac{2}{3}} \leq H \leq Ch^{\frac{1}{2}}, \tag{5.22}$$

where C_0 is a positive constant independent of h, H and u, then Eq. (5.21) leads to

$$\|u - u_h^*\|_1 \leq C(h + H^k). \tag{5.23}$$

This means that under the condition (5.22), the effect of nonconforming coupling does not cause a reduced rate of convergence.

Furthermore, the mesh spacing on S_1 and S_2 should be chosen as optimal coupling:

$$H = O(h^{1/k}). \tag{5.24}$$

Then Eq. (5.23) leads to

$$\|u - u_h^*\|_1 = O(h). \tag{5.25}$$

However, both (5.22) and (5.24) can simultaneously hold for an infinitesimal h if and only if the integer $k = 2$. Therefore we have the following corollary.

Corollary 5.2 *Suppose that the conditions in Theorem 5.1 hold true. Then the necessary and sufficient conditions of an optimal rate, (5.25), of convergence are*

$$k = 2 \quad \text{and} \quad H = O(h^{\frac{1}{2}}),$$

among all combinations (5.12) of the linear and $k(\geq 2)$-order Lagrange FEMs.

From Corollaries 5.1 and 5.2, the asymptotic convergence rates for the combination of the linear and quadratic Lagrange FEMs are

$$\|u - u_h^*\|_1 = O(h^{\frac{1}{2}}) \quad \text{when} \quad H = O(h), \tag{5.26}$$

and

$$\|u - u_h^*\|_1 = O(h) \quad \text{when} \quad H = O(h^{\frac{1}{2}}). \tag{5.27}$$

This shows the significance of the coupling strategies between h and H: different coupling strategies yield different convergence rates.

From (5.26) and (5.27) we also learn the following facts, which seem contrary to intuition. If quadratic elements used in S_2 are too small, like $H = O(h)$, the combination will contrarily have a reduced rate of convergence resulting from the error bounds, $O\left(\frac{h^2}{H^{\frac{3}{2}}} \left|\frac{\partial u}{\partial n}\right|_{0,\Gamma_0}\right)$, in (5.21), a disastrous effect of the nonconforming coupling.

Summarily, the comparison between (5.26) and (5.27) suggest that we should deliberately choose a large length H such that $H = O(h^{\frac{1}{2}})$.

We define a coupling optimal rate of convergence (ORC), by which the errors $\|\varepsilon\|_1$ of numerical solutions have an asymptotically optimal rate of convergence, (5.25). On the basis of Corollary 5.2, we have found a unique coupling ORC: the combination of the linear and quadratic Lagrange FEMs with the coupling, $H = O(h^{\frac{1}{2}})$. We refer to this combination as coupling ORC.

Let us view the assumptions in Theorem 5.1. The uniformly V_h^*- elliptic inequality (5.19) is proven in Li (1991); the inverse assumption (5.20) is easily satisfied provided that the finite elements are quasiuniform and the common boundary Γ_0 consists of piecewise straight lines.

It is worthwhile to point out that we can prove, as in Li (1986, 1990), and Li and Liang (1983), that the condition number of the coefficient matrix $\mathbf{A}$ in (5.15) has bounds, $O(h_{\min}^{-2})$, where $h_{\min}$ is the minimal boundary length of all elements in both S_1 and S_2. These bounds are the same as those in FEM in Strang and Fix (1973). Hence, the stability of numerical solutions obtained by the combinations would not be of asymptotically worse orders than that obtained by FEM.

Remark 5.2 *Although this chapter focus on triangular elements, the combined approaches and analyses can be easily extended to quadrilateral and isoparametric elements. In particular, when Γ_0 consists of piecewise straight lines, we can similarly prove that the combination of bilinear and biquadratic Lagrange FEMs of quadrilateral elements as $H = O(h^{\frac{1}{2}})$ is also a coupling ORC, with an optimal rate of convergence*

$$\|u - \widehat{u}_h^*\|_1 = O(h),$$

where $\hat{u}_h^*$ are the numerical solutions obtained, and H and h are maximal boundary lengths of bilinear and biquadratic elements, respectively.

5.4 COMBINATIONS OF $M(\geq 1)-$ AND $K(> M)-$ORDER LAGRANGE FEMS

In this section, we consider the combinations of all Lagrange FEMs. Without loss of generality, we combine two different methods: the $m(\geq 1)$-order and $k(> m)$-order Lagrange FEMs used in S_1 and S_2, respectively. Note that $m = 1$ has been discussed in Sections 5.1 and 5.3. The admissible functions are chosen:

$$v_h = \begin{cases} V_m, & (x, y) \in S_1, \\ V_k, & (x, y) \in S_2, \end{cases} \tag{5.28}$$

where V_m are piecewise m-order Lagrange interpolation polynomials. Let all nodes of k-order elements be just located at the nodes of m order elements, as in (5.10). In addition, we constrain the admissible functions v_h such that

$$v_h(\tilde{Q}_j) = V_k(\tilde{Q}_j) = V_m(\tilde{Q}_j), \quad \tilde{Q}_j \in \Gamma_0, \tag{5.29}$$

where $\tilde{Q}_j$ are the remaining nodes, on Γ_0, of m-order elements.

Let V_h^* still denote the space of the admissible function v_h (5.28) satisfying (5.29) and (5.6), Eq. (5.12) may, in the form, represent the combination of m and k-order FEMs. Then we have the following theorem.

Theorem 5.2 *Suppose that Eqs. (5.19) and (5.20) hold, where h and h_{ij} are the maximal boundary lengths of $m(\geq 1)$-order element on S_1 and Γ_0, respectively. Also assume $u \in H^{m+1}(S_1)$, $u \in H^{k+1}(S_2)$ and $u \in H^{k+1}(\Gamma_0)$. Then the solutions by the combination of the $m(\geq 1)$- and $k(> m)$-order Lagrange FEMs have the error bounds*

$$\|u - u_h^*\|_1 \leq C \left\{ h^m |u|_{m+1,S_1} + H^k |u|_{k+1,S_2} + \frac{h^{m+1}}{H^{m+1/2}} \left| \frac{\partial u}{\partial n} \right|_{0,\Gamma_0} \tag{5.30} \right.$$

$$\left. + h^{m+1/2} |u|_{m+1,\Gamma_0} + \left(h^{m+1/2} H^{k-m} + \frac{H^{k+1}}{\sqrt{h}} \right) |u|_{k+1,\Gamma_0} \right\}.$$

The proof of Theorem 5.2 is given in Li (1991). We have the following corollary from Theorem 5.2.

Corollary 5.3 *Suppose that all conditions in Theorem 5.2 hold. Then when $H = O(h)$,*

$$\|u - u_h^*\|_1 = O(h^{\frac{1}{2}}),$$

to lead to the same reduced rate of convergence as in Corollary 5.1.

Corollary 5.3 implies that no matter which Lagrange FEMs are combined, we can obtain only the same asymptotically worse order, $O(h^{\frac{1}{2}})$, of convergence rates as long as $H = O(h)$. This disaster is owing to the error bounds

$$O\left(\frac{h^{m+1}}{H^{m+1/2}}\left|\frac{\partial u}{\partial n}\right|_{0,\Gamma_0}\right)$$

in (5.30) resulting from the nonconforming coupling.

Next, we learn from (5.30) that the error bounds

$$\|u - u_h^*\|_1 \leq C(h^m + H^k), \qquad m \leq 1, \ k > m$$

hold if and only if

$$C_0 h^{1/(m+\frac{1}{2})} \leq H \leq C h^{\frac{1}{2}}, \tag{5.31}$$

where C_0 is a constant independent of h, H and u. However, only one integer, $m = 1$, can satisfy (5.31) while $h \to 0$. By recalling the proof of Corollary 5.2, we have the following corollary.

Corollary 5.4 *Suppose that all conditions in Theorem 5.2 hold. Then, among all nonconforming combinations of different Lagrange FEMs, we can find only one coupling with an optimal rate of convergence, i.e., coupling ORC, $m = 1$, $k = 2$, and $H = O(h^{\frac{1}{2}})$.*

5.5   PROOF OF THEOREM 5.1

First, we need several lemmas. By similar arguments in Theorem 4.4 in Chapter 4 we obtain the following lemma.

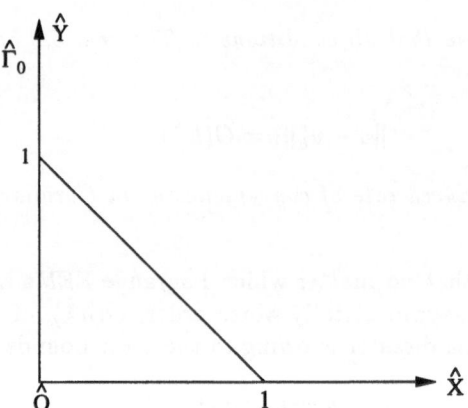

Figure 5.3 The reference element $\widehat{\triangle}$.

Lemma 5.1 *Let (5.19) hold, then*

$$\|u - u_h^*\|_1 \leq C \left\{ \inf_{v_h \in V_h^*} \|u - v_h\|_1 + + \left[\int_{\Gamma_0} \left(\beta \frac{\partial u}{\partial n} \right)^2 d\ell \right]^{\frac{1}{2}} \sup_{w_h \in V_h^*} \frac{\left[\int_{\Gamma_0} (w_h^+ - w_h^-)^2 d\ell \right]^{\frac{1}{2}}}{\|w_h\|_1} \right\},$$

where $w_h^- = w_h|_{\Gamma_0 \cap S_1}$, *and* $w_h^+ = w_h|_{\Gamma_0 \cap S_2}$.

Lemma 5.2 *Let* $\triangle_i$ *be a regular triangular element on* Γ_0, *and* P_k *be the space of k-order polynomials, then*

$$|v|_{1,\Gamma_0 \cap \triangle_i} \leq C \frac{1}{\sqrt{h_i}} |v|_{1,\triangle_i}, \quad \forall v \in P_k, \tag{5.32}$$

where $h_i = Meas(\Gamma_0 \cap \triangle_i)$, *and C is a bounded constant independent of h.*

Proof. There exists an invertible affine transformation:

$$\widehat{z} \in \widehat{\triangle} \rightarrow g(\widehat{z}) = B_i \widehat{z} + b_i \in \triangle_i,$$

where the vector $z = (x, y)^T$, B_i is a 2×2 constant matrix, b_i is a two-dimensional vector, and $\widehat{\triangle}(\widehat{x} \geq 0, \widehat{y} \geq 0,$ and $\widehat{x} + \widehat{y} \leq 1)$ is a reference element shown in Figure 5.3. The k-order polynomial $v \in P_k$ under the inverse transformation is

$$v = \sum_{i+j=0}^{k} a_{ij} x^i y^j \quad \text{on } \triangle_i \rightarrow \widehat{v} = \sum_{i+j=0}^{k} \widehat{a}_{ij} \widehat{x}^i \widehat{y}^j \quad \text{on } \widehat{\triangle},$$

where a_{ij} and $\hat{a}_{ij}$ are constants. We can find from Ciarlet (1978, p.118) that

$$|v|_{1,\Gamma_0 \cap \Delta_i} \leq C\|B_i^{-1}\|\sqrt{Meas\ (\Gamma_0 \cap \Delta_i)}|\hat{v}|_{1,\hat{\gamma} \cap \hat{\Delta}} \leq C\frac{1}{\sqrt{h_i}}|\hat{v}|_{1,\hat{\gamma} \cap \hat{\Delta}}, \quad (5.33)$$

where $\|B_i^{-1}\| = O(h_i^{-1})$, $\hat{Y}$ is the vertical axis in the reference coordinates $\hat{X}\hat{O}\hat{Y}$ in Figure 5.3, and C is a bounded constant independent of h_i.

Since

$$|\hat{v}|_{1,\hat{\gamma} \cap \hat{\Delta}} = \left(\int_0^1 \left[\sum_{j=1}^k j\hat{a}_{0j}\hat{y}^{j-1}\right]^2 d\hat{y}\right)^{\frac{1}{2}},$$

$|\hat{v}|_{1,\hat{\gamma} \cap \hat{\Delta}}$ is, in fact, a norm over a k-dimensional space of the variables $\hat{a}_{0j}$, $j = 1, 2, \ldots, k$. All norms over the finite-dimensional space are equivalent to each other so that there exists a bounded constant C such that

$$|\hat{v}|_{1,\hat{\gamma} \cap \hat{\Delta}} \leq C\left(\sum_{j=1}^k \hat{a}_{0j}^2\right)^{\frac{1}{2}}. \quad (5.34)$$

Similarly, we have

$$\left(\sum_{i+j=1}^k \hat{a}_{ij}^2\right)^{\frac{1}{2}} \leq C|\hat{v}|_{1,\hat{\Delta}}.$$

Now, we obtain,

$$|\hat{v}|_{1,\hat{\gamma} \cap \hat{\Delta}} \leq C\left(\sum_{j=1}^k \hat{a}_{0j}^2\right)^{\frac{1}{2}} \leq C\left(\sum_{i+j=1}^k \hat{a}_{ij}^2\right)^{\frac{1}{2}} \leq C|\hat{v}|_{1,\hat{\Delta}}; \quad (5.35)$$

and from Ciarlet (1978, p.118) again

$$|\hat{v}|_{1,\hat{\Delta}} \leq C\|B_i\|\ |det(B_i)|^{-1/2}\ |v|_{1,\Delta_i} \leq C|v|_{1,\Delta_i}, \quad (5.36)$$

where $\|B_i\| = O(h_i)$ and $|det(B_i)| = O(h_i^2)$. Therefore, Eq. (5.32) follows from (5.33)-(5.36). ∎

Similarly, we can prove the following lemma.

Lemma 5.3 *If the conditions on Lemma 5.2 hold, then*

$$|v|_{2,\Gamma_0 \cap \Delta_i} \leq C \frac{1}{h_i} |v|_{1,\Gamma_0 \cap \Delta_i}, \quad \forall v \in P_k. \tag{5.37}$$

Let u_h on S_1 be the piecewise linear interpolation function of u, and u_h on S_2 be the $k(\geq 2)$-order Lagrange interpolation polynomial of u. Obviously, $u_h \notin V_h^*$. Then it is necessary to construct an auxiliary function $\tilde{u}_h \in V_h^*$ for the error estimates (see Lemma 5.1). The auxiliary function $\tilde{u}_h$ is constructed as follows:

1. $\tilde{u}_h|_{S_2} = u_h$;

2. $\tilde{u}_h$ on S_1 is also a piecewise linear interpolation function, but the values of $\tilde{u}_h$ at the nodes Q_j of the linear elements are, particularly, chosen as:

$$\tilde{u}_h = \begin{cases} u_h^+, & Q_j \in \Gamma_0, \\ u(Q_j), & Q_j \notin \Gamma_0, \end{cases} \tag{5.38}$$

where $u_h^+ = u_h|_{\Gamma_0 \cap S_2}$. Here $\tilde{u}_h$ satisfies the coupling conditions (5.10) so that $\tilde{u}_h \in V_h^*$.

Lemma 5.4 *Let u_h and $\tilde{u}_h$ be the functions defined above, and the linear elements on Γ_0 be quasiuniform. Then*

$$\|u_h - \tilde{u}_h\|_1 \leq C \frac{1}{[\min_{i,j} h_{ij}]^{\frac{1}{2}}} \|u_h^- - \tilde{u}_h^-\|_{0,\Gamma_0},$$

where $u_h^- = u_h|_{\Gamma_0 \cap S_1}$, h_{ij} are the maximal boundary lengths of the linear elements Δ_{ij}^- on Γ_0 in Figure 5.2, and C is a bounded constant independent of h_{ij}.

Proof. A pair of linear elements, Δ_{i1}^- and Δ_{i2}^-, in Figure 5.2 can be transformed from another pair of reference elements in Figure 5.4 by the following affine transformations:

$$\hat{z} \in \hat{\Delta}_1 \longrightarrow g_1(\hat{z}) = B_{i1}\hat{z} + b_{i1} \in \Delta_{i1}^-,$$
$$\hat{z} \in \hat{\Delta}_2 \longrightarrow g_2(\hat{z}) = B_{i2}\hat{z} + b_{i2} \in \Delta_{i2}^-,$$

where the vector $z = (x, y)^T$, B_{i1} and B_{i2} are 2×2 constant matrices, and g_1, g_2, b_{i1} and b_{i2} are two-dimensional vectors.

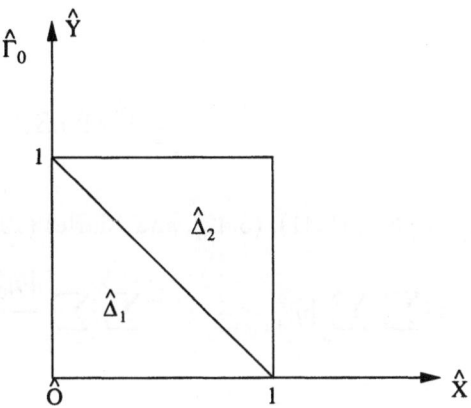

Figure 5.4 The reference elements $\widehat{\triangle}_1$ and $\widehat{\triangle}_2$.

Let $\eta = u_h - \tilde{u}_h$ and $\widehat{\eta}(\widehat{z}) = \eta(z)$. We note that η attached to $\partial\widehat{\triangle}_j$ do not vanish only at the linear elements on Γ_0. Then, from Ciarlet (1978, p.118) we have

$$\|\eta\|_1^2 = \sum_i \sum_{j=1,2} \|\eta\|_{1,\triangle_{\overline{ij}}}^2 \le C \sum_i \sum_{j=1,2} \left\{ h_{ij}^2 |\widehat{\eta}|_{0,\widehat{\triangle}_j}^2 + |\widehat{\eta}|_{1,\widehat{\triangle}_j}^2 \right\}. \tag{5.39}$$

Moreover, the reference function, $\widehat{\eta}$ on $\widehat{\triangle}_i$, is a piecewise linear function:

$$\widehat{\eta} = \widehat{\eta}_i = \widehat{a}_i + \widehat{b}_i \widehat{x} + \widehat{c}_i \widehat{y}, \quad (\widehat{x}, \widehat{y}) \in \widehat{\triangle}_i, \quad i = 1, 2,$$

with the constants, $\widehat{a}_i$, $\widehat{b}_i$ and $\widehat{c}_i$. Also $\widehat{\eta}$ satisfies

$$\widehat{\eta}_i = 0, \quad \text{when } \widehat{X} = 1. \tag{5.40}$$

Then $\widehat{\eta}$ becomes (Figure 5.4)

$$\begin{aligned} \widehat{\eta} = \widehat{\eta}_1 = \overline{a}_1(1 - \widehat{x}) + \overline{c}_1 \widehat{y}, & \quad (\widehat{x}, \widehat{y}) \in \widehat{\triangle}_1, \\ \widehat{\eta} = \widehat{\eta}_2 = \overline{a}_2(1 - \widehat{x}), & \quad (\widehat{x}, \widehat{y}) \in \widehat{\triangle}_2, \end{aligned}$$

with the constants $\overline{a}_1$, $\overline{a}_2$ and $\overline{c}_1 = \overline{a}_2 - \overline{a}_1$.

Similarly to the arguments in Lemma 5.2, we find

$$\sum_{i=1}^2 |\widehat{\eta}|_{1,\widehat{\triangle}_j}^2 \le C(\overline{a}_1^2 + \overline{a}_2^2) \le C \sum_{i=1}^2 |\widehat{\eta}|_{0,\widehat{\Gamma}_0 \cap \widehat{\triangle}_j}^2, \tag{5.41}$$

and

$$\sum_{j=1}^{2} |\hat{\eta}|^2_{0,\hat{\Delta}_j} \leq C(\bar{a}_1^2 + \bar{a}_2^2) \leq C \sum_{j=1}^{2} |\hat{\eta}|^2_{0,\hat{\Gamma}_0 \cap \hat{\Delta}_j}. \tag{5.42}$$

Finally, we can see from (5.39), (5.41), (5.42), and Ciarlet (1978, p.118) that

$$\begin{aligned} \|\eta\|_1^2 &\leq C \sum_i \sum_{j=1,2} |\hat{\eta}|^2_{0,\hat{\Gamma}_0 \cap \hat{\Delta}_j} \leq C \sum_i \sum_{j=1,2} \frac{|\eta|^2_{0,\Gamma_0 \cap \Delta_{ij}}}{h_{ij}} \\ &\leq C \frac{1}{\min_{i,j} h_{ij}} |u_h^- - \tilde{u}_h^-|^2_{0,\Gamma_0}, \end{aligned}$$

where C is the bounded constants independent of h_{ij}. ∎

Remark 5.3 *When the pair of linear elements is at the bottom or the top of the k-order element (Figure 5.2), $\hat{\eta}$ will satisfy (5.40) and one more condition:*

$$\hat{\eta}(0,0) = 0 \quad or \quad \hat{\eta}(0,1) = 0.$$

The conclusions in (5.41) and (5.42) still hold by a similar proof.

Now we prove Theorem 5.1 by estimating all terms in Lemma 5.1, with the help of Lemmas 5.2-5.4. We have

$$\int_{\Gamma_0} (w_h^+ - w_h^-)^2 d\ell \leq C \sum_i (h_{ij})^4 |w_h^+|^2_{2,\Gamma_0 \cap \Delta_i^+},$$

where Δ_i^+ are the k-order elements on Γ_0. Then we see from Lemmas 5.2 and 5.3 and (5.20) that

$$\begin{aligned} |w_h^+|_{2,\Gamma_0 \cap \Delta_i^+} &\leq C \frac{1}{Meas \ (\Gamma_0 \cap \Delta_i^+)} |w_h^+|_{1,\Gamma_0 \cap \Delta_i^+} \\ &\leq C \frac{1}{[Meas \ (\Gamma_0 \cap \Delta_i^+)]^{\frac{3}{2}}} |w_h^+|_{1,\Delta_i^+} \\ &\leq C \frac{1}{H^{\frac{3}{2}}} |w_h^+|_{1,\Delta_i^+}. \end{aligned} \tag{5.43}$$

Therefore, we obtain

$$\int_{\Gamma_0} (w_h^+ - w_h^-)^2 d\ell \leq C \frac{h^4}{H^3} \sum_i |w_h^+|^2_{1,\Delta_i^+} \leq C \frac{h^4}{H^3} \|w_h\|_1^2. \tag{5.44}$$

Since the auxiliary function $\tilde{u}_h \in V_h^*$ (see (5.38)), we have

$$\inf_{v_h \in V_h^*} \|u - v_h\|_1 \le \|u - \tilde{u}_h\|_1 \le \|u - u_h\|_1 + \|u_h - \tilde{u}_h\|_1, \tag{5.45}$$

where u_h on S_1 is the piecewise linear interpolation function of u, and u_h on S_2 is the k-order Lagrange interpolation polynomial of u. Obviously,

$$\|u - u_h\|_1 \le C \left\{ h|u|_{2,S_1} + H^k|u|_{k+1,S_2} \right\}.$$

On the other hand, we find from Lemma 5.4 and (5.20) that

$$\|u_h - \tilde{u}_h\|_1 \le C \frac{1}{h^{\frac{1}{2}}} \|u_h^- - \tilde{u}_h^-\|_{0,\Gamma_0}$$

$$\le C \frac{1}{h^{\frac{1}{2}}} \left\{ \|u - u_h^-\|_{0,\Gamma_0} + \|u - u_h^+\|_{0,\Gamma_0} + \|u_h^+ - \tilde{u}_h^-\|_{0,\Gamma_0} \right\}$$

$$\le C \frac{1}{h^{\frac{1}{2}}} \left\{ h^2|u|_{2,\Gamma_0} + H^{k+1}|u|_{k+1,\Gamma_0} + \|u_h^+ - \tilde{u}_h^-\|_{0,\Gamma_0} \right\}$$

$$\le C \left\{ h^{\frac{3}{2}}|u|_{2,\Gamma_0} + \left[\frac{1}{\sqrt{h}} H^{k+1} + h^{\frac{3}{2}} H^{k-1} \right] |u|_{k+1,\Gamma_0} \right\}. \tag{5.46}$$

In the last step in (5.46), we have used the following inequality:

$$\|u_h^+ - \tilde{u}_h^-\|_{0,\Gamma_0} = \left(\sum_i \|u_h^+ - \tilde{u}_h^-\|_{0,\Gamma_0 \cap \Delta_i^+}^2 \right)^{\frac{1}{2}} \le Ch^2 \left(\sum_i |u_h^+|_{2,\Gamma_0 \cap \Delta_i^+}^2 \right)^{\frac{1}{2}}$$

$$\le Ch^2 \left[\left(\sum_i |u|_{2,\Gamma_0 \cap \Delta_i^+}^2 \right)^{\frac{1}{2}} + \left(\sum_i |u - u_h^+|_{2,\Gamma_0 \cap \Delta_i^+}^2 \right)^{\frac{1}{2}} \right]$$

$$\le Ch^2 (|u|_{2,\Gamma_0} + H^{k-1}|u|_{k+1,\Gamma_0}).$$

Consequently, Eq. (5.21) is finally obtained from Lemma 5.1 and (5.44)-(5.46). This completes the proof of Theorem 5.1. ∎

5.6 REDUCED RATE OF CONVERGENCE

In this section we also consider the elliptic boundary value problem (5.5) and (5.6), and combine the linear finite element method and the $k(\ge 2)-$ order Lagrange FEM in different solution subdomains with the element nodes on the common boundary coinciding with each other, i.e., $L = K$ in (5.13). This will lead to the reduced convergence rates $O(h^{\frac{1}{2}})$. However, the existence of the reduced convergence rate $O(h^{\frac{1}{2}})$ relys on (5.4), which will be proven in Theorem 5.3 below.

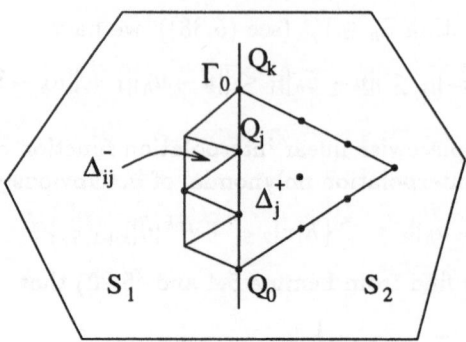

Figure 5.5   A combination of the linear and cubic Lagrange finite element methods.

Note that the admissible functions chosen in combined methods are frequently not continuous along the common boundary of different subdomains. Theorem 5.3 suggests that if we are not careful in coupling, we may have distressing results.

The solution domain, S, is divided into two parts: S_1 and S_2, i.e., $S = S_1 \cup S_2$. The common boundary of S_1 and S_2 is a piecewise straight line, Γ_0, (Figure 5.1). The linear FEM is used in S_1, and the $k(\geq 2)-$ order **Lagrange FEM** is used in S_2. Moreover, $S_1 = \cup_i \triangle_i^-$, $S_2 = \cup_i \triangle_i^+$, where $\triangle_i^-$ and $\triangle_i^+$, are quasiuniform triangular elements in S_1 and S_2, respectively.

We suppose that the nodes of the triangular elements, $\triangle_i^-$ and $\triangle_i^+$ on Γ_0 just coincide with each other. Figure 5.5 represents a combination of the linear and cubic Lagrange FEMs. The admissible functions for such a combined method can be expressed as:

$$v_h = \begin{cases} V_1, & (x,y) \in S_1, \\ V_2, & (x,y) \in S_2, \end{cases}$$

where V_1 are piecewise linear interpolation functions, V_2 are piecewise $k(\geq 2)-$ order Lagrange interpolation polynomials, and v_h satisfy $v_h|_\Gamma = 0$.

The admissible functions, v_h, are continuous in subdomains, S_1 and S_2. However, on the common boundary, Γ_0, of S_1 and S_2, v_h are continuous only at the common nodes, Q_j, of the triangular elements, as shown in Figure 5.5,

$$V_1(Q_j) = V_2(Q_j), \qquad Q_j \in \Gamma_0.$$

Therefore, the combination of the linear and $k(\geq 2)-$ order Lagrange finite elements is nonconforming and is called the nonconforming combination. The above admissible function space is denoted as V_h^0, and

$$V_h^0 \not\subset H^1(S).$$

The combination of using the linear and $k(\geq 2)-$ order Lagrange FEMs is:

To find a solution, $u_h^* \in V_h^0$ such that:

$$a_h(u_h^*, v) = f(v) \quad \forall v \in V_h^0, \tag{5.47}$$

where

$$a_h(u, v) = \sum_i \iint_{\triangle_i^-} \beta(u_x v_x + u_y v_y)ds + \sum_i \iint_{\triangle_i^+} \beta(u_x v_x + u_y v_y)ds,$$

and

$$f(v) = \sum_i \iint_{\triangle_i^-} fvds + \sum_i \iint_{\triangle_i^+} fvds.$$

First, we prove an important theorem.

Theorem 5.3 *Suppose that the linear and $k(\geq 2)-$ order Lagrange FEMs are used in S_1 and S_2, respectively, the nodes of regular triangular elements on Γ_0 just coincide with each other, and the inverse assumption of the triangulation in S_1 holds:*

$$h^- \Big/ Meas(\Gamma_0 \cap \triangle_i^-) \leq C. \tag{5.48}$$

Moreover, assume that $u \in H^2(S)$ and there exists a closed segment Γ_0' of Γ_0 such that:

$$\frac{\partial u}{\partial n}\Big|_{\Gamma_0'} \; always \; \neq 0 \quad and \quad Meas(\Gamma_0') > 0. \tag{5.49}$$

With these assumptions the following inequality always holds:

$$\|u - u_h^*\|_1 \geq C_0(h^-)^{\frac{1}{2}}, \tag{5.50}$$

where $C_0(> 0)$ is a constant, independent of h^- and h^+.

Proof. Strang and Fix (1973, p.179) have proved the following inequality:

$$a_h(u - u_h^*, u - u_h^*)^{\frac{1}{2}} \geq \sup_{w_h \in V_h^0} \frac{|f(w_h) - a_h(u, w_h)|}{a_h(w_h, w_h)^{\frac{1}{2}}}.$$

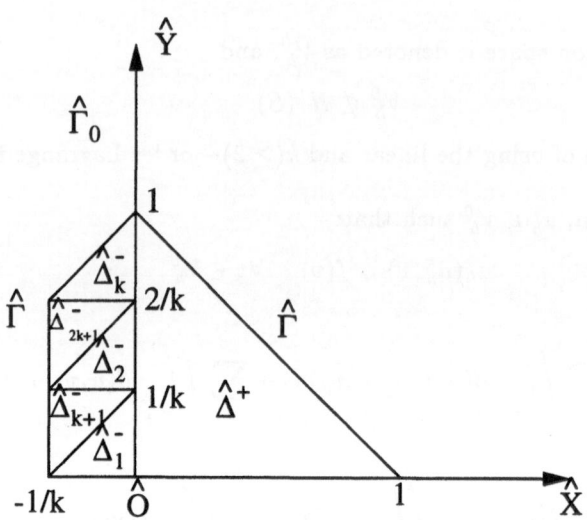

Figure 5.6 The reference element set of $\widehat{\Delta}^+ - \cup_j \widehat{\Delta}_j^-$.

Then we obtain

$$\|u - u_h^*\|_1 \geq \frac{1}{\beta_{max}^{\frac{1}{2}}} \sup_{w_h \in V_h^0} \frac{|f(w_h) - a_h(u, w_h)|}{\|w_h\|_1},$$

because $a_h(w_h, w_h) \leq \beta_{max} \|w_h\|_1^2$, where $\beta_{max} = \max_{(x,y) \in S} \beta(x, y)$. Obviously, β_{max} is a bounded constant independent of h^- and h^+.

We have

$$|f(w_h) - a_h(u, w_h)| = \left| \int_{\Gamma_0} \beta \frac{\partial u}{\partial n} (w_h^+ - w_h^-) \, d\ell \right|.$$

Therefore, in order to prove (5.50), it is sufficient to find a specific function, $\overline{w}_h \in V_h^0$, such that:

$$I = \left| \int_{\Gamma_0} \beta \frac{\partial u}{\partial n} (\overline{w}_h^+ - \overline{w}_h^-) \, d\ell \right| \geq C_0 (h^-)^{\frac{1}{2}} \|\overline{w}_h\|_1. \tag{5.51}$$

Now let us construct $\overline{w}_h$ to satisfy (5.51).

Let $(\widehat{\Delta}^+ - \cup_j \widehat{\Delta}_j^-)$ denote a reference element set of the element, $\widehat{\Delta}^+$, and the elements, $\widehat{\Delta}_j^-$ $(j = 1, 2, \ldots\ldots, 2k-1)$ (see Figure 5.6). A reference function on $(\widehat{\Delta}^+ -$

$\cup_j \widehat{\Delta}_j^-$) is given as follows:

$$
\widehat{w} = \begin{cases} \widehat{P}_k^+ = (\widehat{x} + \widehat{y} - 1)(1 + \widehat{y} + \widehat{y}^2 + \ldots\ldots + \widehat{y}^{k-2})\widehat{y}, & (\widehat{x}, \widehat{y}) \in \widehat{\Delta}^+, \\ \widehat{P}_1^-, & (\widehat{x}, \widehat{y}) \in \cup_j \widehat{\Delta}_j^-, \end{cases}
$$

where $\widehat{P}_1^-$ is a piecewise linear interpolation function, and the values of $\widehat{P}_1^-$ at the nodes of the elements, $\cup_j \widehat{\Delta}_j^-$, are defined as follows:

$$
\begin{aligned}
\widehat{P}_1^- \left(-\tfrac{1}{k}, \tfrac{i}{k}\right) &= 0, & j &= 0, 1, 2, \ldots\ldots, k-1, \\
\widehat{P}_1^- \left(0, \tfrac{i}{k}\right) = P_k^+ \left(0, \tfrac{i}{k}\right) &= \left(\tfrac{i}{k}\right)\left[\left(\tfrac{i}{k}\right)^{k-1} - 1\right], & j &= 0, 1, 2, \ldots\ldots, k.
\end{aligned} \tag{5.52}
$$

Obviously, $\widehat{P}_k^+$ is a $k (\geq 2)$-order polynomial function. The reference function, $\widehat{w}$, is continuous on the common nodes of $\widehat{\Delta}^+$ and $\cup_j \widehat{\Delta}_j^-$, based on (5.52). Moreover, we have

$$
\widehat{w}|_{\widehat{\Gamma}} = 0,
$$

where $\widehat{\Gamma}$ is the boundary of the reference element set, $(\widehat{\Delta}^+ - \cup_j \widehat{\Delta}_j^-)$.

An element set, $(\Delta_i^+ - \cup_j \Delta_{ij}^-)$, on Γ_0 in Figure 5.5 can be transformed from the reference element set, $(\widehat{\Delta}^+ - \cup_j \widehat{\Delta}_j^-)$, in Figure 5.6 by the following affine transformations:

$$
\begin{aligned}
\widehat{z}^+ \in \widehat{\Delta}^+ \to g(\widehat{z}^+) &= \mathbf{B}_i^+ \widehat{z}^+ + \mathbf{b}_i^+ \in \Delta_i^+, & (5.53) \\
\widehat{z}_j^- \in \widehat{\Delta}_j^- \to g_j(\widehat{z}_j^-) &= \mathbf{B}_{ij}^- \widehat{z}_j^- + \mathbf{b}_{ij}^- \in \Delta_{ij}^+, & (5.54)
\end{aligned}
$$

where the vector $\mathbf{z} = (x, y)^T$, $\mathbf{B}_i^+$ and $\mathbf{B}_{ij}^-$ are 2×2 constant matrices, and g, g_j, $\mathbf{b}_i^+$ and $\mathbf{b}_{ij}^-$ are 2-dimensional vectors. We can easily prove the existence of the affine transformations, (5.53) and (5.54).

Because of the assumption (5.49), we may simply suppose that

$$
\left.\frac{\partial u}{\partial n}\right|_{\Gamma_0'} \geq \sigma > 0, \tag{5.55}
$$

where σ is a constant. The interpolation functions, $\overline{w}_h$, can be constructed as follows:

1. $\overline{w}_h \neq 0$ only in the element sets, $(\triangle_i^+ - \cup_j \triangle_{ij}^-)$, on Γ_0'.

2. $\overline{w}_h$ is transformed from the reference function, $\hat{w}$, by the affine transformations, (5.53) and (5.54).

Obviously, $\overline{w}_h|_{S_1}$ is the piecewise linear interpolation function, and $\overline{w}_h|_{S_2}$ is the piecewise $k(\geq 2)-$ order Lagrange interpolation polynomial. The function $\overline{w}_h$ is continuous on the common nodes, Q_j, of $\triangle_i^+$ and $\cup_j \triangle_{ij}^-$ on Γ_0. Moreover, $\overline{w}_h|_\Gamma = 0$. Then the function $\overline{w}_h \in V_h^0$.

Below, let us prove (5.51). We have

$$\overline{w}_h^+\big|_{\Gamma_0'} \leq \overline{w}_h^-\big|_{\Gamma_0'} \leq 0, \quad (x,y) \in \Gamma_0', \tag{5.56}$$

because

$$\hat{w}^+\big|_{\hat{\Gamma}_0} \leq \hat{w}^-\big|_{\hat{\Gamma}_0} \leq 0, \quad (x,y) \in \hat{\Gamma}_0.$$

Hence, it follows from (5.55) and (5.56) and $\beta \geq \beta_0 > 0$ that

$$I \geq \left| \int_{\Gamma_0'} \beta \frac{\partial u}{\partial n} \left(\overline{w}_h^+ - \overline{w}_h^-\right) d\ell \right| \geq \beta_0 \sigma \sum_i \int_{\Gamma_0' \cap \triangle_i^+} \left(\overline{w}_h^- - \overline{w}_h^+\right) d\ell$$

$$= \beta_0 \sigma \sum_i Meas\left(\Gamma_0' \cap \triangle_i^+\right) \int_{\hat{\Gamma}_0 \cap \hat{\triangle}_i^+} \left(\hat{w}^- - \hat{w}^+\right) d\hat{l}.$$

We can obtain

$$\int_{\hat{\Gamma}_0 \cap \hat{\triangle}_i^+} \left(\hat{w}^- - \hat{w}^+\right) d\hat{l} \geq \mu > 0$$

with a constant μ. Therefore it follows that

$$I \geq \beta_0 \sigma \mu\, Meas(\Gamma_0'). \tag{5.57}$$

On the other hand, we have from Ciarlet (1978):

$$\|\overline{w}_h\|_{1,\triangle_i^+} \leq C\left[det(B_i^+)\right]^{\frac{1}{2}} \left\{ \|\, [B_i^+]^{-1} \,\|\, |\hat{w}|_{1,\hat{\triangle}_i^+} + |\hat{w}|_{0,\hat{\triangle}_i^+} \right\}$$

$$\leq C\|\hat{w}\|_{1,\hat{\triangle}_i^+} = C\|\hat{P}_k^+\|_{1,\hat{\triangle}_i^+},$$

where C is a bounded constant independent of h^- and h^+. Because of the bounded norm, $\|\hat{P}_k^+\|_{1,\hat{\triangle}_i^+}$, we obtain

$$\|\overline{w}_h\|_{1,\triangle_i^+} \leq C.$$

Similarly, we have

$$\sum_{j=1}^{2k-1} \|\overline{w}_h\|_{1,\Delta_{ij}^-}^2 \leq C \sum_{j=1}^{2k-1} \|\widehat{P}_1^-\|_{1,\widehat{\Delta}_j}^2 \leq C.$$

It is worth noting that the specific function $\overline{w}_h \neq 0$ only on the element sets, $(\Delta_i^+ - \cup_j \Delta_{ij}^-)$ on Γ_0'. Therefore, we obtain:

$$\|\overline{w}_h\|_1^2 = \sum_{i=1}^{M} \left\{ \|\overline{w}_h\|_{1,\Delta_i^+}^2 + \sum_{j=1}^{2k-1} \|\overline{w}_h\|_{1,\Delta_{ij}^-}^2 \right\} \leq CM,$$

where M is the number of the element sets, $(\Delta_i^+ - \cup_j \Delta_{ij}^-)$.

Moreover, we have from the inverse assumption, (5.48),

$$kM \leq \frac{Meas(\Gamma_0)}{\text{Min } Meas(\Gamma_0 \cap \Delta_i^-)} \leq \frac{C_2}{h^-}.$$

Therefore, it follows that

$$(h^-)^{\frac{1}{2}} \|\overline{w}_h\|_1 \leq C. \tag{5.58}$$

Finally, we obtain from (5.57) and (5.58):

$$I = \left| \int_{\Gamma_0} \beta \frac{\partial u}{\partial n} \left(\overline{w}_h^+ - \overline{w}_h^- \right) d\ell \right| \geq \beta_0 \sigma \mu \, Meas(\Gamma_0') \geq C_0 \left(h^- \right)^{\frac{1}{2}} \|\overline{w}_h\|_1.$$

Eq. (5.51) has been proved. This completes the proof of Theorem 5.3. ■

Remark 5.4 *Theorem 5.3 implies a significance of coupling techniques in combinations. In the coupling as shown in Figure 5.5 with $H = kh$, the k-order FEM works so superfluously that there incurs the reduced convergence rates. This clearly displays how important the harmonious collaboration in combinations is!. Such a concordant collaboration is already shown in the coupling ORC, the way that the quadratic FEM is devoted itself just to balance the linear FEM, as that $H = O(h^{\frac{1}{2}})$. From other analysis in this book, the optimal coupling lies that all methods in combinations should just balance to each other.*

Be careful in coupling. The overwork of one method not only consume more CPU time and computer storage, but also may result in the reduced convergence rates of the entire combinations! This result is unexpected.

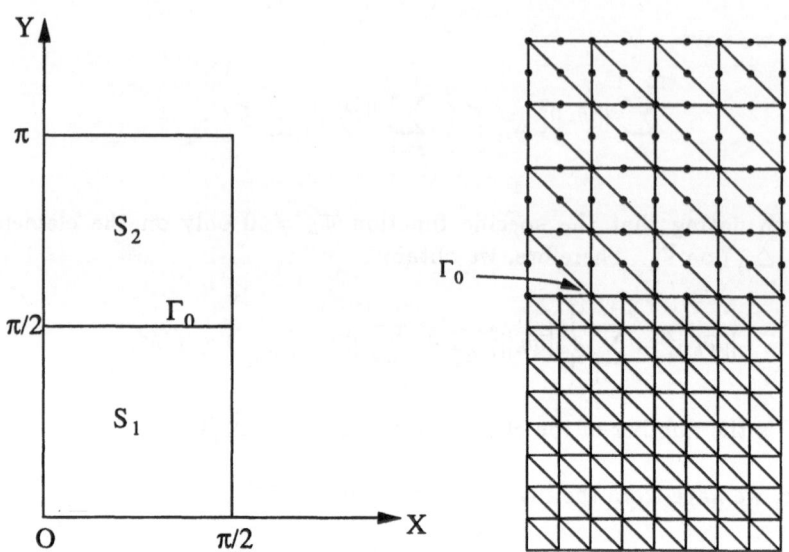

Figure 5.7 A model problem.

Figure 5.8 The coupling ($N = 8$ and $M = 4$).

5.7 NUMERICAL EXPERIMENTS OF RRC COUPLING

We will test the optimal rates (5.2) of convergence (ORC) and the reduced rate (5.3) of convergence (RRC) of couplings for linear and quadratic FEMs. In fact, the verification of ORC coupling is easy (refer Li (1991)); and the verification of RRC coupling is much more difficult.

Consider first the Laplace equation on a rectangle domain S ($0 < x < \pi/2$, $0 < y < \pi$), with the boundary conditions (see Figure 5.7)

$$
\begin{aligned}
u &= 0, & x &= 0, & 0 &< y < \pi, \\
\tfrac{\partial u}{\partial x} &= 0, & x &= \pi/2, & 0 &< y < \pi, \\
u &= 0, & 0 &< x < \pi/2, & y &= \pi, \\
u &= \sinh \pi \cdot \sin x, & 0 &< x < \pi/2, & y &= 0.
\end{aligned}
\tag{5.59}
$$

The true solution has been known as $u(x, y) = \sin x \cdot \sinh(\pi - y)$.

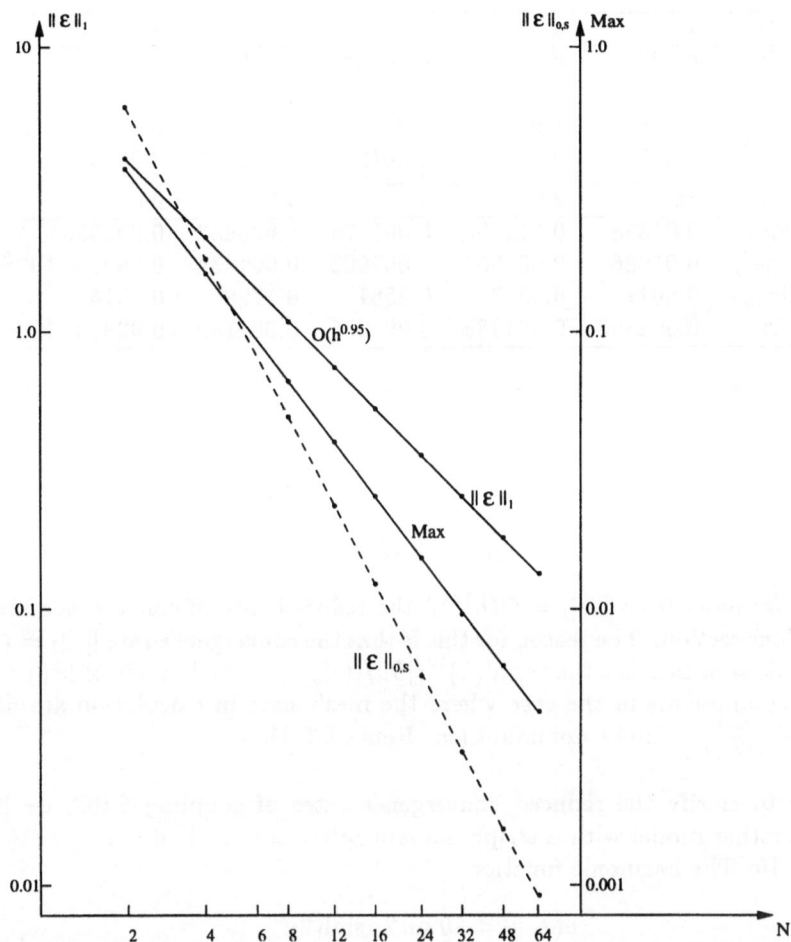

Figure 5.9 Error norm curves of the numerical solutions by coupling RRC with $N = 2M$.

Let Γ_0 be the straight line, $y = \pi/2$, and let both linear elements on S_1 and quadratic elements on S_2 be uniform. Also N denotes the number of linear elements along x; and M the number of quadratic elements along x. Let

$$H = 2h, \quad \text{i.e.,} \quad N = 2M, \tag{5.60}$$

in Figure 5.8; this combination is defined as coupling RRC because of a reduced rate of convergence in Corollary 5.1. The calculated results have been provided in Table 5.1 and Figure 5.9.

N	2	4	6	8	12
Max	0.3871	0.1274	0.06948	0.04637	0.02715
$\|\varepsilon\|_{0,S}$	0.5928	0.1613	0.7320	0.04171	0.01893
$\|\varepsilon\|_h$	3.878	1.9722	1.3212	0.9940	0.6659
Δ	0.07813	0.01098	0.001262	0.001331	0.002418

N	16	24	32	48	64
Max	0.01898	0.01175	0.008479	0.005433	0.003993
$\|\varepsilon\|_{0,S}$	0.01086	0.005031	0.002953	0.001434	0.8827×10^{-3}
$\|\varepsilon\|_h$	0.5017	0.3373	0.2551	0.1728	0.1316
Δ	0.002393	0.001985	0.001635	0.001189	0.9287×10^{-3}

Table 5.1 Error norms for the numerical solutions by coupling RRC with $N = 2M$.

We notice from Figure 5.9 that when $h \longrightarrow \pi/64\sqrt{2}$, i.e., $N \longrightarrow 64$,

$$\|\varepsilon\|_1 \longrightarrow 3.2h^{0.95}, \tag{5.61}$$

which is far away from $\|\varepsilon\|_1 = O(h^{1/2})$, the reduced rate of convergence proven in the previous section. The reason for this is that the convergence rate $\|\varepsilon\|_1 = O(h^{1/2})$ may not be seen because the term $(h)^{1/2}|\partial u/\partial n|_{0,\Gamma_0}$ in the bounds of $\|\varepsilon\|_1$ may not be very troublesome in the case where the mesh sizes in calculation are not very small and $|\frac{\partial u}{\partial n}|_{0,\Gamma_0}$ is not dominant (see Remark 5.1).

In order to clarify the reduced convergence rates of coupling RRC, we have to choose another model with a stripe domain $S(0 \le x \le \pi/2,\ 0 \le y \le \pi/16)$, as in Figure 5.10. The harmonic function

$$u(x, y) = 10 \sin 2x \sinh 2y \tag{5.62}$$

is governed by the Laplace equation and boundary conditions:

$$\triangle u = 0, \quad (x, y) \in S,$$

$$u|_{x=0} = u|_{x=\pi/2} = u|_{y=0} = 0, \tag{5.63}$$

$$u|_{y=\pi/16} = 10 \sin 2x \sinh \pi/8.$$

The midline, $y = \pi/32$, of the stripe is employed as the interface Γ_0 so that the solution domain S is divided into two subdomains: S_1 (the lower layer) and S_2 (the upper layer). The linear and quadratic FEM are used in S_1 and S_2 respectively, and the coupling relations of (5.60), (5.10) and (5.11) are then chosen to match two FEMs.

N	M	L_1	L_2	Max	$\|\varepsilon\|_{0,S}$	$\|\varepsilon\|_1$	Δ
32	16	2	1	0.06285	0.007248	0.6140	0.01848
64	32	4	2	0.03089	0.002570	0.4052	0.008836
96	48	6	3	0.02049	0.001434	0.3224	0.005804
128	64	8	4	0.01533	0.9582×10^{-3}	0.2756	0.004321
192	96	12	6	0.01020	0.5524×10^{-3}	0.2219	0.002859
256	128	16	8	0.007642	0.3786×10^{-3}	0.1908	0.002137
384	192	24	12	0.005089	0.2264×10^{-3}	0.1547	0.001419
512	256	32	16	0.003815	0.1593×10^{-3}	0.1335	0.001063
768	384	48	24	0.002542	0.9876×10^{-4}	0.1086	0.7071×10^{-3}

Table 5.2   Error norms of the numerical solutions by coupling RRC for the stripe model.

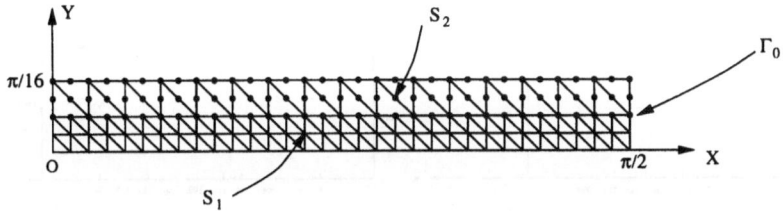

Figure 5.10   A stripe domain.

The numerical results by coupling RRC have been provided in Table 5.2, where N and $M(= N/2)$ are the numbers of linear and quadratic elements along the $x-$ direction, and L_1 and L_2 are those along $y-$ direction. For the division of Figure 5.10,

$$N = 32, \quad M = 16, \quad L_1 = 2, \quad L_2 = 1. \tag{5.64}$$

Based on the error norms obtained in Table 5.2, we can find the following asymptotic relations (see Figures 5.11 and 5.12):

$$\|\varepsilon\|_1 \longrightarrow 2.02h^{1/2}, \tag{5.65}$$

$$\|\varepsilon\|_{0,S} \longrightarrow 0.11h^{7/6}, \tag{5.66}$$

$$Max = \max_{i,j} |\varepsilon_{i,j}| \longrightarrow 0.88h, \tag{5.67}$$

$$\Delta = |(u_h^*)^+ - (u_h^*)^-|_{0,\Gamma_0} \longrightarrow 0.245h. \tag{5.68}$$

Eq. (5.65) perfectly verifies the conclusions on the reduced convergence rates in Corollaries 5.1 and 5.3; Eqs. (5.65) and (5.68) agree with each other on the basis

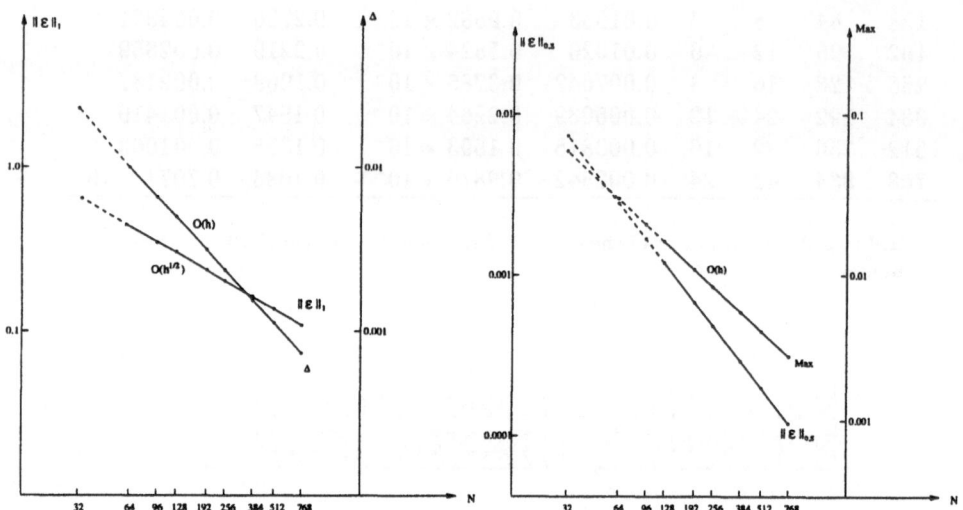

Figure 5.11 The curves of $\|\varepsilon\|_h$ and Δ of the numerical solutions by coupling RRC for the stripe model.

Figure 5.12 The curves of $\|\varepsilon\|_{0,S}$ and Max of the numerical solutions by coupling RRC for the stripe model.

of Lemma 5.1. it is interesting to note that the convergence rates in (5.65) are order $O(h^{1/2})$. In fact, for the model of Figure 5.10 with the long interface Γ_0, the error term, $O((h^{1/2})|\partial u/\partial n|_{0,\Gamma_0})$, resulting from matching on Γ_0, will play a dominant role in the errors $\|\varepsilon\|_1$. This is why the reduced convergence rates of $O(h^{1/2})$ can be observed by the numerical experiments with computable values of small h (but not as $h \longrightarrow 0$).

To summarize two kinds of combinations in Chapters 4 and 5, we conclude that the combined methods may have the optimal performance if all errors resulting from two methods and in particular from their coupling are fully estimated, and if the optimal coupling strategies are undertaken.

Following the lines given in this part, many other kinds of combinations may be established, see Li (1990) and Part III, and more study on error estimates in L_2 and L_∞ norms may also be developed.

There is no more lovely, friendly
and charming relationship communion or company
than a good marriage.
——— *Table talk (1596)* ———

Martin Luther
(1483 – 1546)

PART III
COUPLING TECHNIQUES

This part consists of six chapters describing different coupling techniques:

Chapter 6: Lagrange Multipliers and other Coupling Techniques which do not introduce extra variables.

Chapter 7: Penalty Techniques, for combinations of RGM-FEM.

Chapter 8: Simplified Hybrid Methods, for combinations of RGM-FEM.

Chapter 9: Penalty plus Hybrid Techniques, for combinations of RGM-FEM.

Chapter 10: Optimal Combinations for Various FEMs, using the coupling techniques in Chapter 9.

Chapter 11: Combinations of RGM and FDM, using the coupling techniques in Chapters 7–9.

The Lagrange multipliers are described in Chapter 6, by introducing extra variables; a linkage between the Lagrange multipliers and the nonconforming constants is also discovered. Different coupling techniques are explored in Chapters 7–9, in detail for the combinations of RGM-FEM, whose overview and summary are provided in Sections 6.4 and 9.4. These effective techniques are also suited to other kinds of combinations, demonstrated by the samples given in Chapters 10 and 11. The contents of this part are adapted mainly from Babuska (1973a), Liang and He (1992), Li (1992a, b, 1996a, 1997a, b), Li and Liang (1983), and Li and Bui (1988a–c, 1990a, b, 1992a, b).

The coupling techniques are critical to combinations so that we develop different effective coupling techniques carefully in this part. The nonconforming techniques in Part II need special treatments in programming; the techniques using additional integrals on Γ_0 in this part may greatly simplify the combined algorithms. Therefore, several coupling techniques that use penalty, hybrid, and penalty plus hybrid integrals are explored in Chapters 7, 8, and 9 separately. The new coupling techniques are different from the Lagrange multipliers in that no extra variables are required. Note that the techniques in Chapters 7 and 9 may approach the nonconforming techniques in Chapter 4 if the penalty constant or the penalty power increases to infinity.

In fact, the additional penalty plus hybrid integrals have been employed in FEMs already (see Fairweather (1978)), but they do not draw much attention because of algorithm complexity. By applying the penalty plus hybrid techniques, the optimal combinations of various FEMs can be achieved in Chapter 10, as contrasted to the combinations in Chapter 5 using nonconforming techniques, where the reduced convergence rates often occur. The complexity on the algorithms of penalty plus hybrid techniques is not severe, since the additional integrals involve Γ_0 only, and since the associated matrix obtained is positive definite and symmetric.

The simplified hybrid techniques in Chapter 8 are also attractive because they are analogous to the hybrid method in FEM, but do not require extra variables either. Moreover, a linkage between the Lagrange multiplier and the simplified hybrid technique is also discovered. However, the associated matrix obtained is positive definite but not symmetric and it can be solved by special algorithms in Golub and Loan (1989).

Based on analysis and numerical experiments, the penalty combinations that stem from the nonconforming combinations are strongly recommended due to algorithm simplicity. Note that the penalty techniques are pessimistic in FEMs, which cause the reduced convergence rates (see Ciarlet (1978)).

In Chapter 11, different coupling techniques are employed to combinations of RGM-FDM to gain not only optimal convergence, but also superconvergence by means of the suitable norms that involve discrete summation. Chapter 11 is also regarded as the representative of generalized combinations, to display what are described in Chapters 3–10 may also be applied to create other kinds of combinations. For instance, we may also develop new combinations of FVM with other methods, based on Section 1.6.6 of Chapter 1, where the FVM may be viewed as special Galerkin methods.

6

LAGRANGE MULTIPLIERS AND OTHER COUPLING TECHNIQUES

6.1 INTRODUCTION

In part II, the nonconforming constraints are adopted to couple different methods. Now, we solicit the additional integrals along the common boundary to play the same matching role. The traditional Lagrangian multiplier and the hybrid method by introducing extra variables on Γ_0 are described in this chapter. In the rest of this part we develop the coupling techniques without introducing extra variables. Using fewer variables is significant in computations. The overview of coupling techniques is given in this chapter; details of comparisons defer in Section 9.5.

The Lagrange multiplier method was first introduced by Babuska (1973a) to treat the constraint Dirichlet boundary condition as a natural boundary condition, and to relax the limitation on the admissible functions used. Since then the techniques of Lagrange multipliers have drawn much attention. Such techniques have been adopted to mixed and hybrid methods, see Brezzi and Fortin (1991) and Raviart and Thomas (1977).

The boundary condition using Lagrange multipliers is also extended to that involving flux in Bramble (1981). More analysis and applications are given in Fix (1976), Pitkaranta (1979, 1981), Lee (1993), Barbosa and Hughes (1992), in particular for domain decomposition methods in Liang and Liang (1990), Liang and He (1992), and Mandel and Tezaur (1996).

In the next section, we give in Corollary 6.1 an interpretation of the nonconforming coupling in Chapter 4 and 5 as Lagrange multipliers, and in Section 6.3 we provide a brief analysis on the generalized Lagrange multipliers. By using extra variables, optimal error estimates are given in Theorem 6.1, and an effort has been made to verify the nontrivial Ladyzhenskaya-Babuska-Brezzi (LBB) condition given in Lemmas 6.1–6.3. In the last section, we give an overview on the new couplings without extra variables that are embedded in the sequential chapters.

6.2 INTERPRETATION OF NONCONFORMING COMBINATIONS AS LAGRANGE MULTIPLIERS

In this section, we adopt Lagrange multipliers to couple different methods along the interior boundary Γ_0: to accommodate

$$u^+ = u^-, \quad \text{on } \Gamma_0, \tag{6.1}$$

for the constraint in nonconforming combination in Part II

$$v^+(Q_i) = v^-(Q_i), \quad \forall Q_i \in \Gamma_0. \tag{6.2}$$

Consider the elliptic equations with the Neumann conditions

$$\begin{cases} -\nabla(p\nabla u) + cu = f, & \text{in } S, \\ \frac{\partial u}{\partial n} = 0, & \text{on } \partial S, \end{cases} \tag{6.3}$$

where $c > 0$. The solutions of the nonconforming combinations in Chapters 4 and 5 are also described by

$$I(u_h) = \min_{v \in \overline{V}_h} I(v), \tag{6.4}$$

where $\overline{V}_h$ denotes the space of admissible functions satisfying (6.2), and

$$I(v) = \frac{1}{2} \iint_{S_1} \left[p(\nabla v)^2 + cv^2\right] ds + \frac{1}{2} \iint_{S_2} \left[p(\nabla v)^2 + cv^2\right] ds - \iint_S fv ds.$$

The minimization (6.4) under the condition (6.2) can also be represented equivalently by the Lagrange multiplier (see Courant and Hilbert (1953))

$$I^*(u_h) = \min_{\lambda_i, v \in V_h} I^*(v),$$

where

$$I^*(v) = \min_{\lambda_i, v \in V_h} \left\{ I(v) - \sum_{Q_i \in \Gamma_0} \lambda_i(v^+(Q_i) - v^-(Q_i)) \right\}. \tag{6.5}$$

The Lagrange multipliers λ_i are regarded as new variables. The corresponding algebraic equations are

$$\begin{cases} \mathbf{Ax} - \mathbf{By} = \mathbf{b}, \\ \mathbf{B}^T \mathbf{x} = 0. \end{cases} \tag{6.6}$$

Note that the solutions of (6.6) are unique and exactly the same as those obtained from the nonconforming combinations (6.4), where a simple prior substitution for eliminating λ_i is described in Section 5.2. Therefore, the associated matrix of the nonconforming combination has symmetric and positive definite properties, to make computation much easy. Moreover, the theoretical analysis in Chapters 4 and 5 is also valid for the discrete formulation of Lagrange multipliers. Also note that the analysis therein is rather easier since the standard FEM analysis follows, to circumvent the troublesome LBB condition that is shown below.

Now we turn out the continuity requirement (6.1) by introducing an additional integral $\int_{\Gamma_0} \lambda(v^+ - v^-)d\ell$, where λ is a continuous function of Lagrange multipliers. We describe the Lagrange multiplier coupling as:

To seek (u_h, λ) such that

$$A(u_h, v) - D(u_h v; \lambda, \mu) = f(v), \tag{6.7}$$

where

$$A(u, v) = \iint_{S_1} p \, \nabla u \, \nabla v ds + \iint_{S_2} p \, \nabla u \, \nabla v ds, \tag{6.8}$$

$$D(u, v; \lambda, \mu) = \int_{\Gamma_0} \lambda(v^+ - v^-)d\ell + \int_{\Gamma_0} \mu(u^+ - u^-)d\ell,$$

$$f(v) = \iint_S f v ds.$$

By the Green theory, it is easy to verify the true solution is $\lambda = p\frac{\partial u}{\partial n}$, where n is the outward normal to ∂S_1.

Now we show that a suitable approximation of the boundary integration also leads to the solution of the nonconforming combinations (6.4).

Choose the trapezoidal rule

$$\widehat{\int_{\Gamma_0}} f d\ell = \frac{1}{2} \sum_{i=1}^{N} \overline{Z_{i-1}Z_i} \left[f(Z_i) + f(Z_{i-1}) \right],$$

then the solution involving integration approximation is given by:

$$A(\tilde{u}_h, v) - \hat{B}(\tilde{u}_h, v; \tilde{\lambda}, \mu) = f(v), \tag{6.9}$$

where $A(u, v)$ is the same as (6.8), and

$$\hat{B}(u, v, \lambda, \mu) = \widehat{\int_{\Gamma_0}} \lambda(v^+ - v^-)d\ell + \widehat{\int_{\Gamma_0}} \mu(u^+ - u^-)d\ell, \tag{6.10}$$

$$\widehat{\int_{\Gamma_0}} \lambda(v^+ - v^-) d\ell = \sum_{Q_i \in \Gamma_0} \lambda_i \left[\frac{h_i + h_{i+1}}{2} \right] (v^+(Q_i) - v^-(Q_i)), \quad (6.11)$$

where $h_i = \overline{Z_i Z_{i-1}}$, and $h_0 = h_N = 0$.

Letting $\lambda_i^* = \lambda_i \left[\frac{h_i + h_{i+1}}{2} \right]$, the discrete formation of (6.11) becomes

$$\sum_{Q_i \in \Gamma_0} \lambda_i^* (v^+(Q_i) - v^-(Q_i)),$$

exactly the same as those in (6.5). We conclude this as an important corollary.

Corollary 6.1 *The solutions of the nonconforming combination are the same as those obtained from the Lagrange multipliers in the discrete form (6.5), as well as in (6.10) using the trapezoidal rule or other integration rules that choose only $Q_i \in \Gamma_0$ as the integration nodes.*

6.3   GENERAL APPROACHES OF LAGRANGE MULTIPLIERS

The Lagrange multiplier is typical in minimization under constraints. For some integration rules in which the element nodes $Q_i \in \Gamma_0$ are not chosen as the integration nodes, the Lagrange multipliers may result in a different solution from that by the nonconforming combination. Hence we will give a brief analysis for the Lagrange multiplier combination. The following presentation develops from Babuska (1973b) and Quarteroni and Valli (1994).

Choose

$$v = \begin{cases} v^- = v_k, & \text{in } S_1, \\ v^+, & \text{in } S_2, \end{cases} \quad (6.12)$$

where v^+ and v^- are two different admissible functions, and v_k is the piecewise k-order Lagrange polynomials.

For simplicity, consider $p = 1$ and $c = 1$ in (6.3), i.e.,

$$\begin{cases} -\triangle u + u = f, & \text{in } S, \\ \frac{\partial u}{\partial n} = 0, & \text{on } \partial S, \end{cases} \quad (6.13)$$

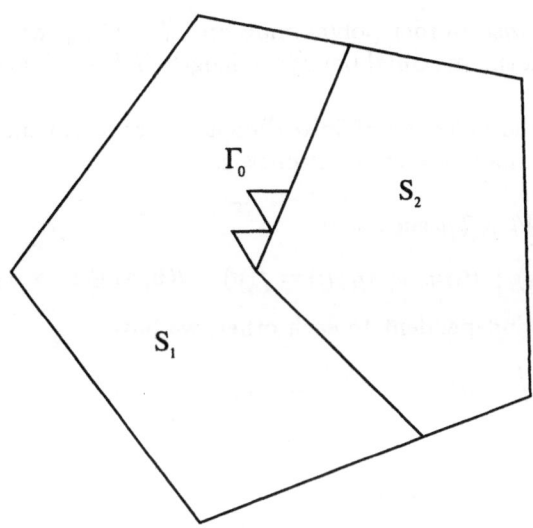

Figure 6.1   A partition of S.

where S is a polygon and Γ_0 is piecewise straight lines (see Figure 6.1). Then the true solutions u and $\lambda = \frac{\partial u}{\partial n}$ are defined by $u \in H^1(S)$ and $\lambda \in H^{\frac{1}{2}}(\Gamma_0)$ such that

$$A_h(u,v) + B(u,v;\lambda,\mu) = f(v), \tag{6.14}$$

where

$$A_h(u,v) = \iint_{S_1} (\nabla u \, \nabla v + uv)\,ds + \iint_{S_2} (\nabla u \, \nabla v + uv)\,ds,$$

$$B(u,v;\lambda,\mu) = -\int_{\Gamma_0} \lambda(v^+ - v^-)\,d\ell - \int_{\Gamma_0} \mu(u^+ - u^-)\,d\ell.$$

Define the error norms

$$\|(v,\mu)\|_H = (\|v\|_1^2 + \|\mu\|_{-\frac{1}{2},\Gamma_0}^2)^{1/2}, \quad \|v\|_1 = (\|v\|_{1,S_1}^2 + \|v\|_{1,S_2}^2)^{1/2},$$

$$\|\mu\|_{-\frac{1}{2},\Gamma_0} = \sup_v \frac{\left|\int_{\Gamma_0} \mu(v^+ - v^-)\,d\ell\right|}{|v^+|_{\frac{1}{2},\Gamma_0} + |v^-|_{\frac{1}{2},\Gamma_0}}, \tag{6.15}$$

$$\|v\|_{\frac{1}{2},\Gamma_0} = \left(\int_{\Gamma_0} \int_{\Gamma_0} \frac{(v(P) - v(Q))^2}{(P-Q)^2}\,d\ell(Q)\,d\ell(P) \right.$$

$$\left. + \frac{1}{2}(d_1^{-1} + d_2^{-1})\|v\|_{0,\Gamma_0}^2 \right)^{\frac{1}{2}}, \tag{6.16}$$

where d_i is roughly the diameter of S_i.

Let λ_ℓ^H be the piecewise ℓ-order polynomials on $\Gamma_0^H = \cup_i \Gamma_0^i$, where Γ_0^H are quasiuniform sections with the maximal boundary length H (see Figure 6.1).

Denote by $V_h \times T_H$ the collection of finite dimensions of (6.12) and λ_ℓ^H on Γ_0^H. Then the combination by the Lagrange multiplier is:

To seek $(u_h, \lambda_H) \in V_h \times T_H$ such that

$$A_h(u_h, v_h) + B(u_h, v; \lambda_H, \mu) = f(v), \quad \forall (v, \mu) \in V_h \times T_H. \tag{6.17}$$

Since v^+ and v^- are independent to each other, we have

$$A_h(u_h, v) = \int_{\Gamma_0} \lambda_H(v^+ - v^-)d\ell + f(v), \quad \int_{\Gamma_0} \mu(u_h^+ - u_h^-)d\ell = 0. \tag{6.18}$$

We need the following assumptions.

(A1) For $A_h(u, v)$ there exist the bounds

$$C_0\|v\|_1^2 \leq A_h(v, v), \quad |A_h(u, v)| \leq C_1\|u\|_1\|v\|_1, \quad \forall v \in V_h^0.$$

(A2) For $\int_{\Gamma_0} \mu(v^+ - v^-)d\ell$, the Ladyzhenskaya-Babuska-Brezzi (LBB) condition holds:

$\forall \mu_H \in T_H, \exists v_h \in V_h, v_h \neq 0$, such that

$$\left| \int_{\Gamma_0} \mu_H(v_h^+ - v_h^-)d\ell \right| \geq \beta\|v_h\|_1\|\mu_H\|_{-\frac{1}{2},\Gamma_0}. \tag{6.19}$$

(A3) Also the following bounds hold

$$\left| \int_{\Gamma_0} \lambda(v^+ - v^-)d\ell \right| \leq C\|\lambda\|_{-\frac{1}{2},\Gamma_0}\|v\|_1. \tag{6.20}$$

Now we give a theorem.

Theorem 6.1 *Let* **(A1)**–**(A3)** *hold. There exist the error bounds*

$$\|\lambda - \lambda_H\|_{-\frac{1}{2},\Gamma_0} \leq C \left\{ \inf_{v \in V_h} \|u - v\|_1 + \inf_{\eta \in T_H} \|\lambda - \eta\|_{-\frac{1}{2},\Gamma_0} \right\}, \tag{6.21}$$

and

$$\|u - u_h\|_1 \leq C \left\{ \inf_{v \in V_h} \|u - v\|_1 + \inf_{\eta \in T_H} \|\lambda - \eta\|_{-\frac{1}{2},\Gamma_0} \right\}. \tag{6.22}$$

Proof: Since the true solutions u and λ also satisfy (6.17) and (6.18), then

$$A_h(u - u_h, v) = \int_{\Gamma_0} (\lambda - \lambda_H)(v^+ - v^-)d\ell. \tag{6.23}$$

Let $w = u_h - v$ where $v \in V_h$, and $\delta = \lambda_H - \eta$ where $\eta \in T_H$. We have from **(A1)** and (6.23)

$$C_0\|u_h - v\|_1^2 \le A_h(u_h - v, w) \tag{6.24}$$

$$= A_h(u - v, w) - \int_{\Gamma_0}(\lambda - \eta)(w^+ - w^-)d\ell - \int_{\Gamma_0}(\eta - \lambda_H)(w^+ - w^-)d\ell$$

$$= A_h(u - v, w) - \int_{\Gamma_0}(\lambda - \eta)(w^+ - w^-)d\ell + \int_{\Gamma_0}\delta[(u_h^+ - v^+) - (u_h^- - v^-)]d\ell.$$

From (6.18), we obtain

$$\int_{\Gamma_0} \delta(u_h^+ - u_h^-)d\ell = \int_{\Gamma_0} \delta(u^+ - u^-)d\ell = 0. \tag{6.25}$$

We have from **(A3)**

$$\int_{\Gamma_0} \delta[(u_h^+ - v^+) - (u_h^- - v^-)]d\ell = \int_{\Gamma_0} \delta[(u^+ - v^+) - (u^- - v^-)]d\ell$$

$$\le C\|\lambda_H - \eta\|_{-\frac{1}{2}}\|u - v\|_1. \tag{6.26}$$

Hence, we obtain from (6.24), (6.26) and **(A3)**

$$C_0\|w\|_1^2 = C_0\|u_h - v\|_1^2$$

$$\le C\left\{\left(\|u - v\|_1 + \|\lambda - \eta\|_{-\frac{1}{2},\Gamma_0}\right)\|w\|_1 + \|\lambda_H - \eta\|_{-\frac{1}{2},\Gamma_0}\|u - v\|_1\right\}.$$

This is an inequality equation of order 2,

$$C_0 x^2 \le bx + d, \quad b, \ d > 0, \tag{6.27}$$

where

$$x = \|u_h - v\|_1 = \|w\|_1, \quad b = C(\|u - v\|_1 + \|\lambda - \eta\|_{-\frac{1}{2},\Gamma_0}),$$

$$d = C\|\lambda_H - \eta\|_{-\frac{1}{2},\Gamma_0}\|u - v\|_1 \le \frac{C}{2}\left(\varepsilon\|\lambda_H - \eta\|_{-\frac{1}{2},\Gamma_0} + \varepsilon^{-1}\|u - v\|_1\right)^2,$$

with $\varepsilon > 0$. Therefore, we obtain the bounds of $\|w\|_1$ by

$$\|w\|_1 = x \le \frac{b + \sqrt{b^2 + 4C_0 d}}{2C_0} \tag{6.28}$$

$$\le C\left\{\|u - v\|_1 + \|\lambda - \eta\|_{-\frac{1}{2},\Gamma_0} + \varepsilon^{-1}\|u - v\|_1\right\} + C\varepsilon\|\lambda_H - \eta\|_{-\frac{1}{2},\Gamma_0}.$$

Next, we adopt the LBB condition for each $\eta \in T_H$, $\lambda_H - \eta \in T_H$, $\exists v_h \in V_h$ such that

$$\|\lambda_H - \eta\|_{-\frac{1}{2},\Gamma_0} \leq \frac{1}{\beta} \sup \frac{\int_{\Gamma_0}(\lambda_H - \eta)(v_h^+ - v_h^-)d\ell}{\|v_h\|_1}. \tag{6.29}$$

Also it follows from (6.23) that

$$\begin{aligned}
\int_{\Gamma_0}(\lambda_H - \eta)(v^+ - v^-)d\ell &= \int_{\Gamma_0}(\lambda_H - \lambda)(v^+ - v^-)d\ell + \int_{\Gamma_0}(\lambda - \eta)(v^+ - v^-)d\ell \\
&= -A_h(u - u_h, v) + \int_{\Gamma_0}(\lambda - \eta)(v^+ - v^-)d\ell \\
&\leq C\|u - u_h\|_1\|v\|_1 + \|\lambda - \eta\|_{-\frac{1}{2},\Gamma_0}\|v\|_1. \tag{6.30}
\end{aligned}$$

From (6.28) – (6.30) we obtain

$$\begin{aligned}
\|\lambda_H - \eta\|_{-\frac{1}{2},\Gamma_0} &\leq C(\|u - u_h\|_1 + \|\lambda - \eta\|_{-\frac{1}{2},\Gamma_0}) \\
&\leq C\left\{\|u - v\|_1 + \|v - u_h\| + \|\lambda - \eta\|_{-\frac{1}{2},\Gamma_0}\right\} \tag{6.31} \\
&\leq C_1\left\{\|u - v\|_1 + \|\lambda - \eta\|_{-\frac{1}{2},\Gamma_0} + + \varepsilon^{-1}\|u - v\|_1\right\} + C_1\varepsilon\|\lambda_H - \eta\|_{-\frac{1}{2},\Gamma_0},
\end{aligned}$$

where $C_1(> 0)$ is also a bounded constant independent of v, η and ε. Letting $C_1\varepsilon \leq \frac{1}{2}$, Eq. (6.31) leads to

$$\|\lambda_H - \eta\|_{-\frac{1}{2},\Gamma_0} \leq C\left\{\|u - v\|_1 + \|\lambda - \eta\|_{-\frac{1}{2},\Gamma_0}\right\}. \tag{6.32}$$

The first desired result (6.21) follows from the triangular inequality

$$\|\lambda - \lambda_H\|_{-\frac{1}{2},\Gamma_0} \leq \|\lambda - \eta\|_{-\frac{1}{2},\Gamma_0} + \|\lambda_H - \eta\|_{-\frac{1}{2},\Gamma_0}.$$

Moreover, from (6.28) and (6.32)

$$\|u_h - v\|_1 \leq C\left\{\|u - v\|_1 + \|\lambda - \eta\|_{-\frac{1}{2},\Gamma_0}\right\}.$$

Then the second desired result (6.22) follows from

$$\|u - u_h\|_1 \leq \|u - v\|_1 + \|u_h - v\|_1. \quad \blacksquare$$

Let us examine the assumptions (A1) – (A3). (A1) is valid obviously, and for (A3) we have

$$\left|\int_{\Gamma_0}\lambda(v^+ - v^-)d\ell\right| \leq |\lambda|_{-\frac{1}{2},\Gamma_0}\|v^+ - v^-\|_{\frac{1}{2},\Gamma_0}$$

$$
\begin{aligned}
&\leq \ \|\lambda\|_{-\frac{1}{2},\Gamma_0}(\|v^+\|_{\frac{1}{2},\Gamma_0} + \|v^-\|_{\frac{1}{2},\Gamma_0}) \\
&\leq \ C\|\lambda\|_{-\frac{1}{2},\Gamma_0}(\|v\|_{1,S_1} + \|v\|_{1,S_2}) \\
&\leq \ C\|\lambda\|_{-\frac{1}{2},\Gamma_0}\|v\|_1.
\end{aligned}
$$

The challenging task is to verify the LBB condition (6.19). For this we shall prove two lemmas. First let us prove the original LBB condition, then give an extension of Fortin's lemma.

Lemma 6.1 *There exists a constant* $\beta(> 0)$ *independent of* u *and* v *such that* $\forall \mu \in H^{-\frac{1}{2}}(\Gamma_0), \exists v \in H^1(S), v \neq 0$ *such that*

$$
\int_{\Gamma_0} \mu(v^+ - v^-)d\ell \geq \beta\|\mu\|_{-\frac{1}{2},\Gamma_0}\|v\|_1. \tag{6.33}
$$

Proof: Consider an auxiliary problem

$$
\begin{cases}
-\triangle v^+ + v^+ = 0, & \text{in } S^+, \\
-\triangle v^- + v^- = 0, & \text{in } S^-, \\
\frac{\partial v}{\partial n}\big|_{\Gamma} = 0, & \\
\frac{\partial v^+}{\partial n}\big|_{\Gamma_0} = \mu, \ \frac{\partial v^-}{\partial n}\big|_{\Gamma_0} = -\mu.
\end{cases} \tag{6.34}
$$

We have

$$
\begin{aligned}
\int_{\Gamma_0} \mu v^+ d\ell &= \int_{\Gamma_0} \frac{\partial v^+}{\partial n} v^+ d\ell = \iint_{S_2}((\triangle v)^2 + v^2)ds = \|v\|_{1,S_2}^2, \\
-\int_{\Gamma_0} \mu v^- d\ell &= \int_{\Gamma_0} \frac{\partial v^-}{\partial n} v^- d\ell = \iint_{S_1}((\triangle v)^2 + v^2)ds = \|v\|_{1,S_1}^2.
\end{aligned}
$$

Then

$$
\int_{\Gamma_0} \mu(v^+ - v^-)d\ell = \|v\|_1^2. \tag{6.35}
$$

From (6.34)

$$
\|\mu\|_{-\frac{1}{2},\Gamma_0} = \left\|\frac{\partial v^+}{\partial n}\right\|_{-\frac{1}{2},\Gamma_0} \leq C\|v^+\|_{\frac{1}{2},\Gamma_0} \leq C\|v\|_1. \tag{6.36}
$$

Then we obtain from (6.35) and (6.36),

$$
\int_{\Gamma_0} \mu(v^+ - v^-)d\ell = \|v\|_1^2 \geq \beta\|\mu\|_{-\frac{1}{2},\Gamma_0}\|v\|_1.
$$

This is the LBB condition. ■

Now we give an extension of Fortin's lemma.

Lemma 6.2 *Let the original LBB condition (6.33) hold. Also $\forall \mu_h \in T_h$, $\exists v_h$ to approximate v such that*

$$
\begin{cases}
- \triangle v_h^+ + v_h^+ = 0 & \text{in } S^+, \\
- \triangle v_h^- + v_h^- = 0 & \text{in } S^-, \\
\frac{\partial v_h}{\partial n}\big|_\Gamma = 0, \\
\frac{\partial v_h^+}{\partial n} = \mu_h, \frac{\partial v_h^-}{\partial n} = -\mu_h,
\end{cases}
\tag{6.37}
$$

and the following inequalities hold,

$$
C_0\|v\|_1 \le \|v_h\|_1 \le C\|v\|_1, \quad C_0 > 0,
\tag{6.38}
$$

$$
\left| \int_{\Gamma_0} \mu_h \left[(v^+ - v_h^+) - (v^- - v_h^-) \right] d\ell \right| \le o(1)\|v\|_1^2.
\tag{6.39}
$$

With these assumptions, the discrete LBB condition (6.19) holds.

Proof: We have

$$
\left| \int_{\Gamma_0} \mu_h (v_h^+ - v_h^-) d\ell \right|
$$

$$
\ge \left| \int_{\Gamma_0} \mu_h (v^+ - v^-) d\ell \right| - \left| \int_{\Gamma_0} \mu_h (v^+ - v^-) d\ell - \int_{\Gamma_0} \mu_h (v_h^+ - v_h^-) d\ell \right|
$$

$$
\ge \left\{ \beta\|\mu_h\|_{-\frac{1}{2},\Gamma_0} \|v\|_1 - o(1)\|v\|_1^2 \right\}.
$$

From (6.37)

$$
\|v_h\|_1^2 = \int_{\Gamma_0} \mu_h (v_h^+ - v_h^-) d\ell \le C\|\mu_h\|_{-\frac{1}{2},\Gamma_0} \|v_h\|_1,
$$

to lead to

$$
\|v_h\|_1 \le C\|\mu_h\|_{-\frac{1}{2},\Gamma_0}.
$$

Also from (6.38) and the above bounds

$$
\|v\|_1^2 \le C\|v_h\|_1 \|v\|_1 \le C_1\|\mu_h\|_{-\frac{1}{2},\Gamma_0} \|v\|_1.
$$

Finally, we obtain

$$\left| \int_{\Gamma_0} \mu_h(v_h^+ - v_h^-) d\ell \right| \geq (\beta - o(1)) \|\mu_h\|_{-\frac{1}{2},\Gamma_0} \|v\|_1 \geq \frac{(\beta - o(1))}{C} \|\mu_h\|_{-\frac{1}{2},\Gamma_0} \|v_h\|_1.$$

The discrete LBB condition (6.19) is obtained by replacing β by $(\beta - o(1))/C$.   ∎

Note that when $o(1) \longrightarrow 0$ as $h \longrightarrow 0$, Lemma 6.2 leads to Fortin (1977). However, since the different admissible functions chosen in combinations may not satisfy Fortin's lemma exactly, Lemma 6.2 is more useful in our analysis. The general Lagrange multipliers offer a flexibility in matching different methods, in particular for the domain decomposition method, see Liang and Liang (1990). However, the verification on the discrete LBB condition is nontrivial. Below we cite some important results in Liang and He (1992).

Let k- and m-order FEMs be chosen in S_1 and S_2, h be the maximal boundary length of the elements in both S_1 and S_2, and H be that of piecewise ℓ-order polynomials in Γ_0. Then we have the following lemma.

Lemma 6.3 *Let* $\lim_{h \to 0} \ell \sqrt{\frac{h}{H}} = 0$, *then* $\forall \mu_h \in T_h$, $\exists v_h$ *satisfying (6.38) and (6.39) such that*

$$\left| \int_{\Gamma_0} \mu_h \left((v^+ - v_h^+) - (v^- - v_h^-) \right) d\ell \right| = o(1) \|v\|_1.$$

Proof: There is an inequality in Liang and Liang (1990)

$$|\mu_H|_{0,\Gamma_0} \leq C \frac{\ell}{\sqrt{H}} |\mu_H|_{-\frac{1}{2},\Gamma_0}. \tag{6.40}$$

Denote by $v_h^\pm$ the interpolants of $v^\pm$, we have from (6.40) and interpolation errors

$$\left| \int_{\Gamma_0} \mu_H(v^\pm - v_h^\pm) d\ell \right| \leq \|\mu_H\|_{0,\Gamma_0} \|v^\pm - v_h^\pm\|_{0,\Gamma_0}$$

$$\leq C \frac{\ell}{\sqrt{H}} \|\mu_H\|_{-\frac{1}{2},\Gamma_0} \left(\sqrt{h} \|v^\pm\|_{\frac{1}{2},\Gamma_0} \right)$$

$$\leq C_1 \ell \sqrt{\frac{h}{H}} \|v_h\|_1 \|v\|_1 \leq C_2 \ell \sqrt{\frac{h}{H}} \|v\|_1^2 = o(1) \|v\|_1^2.$$

The discrete LBB condition follows from Lemma 6.2.   ∎

For the integration approximation, $\widehat{\int_{\Gamma_0}}$ of $\int_{\Gamma_0}$, the combination (6.17) leads to

$$A_h(u_h, v_h) - \widehat{\int_{\Gamma_0}} \lambda_h(v^+ - v^-)d\ell - \widehat{\int_{\Gamma_0}} \mu(u_h^+ - u_h^-)d\ell = f(v).$$

The error bounds in Theorem 6.1 are extended as follows (see Roberts and Thomas (1991))

$$\|u - u_h\|_1 + \|\lambda - \lambda_H\|_{-\frac{1}{2}, \Gamma_0}$$

$$\leq C \left\{ \inf_{v \in V_h} \|u - v\|_1 + \inf_{\eta \in T_H} \|\lambda - \eta\|_{-\frac{1}{2}, \Gamma_0} + \sup_{w \in V_h} \frac{\left| \left(\int_{\Gamma_0} - \widehat{\int_{\Gamma_0}} \right) \lambda_H(w^+ - w^-)d\ell \right|}{\|w\|_1} \right\}.$$

The error estimates from Theorem 6.1 and the above bounds are optimal. In (6.14) we introduce the extra variables λ, which play a coupling rule. Liang and Liang (1990) also introduces other extra variables that play the same coupling role:

For solving (6.13), to seek $u_h \in V_h$ and $\phi_H \in T_H^*$ such that

$$A_h(u_h, v) + B^*(u_h, v; \phi_H, \psi) = f(v),$$

where $f(v)$ and $A_h(u_h, v)$ are given in (6.14), but

$$
\begin{aligned}
B^*(u, v; \phi, \psi) = & -\int_{\Gamma_0} \frac{\partial u^+}{\partial n}(v^+ - \psi)d\ell - \int_{\Gamma_0} \frac{\partial u^-}{\partial n}(v^- - \psi)d\ell \\
& -\int_{\Gamma_0} \frac{\partial v^+}{\partial n}(u^+ - \phi)d\ell - \int_{\Gamma_0} \frac{\partial v^-}{\partial n}(u^- - \phi)d\ell.
\end{aligned}
$$

The space of T_H^* consists of piecewise ℓ-order polynomials on Γ_0^H satisfying one more continuity condition on ψ at the corners of Γ_0. Detailed analysis is given in Liang and Liang (1990).

6.4 OVERVIEW OF COUPLING TECHNIQUES WITHOUT EXTRA VARIABLES

The coupling techniques using additional integrals are very promising, as shown in Chapters 7-9. These techniques include the penalty technique, the simplified hybrid techniques, and the mixed types of both penalty and hybrid techniques.

Six combinations based on coupling techniques have appeared: the nonconforming combination, the penalty combination, the simplified hybrid combination, Combinations I and II, and Symmetric Combination; all of them are first designed for coupling the RGM-FEM in solving elliptic equations, especially with singularities. An overview of coupling strategies in this subsection is an introduction to the rest of this part.

Let us consider a general elliptic equation

$$-\frac{\partial}{\partial x}\left(p\frac{\partial u}{\partial x}\right) - \frac{\partial}{\partial y}\left(p\frac{\partial u}{\partial y}\right) + cu = f, \qquad (x,y) \in S, \tag{6.41}$$

with the mixed type of boundary conditions

$$u = g_1 \quad \text{in } \Gamma_D,$$
$$p\frac{\partial u}{\partial n} + qu = g_2 \quad \text{in } \Gamma_N,$$

where S is a bounded, polygonal domain, Γ is the exterior boundary of S such that $\Gamma = \Gamma_D \bigcup \Gamma_N$, $Meas(\Gamma_D) > 0$ while $q \equiv 0$, and the functions p, c, f, q, g_1 and g_2 are sufficiently smooth. Moreover, the functions in (6.41) satisfy

$$p = p(x,y) \geq p_0 > 0, \quad c = c(x,y) > 0, \quad q = q(x,y) \geq 0.$$

Let the Sobolev spaces be defined by

$$H_*^1(S) = \{v|v, \ v_x, \ v_y \in L^2(S) \ \text{and} \ v|_{\Gamma_D} = g_1\},$$
$$H_0^1(S) = \{v|v, \ v_x, \ v_y \in L^2(S) \ \text{and} \ v|_{\Gamma_D} = 0\}.$$

Then, the well-posed problem for (6.41) can be equivalently written as:

$$I(u,v) = f(v), \quad \forall v \in H_0^1(S),$$

where $u \in H_*^1(S)$ and the notations are

$$I(u,v) = \iint_S (p\,\nabla u\,\nabla v + cuv)\,ds + \int_{\Gamma_N} quvd\ell,$$
$$f(v) = \iint_S fvds + \int_{\Gamma_N} g_2vd\ell.$$

Let the solution domain S be divided by a common boundary Γ_0 into two subdomains S_1 and S_2, as in Figure 6.1. The Ritz-Galerkin method is used in S_2 where there may occur a singular point; and the $k(\geq 1)-$ order Lagrange FEM is used in S_1 where the true solution u is assumed to be smooth enough such that $u \in H^{k+1}(S_1)$.

Consequently, the admissible functions used can be expressed in the form

$$v = \begin{cases} v^- = v_1^k, & \text{in } S_1, \\ v^+ = \sum_{i=1}^{L} a_i \Psi_i, & \text{in } S_2, \end{cases} \tag{6.42}$$

where v_1^k are piecewise k-order Lagrange interpolation polynomials on a quasiuniform triangulation of S_1 with a maximal boundary length h of the elements, and $\{\Psi_i\}$ are the known singular or analytic functions, and a_i are the unknown coefficients to be sought.

The basis functions $\{\Psi_i\}_{i=1}^{\infty}$ are complete and linearly independent in $L(S_2)$. We denote that the admissible functions v in (6.42) are not continuous along the common boundary

$$v^+ \neq v^- \quad \text{on } \Gamma_0.$$

Therefore, $v \notin H^1(S)$. We then need to define a new space H such that

$$H = \{v | v \in L^2(S), \ v \in H^1(S_1), \ v \in H^1(S_2)\}.$$

Let $V_h^* \in H$ be a finite-dimensional collection of admissible functions (6.42). The coupling techniques using additional integrals may fall into two categories:

I Extra unknowns introduced, such as Lagrange multipliers;

II No extra unknowns introduced.

The second category of additional integrands is our focus, five different combinations are presented in the rest of this part for combinations of RGM-FEM, RGM-FDM and various FEMs. More comparisons and relations are given in Section 9.4 later.

Define a bilinear functional with additional integrals along Γ_0:

$$B^{(P_c, \alpha, \beta)}(u, v) = \iint_{S_1} (p \nabla u \nabla v + cuv)\, ds + \iint_{S_2} (p \nabla u \nabla v + cuv)\, ds + D(u, v),$$

where the general coupling strategies without additional unknowns are given by

$$D(u, v) = \int_{\Gamma_N} quv d\ell + \frac{P_c}{h^{*\sigma}} \int_{\Gamma_D} uv d\ell - \int_{\Gamma_D} p\frac{\partial u}{\partial n} v d\ell - \int_{\Gamma_D} p\frac{\partial v}{\partial n} u d\ell$$

$$+ \frac{P_c}{h^{*\sigma}} \int_{\Gamma_0} (u^+ - u^-)(v^+ - v^-)d\ell - \int_{\Gamma_0} p\left(\alpha\frac{\partial u^+}{\partial n} + \beta\frac{\partial u^-}{\partial n}\right)(v^+ - v^-)d\ell$$

$$- \int_{\Gamma_0} p\left(\alpha\frac{\partial v^+}{\partial n} + \beta\frac{\partial v^-}{\partial n}\right)(u^+ - u^-)d\ell,$$

and a linear functional

$$f^{(P_c)}(v) = \iint_S fvds + \int_{\Gamma_N} g_2 vd\ell + \frac{P_c}{h^{*\sigma}} \int_{\Gamma_D} g_1 vd\ell - \int_{\Gamma_D} p\frac{\partial v}{\partial n} g_1 d\ell,$$

where n is the outward normal to ∂S_2 or Γ_D, P_c is a penalty constant, $\sigma(> 0)$ is a penalty power, h^* is the average (or maximal) boundary length of triangular elements along Γ_0, and α and β are bounded constants. Therefore, an approximate solution u_h can be obtained by

$$B^{(P_c, \alpha, \beta)}(u_h, v) = f^{(P_c)}(v), \quad \forall v \in V_h^*. \tag{6.43}$$

In fact, the combinations as (6.43) are general approaches using additional integrals, which contain the following five important combined methods:

1. **Combination I**
   $P_c > 0$, $\alpha = 0$, and $\beta = 1$ (Li and Bui (1988a, b)).

2. **Combination II**
   $P_c > 0$, $\alpha = 1$, and $\beta = 0$ (Li and Bui (1992a, 1988a)).

3. **Symmetric Combination**
   $P_c > 0$, and $\alpha = \beta = \frac{1}{2}$ (Li and Bui (1992a, 1988a)).

4. **The Simplified Hybrid Combination**
   $P_c = 0$, $\alpha = 1$, and $\beta = 0$
   (Li and Liang (1983), and Li and Bui (1988a, 1990a)).

5. **The Penalty Combination**
   $P_c > 0$, and $\alpha = \beta = 0$ (Li (1992b) and Li and Bui (1992b)).

   Note that the above coupling techniques described may be applied to combinations under the LSM frame work, because they all play a role in amending the nonconforming of different admissible functions along Γ_0.

   It should also be noted that using the techniques of additional penalty integrals, simplified hybrid integral, or both can also be found in the penalty method of Barrett and Elliott (1986) for dealing with the Dirichlet boundary condition; in the global element method of Delves (1979), and Delves and Hall

(1979) for coupling different expansions; and in the finite element method for the boundary integral and finite element methods of Zienkiewicz, Kelley and Bettess (1977).

Besides the coupling of RGM and FEM, there have appeared the coupling or combining of other different numerical methods, e.g., the coupling of spectral and FEM in Bernardi, Maday and Landriani (1989/90), Bernardi, Debit and Maday (1990), Kirsch and Monk (1990), and Canuto, Maday and Quarteroni (1982); the coupling of FEM and Marker and cell methods in Girault (1976); the coupling of Galerkin and characteristic methods in Bermejo (1991); and the coupling of the FEM and the perturbation method in Rangogni (1986). The imbedding technique in Borgers and Widlund (1990) is also a kind of coupling. It is expected that more couplings on different methods will be reported in the future.

7

PENALTY TECHNIQUES

In Chapter 6, the Lagrange multipliers need extra variables λ, and lead to an associated matrix that is either non-symmetric or non-positive definite. In this chapter, we will develop new coupling techniques by using additional integrals on Γ_0. The advantages of the new coupling techniques in Chapters 7 and 9 are twofold:

(1) No need of extra variables.

(2) The associated matrix is positive definite and symmetric.

Let us follow the lines in Part II, where a continuity condition for admissible functions is constrained only at the element nodes on the common boundary of the RGM and the FEM. The admissible functions are still discontinuous elsewhere along the entire common boundary. Evidently, additional integrals along the common boundary can be employed to couple the Ritz-Galerkin and finite element methods, instead of the direct constraint conditions in Chapters 4 and 5. The penalty technique is very interesting because of its simplicity. As is well known, if this technique is applied to all element boundaries for the finite element methods, a reduced convergence rate is obtained (see Ciarlet (1978)). However, if this technique is applied only to an exterior boundary of the Dirichlet condition, an optimal convergence rate can be achieved. This result was first reported by Babuška (1973b), and subsequently by many other authors, such as King (1974), King and Serbin (1978), Utku and Carey (1982), Shi (1984), Barrett and Elliott (1986), and Dussult (1995).

The important results in this chapter are given in Theorems 7.2 and 7.3. Under the norms defined in (7.7), the solutions by the penalty combination of RGM– linear FEM have the optimal convergence $O(h)$, even when some variational crimes are committed. Note that the same coupling strategies between L and h can be chosen as in Chapter 4. Another important result is given in Propositions 7.1 and 7.2 in the last section, to reveal that the nonconforming combinations in Chapter 4 can be regarded as a limitation of the penalty combinations in this chapter, when the penalty constant $P_c \longrightarrow \infty$ or the penalty power $\sigma \longrightarrow \infty$. Consequently, it is valid and beneficial to replace the nonconforming combinations by the penalty combinations if we can.

The penalty combinations will be studied again in Chapters 11 and 14. Under the different norms $\overline{\overline{\|v\|}}_h$ defined in Chapter 11 involving discrete summations, not only can the superconvergence rates $O(h^{2-\delta})$, $0 < \delta << 1$ be obtained, but also the FEMs used in combinations may not be confined to be linear elements.

7.1   DESCRIPTION OF COUPLING TECHNIQUES

In this chapter, we shall use the penalty technique along the interface. This technique can also yield an optimal convergence rate for linear FEM, contrasted to that the optimal convergence rate cannot be obtained from Babŭska (1970).

Consider the elliptic boundary value problem of two dimensions

$$-\frac{\partial}{\partial x}\left(\beta\frac{\partial u}{\partial x}\right) - \frac{\partial}{\partial y}\left(\beta\frac{\partial u}{\partial y}\right) = f, \quad \text{in } S, \tag{7.1}$$

$$u = 0, \quad \text{in } \partial S, \tag{7.2}$$

where S is a polygon.

Let the solution domain S be divided into two subdomains S_1 and S_2 by a piecewise straight line Γ_0, i.e., $S = S_1 \cup S_2$. The RGM is used in S_2, and the k-order Lagrange FEM is used in S_1, where $k \geq 1$. We choose the admissible functions v as

$$v = \begin{cases} v^+ = \sum_{l=0}^{L} \tilde{D}_l \varphi_l, & \text{in } S_2, \\ v^- = v_k^h, & \text{in } S_1, \end{cases} \tag{7.3}$$

where $\tilde{D}_l$ are unknown coefficients, $\{\varphi_l\}$ are known and complete and linearly independent functions in $L(S_2)$, and v_k^h are piecewise k-order Lagrange interpolation polynomials on the triangulation of k-order quasiuniform elements with a maximal boundary length h.

Denote V_h^0 as the above admissible function space satisfying the Dirichlet boundary condition (7.2), and

$$V_h^0 \not\subset H^1(S), \tag{7.4}$$

where $H^1(S)$ is the Sobolev space: $H^1(S) = \{v | v, \; v_x, \; v_y \in L^2(S)\}$. The relation (7.4) results from $v^+ \neq v^-$ on Γ_0, where $v^+ = v|_{S_2}$ and $v^- = v|_{S_1}$.

Define a quadratic functional

$$a_h(u,v) = \iint_{S_1} \beta\left(u_x v_x + u_y v_y\right) ds + \iint_{S_2} \beta\left(u_x v_x + u_y v_y\right) ds +$$
$$+ \frac{P_c}{h^{*\sigma}} \int_{\Gamma_0} \left(u^+ - u^-\right)\left(v^+ - v^-\right) d\ell, \tag{7.5}$$

where P_c and σ are nonnegative, bounded constants independent of L, h, u and v, and h^* is the maximal boundary length of k-order elements on Γ_0. Then the penalty combinations are designed to seek an approximate solution $u_h \in V_h^0$ such that

$$a_h(u_h, v) = f(v), \quad \forall v \in V_h^0. \tag{7.6}$$

Here, the penalty integral

$$\frac{P_c}{h^{*\sigma}} \int_{\Gamma_0} \left(v^+ - v^-\right)^2 d\ell$$

will play the role of matching the RGM and the FEM across their common boundary Γ_0. We refer Γ_0 to the coupling boundary, P_c and σ to the penalty constant and the penalty power, respectively.

In terms of (7.4), we define a new norm over V_h^0:

$$\|v\|_{h,P_c} = \left\{ \|v\|_{1,S_1}^2 + \|v\|_{1,S_2}^2 + \frac{P_c}{h^{*\sigma}} \|v^+ - v^-\|_{0,\Gamma_0}^2 \right\}^{\frac{1}{2}}. \tag{7.7}$$

We here give the following theorem.

Theorem 7.1 *Assume that*

$$C_0\|v\|_{h,P_c}^2 \leq a_h(v,v), \quad v \in V_h^0,$$
$$|a_h(u,v)| \leq C\|u\|_{h,P_c}\|v\|_{h,P_c}.$$

There exists a bounded constant C independent of L, h, and u such that

$$\|u - u_h\|_{h,P_c} \leq C\left\{ \inf_{v \in V_h^0} \|u - v\|_{h,P_c} + h^{*\frac{\sigma}{2}} \left|\frac{\partial u}{\partial n}\right|_{0,\Gamma_0} \right\},$$

where n is the outward normal to ∂S_2 and h is the maximal boundary length of k-order elements in S_1.

Proof: For the true solution of (7.1) and (7.2) the bilinear form

$$a_h(u,v) = \iint_{S_1} \beta \nabla u \nabla v \, ds + \iint_{S_2} \beta \nabla u \nabla v \, ds = \int_{\Gamma_0} \beta \frac{\partial u}{\partial n}(v^+ - v^-)d\ell + \iint_S fv \, ds.$$

Let $w = u_h - v$ where $v \in V_h^0$, then $w \in V_h^0$. The uniformly V_h^0-elliptic inequality gives

$$C_0\|u_h - v\|_{h,P_c}^2 \leq a_h(u_h - v, w) = \left| a_h(u - v, w) - \int_{\Gamma_0} \beta \frac{\partial u}{\partial n}(v^+ - v^-)d\ell \right|$$

$$\leq C\|u - v\|_{h,P_c}\|w\|_{h,P_c} + C \left\| \frac{\partial u}{\partial n} \right\|_{0,\Gamma_0} \|v^+ - v^-\|_{0,\Gamma_0}$$

$$\leq C \left(\|u - v\|_{h,P_c} + h^{*\frac{g}{2}} \left\| \frac{\partial u}{\partial n} \right\|_{0,\Gamma_0} \right) \|w\|_{h,P_c}.$$

Hence we have

$$\|u_h - v\|_{h,P_c} \leq \frac{C}{C_0} \left(\|u - v\|_{h,P_c} + h^{*\frac{g}{2}} \left\| \frac{\partial u}{\partial n} \right\|_{0,\Gamma_0} \right).$$

The desired result (7.7) follows from the inequality

$$\|u - u_h\|_{h,P_c} \leq \|u - v\|_{h,P_c} + \|v - u_h\|_{h,P_c}. \qquad \blacksquare$$

Assume that the solutions on S_1 and Γ_0 (but not in S_2 where there may exist singularities) are smooth enough such that

$$u \in H^{k+1}(S_1) \text{ and } u \in H^{k+1}(\Gamma_0), \ k \geq 1. \qquad (7.8)$$

On the basis of the Sobolev imbedding theorem (Sobolev (1963), p.64), we have

$$C_0 \left\| \frac{\partial u}{\partial n} \right\|_{0,\Gamma_0} \leq \|u\|_{2,S_1} \leq C.$$

From $h^* < h$ and Theorem 7.1, the error bounds of approximate solutions u_h are

$$\|u - u_h\|_{h,P_c} \leq C \left(\inf_{v \in V_h^0} \|u - v\|_{h,P_c} + h^{\frac{g}{2}} \right)$$

$$\leq C \left[\inf_{v \in V_h^0} \left(\|u - v\|_{1,S_2} + \|u - v\|_{1,S_1} + \frac{\sqrt{P_c}}{h^{*\frac{g}{2}}} \|v^+ - v^-\|_{0,\Gamma_0} \right) + h^{\frac{g}{2}} \right]$$

$$\leq C \inf_{v \in V_h^0} \|u - v\|_{1,S_2} + O\left(h^k\right) + O\left(\frac{h^{k+1}}{h^{*\frac{g}{2}}}\right) + O\left(h^{\frac{g}{2}}\right)$$

$$\leq C \inf_{v \in V_h^0} \|u - v\|_{1,S_2} + O\left(h^k\right) + O\left(h^{k+1-\frac{g}{2}}\right) + O\left(h^{\frac{g}{2}}\right). \qquad (7.9)$$

In the last step of (7.9), we have used $h/h^* \leq C$ for quasiuniform k-order elements.

To reach the optimal convergence rates $O\left(h^k\right)$ of solutions by the penalty combinations, the following identities must be fulfilled:

$$k = k + 1 - \frac{\sigma}{2} = \frac{\sigma}{2}, \tag{7.10}$$

$$\inf_{v \in V_h^o} \|u - v\|_{1, S_2} = O\left(h^k\right).$$

Obviously, Eq. (7.10) reduces to $\sigma = 2$ and $k = 1$. We now write this important conclusion as a corollary.

Corollary 7.1 *Let (7.8) be given, and suppose that all k-order elements are quasiuniform. Then necessary conditions of the penalty combinations having optimal convergence rates of the solutions are:*

1. *The penalty power $\sigma = 2$;*

2. *$k = 1$, that is, the linear FEM only.*

7.2 ERROR BOUNDS IN PRESENCE OF VARIATIONAL CRIMES

From Corollary 7.1, we only discuss the combination of the RGM and the linear FEM with the penalty power $\sigma = 2$. Such penalty combinations will be applied to solving the Laplace equation with singularity (Figure 4.2 in Chapter 4):

$$\Delta u = \frac{\partial^2 u}{\partial x^2} + \frac{\partial^2 u}{\partial y^2} = 0, \text{ in } S,$$

$$u = 0, (x, y) \in \overline{AC}, \tag{7.11}$$

$$\frac{\partial u}{\partial n} = 0, (x, y) \in \overline{AB}, \tag{7.12}$$

$$\frac{\partial u}{\partial n} = g, (x, y) \in \overline{BF} \cup \overline{FG} \cup \overline{GC}.$$

Let $\Theta = \angle CAB$. Choose a circular arc $l_{R^*} (r = R^*, 0 \leq \theta < \Theta)$ as the coupling boundary Γ_0, i.e., $\Gamma_0 = l_{R^*}$. In this case, $S_2 = \{(r, \theta), 0 \leq r \leq R^*, 0 \leq \theta \leq \Theta\}$, and S_1 is the rest of S.

The admissible functions can be chosen as in Chapter 4,

$$v = \begin{cases} v^+ = \sum_{l=0}^{L} \tilde{D}_l (r/R)^{\tau\left(l+\frac{1}{2}\right)} \cos \tau \left(l+\frac{1}{2}\right)\theta, & \text{in } S_2, \\ v^- = v_1^h, & \text{in } \hat{S}_1^h, \end{cases} \tag{7.13}$$

where $\tau = \pi/\Theta$, R is the radius of a circular arc $l_R(r = R, \ 0 \le \theta \le \Theta)$ within S, $\hat{S}_1^h$ is the triangulation domain of S_1, and $S_1 \approx \hat{S}_1^h$. In (7.13), only linear piecewise interpolation functions v_1^h are used because of Corollary 7.1. The space of the functions (7.13) satisfying the boundary condition (7.11) is still denoted by V_h^0.

The penalty combinations can be described from (7.6) as:

To seek $\tilde{u}_h \in V_h^0$ such that

$$\tilde{a}(\tilde{u}_h, v) = g(v), \qquad v \in V_h^0, \tag{7.14}$$

where

$$\tilde{a}_h(u, v) = \iint_{\hat{S}_1^h} (u_x v_x + u_y v_y)\, ds + \iint_{S_2} (u_x v_x + u_y v_y)\, ds$$
$$+ \frac{P_c}{h^{*2}} \widehat{\int_{\Gamma_0}} (u^+ - u^-)(v^+ - v^-)\, d\ell, \tag{7.15}$$

$$g(v) = \int_{\overline{BF \cup FG \cup GC}} gv d\ell,$$

where $\widehat{\int}_{\Gamma_0}$ denotes a numerical approximation of $\int_{\Gamma_0}$. In (7.15), the penalty power, $\sigma = 2$, is used.

Compared with (7.6), there exist two kinds of variational crimes in (7.15) (the terminology in Strang and Fix (1973)):

(1) the integration approximation along Γ_0 and

(2) the noncoincidence of $\hat{S}_1^h$ and S_1.

The approximate solution is denoted by $\tilde{u}_h$ for distinguishing from u_h; and further estimates for error bounds have to be done.

For the penalty integral $\int_{\Gamma_0} \xi \eta d\ell$, where $\xi = u^+ - u^-$ and $\eta = v^+ - v^-$, if ξ and η are substituted by their piecewise linear interpolation functions $\hat{\xi}$ and $\hat{\eta}$ along Γ_0, approximate integration values can be easily obtained:

$$\int_{\Gamma_0} \xi\eta d\ell \approx \widehat{\int_{\Gamma_0} \xi\eta d\ell} = \int_{\Gamma_0} \hat{\xi}\hat{\eta} d\ell \qquad (7.16)$$

$$= \sum_i \frac{\Delta z_i}{6} \left[2\xi(z_i)\eta(z_i) + \xi(z_i)\eta(z_{i+1}) + \xi(z_{i+1})\eta(z_i) + 2\xi(z_{i+1})\eta(z_{i+1}) \right],$$

where $\Delta z_i = z_{i+1} - z_i$ and z is the coordinate along Γ_0.

Below we shall analyze the effects of, especially, the integration approximation (7.16) on error bounds.

Define the following norms and let $w = v^+ - v^-$:

$$\|v\|_{h,P_c} = \left(\|v\|_{1,\hat{S}_1^h}^2 + \|v\|_{1,S_2}^2 + \frac{P_c}{h^{*\sigma}} \|w\|_{0,\Gamma_0}^2 \right)^{\frac{1}{2}}, \qquad (7.17)$$

$$\overline{\|v\|}_{0,\Gamma_0} = \left(\widehat{\int_{\Gamma_0} v^2 d\ell} \right)^{\frac{1}{2}} = \left(\int_{\Gamma_0} \hat{v}^2 d\ell \right)^{\frac{1}{2}} = \|\hat{v}\|_{0,\Gamma_0}. \qquad (7.18)$$

Then we have the following two lemmas.

Lemma 7.1 *Suppose that the quasiuniform triangular elements and the integration approximation (7.16) are used, there exist the following bounds for any $v \in V_h^0$:*

$$\|w\|_{0,\Gamma_0} \le \overline{\|w\|}_{\Gamma_0} + C \left(h^2 \left[\tau \left(L + \frac{1}{2} \right) \right]^{\frac{3}{2}} + h^{3/2} \right) \|v\|_{h,P_c}, \qquad (7.19)$$

$$\overline{\|w\|}_{\Gamma_0} \le \|w\|_{0,\Gamma_0} + C \left(h^2 \left[\tau \left(L + \frac{1}{2} \right) \right]^{\frac{3}{2}} + h^{3/2} \right) \|v\|_{h,P_c}, \qquad (7.20)$$

where $w = v^+ - v^-$, and h is the maximal boundary length of triangular elements.

Proof: From (7.18), we have from the triangular inequality

$$\|w\|_{0,\Gamma_0} - \overline{\|w\|}_{0,\Gamma_0} = \|w\|_{0,\Gamma_0} - \|\hat{w}\|_{0,\Gamma_0} \le \|w - \hat{w}\|_{0,\Gamma_0}.$$

Let $w = v^+ - v^-$, we obtain

$$\|v^+ - v^-\|_{0,\Gamma_0} - \overline{\|v^+ - v^-\|}_{0,\Gamma_0} \le \|v^+ - \hat{v}^+\|_{0,\Gamma_0}. \qquad (7.21)$$

Note that $\widehat{v}^+$ is the piecewise linear interpolation of v^+. From the arguments of Lemma 4.3 in Chapter 4 we obtain

$$
\begin{aligned}
\|v^+ - \widehat{v}^+\|_{0,\Gamma_0} &= \left(\int_{\Gamma_0} (v^+ - \widehat{v}^+)^2 d\ell \right)^{1/2} \\
&\leq C \left(h^{3/2} + \left[\tau \left(L + \frac{1}{2} \right)^{3/2} h^2 \right] \right) \|v\|_{1,S_2} \\
&\leq C \left(h^{3/2} + \left[\tau \left(L + \frac{1}{2} \right)^{3/2} h^2 \right] \right) \|v\|_{h,P_c}.
\end{aligned}
\tag{7.22}
$$

The first result (7.19) follows from (7.21). The proof of (7.20) is similar. ∎

Lemma 7.2 *Let $Meas(S_1 \cap \overline{AC}) \neq 0$, $u \in H^2(\Gamma_0)$, and rule (7.16) be used. Assume that the triangular elements in S_1 are quasiuniform, and*

$$
h \left[\tau \left(L + \frac{1}{2} \right) \right]^{3/2} \leq C.
\tag{7.23}
$$

Then there exist the bounds

$$
\widetilde{a}_h(u,v) \leq C \left(\|u\|_{1,P_c} + h^{\frac{3}{2}} |u|_{2,\Gamma_0} \right) \|v\|_{1,P_c}, \quad \forall v \in V_h^0,
\tag{7.24}
$$

and

$$
C_0 \|v\|_{h,P_c}^2 \leq \widetilde{a}_h(v,v), \quad \forall v \in V_h^0,
\tag{7.25}
$$

where C_0 and C are two bounded constants independent of h, L, u and v.

Proof: From the assumption $Meas(S_1 \cap \overline{AC}) \neq 0$ and the Poincare-Friedrichs inequality, the uniformly V_h^0–elliptic inequality (7.25) holds. Now we show (7.24) only. By the Schwarz inequality,

$$
\begin{aligned}
\widehat{\int_{\Gamma_0}} (u^+ - u^-)&(v^+ - v^-) d\ell \\
&\leq \left(\widehat{\int_{\Gamma_0}} (u^+ - u^-)^2 d\ell \right)^{1/2} \times \left(\widehat{\int_{\Gamma_0}} (v^+ - v^-)^2 d\ell \right)^{1/2}.
\end{aligned}
\tag{7.26}
$$

For $v \in V_h^0$, Lemma 7.1 gives

$$
\left(\widehat{\int_{\Gamma_0}} (v^+ - v^-)^2 d\ell \right)^{1/2}
\tag{7.27}
$$

$$\leq \left(\int_{\Gamma_0}(v^+ - v^-)^2 d\ell\right)^{1/2} + C\left(h^2\left[\tau\left(L+\frac{1}{2}\right)\right]^{\frac{3}{2}} + h^{\frac{3}{2}}\right)\|v\|_{h,P_c}.$$

However for $u \notin V_h^0$, we obtain by a similar argument in Lemma 7.1

$$\left(\widehat{\int_{\Gamma_0}}(u^+ - u^-)^2 d\ell\right)^{1/2} \leq \left(\int_{\Gamma_0}(u^+ - u^-)^2 d\ell\right)^{1/2} + Ch^2|u^+|_{2,\Gamma_0}. \qquad (7.28)$$

Therefore, combining (7.26), (7.28) and (3.23) leads to

$$\frac{P_c}{h^{*2}}\widehat{\int_{\Gamma_0}}(u^+ - u^-)(v^+ - v^-)d\ell$$

$$\leq C\left(\frac{1}{h^*}h^2\left[\tau\left(L+\frac{1}{2}\right)\right]^{3/2} + \frac{h^{\frac{3}{2}}}{h^*}\right)\|v\|_{h,P_c} \times \left(\|u\|_{h,P_c} + \frac{h^2}{h^*}|u^+|_{2,\Gamma_0}\right)$$

$$\leq C\left(\|u\|_{1,P_c} + h^{\frac{3}{2}}|u^+|_{2,\Gamma_0}\right)\|v\|_{h,P_c}. \qquad (7.29)$$

This is the desired result (7.24). ∎

Eq. (7.23) gives a necessary coupling relation between $(L+1)$ and h, the same as (4.41) in Chapter 4. Now we prove the error theorem.

Theorem 7.2 *Let $Meas\,(\partial S_1 \cap \overline{AC} \neq 0)$, and $u \in H^2(S_1)$ be given. Also suppose that the conditions in Lemma 7.2 hold, and the integral approximation (7.16) is used. Then when $\sigma = 2$, the solutions by the penalty combinations of the RGM and the linear FEM have the error bounds*

$$\|u - \tilde{u}_h\|_{h,P_c} \leq C\left\{\inf_{v \in V_h^0}\|u - v\|_{h,P_c} + h\right\}. \qquad (7.30)$$

Proof. First for a true solution u, we have

$$\tilde{a}_h(u,v) = \iint_{\hat{S}_1^h}(u_x v_x + u_y v_y)\,ds + \iint_{S_2}(u_x v_x + u_y v_y)\,ds \qquad (7.31)$$

$$= \int_{l_{R^*}}\frac{\partial u}{\partial r}(v^+ - v^-)\,d\ell - \int_{\hat{l}_{R^*}}\frac{\partial u}{\partial n}v^-\,d\ell + \int_{l_{R^*}}\frac{\partial u}{\partial r}v^-\,d\ell + g(v),$$

where n is the inward normal to $\partial\hat{S}_1^h$.

Since Lemma 7.2 holds, using (7.14) and (7.31) gives

$$
\begin{aligned}
C_0 \|\tilde{u}_h - v\|_{h,P_c}^2 &\leq \tilde{a}_h(\tilde{u}_h - v, \tilde{u}_h - v) \\
&= \tilde{a}_h(\tilde{u}_h - v, u - v) - \int_{l_{R^*}} \frac{\partial u}{\partial r}(w^+ - w^-)d\ell \\
&\quad + \left(\int_{\hat{l}_{R^*}} \frac{\partial u}{\partial n} w^- \, d\ell - \int_{l_{R^*}} \frac{\partial u}{\partial r} w^- \, d\ell \right),
\end{aligned}
\tag{7.32}
$$

with $w = v - \tilde{u}_h$ and $v \in V_h^0$. Moreover from the Green theorem

$$
\left| \int_{\hat{\ell}_{R^*}} \frac{\partial u}{\partial n} w^- \, d\ell - \int_{\ell_{R^*}} \frac{\partial u}{\partial r} w^- \, d\ell \right| = \left| \int_{\hat{S}_1^h \cap S_2} \nabla u \, \nabla w^- \, ds \right|
$$

$$
\leq \|u\|_{1,\hat{S}_1^h \cap S_2} \|w^-\|_{1,\hat{S}_1^h \cap S_2}.
\tag{7.33}
$$

Since the area of $\hat{S}_1^h \cap S_2$ is $O(h^3)$, and since the solution derivatives on $\hat{S}_1^h \cap S_2$, far from the origin, are bounded, we obtain, $\|u\|_{1,\hat{S}_1^h \cap S_2} \leq Ch^{\frac{3}{2}}$, to lead to

$$
\left| \int_{\hat{\ell}_{R^*}} \frac{\partial u}{\partial n} w^- \, d\ell - \int_{\ell_{R^*}} \frac{\partial u}{\partial r} w^- \, d\ell \right| \leq Ch^{\frac{3}{2}} \|w^-\|_{1,\hat{S}_1^h}.
\tag{7.34}
$$

Therefore, we have from Lemma 7.1

$$
\begin{aligned}
\left| \int_{\ell_{R^*}} \frac{\partial u}{\partial r}(w^+ - w^-)d\ell \right| &\leq \left\| \frac{\partial u}{\partial r} \right\|_{0,\Gamma_0} \|w^+ - w^-\|_{0,\Gamma_0} \\
&\leq C \left\| \frac{\partial u}{\partial r} \right\|_{0,\Gamma_0} \left\{ \overline{\|w^+ - w^-\|}_{0,\Gamma_0} + Ch\|w\|_{h,P_c} \right\} \\
&\leq Ch\|u\|_{2,S_1}\|w\|_{h,P_c}.
\end{aligned}
\tag{7.35}
$$

We can see from (7.24), (7.32), (7.34) and (7.35)

$$
\|\tilde{u}_h - v\|_{h,P_c}^2 \leq C \left(\|u - v\|_{h,P_c} + h^{\frac{3}{2}} + h \right) \|v - \tilde{u}_h\|_{h,P_c}.
$$

Therefore

$$
\|\tilde{u}_h - v\|_{h,P_c} \leq C \left(\|u - v\|_{h,P_c} + h \right),
\tag{7.36}
$$

and the desired result (7.30) follows from

$$
\|u - \tilde{u}_h\|_{h,P_c} \leq \|u - v\|_{h,P_c} + \|v - \tilde{u}_h\|_{h,P_c}. \quad \blacksquare
$$

7.3 COUPLING STRATEGY BETWEEN $(L+1)$ AND H

Besides the conditions in Theorem 7.2, we need more assumptions:

$$R^* < R, \tag{7.37}$$

where $l_R \in S_1$, and

$$u \in H^2(l_R). \tag{7.38}$$

Below, we first employ Theorem 7.2 to provide explicitly the error bounds, and then discuss the coupling strategy between $(L+1)$ and h. Choose an auxiliary function w_h as in Chapter 4 as the trial function v, for applying (7.30) as $v = \overline{w}_h$. An analytic cut-off function $\phi(r)$ is given by $\phi(r) \in C^\infty(R^*, R)$ and

$$\phi(r) = 0 \text{ for } r \leq R_1, \quad \phi(r) = 1 \text{ for } r \geq R_2,$$

where the radii satisfy: $R^* < R_1 < R_2 < R$. A special function v is designed as

$$v = \begin{cases} u, & \text{for } r \geq R, \\ f_L(D_\ell) + \phi(r) \sum_{\ell=L+1}^{L} D_\ell \left(\frac{r}{R}\right)^{\tau\left(\ell+\frac{1}{2}\right)} \cos\left[\tau\left(\ell+\frac{1}{2}\right)\theta\right], & \text{for } R^* \leq r \leq R, \\ f_L(D_\ell), & \text{for } r \leq R^*, \end{cases} \tag{7.39}$$

where D_l are expansion coefficients of u, and

$$f_L(D_l) = \sum_{l=0}^{L} D_\ell \left(\frac{r}{R}\right)^{\tau\left(\ell+\frac{1}{2}\right)} \cos\left[\tau\left(\ell+\frac{1}{2}\right)\theta\right].$$

Then the trial function $w_h \in V_h^0$ is defined by

$$w_h = \begin{cases} w_h^+ = f_L(D_\ell), & \text{in } S_2, \\ w_h^- = \overline{v}_1, & \text{in } \hat{S}_1^h, \end{cases} \tag{7.40}$$

where $\overline{v}_1$ are piecewise linear interpolation functions of v (7.39), on the the triangulation domain $\hat{S}_1^h$. Such a trial function w_h also satisfies the continuity condition on the element nodes (R^*, θ_j):

$$w_h^+(R^*, \theta_j) = w_h^-(R^*, \theta_j), \quad \forall j, \text{ i.e.,}$$

$$\overline{v}_1(R^*, \theta_j) = \sum_{\ell=0}^{L} D_\ell (R^*/R)^{\tau\left(\ell+\frac{1}{2}\right)} \cos\left[\tau\left(\ell+\frac{1}{2}\right)\theta_j\right], \quad \forall j. \tag{7.41}$$

We also obtain a lemma from Lemma 4.2 in Chapter 4:

Lemma 7.3 *Let (7.37) and $u \in H^2(S_1)$ be given. Then for quasiuniform elements in $\hat{S}_1^h$, there exist the bounds*

$$\|u - w_h^-\|_{1,\hat{S}_1^h} + \|u - w_h^+\|_{1,S_2} \le C \left[h + \left(\frac{R_2}{R} \right)^{\tau(L + \frac{3}{2})} \right],$$

with $R^* < R_2 < R$.

In addition, we need two more lemmas.

Lemma 7.4 *Let (7.38) be given, then the true expansion coefficients D_l in*

$$u(R, \theta) = \sum_{\ell=0}^{\infty} D_\ell \left(\frac{r}{R} \right)^{\tau(\ell + \frac{1}{2})} \cos \left[\tau(\ell + \frac{1}{2})\theta \right]$$

are bounded by

$$|D_\ell| \le \frac{C}{\left[\tau \left(\ell + \frac{1}{2} \right) \right]^2}, \tag{7.42}$$

with a bounded constant C independent of ℓ.

Proof. Since

$$D_\ell = \frac{2}{\Theta} \int_0^{\Theta} u(R, \theta) \cos \left[\tau \left(l + \frac{1}{2} \right) \theta \right] d\theta,$$

by performing integration by parts and using the boundary conditions (7.11) and (7.12), we obtain

$$
\begin{aligned}
D_\ell &= \frac{-2}{\Theta \tau \left(\ell + \frac{1}{2} \right)} \int_0^{\Theta} \frac{\partial u(R, \theta)}{\partial \theta} \sin \left[\tau \left(\ell + \frac{1}{2} \right) \theta \right] d\theta \\
&= \frac{-2}{\Theta \left[\tau \left(\ell + \frac{1}{2} \right) \right]^2} \int_0^{\Theta} \frac{\partial^2 u(R, \theta)}{\partial \theta^2} \cos \left[\tau \left(\ell + \frac{1}{2} \right) \theta \right] d\theta.
\end{aligned}
$$

Then

$$|D_\ell| \le \frac{C}{\left[\tau \left(\ell + \frac{1}{2} \right) \right]^2} \|u\|_{2, \ell_{R^*}}, \tag{7.43}$$

and the desired result (7.42) follows from the assumption (7.38). ∎

Lemma 7.5 *Let (7.37), (7.38), and $u \in H^2(S_1)$ be given. Then there exist the bounds*

$$\|w_h^+ - w_h^-\|_{0,l_{R^*}} \leq Ch^2, \tag{7.44}$$

where $w_h^{\pm}$ are defined in (7.40), and h is the maximal boundary length of the elements.

Proof. In terms of (7.41), we construct a linear interpolation function $P(\theta)$ of both w_h^+ and w_h^-, with respect to θ:

$$P(\theta) = w_h(R^*, \theta_j) + \frac{w_h(R^*, \theta_{j+1}) - w_h(R^*, \theta_j)}{\theta_{j+1} - \theta_j}(\theta - \theta_j),$$

where $\theta_j \leq \theta \leq \theta_{j+1}$. Then we have

$$\begin{aligned}
\|w_h^+ - w_h^-\|_{0,l_{R^*}} &\leq \|w_h^+ - P(\theta)\|_{0,l_{R^*}} + \|w_h^- - P(\theta)\|_{0,l_{R^*}} \\
&\leq Ch^2 \left(|w^+|_{2,l_{R^*}} + T \right),
\end{aligned} \tag{7.45}$$

with the notation

$$T = \left[\sum_i \int_{\theta_i}^{\theta_{i+1}} \left(\frac{\partial^2 \bar{v}_1}{\partial \theta^2} \right)^2 R^* d\theta \right]^{\frac{1}{2}}.$$

Through some integration calculations, we obtain

$$|w_h^+|_{2,l_{R^*}}^2 = |f_L(D_l)|_{2,l_{R^*}}^2 = \frac{\Theta}{2} \sum_{l=0}^{L} D_l^2 \left[\tau \left(l + \frac{1}{2} \right) \right]^4 \left(\frac{R^*}{R} \right)^{\tau(2l+1)}.$$

Hence Eq. (7.37) and Lemma 7.4 grant a constant bound of $|w_h^+|_{2,l_{R^*}}$:

$$|w_h^+|_{2,l_{R^*}} \leq C. \tag{7.46}$$

On the other hand, since $w_h^-(= \bar{v}_1)$ are piecewise linear interpolation functions, we may represent them on the element $\triangle_i$ as

$$w_h^- = a_i + b_i x + c_i y = a_i + b_i r \cos\theta + c_i r \sin\theta,$$

where a_i, b_i, and c_i are constants within $\triangle_i$. Therefore, we have

$$T^2 = \sum_i \int_{\theta_i}^{\theta_{i+j}} \left(\frac{\partial^2 \bar{v}_1}{\partial \theta^2} \right)^2 R^* d\theta \tag{7.47}$$

$$= \sum_i \int_{\theta_i}^{\theta_{i+1}} \left(b_i R^* \cos\theta + c_i R^* \sin\theta \right)^2 R^* d\theta$$

$$\leq C \left(\left\| \frac{\partial \overline{v}_1}{\partial x} \right\|_{0,l_{R^*}}^2 + \left\| \frac{\partial \overline{v}_1}{\partial y} \right\|_{0,l_{R^*}}^2 \right) \leq C' \left(\left\| \frac{\partial v}{\partial x} \right\|_{0,l_{R^*}}^2 + \left\| \frac{\partial v}{\partial y} \right\|_{0,l_{R^*}}^2 \right),$$

with a bounded constant C' independent of h, L and u. Therefore, the condition $u \in H^2(S_1)$ and the Sobolev imbedding theorem also grant a boundedness of T:

$$T \leq C. \tag{7.48}$$

Finally, combining (7.45), (7.46) and (7.48) gives the desired result (7.44).

Let w_h in (7.40) be the trial function v in Theorem 7.2, i.e., $v = w_h \in V_h^0$. The following theorem can be easily proved from Theorem 7.2, Lemmas 7.3, 7.5, and the norm definition (7.17).

Theorem 7.3 *Let (7.37), (7.38), and all assumptions in Theorem 7.2 be given, then*

$$\| u - \overline{u}_h \|_{h,P_c} \leq C \left[h + \left(\frac{R_2}{R} \right)^{\tau\left(L + \frac{3}{2}\right)} \right] \tag{7.49}$$

with $R^* < R_2 < R$.

Below, based on Theorem 7.3, we discuss the coupling relation between $(L+1)$ and h. For optimal convergence rates, the following relation must be fulfilled:

$$\left(\frac{R_2}{R} \right)^{\tau\left(L + \frac{3}{2}\right)} = \overline{C}h, \tag{7.50}$$

with a constant $\overline{C}$ independent of L, h and u. Then, we have

$$L + 1 = \frac{\ln \overline{C} + \ln h}{\tau \left[\ln \left(\frac{R_2}{R} \right) \right]} - \frac{1}{2}, \tag{7.51}$$

and

$$L + 1 = O\left(|\ln h|\right) \quad \text{when} \quad h \longrightarrow 0. \tag{7.52}$$

In this case, the assumption (7.23) is automatically satisfied. We write down this important conclusion as a corollary.

Corollary 7.2 *Let all assumption in Theorem 7.3 hold. There then exist the optimal convergence rates*

$$\|u - \tilde{u}_h\|_{h,P_c} = O(h),$$

provided that the coupling strategy (7.50)–(7.52) between $(L+1)$ and h is chosen.

We can also obtain from Chapter 4 a more useful corollary for practical application.

Corollary 7.3 *Let all assumptions in Theorem 7.3 hold. There then exist the optimal convergence rates*

$$\|u - \tilde{u}_h\|_{h,P_c} = O(h),$$

provided that the coupling strategy between $(L_h + 1)$ and h is chosen as:

$$L_h + 1 = (L_{h_0} + 1) + \frac{\left|\ln\left(\frac{h}{h_0}\right)\right|}{\left|\tau \ln\left(\frac{R_2}{R}\right)\right|}, \qquad R^* < R_2 < R, \tag{7.53}$$

or

$$L_{\frac{h}{2}} + 1 \approx (L_h + 1) + 1, \tag{7.54}$$

for a special case of $\Theta = \pi (i.e., \sigma = \pi/\Theta = 1)$ and $R^*/R = \frac{1}{2}$, where $L_h + 1$ is the total number of expansion functions in S_2, with respect to a fixed, maximal boundary length h of elements in S_1^h.

Remark 7.1 *A necessary constraint (4.41) in Chapter 4 exists between L and h in the nonconforming combination,*

$$h\left[\tau\left(L + \frac{1}{2}\right)\right]^{\frac{3}{2}} \le C,$$

which results from a nonconforming coupling on Γ_0. In this section, the same constraint relation (7.23) between L and h is also required by the penalty combinations of (7.14). Second, the admissible functions (7.13) do not indeed have to satisfy the constraint conditions (7.41) for w_h; the special auxiliary function w_h of (7.40) satisfying (7.41) is employed here only for obtaining the optimal error bounds (7.49) in theoretical analysis.

$L+1$	δ_1	δ_2	Max	$\|\varepsilon\|_{0,S}$	$\|\varepsilon\|_{h,P_c}$	$\|\varepsilon\|_{h,1}$
2	1.984	1.377	4.767	2.828	45.50	37.83
3	1.752	1.345	4.093	2.444	43.99	36.40
4	1.757	1.344	4.171	2.404	43.80	36.19
5	1.773	1.346	4.166	2.401	43.82	36.19
6	1.749	1.347	4.165	2.400	43.84	36.19
8	1.729	1.348	4.163	2.400	43.84	36.19
10	1.742	1.347	4.163	2.399	43.83	36.19
14	2.320	1.517	4.214	2.408	45.86	36.44
16	23.76	12.75	3.699	3.697	250.0	87.68

Table 7.1   The error norms of numerical solutions for $N = M = 2, \sigma = 2, R^* = 0.5$, and $P_c = 14$ while changing $L + 1$. In this table and Table 7.3, $\delta_1 = \|\varepsilon^+ - \varepsilon^-\|_{\infty, l_{R^*}}$ and $\delta_2 = \|\varepsilon^+ - \varepsilon^-\|_{0, l_{R^*}}$.

$L+1$	$\bar{D}_0$	$\bar{D}_1$	$\bar{D}_2$	$\bar{D}_3$	$\bar{D}_7$	$\bar{D}_{11}$	$\bar{D}_{15}$
2	398.689	87.9558					
3	398.413	87.7242	18.5167				
4	398.415	87.7508	18.5609	−11.7765			
5	398.413	87.7518	18.5624	−11.7726			
6	398.412	87.7569	18.5593	−11.7742			
8	398.412	87.7556	18.5650	−11.7705	3.23369		
10	398.412	87.7553	18.5662	−11.7695	2.16188		
14	398.391	87.7417	16.8299	−10.1324	1.97906	−51.6011	
16	389.687	82.3472	16.4866	−10.1902	1.62508	−44.2696	411911
True Coeffs.	401.162	87.6559	17.2379	−8.07122	−.086933	−.003184	−.00012

Table 7.2   The approximate coefficients for $M = N = 2$, $\sigma = 2$, $R^* = 0.5$, and $P_c = 14$ while changing $L + 1$.

7.4 NUMERICAL EXPERIMENTS

To solve Motz's problem, we use the penalty combinations (7.14) for the division as shown in Figure 4.4 of Chapter 4. Letting $\sigma = 2$, we have obtained the numerical solutions and evaluated their errors in two cases.

(I) The division of Figure 4.4 is taken as an example for testing different values of P_c. After numerical solutions have been obtained, we compute the error norms

$$\|\varepsilon\|_{h,P_c}, \quad \text{defined as (7.17)},$$

$$\|\varepsilon\|_{h,1} = \|\varepsilon\|_{h,P_c}, \quad \text{when } P_c = 1,$$

$$\|\varepsilon\|_{0,S} = \left[\iint_S \varepsilon^2 ds\right]^2, \quad \max = \max\left[\max_{\widehat{S}_1^h} |\varepsilon_{i,j}|, \max_{S_2} |\varepsilon|\right],$$

$$\delta_1 = \|\varepsilon^+ - \varepsilon^-\|_{\infty, \ell_{R^*}} = \max_{\ell_{R^*}} |\varepsilon^+ - \varepsilon^-| = \max_{\ell_{R^*}} |u_h^+ - u_h^-|,$$

$$\delta_2 = \|\varepsilon^+ - \varepsilon^-\|_{0, \ell_{R^*}} = \left[\int_{\ell_{R^*}} (\varepsilon^+ - \varepsilon^-)^2 d\ell\right]^{1/2} = \left[\int_{\ell_{R^*}} (u_h^+ - u_h^-)^2 d\ell\right]^{\frac{1}{2}},$$

where $\varepsilon = u - u_h$, and ε_{ij} are the errors at element nodes in $\widehat{S}_1^h$.

First for the partition of Figure 4.2, we choose the good matches $L + 1 = 4$, $M = N = 2$ as given in Section 4.8 of Chapter 4. Let $h^* = \frac{\pi R^*}{8} = \frac{\pi}{16}$. The penalty constant P_c is tested to discover a good value as $P_c = 14$. By following Section 4.8, we also change L. Error norms and approximation coefficients are provided as Tables 7.1 and 7.2.

From Table 7.1 we can see that $L + 1 = 4$ is a good choice for matching $M = N = 2$ because all error norms are small and because the calculation work is also small. When $L + 1 = 2$, the error norms are large owing to too few terms of expansions used. However, when $L + 1$ is large, e.g., $L + 1 = 16$, the error norms become large. We recall that too large a number, $L + 1$, will violate the condition (7.23) of (7.24).

(II) We refine the division of Figure 4.4 by increasing $M(= N)$. Since we have found from (I) an optimal match $L + 1 = 4$ for $M = N = 2$, on the basis of other optimal matches expected, we choose $L + 1 = 5$ for $M = N = 4$ and $L + 1 = 6$ for $M = N = 8$. In calculation, the penalty power $\sigma = 2$ and the penalty constant $P_c = 14$ are used, see the next subsection. The results are given in Tables 7.3 and 7.4.

Divisions	δ_1	δ_2	Max	$\|\varepsilon\|_{0,S}$	$\|\varepsilon\|_{h,P_c}$	$\|\varepsilon\|_{h,1}$
$M = N = 2$	1.757	1.344	4.171	2.404	43.80	36.19
$L+1 = 4$						
$M = N = 3$	0.7818	0.6207	1.918	1.024	29.29	23.78
$L+1 = 5$						
$M = N = 4$	0.4406	0.3572	1.088	0.5681	22.09	17.77
$L+1 = 5$						
$M = N = 6$	0.1973	0.1628	0.4779	0.2501	14.84	11.83
$L+1 = 6$						
$M = N = 8$	0.1115	0.09280	0.2708	0.1402	11.19	8.869
$L+1 = 6$						

Table 7.3 The error norms of numerical solutions for $M = N, \sigma = 2, P_c = 14$, and $R^* = 0.5$.

Divisions	$\tilde{D}_0$	$\tilde{D}_1$	$\tilde{D}_2$	$\tilde{D}_3$	$\tilde{D}_4$	$\tilde{D}_5$
$M = N = 2$	398.415	87.7508	18.5609	−11.7765		
$L+1 = 4$						
$M = N = 3$	399.946	87.6630	17.8865	−10.3156	1.39370	
$L+1 = 5$						
$M = N = 4$	400.476	87.6568	17.6280	−9.54075	1.79517	
$L+1 = 5$						
$M = N = 6$	400.856	87.6563	17.4225	−8.81746	1.78080	−0.091906
$L+1 = 6$						
$M = N = 8$	400.989	87.6563	17.3445	−8.51435	1.68054	0.044985
$L+1 = 6$						
True Coeffs.	401.162	87.6559	17.2379	−8.07122	1.44027	0.331055

Table 7.4 The approximate coefficients for $M = N, \sigma = 2, P_c = 14$ and $R^* = 0.5$.

It appears that the following asymptotic relations exist:

$$\|\varepsilon\|_{h,P_c} = O(h), \qquad \|\varepsilon\|_{h,1} = O(h), \tag{7.55}$$

$$\|\varepsilon\|_{0,S} = O(h^2), \qquad \text{max} = O(h^{2-\delta}), \tag{7.56}$$

$$\|\varepsilon^+ - \varepsilon^-\|_{0,\ell_{R^*}} = O(h^{2-\delta}), \qquad \|\varepsilon^+ - \varepsilon^-\|_{\infty,\ell_{R^*}} = O(h^{2-\delta}), \tag{7.57}$$

$$\delta D_0 = |D_0 - \tilde{D}_0| = O(h^2), \tag{7.58}$$

where δ is a very small, positive number. All these asymptotic relations (7.55)–(7.58) are, almost, the same as those in the FEM and in Chapter 4. In particular, Eq. (7.55) coincides with the analysis in Sections 7.2 and 7.3.

We use only six terms of expansions in S_2 to match the finest division, $M = N = 8$, in S_1, and to achieve optimal convergence rates. And it is also only six terms of expansions in S_2 that can yield a good approximation of the first coefficient: $\widetilde{D}_0 = 400.989$ with a relative error of 0.0004. This clearly displays high efficiency of the penalty combinations for singularity problems.

In Li (1992b), we have carried out the computations with $\sigma = 1$ and 3 under $P_c = 14$, to show

$$\|\varepsilon\|_{h,P_c} = O(h^{0.9}) \qquad \text{for } \sigma = 1,$$
$$\|\varepsilon\|_{h,P_c} = O(h) \qquad \text{for } \sigma = 2,$$
$$\|\varepsilon\|_{h,P_c} = O(h^{0.83}) \qquad \text{for } \sigma = 3,$$

where $\|\varepsilon\|_{h,P_c}$ is defined as (7.17). Only the penalty power, $\sigma = 2$, can reach the optimal convergence rates of $\|\varepsilon\|_{h,P_c}$. This fact also verifies the analysis in Corollary 7.1.

7.5 RELATION TO NONCONFORMING COMBINATIONS

To explore the relations between the penalty combinations in this chapter and the nonconforming combinations in Chapter 4, we first choose the trapezoidal rule

$$\int_x^{x+\Delta x} f(x)dx \approx \frac{1}{2} \Delta x \left[f(x) + f(x + \Delta x) \right], \tag{7.59}$$

for the penalty integration

$$\int_{\Gamma_0} (u^+ - u^-)(v^+ - v^-)d\ell \approx \widetilde{\int_{\Gamma_0} (u^+ - u^-)(v^+ - v^-)d\ell}$$

$$= \sum_{i=0}^N (u^+(R^*,\theta_j) - u_{0j})(v^+(R^*,\theta_j) - v_{0j})\frac{R^*}{2}(\Delta\theta_j + \Delta\theta_{j-1}), \tag{7.60}$$

where $\Delta\theta = \theta_{j+1} - \theta_j$ and $\Delta\theta_N = \Delta\theta_{-1} = 0$. Let (R^*,θ_j) in (7.60) denote the element nodes on Γ_0 with

$$v^+(R^*,\theta_j) = \sum_{\ell=0}^L \widetilde{D}_\ell (R^*)^{\ell+1/2} \cos[(\ell+1/2)\theta];$$

$$v_{ij} = v^-(r_i, \theta_j), \quad r_i > R^*(i > 0); \quad v_{0j} = v^-(R^*, \theta_j), \quad (R^*, \theta_j) \in \Gamma_0.$$

Through manipulation, from the nonconforming combination we can easily obtain the following difference equations on Γ_0

$$
\begin{aligned}
0 = & \frac{1}{2}\pi \left(\ell + \frac{1}{2}\right) \tilde{D}_\ell (R^*)^{2\ell+1} - \frac{P_c}{h^2} \sum_{j=0}^{N} \left[v_{0j} - \sum_{\ell=0}^{L} \tilde{D}_\ell (R^*)^{\ell+\frac{1}{2}} \cos \left(\ell + \frac{1}{2}\right)\theta_j \right] \\
& \times (R^*)^{\ell+\frac{3}{2}} \cos \left(\left(\ell + \frac{1}{2}\right)\theta_j\right) \frac{R^*}{2}(\triangle\theta_j + \triangle\theta_{j-1}), \quad \text{for } \ell = 0, 1, \ldots, L, (7.61)
\end{aligned}
$$

$$0 = E(v_{ij}), \quad \text{for } i > 0, \tag{7.62}$$

$$
\begin{aligned}
0 = & E(v_{0j}) + \frac{P_c}{h^2} \sum_{i=0}^{N} \left[v_{0j} - \sum_{\ell=0}^{L} \tilde{D}_\ell (R^*)^{\ell+\frac{1}{2}} \cos \left(\ell + \frac{1}{2}\right)\theta_j \right] \times \\
& \times \frac{R^*}{2}(\triangle\theta_j + \triangle\theta_{j-1}), \quad \text{for } i = 0, \tag{7.63}
\end{aligned}
$$

where $E(v_{ij})$ (or $E(v_{0j})) = 0$ denotes the algebraic equations resulted purely from the FEM in S_1. In fact, Eqs. (7.61)–(7.63) can also be obtained from Section 9.4 in Chapter 9.

Naturally, the solutions, v^+ and v^-, obtained here are not exactly the same at the element nodes (R^*, θ_j), i.e.,

$$v^+(R^*, \theta_j) \neq v^-(R^*, \theta_j). \tag{7.64}$$

This is a key difference between the penalty combination and the nonconforming combination. Furthermore, when the penalty constant, $P_c \longrightarrow \infty$, the following coupling condition follows from (7.61) and (7.63):

$$v^+(R^*, \theta_j) = v^-(R^*, \theta_j). \tag{7.65}$$

This is the very constraint condition in Chapter 4. In this case, the solutions from the penalty combinations will, indeed, approach those from the nonconforming combination, provided that the integral approximations of (7.59) and (7.60) are chosen. In later practical calculations, the penalty constant will be suitably large but not infinite, based on computer limitation and solution stability. In fact, a similar technique can be found in Zienkiewicz (1977) while dealing with the Dirichlet boundary condition.

Let us write down this result as the following proposition.

P_c	Δ	α_1	α_2	α_3	Max	$\|\varepsilon\|_{0,S}$	$\|\varepsilon\|_{h,1}$
0	454.3	317.1	252.9	366.3	137.3	132.1	340.4
0.1	107.8	75.34	54.67	84.94	32.44	25.63	81.03
1	14.34	11.12	7.948	12.36	3.580	3.740	36.68
4	3.697	3.866	2.941	4.722	4.062	2.454	35.59
12	1.242	2.197	1.836	3.462	4.164	2.403	35.53
14	1.065	2.076	1.759	3.396	4.171	2.404	35.53
16	0.9320	1.986	1.702	3.349	4.176	2.405	35.53
64	0.2335	1.651	1.409	3.148	4.205	2.416	35.53
512	0.02921	1.603	1.326	3.106	4.213	2.421	35.53
4×10^4	.000374	1.597	1.314	3.100	4.214	2.422	35.53
4×10^5	$.37 \times 10^{-4}$	1.597	1.314	3.100	4.214	2.422	35.53
4×10^6	$.37 \times 10^{-5}$	1.597	1.314	3.100	4.214	2.422	35.53
4×10^7	$.6 \times 10^{-6}$	1.598	1.315	3.100	4.214	2.427	35.53
4×10^8	$< .5 \times 10^{-7}$	1.605	1.324	3.109	4.220	2.428	35.54
4×10^9	$< .5 \times 10^{-7}$	1.660	1.374	3.160	4.255	2.459	35.54
4×10^{10}	$< .5 \times 10^{-7}$	2.639	2.380	4.416	4.983	3.085	35.69
Nonconf. Comb.	0	1.597	1.314	3.100	4.214	2.422	35.53

Table 7.5   The error norms of calculated solutions for $M = N = 2, L+1 = 4$ and $R^* = 0.5$ while changing the penalty constant P_c In this table, $\alpha_1 = \|\varepsilon^+\|_{\infty,l_{R^*}}$, $\alpha_2 = \|\varepsilon^+\|_{0,l_{R^*}}$ and $\alpha_3 = \|\varepsilon^+\|_{1,l_{R^*}}$.

Proposition 7.1 *Let the penalty integral be approximated by the trapezoidal rule (7.59) and (7.60). Then when the penalty constant P_c is chosen properly large, the numerical solution obtained from the penalty combinations will approach that from the nonconforming combination.*

Below, we provide the numerical results.

First, take the division of Figure 4.4 as a sample to be calculated where $M = N = 2$ and $R^* = 0.5$. A good penalty constant $P_c = 14$ is observed from the calculation in Table 7.5 where $\varepsilon = u - u_h$, $\Delta = \max_j |v_{0j}^+ - v_{0j}^-|$, and $v_{0j}^\pm$ are the solutions on the nodes (R^*, θ_j) of the common boundary ℓ_{R^*} for $v^\pm$. The norm $\Delta \equiv 0$ is for the nonconforming combination.

P_c	$\tilde{D}_0$	$\tilde{D}_1$	$\tilde{D}_2$	$\tilde{D}_3$
0	0	0	0	0
0.1	316.135	48.9362	7.38716	−4.81281
1	388.794	82.0811	16.5068	−10.3467
4	396.524	86.6082	18.1326	−11.4704
12	398.288	87.6734	18.5318	−11.7555
14	398.415	87.7508	18.5609	−11.7765
16	398.510	87.8086	18.5827	−11.7922
64	399.010	88.1132	18.6983	−11.8757
512	399.156	88.2024	18.7322	−11.9003
4×10^4	399.177	88.2150	18.7370	−11.9037
4×10^5	399.177	88.2152	18.7371	−11.9037
4×10^6	399.177	88.2152	18.7371	−11.9037
4×10^7	399.176	88.2153	18.7373	−11.9043
4×10^8	399.161	88.2164	18.7335	−11.9072
4×10^9	399.081	88.2733	18.7385	−11.9019
4×10^{10}	397.466	88.7345	18.9113	−12.3341
Nonconforming	399.177	88.2152	18.7371	−11.9038
Exact Coeffs.	401.162	87.6559	17.2379	−8.07122

Table 7.6 The calculated coefficients for $M = N = 2, L + 1 = 4$ and $R^* = 0.5$ for changing the penalty constant P_c.

The coefficients obtained are listed in Table 7.6. Moreover, Table 7.7 shows the solution v^- on the nodes (R^*, θ_j), where $\theta_j = j\pi/8$, $j = 0 \sim 7$. For comparison, we also provide the results of nonconforming combination and the true solutions.

(1) Obviously, the choice of $P_c = 0$ causes a wrong solution since nothing is related to the two methods. Also when $P_c = 12 \sim 14$, the error norms obtained in Table 7.5 are small.

(2) When the coupling constant P_c again increases from 512 to 4×10^6, the error norm $\triangle = \max_j |v_{0j}^+ - v_{0j}^-|$ diminishes to 0.37×10^{-5}. This shows that the continuity constraints (7.65) as $v_{0j}^+ = v_{0j}^-$ hold within very small errors in the penalty combinations, to have verified Proposition 7.1.

Looking at the solutions while $P_c = 4 \times 10^4 \sim 4 \times 10^6$, the approximate coefficients $\tilde{D}_\ell$ and the solutions $v^-(R^*, \theta_j)$ in Tables 7.6 and 7.7 are the same as those by the nonconforming combination, within a relative error 1×10^{-6}.

P_c	θ					
	0	$\pi/8$	$\pi/4$	$\pi/2$	$3\pi/4$	$7\pi/8$
0	454.320	442.024	405.918	282.482	141.950	73.0286
0.1	349.500	336.506	300.315	188.682	88.2723	44.3517
1	320.281	308.599	275.942	175.645	83.0461	41.6915
4	316.894	305.477	273.322	174.729	82.7550	41.5593
12	316.108	304.764	272.719	174.540	82.6945	41.5356
14	316.052	304.713	272.676	174.527	82.6902	41.5340
16	316.009	304.674	272.644	174.517	82.6870	41.5328
64	315.785	304.472	272.472	174.466	82.6703	41.5268
512	315.719	304.413	272.422	174.451	82.6654	41.5251
4×10^4	315.710	304.405	272.415	174.449	82.6647	41.5248
4×10^5	315.710	304.405	272.415	174.449	82.6647	41.5248
4×10^6	315.710	304.405	272.415	174.449	82.6647	41.5248
4×10^7	315.709	304.404	272.414	174.448	82.6643	41.5247
4×10^8	315.698	304.393	272.405	174.440	82.6594	41.5221
4×10^9	315.662	304.355	272.359	174.386	82.6197	41.5001
4×10^{10}	314.676	303.381	271.391	173.415	82.0751	41.2497
Noncon.	315.710	304.405	272.415	174.449	82.6647	41.5248
Exact Sol.	317.063	305.517	273.364	176.045	82.9895	41.4140

Table 7.7 The calculated solutions $v^-, (R^*, \theta)$ on the common semi-circle l_{R^*} for $M = N = 2, L + 1 = 4$ and $R^* = 0.5$ while changing the penalty constant P_c.

Therefore, the same values of the error norms, from $\|\varepsilon^+\|_{\infty, \ell_{R^*}}$ to $\|\varepsilon\|_h$ in Table 7.5, are obtained by two combinations, to coincide with the above analysis. This fact also gives us an idea that the solutions of the nonconforming combination may even be provided by the penalty combinations with a suitable, large penalty constant P_c, i.e., $P_c = 4 \times 10^4 \sim 4 \times 10^6$ in the Cyber 835 computer, with 14 significant digits in double precision (48 bits mantissa). Of course, the upper bound of the suitable values of P_c depends on the precision of a computer, and the lower bound of P_c depends on the partition of S.

(3) When the penalty constant P_c becomes extremely large, such as $4 \times 10^7 \sim 4 \times 10^{10}$, the error norms, $\|\varepsilon\|_h$, $\|\varepsilon\|_{0,S}$, etc., except $\triangle$, become larger. In this case the solutions obtained are different from those from the nonconforming combination. The reason is that an extremely large P_c, i.e., $P_c = 4 \times 10^{10}$, causes instability of the numerical solutions.

Remark 7.2 *In summary, if the penalty constant P_c is chosen as a bounded value, e.g., $P_c \approx 14$, the penalty method (7.17) can be regarded as an independent kind of combined method, yielding good approximate solutions. However, if P_c is chosen as a suitably, large value, e.g., $P_c = 4 \times 10^4 \sim 4 \times 10^6$ for Cyber 835, the approximate solutions will approach those from the nonconforming combination.*

Now we consider the integration rule (7.16)

$$\widetilde{\int_{\Gamma_0}} v^2 d\ell = \frac{1}{3} \sum_{i=0}^{N_i-1} \Delta z_i \left[v^2(z_i) + v(z_i)v(z_{i+1}) + v^2(z_{i+1}) \right].$$

The trapezoidal rule (7.60) is

$$\widetilde{\int_{\Gamma_0}} v^2 d\ell = \frac{1}{2} \sum_{i=0}^{N_i-1} \Delta z_i \left[v^2(z_i) + v^2(z_{i+1}) \right].$$

Since $x^2 + y^2 \leq 2(x^2 + xy + y^2)$ and $xy \leq \frac{1}{2}(x^2 + y^2)$, we have

$$\frac{1}{3} \widetilde{\int_{\Gamma_0}} v^2 d\ell \leq \widehat{\int_{\Gamma_0}} v^2 d\ell \leq \widetilde{\int_{\Gamma_0}} v^2 d\ell. \tag{7.66}$$

Let

$$\widehat{\int_{\Gamma_0}} v^2 d\ell = \beta \widetilde{\int_{\Gamma_0}} v^2 d\ell, \quad \beta \in [\tfrac{1}{3}, 1].$$

The difference equations in (7.61) and (7.63) are modified by the rather different penalty constants $P_c^* = \beta P_c$. Also when $P_c \longrightarrow \infty$, then $P_c^* \longrightarrow \infty$, and the same nonconforming constants (7.64) are obtained. We reach the following proposition; a further analysis between the penalty and nonconforming combinations is given in Chapters 11 and 14.

Proposition 7.2 *Proposition 7.1 holds for the integration rule (7.16).*

Remark 7.3 *For the elliptic equation (7.1), the solutions across an interior boundary Γ_0 have to satisfy two continuity conditions:*

$$u^+ = u^-, \quad \text{on } \Gamma_0, \tag{7.67}$$

and

$$\beta^+ \frac{\partial u^+}{\partial n} = \beta^- \frac{\partial u^-}{\partial n}, \quad \text{on } \Gamma_0. \tag{7.68}$$

If the functions (7.3) are used for solving (7.1), the combinations of two methods should also be combined with the additional integrals on Γ_0, concerning both (7.67) and (7.68). It is, however, remarkable that the penalty combination of the RGM and the linear FEM is designed only by (7.67) in this chapter, and can also achieve optimal convergence rates by using the significant coupling relations (7.53) and (7.54). This progress shows a great potential for the penalty methods in combining different numerical approaches. The reason for doing this is that the conforming requirement of admissible functions is only (7.67) for 2-order elliptic equations.

Remark 7.4 *We should point out here that the limitation $k = 1$ and $\sigma = 2$ for optimal convergence rates is undertaken in the norm $\|\cdot\|_{h,P_c}$ defined in (7.7). Such a limitation may be removed by using the discrete norm, $\overline{\overline{\|v\|}}_h$ (or $\|v\|_1$):*

$$\overline{\overline{\|v\|}}_h = \left\{ \|v\|_{1,S_1^2}^2 + \|v\|_{1,S_2^2}^2 + \frac{P_c}{h^\sigma} \overline{\overline{\|v^+ - v^-\|}}_{0,\Gamma_0}^2 \right\}^{\frac{1}{2}},$$

where the discrete norm on Γ_0 is defined by the integral approximation

$$\overline{\overline{\|v\|}}_{0,\Gamma_0}^2 = \widehat{\int_{\Gamma_0}} v^2 d\ell.$$

Take the trapezoidal rule (7.59) as an example,

$$\widehat{\int_{\Gamma_0}} v^2 d\ell = \frac{1}{2} \sum_{i=0}^{N-1} \overline{z_i z_{i+1}} \left[v^2(z_i) + v^2(z_{i+1}) \right].$$

When σ is large, the nonconforming constraint (7.65) is approximately satisfied. Also the norm $\|v^+ - v^-\|_{0,\Gamma}$ leads nearly to zero, thus to maintain the optimal (or superconvergence) rates. Details are given in Chapter 11.

8

SIMPLIFIED HYBRID METHODS

To solve the boundary value problems of elliptic equations, especially with singularities and unbounded domains, the simplified hybrid combinations are presented in this chapter, which also adopt the solution flux on the side ∂S_2 of S_2 where the Ritz-Galerkin method applies. These couplings are simplified to the general hybrid techniques here, where the flux on both ∂S_2 and ∂S_1 must be employed. An equivalence between this coupling and Zienkiewicz, Kelley and Bettess (1977) is revealed in Section 8.2. Note that the coupling techniques here do not require extra variables on the interface of two methods. The important results are given in Theorems 8.1, 8.2 and 8.5 to show that the optimal convergence rates $O(h)$ can also be reached. In Sections 8.4 and 8.8, the numerical stability is proven to be also optimal, although the associate matrix is positive definite but not symmetric. The numerical results given in the last section verify the theoretical analysis made.

8.1 DESCRIPTION OF COUPLING

Consider the elliptic equation

$$Lu = -\frac{\partial}{\partial x}\left(\beta\frac{\partial u}{\partial x}\right) - \frac{\partial}{\partial y}\left(\beta\frac{\partial u}{\partial y}\right) + cu = f, \qquad (x,y) \in S, \tag{8.1}$$

with the Dirichlet boundary condition

$$u = g, \qquad (x,y) \in \Gamma, \tag{8.2}$$

where S is bounded and simply connected domain with the boundary Γ, and the operator

$$L = -\frac{\partial}{\partial x}\beta\frac{\partial}{\partial x} - \frac{\partial}{\partial y}\beta\frac{\partial}{\partial y} + c.$$

The function β, c and f are sufficiently smooth, $c = c(x,y) \geq 0$, $\beta = \beta(x,y) \geq \beta_0 \geq 0$, and β_0 is a constant.

The problem (8.1) and (8.2) can be expressed in a weak form

$$a(u,v) = f(v), \qquad \forall v \in H_0^1(S), \tag{8.3}$$

where the true solution $u \in H^1_*(S)$, the notations are

$$a(u,v) = \iint_S [\beta(u_x v_x + u_y v_y) + cuv]\,ds, \quad f(v) = \iint_S fv\,ds, \qquad (8.4)$$

and the spaces are defined as

$$H^1_0(S) = \{v|v,\ v_x,\ v_y \in L^2(S) \text{ and } v|_\Gamma = 0\},$$

$$H^1_*(S) = \{v|v,\ v_x,\ v_y \in L^2(S) \text{ and } v|_\Gamma = g\}.$$

Let S be divided into two subdomains S_1 and S_2 with a common boundary Γ_0. A combination of the RGM-FEM is obtained if on the solution subdomain S_1, piecewise low-order interpolation polynomials are chosen as admissible functions, but on the other subdomain S_2, analytic functions or singular functions are chosen as admissible functions. Here, the key issue in combinations is how to couple two quite different methods on their common boundary Γ_0. Recall that the nonconforming and the penalty techniques have been provided in Chapters 4 and 7.

In this chapter we will develop another kind of coupling using a hybrid term but without introducing extra variables. As for the combination with the simplified hybrid trick, a condition is

$$f \equiv 0 \quad \text{on } S_2. \qquad (8.5)$$

Obviously, it holds for homogeneous equation $Lu = 0$ in S_2. Even for the nonhomogeneous equation (8.1) which is satisfied by a particular solution u^* on S_2, if such a particular solution can be found, Eq. (8.1) may be reduced to

$$Lw = 0 \quad \text{on } S_2,$$

with a new variable $w = u - u^*$. Hence we assume that (8.5), i.e.,

$$Lu = 0 \quad \text{on } S_2 \qquad (8.6)$$

always holds in this chapter. Note that such a limitation can be removed formally by integrating u^* into the admissible functions.

Define a space

$$H = \{v \in L_2(S),\ v \in H^1(S_1),\ v \in H^1(S_2) \text{ and } Lv = 0 \text{ on } S_2\}, \qquad (8.7)$$

and its subspace

$$H_0 = \{v \in H \text{ and } v|_\Gamma = 0\}.$$

Let $V_h^0 \in H_0$ be a finite-dimensional collection of the functions such that, for $v \in V_h^0$,

(I) $v|_{S_1}$ are piecewise low-order interpolation polynomials on a regular triangulation on S_1 with the maximal boundary length h;

(II) $v|_{S_2} = \sum_{i=1}^{N} a_i \psi_i$, where $L\psi_i = 0$, a_i are unknown coefficients, and $\{\psi_i\}$ are complete basis functions of linear independence. Such basis functions can be found in Bergman (1969) and Vekua (1967).

Moreover, let $V_h^* \in H$ be a finite-dimensional collection of the functions satisfying **(I)** and **(II)** described above as well as $v|_\Gamma = g$.

Remark 8.1 *For simplicity in analyses, we suppose that the functions $v \in V_h^*$ strictly satisfy the boundary condition (8.2); otherwise, the analysis may follow Ciarlet (1978) and Strang and Fix (1973).*

Under the condition of (8.5), the simplified hybrid combination is the procedure to find an approximate solution $u_h^* \in V_h^*$ such that

$$B(u_h^*, w) = f(w), \quad w \in V_h^0, \tag{8.8}$$

where the bilinear form is (Figure 8.1)

$$
\begin{aligned}
B(v, w) =\ & \int_{S_1} [\beta \,\nabla v \,\nabla w + cvw]\, ds + \int_{S_2} [\beta \,\nabla v \,\nabla w + cvw]\, ds \\
& + \int_{\Gamma_0} \beta \left[\frac{\partial v_2}{\partial n} w_1 - \frac{\partial w_2}{\partial n} v_1 \right] d\ell,
\end{aligned}
\tag{8.9}
$$

i.e.,

$$
\begin{aligned}
B(v, w) =\ & \int_{S_1} [\beta \,\nabla v \,\nabla w + cvw]\, ds \\
& + \int_{\Gamma_0} \beta \frac{\partial v_2}{\partial n} w_2 + \int_{\Gamma_0} \beta \left[\frac{\partial v_2}{\partial n} w_1 - \frac{\partial w_2}{\partial n} v_1 \right] d\ell.
\end{aligned}
\tag{8.10}
$$

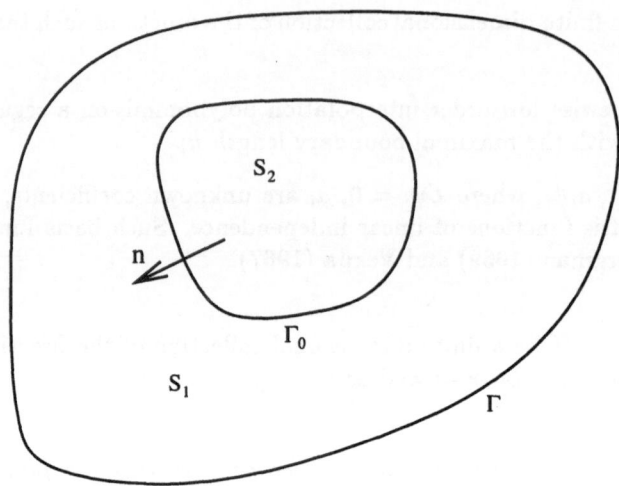

Figure 8.1   The division of a solution domain.

and the linear functional is $f(w) = \int_{S_1} fwds$. We use the notations $v_1 = v|_{S_1}$, $v_2 = v|_{S_2}$, and n is the normal of S_2 to Γ_0 in Figure 8.1.

The equivalence of (8.9) and (8.10) is derived from the following important equalities:

$$\int_{S_2} [\beta \nabla v \nabla w + cvw]\, ds = \int_{\Gamma_0} \beta \frac{\partial v_2}{\partial n} w_2 d\ell = \int_{\Gamma_0} \beta \frac{\partial w_2}{\partial n} v_2 d\ell \qquad (8.11)$$

for $v \in H$ and $w \in H$. Eqs. (8.11) are easily proved from Green's theorem and the homogeneous equation (8.6) which is satisfied by the function $v \in H$ and $w \in H$.

In (8.9) and (8.10), there is an additional integral on Γ_0:

$$\int_{\Gamma_0} \beta \left[\frac{\partial v_2}{\partial n} w_1 - \frac{\partial w_2}{\partial n} v_1 \right] d\ell \qquad (8.12)$$

which plays a role in coupling the RGM and the FEM on Γ_0. Eq. (8.8) is called the simplified hybrid combination because the integral from (8.12) is somewhat like that in the simplified hybrid-FEM of Fix (1976), Fix, Liang and Lee (1982), Raviart and Thomas (1977), Kozlov, Maz'ya and Rozin (1994) and Tong, Pian and Lasry (1973).

8.2 RELATION TO LAGRANGE MULTIPLIER COUPLING

Now, let us prove the equivalence of (8.8) and the method of Lagrange multiplier in Chapter 6, also see Zienkiewciz, Kelley and Bettess (1977).

Define a potential energy on H for (8.1) and (8.2):

$$\Pi(v) = \frac{1}{2}\int_{S_1}\left[\beta\left(\nabla v\right)^2 + cv^2\right]ds + \frac{1}{2}\int_{S_2}\left[\beta\left(\nabla v\right)^2 + cv^2\right]ds$$
$$- \int_{\Gamma_0}\lambda[v_2 - v_1]d\ell - \int_{S_1}fvds,$$

with a Lagrange multiplier $\lambda = \beta\frac{\partial v_2}{\partial n}$ because of the true value $\lambda = \beta(\partial u/\partial n)$. Hence, an approximate solution $\bar{v}$ is obtained by having the stationary, $\Pi^*(\bar{v})$ for $\Pi^*(v)$ under $v \in V_h^*$, where the notation

$$\Pi^*(v) = \frac{1}{2}\int_{S_1}\left[\beta\left(\nabla v\right)^2 + cv^2\right]ds + \frac{1}{2}\int_{S_2}\left[\beta\left(\nabla v\right)^2 + cv^2\right]ds$$
$$- \int_{\Gamma_0}\beta\frac{\partial v_2}{\partial n}[v_2 - v_1]d\ell - \int_{S_1}fvds, \quad v \in V_h^*. \quad (8.13)$$

Performing the variation on (8.13), we obtain

$$\int_{S_1}[\beta\,\nabla\bar{v}\,\nabla w + c\bar{v}w]ds + \int_{S_2}[\beta\,\nabla\,\bar{v}\,\nabla\,w + c\bar{v}w]ds - \qquad (8.14)$$
$$- \left\{\int_{\Gamma_0}\beta\left[\frac{\partial\bar{v}_2}{\partial n}(w_2 - w_1)\right]d\ell + \int_{\Gamma_0}\beta\left[\frac{\partial w_2}{\partial n}(\bar{v}_2 - \bar{v}_1)\right]d\ell\right\} = \int_{S_1}fwds$$

for $\bar{v} \in V_h^*$ and $\forall w \in V_h^0$.

The functions w_1 and w_2 in (8.14) are arbitrary and independent of each other. We may then let them be equal to zero, respectively, so that we obtain two equalities:

$$\int_{S_1}[\beta\,\nabla\bar{v}\,\nabla w + c\bar{v}w]ds + \int_{\Gamma_0}\beta\frac{\partial\bar{v}_2}{\partial n}w_1 d\ell = \int_{S_1}fwds \qquad (8.15)$$

and

$$\int_{S_2}[\beta\,\nabla\bar{v}\,\nabla w + c\bar{v}w]ds - \int_{\Gamma_0}\beta\frac{\partial\bar{v}_2}{\partial n}w_2 d\ell - \int_{\Gamma_0}\beta\frac{\partial w_2}{\partial n}(\bar{v}_2 - \bar{v}_1)d\ell = 0. \qquad (8.16)$$

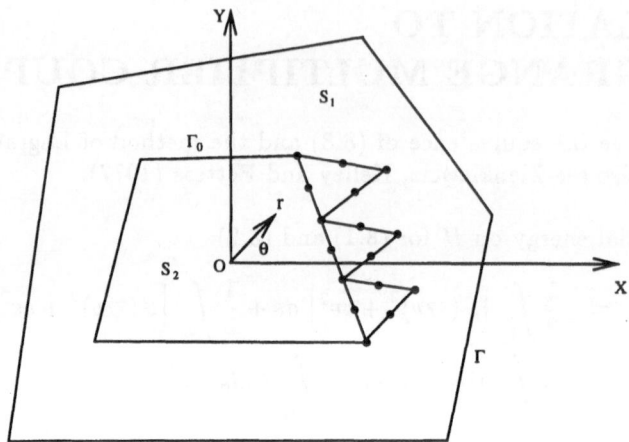

Figure 8.2 The division in the combined method.

Eq. (8.16) is written by applying (8.11) as

$$-\int_{\Gamma_0} \beta \frac{\partial \overline{v}_2}{\partial n} w_2 d\ell + \int_{\Gamma_0} \beta \frac{\partial w_2}{\partial n} \overline{v}_1 d\ell = 0. \tag{8.17}$$

Therefore, the combined description (8.8) is obtained from (8.15) and (8.17). The equivalence of the simplified hybrid description (8.8) and the method of Lagrange multiplier is thus proved.

Proposition 8.1 *The solution of the simplified hybrid description is also the solution of the Lagrange multiplier description.*

8.3   ERROR ANALYSIS

Define a norm on H as

$$\|v\|_1 = \left(\|v\|_{1,S_1}^2 + \|v\|_{1,S_2}^2 \right)^{\frac{1}{2}}.$$

For simplicity in analyses, suppose that S is a convex polygon, and Γ_0 is a piecewise straight line (Figure 8.2). Then we have the following theorem.

Theorem 8.1 *Let (8.5) be given, and suppose that the bilinear form in (8.8) is uniform V_h^0-elliptic, i.e., there exists a positive constant α independent of h and N*

such that

$$\alpha\|v\|_1^2 \leq B(v,v), \quad v \in V_h^0. \tag{8.18}$$

Then the solution u_h^* of (8.8) has the error bounds

$$\|u - u_h^*\|_1 \leq C \inf_{\tilde{w} \in V_h^*} \|u - \tilde{w}\|_1. \tag{8.19}$$

Proof. Let u be the solution of (8.1) and (8.2) under the condition (8.5). Since u and $\partial u/\partial n$ are continuous on Γ_0. We see from Green's theorem and (8.11) that, $\forall v \in H_0$,

$$
\begin{aligned}
B(u,v) &= \int_{S_1} fv\, ds + \int_{\Gamma_0} \left[-\beta \frac{\partial u}{\partial n} v_1 + \beta \frac{\partial u}{\partial n} v_2 \right] d\ell + \int_{\Gamma_0} \left[\beta \frac{\partial u}{\partial n} v_1 - \beta \frac{\partial v_2}{\partial n} u \right] d\ell \\
&= \int_{S_1} fv\, ds + \int_{\Gamma_0} \beta \left[\frac{\partial u}{\partial n} v_2 - \frac{\partial v_2}{\partial n} u \right] d\ell = \int_{S_1} fv\, ds.
\end{aligned}
$$

Hence, the solution u also satisfies (8.8). We have

$$B(u - u_h^*, w) = 0, \quad \forall w \in V_h^0. \tag{8.20}$$

Since v_2 satisfies the homogeneous equation (8.6) we find from the trace theorem of Lions and Magènes (1968) (also see Babuška and Aziz (1972, p. 32-33)) that

$$\left\| \frac{\partial v_2}{\partial n} \right\|_{-\frac{1}{2}, \Gamma_0} \leq C\|v\|_{1, S_2}, \tag{8.21}$$

with a bounded constant C. Also, we see from the imbedding theorem of Sobolev (1963) that

$$\|v_1\|_{\frac{1}{2}, \Gamma_0} \leq C\|v\|_{1, S_1}. \tag{8.22}$$

Thus we have from (8.9), (8.21) and (8.22) that, for $v \in H_0$ and $w \in H_0$,

$$
\begin{aligned}
|B(v,w)| &\leq \|v\|_1 \|w\|_1 + \max \beta \int_{\Gamma_0} \left[\left| \frac{\partial v_2}{\partial n} w_1 \right| + \left| \frac{\partial w_2}{\partial n} v_1 \right| \right] d\ell \\
&\leq \|v\|_1 \|w\|_1 + \max \beta \left\{ \left\| \frac{\partial v_2}{\partial n} \right\|_{-\frac{1}{2}, \Gamma_0} \|w_1\|_{\frac{1}{2}, \Gamma_0} + \left\| \frac{\partial w_2}{\partial n} \right\|_{-\frac{1}{2}, \Gamma_0} \|v_1\|_{\frac{1}{2}, \Gamma_0} \right\} \\
&\leq C\|v\|_1 \|w\|_1.
\end{aligned}
\tag{8.23}
$$

So, the bilinear form $B(v,w)$ is bounded on H_0.

Let $\tilde{w} \in V_h^*$ be arbitrary. We then see from (8.18), (8.20) and (8.23) that

$$\alpha \|u_h^* - \tilde{w}\|_1^2 \leq B(u_h^* - \tilde{w}, u_h^* - \tilde{w}) = B(u - \tilde{w}, u_h^* - \tilde{w}) \leq C\|u - \tilde{w}\|_1 \|u_h^* - \tilde{w}\|_1.$$

Hence, $\|u_h^* - \tilde{w}\|_1 \leq \frac{C}{\alpha}\|u - \tilde{w}\|_1$. Thus

$$\|u - u_h^*\|_1 \leq \|u - \tilde{w}\|_1 + \|\tilde{w} - u_h^*\|_1 \leq \left(1 + \frac{C}{\alpha}\right)\|u - \tilde{w}\|_1.$$

Consequently, Eq. (8.19) is obtained. ∎

Theorem 8.1 is an optimal error estimate of the solutions for the simplified hybrid combination (8.8).

Theorem 8.2 *Let (8.5) be given, and suppose that $u \in H^{k+1}(S_1)$, the k-order Lagrange FEM is used on S_1 and the uniformly V_h^0-elliptic inequality (8.18) holds. Then*

$$\|u_h^* - \tilde{w}\|_1 \leq C \left\{ h^k |u|_{k+1,S_1} + \left[\|R_L\|_{0,\Gamma_0} \left\| \frac{\partial R_L}{\partial n} \right\|_{0,\Gamma_0} \right]^{\frac{1}{2}} \right\}, \tag{8.24}$$

where R_L is the remainder of an approximate expansion u_L of u, which is expressed as $u_L = \sum_{i=1}^{L} \bar{a}_i \psi_i$ with the expansion coefficients $\bar{a}_i$.

Proof. Let u_h be the piecewise k-order Lagrange interpolation polynomial of u on the triangulation of S_1. An auxiliary function $\overline{w}$ is constructed such that

$$\overline{w} = \begin{cases} u_h, & (x,y) \in S_1, \\ u_L, & (x,y) \in S_2. \end{cases}$$

Then $\overline{w} \in V_h^*$. We obtain from Theorem 8.1 that

$$\|u - u_h^*\|_1 \leq C \inf_{\tilde{w} \in V_h^*} \|u - \tilde{w}\|_1 \leq C\|u - \overline{w}\|_1.$$

Moreover, we see from (8.18) that, for $v = u - \overline{w}$,

$$\begin{aligned} \alpha\|u - \overline{w}\|_1^2 &= \alpha\|v\|_1^2 \leq B(v,v) \\ &= \int_{S_1} [\beta(\nabla v)^2 + cv^2]\,ds + \int_{S_2} [\beta(\nabla v)^2 + cv^2]\,ds \\ &\leq C\|u - u_h\|_{1,S_1}^2 + \int_{S_2} [\beta(\nabla R_L)^2 + cR_L^2]\,ds. \end{aligned}$$

Then

$$\|u - u_h^*\|_1 \leq C \left(\|u - u_h\|_{1,S_1} + \left\{ \int_{S_2} [\beta(\nabla R_L)^2 + cR_L^2] \, ds \right\}^{\frac{1}{2}} \right). \qquad (8.25)$$

Note the error bounds for the FEM,

$$\|u - u_h\|_{1,S_1} \leq Ch^k |u|_{k+1,S_1}.$$

Next, the remainder R_L satisfies the homogeneous equation (8.6). Then we have from (8.11) that

$$\int_{S_2} [\beta(\nabla R_L)^2 + cR_L^2] \, ds = \int_{\Gamma_0} \beta R_L \frac{\partial R_L}{\partial n} d\ell \leq C \|R_L\|_{0,\Gamma_0} \left\| \frac{\partial R_L}{\partial n} \right\|_{0,\Gamma_0}. \qquad (8.26)$$

Inequality (8.24) is obtained by combining (8.25) and (8.26). ∎

Now, let us examine the uniformly V_h^0-elliptic inequality (8.18). The bilinear form for $u = v \in V_h^0$ is

$$B(v, v) = \int_{S_1} [\beta(\nabla v)^2 + cv^2] \, ds + \int_{S_2} [\beta(\nabla v)^2 + cv^2] \, ds.$$

Since $v|_{\Gamma \cap S_1} = 0$ for $v \in V_h^0$ and $Meas(\Gamma \cap S_1) > 0$, we see from the Poincare-Friedrichs inequality that there exists a positive constant α_1 independent of h and L such that

$$\alpha_1 \|v\|_{1,S_1}^2 \leq \int_{S_1} [\beta(\nabla v)^2 + cv^2] \, ds. \qquad (8.27)$$

Similarly, if $Meas(\Gamma \cap S_2) > 0$, we also have the inequality

$$\alpha_2 \|v\|_{1,S_2}^2 \leq \int_{S_2} [\beta(\nabla v)^2 + cv^2] \, ds, \qquad (8.28)$$

where α_2 is a positive constant. Hence the uniformly V_h^0-elliptic inequality (8.18) holds from (8.27) and (8.28).

However, in the general case, S_2 might be all inside S, i.e., $Meas(\Gamma \cap S_2) = 0$. Eq. (8.28) can also hold provided that the function c in (8.1) satisfies

$$c = c(x, y) \geq c_0 > 0 \quad \text{on } S_2. \qquad (8.29)$$

Lemma 8.1 *The uniformly V_h^0-elliptic inequality (8.18) holds provided that either $Meas(\Gamma \cap S_2) > 0$ or (8.29) holds.*

Remark 8.2 *For the case where neither of the conditions in Lemma 8.1 holds, for example, if*

$$L^* u = -\frac{\partial}{\partial x}\left(\beta \frac{\partial u}{\partial x}\right) - \frac{\partial}{\partial y}\left(\beta \frac{\partial u}{\partial y}\right) = 0, \quad (x, y) \in S_2,$$

with S_2 inside S, the uniformly V_h^0-elliptic inequality (8.18) does not hold because an arbitrary constant is permitted for admissible functions v_h on S_2. In this case, the spaces H and H_0 shall again satisfy a constraint condition, for example, $\int_{S_2} v ds = 0$. Then, we may define the spaces

$$H^* = \left\{ v \in L_2(S), \ v \in H^1(S_1), \ v \in H^1(S_2), L^* v = 0 \text{ on } S_2 \text{ and } \int_{S_2} v ds = 0 \right\},$$

and $H_0^* = \{ v \in H^* \text{ and } V|_\Gamma = 0 \}$. Moreover, let the subspaces $V_h^* \in H^*$ and $V_h^0 \in H_0^*$. Therefore, the corresponding uniformly elliptic inequality on $V_h^0 \in H_0^*$ for the simplified hybrid combination still holds so that the combined method (8.8) and Theorems 8.1 and 8.2 are all valid.

8.4 SOLVING ALGEBRAIC EQUATION SYSTEMS

An algebraic equation system is obtained from (8.8)

$$\mathbf{Av} + \mathbf{Ed} = \mathbf{b}_1, \tag{8.30}$$

$$-\mathbf{E}^T \mathbf{v} + \mathbf{Dd} = \mathbf{b}_2, \tag{8.31}$$

where $\mathbf{v}$ is the unknown vector with the elements $v_{i,j}$, $\mathbf{d}$ is the unknown vector with the coefficients a_j, $\mathbf{b}_1$ and $\mathbf{b}_2$ are known vectors, $\mathbf{A}$, $\mathbf{D}$, $\mathbf{E}$ and $\mathbf{E}^T$ are matrices, and $\mathbf{E}^T$ is the transposed matrix of $\mathbf{E}$. The matrix $\mathbf{A}$ is positive definite, symmetric and sparse because it comes from the integral $\int_{S_1} \beta[\nabla v \, \nabla w + cvw] ds$, the matrix $\mathbf{D}$ is also positive definite and symmetric because it comes from the integral $\int_{S_2} \beta[\nabla v \, \nabla w + cvw] ds$, (see (8.28)), and the matrices $\mathbf{E}$ and $\mathbf{E}^T$ are from the integrals $\int_{\Gamma_0} \beta(\partial v_2 / \partial n) w_1 d\ell$ and $\int_{\Gamma_0} \beta(\partial w_2 / \partial n) v_1 d\ell$, respectively.

Since the coefficient matrix

$$\begin{pmatrix} \mathbf{A} & \mathbf{E} \\ -\mathbf{E}^T & \mathbf{D} \end{pmatrix}$$

in (8.30) and (8.31) is positive definite but nonsymmetric, the following strategy for solving them is recommended.

We see from (8.31) that

$$d = D^{-1}[E^T v + b_2]. \tag{8.32}$$

Then we obtain, by substituting d into (8.30), that

$$F v = b \tag{8.33}$$

with the matrix

$$F = A + E D^{-1} E^T, \tag{8.34}$$

and the known vector, $b = b_1 - E D^{-1} b_2$. Obviously, the solution v is easily evaluated from (8.33) because the matrix F is also positive definite, symmetric and sparse. Then the solution d is obtained from (8.32).

Now, let us consider the stability of (8.33), which is measured by the bounds of the following condition number of the matrix F:

$$Cond.(F) = \lambda_{max}(F) \Big/ \lambda_{min}(F),$$

where $\lambda_{max}(F)$ and $\lambda_{min}(F)$ are the maximal and minimal eigenvalues of F, respectively.

Theorem 8.3 *Let there be given either (8.29) or $Meas(\Gamma \cap S_2) > 0$, then*

$$Cond.(F) \leq C_1 h_{min}^{-2} \left\{ 1 + \lambda_{max}(E E^T) \Big/ \lambda_{min}(D) \right\}, \tag{8.35}$$

where h_{min} is the minimal boundary length of triangular elements on S_1.

Proof. Since D is a positive definite and symmetric matrix, we have

$$\lambda_{max}(E D^{-1} E^T) \leq \lambda_{max}(E E^T) \Big/ \lambda_{min}(D).$$

Also $\lambda_{min}(E D^{-1} E^T) \geq 0$. Then we see that

$$
\begin{aligned}
Cond.(F) \;&\leq\; \frac{\lambda_{max}(A) + \lambda_{max}(E D^{-1} E^T)}{\lambda_{min}(A) + \lambda_{min}(E D^{-1} E^T)} \\[2mm]
&\leq\; \left[\lambda_{max}(A) + \lambda_{max}(E D^{-1} E^T) \right] \Big/ \lambda_{min}(A) \\[2mm]
&\leq\; \left[\lambda_{max}(A) + \frac{\lambda_{max}(E E^T)}{\lambda_{min}(D)} \right] \Big/ \lambda_{min}(A).
\end{aligned}
$$

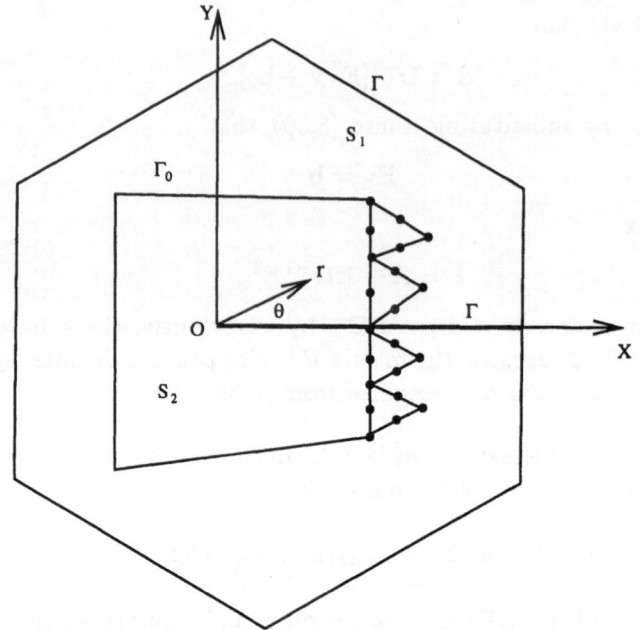

Figure 8.3 The crack problem.

Hence, (8.35) is obtained from the following estimate in the FEM:

$$\lambda_{max}(A) \le C_1, \quad \lambda_{min}(A) = O(h^2_{min}). \quad \blacksquare$$

It is shown in Theorem 8.3 that the condition number $Cond.(\mathbf{F})$ will not be too large if the ratio of $\lambda_{max}(\mathbf{EE}^T)/\lambda_{min}(\mathbf{D})$ is not too large. As to (8.32), the analysis of stability is obvious.

8.5 FOR EQUATION $- \triangle U + U = F$

In this section, we take the following model problem as an example for the application of the simplified hybrid combination (8.8):

$$
\begin{aligned}
- \triangle u + u &= f, & (x,y) \in S_1, \\
- \triangle u + u &= 0, & (x,y) \in S_2, \\
u &= g, & (x,y) \in \Gamma.
\end{aligned}
\tag{8.36}
$$

The condition (8.29) holds because of $c \equiv 1$ on S.

The solution u on S_2 can be expanded, with the help of the method of separation of variables (see Tikhonov and Samarskii (1973)), as

$$u = \bar{a}_0 I_0(r) + \sum_{n=1}^{L} I_n(r) \left[\bar{a}_n \cos n\theta + \bar{b}_n \sin n\theta \right] + R_L, \qquad (8.37)$$

where $\bar{a}_n$ and $\bar{b}_n$ are the expansion coefficients, R_L is the remainder, and $I_n(r)$ is the Bessel function for a purely imaginary argument, defined by

$$I_n(r) = \sum_{k=0}^{\infty} \frac{1}{\Gamma(k+1)\Gamma(k+n+1)} \left(\frac{r}{2} \right)^{2k+n}.$$

Then the admissible functions should be chosen as (Figure 8.2):

$$v = \begin{cases} v_k, & (x,y) \in S_1, \\ a_0 I_0(r) + \sum_{n=1}^{L} I_n(r)[a_n \cos n\theta + b_n \sin n\theta], & (r,\theta) \in S_2, \end{cases}$$

where a_n and b_n are unknown coefficients, and v_k are piecewise k-order Lagrange polynomials on the triangulation of S_1.

The basic functions $I_0(r)$, $I_n(r) \cos n\theta$ and $I_n(r) \sin n\theta$ all satisfy (8.36) in S_2. So, the space V_h^* consisting of (8.36) does belong to H, as defined by (8.7); similarly $V_h^0 \in H_0$. Therefore, an approximate solution can be evaluated from the simplified hybrid combined method (8.8), and Theorems 8.1-8.3 also hold.

Next, consider a singularity problem of a crack lying on the axis x, with the following boundary condition on the crack (Figure 8.3).

$$u|_\Gamma = 0 \quad (y = 0 \text{ and } x \geq 0). \qquad (8.38)$$

There exists a singularity at the origin, which is placed on S_2 (see Figure 8.3). The solution on S_2 can be similarly found as

$$u = \sum_{n=1}^{L} \bar{a}_n I_{\frac{n}{2}}(r) \sin \frac{n}{2}\theta + R_L, \qquad (8.39)$$

with the coefficients $\bar{a}_n$ and the remainder R_L. Obviously, the derivative $\partial u / \partial r$ is unbounded when $r \to 0$, with a singularity at the origin.

Here, we use the combined method for solving the crack problem and choose the admissible functions:

$$v = \begin{cases} v_k, & (x,y) \in S_1, \\ \sum\limits_{n=1}^{L} a_n I_{\frac{n}{2}}(r) \sin \frac{n}{2}\theta, & (r,\theta) \in S_2, \end{cases}$$

with the unknown coefficients a_n and the basic functions, $I_{\frac{n}{2}}(r) \sin \frac{n}{2}\theta$, which satisfy (8.36) and (8.38). It is worth pointing out that even for the singularity problems, Theorem 8.1-8.3 still hold. We have the following corollary.

Corollary 8.1 *Suppose that the conditions in Theorem 8.3 hold, and $u|_{\Gamma_0}$ has the bounded partial derivatives of order μ. Then, the solution u_h^* of (8.8) satisfies the following error bounds:*

$$\|u - u_h^*\|_1 \leq C \left[h^k |u|_{k+1,S_1} + \frac{1}{L^{\mu - \frac{1}{2}}} \right]. \tag{8.40}$$

Proof. For the remainders R_L in (8.37) and (8.39), we see from Eisentstat (1974) that

$$\|R_L\|_{0,\Gamma_0} \leq C \frac{1}{L^\mu} \quad and \quad \left\| \frac{\partial R_L}{\partial n} \right\|_{0,\Gamma_0} \leq C \frac{1}{L^{\mu-1}}.$$

Then, Eq. (8.40) is obtained from Theorem 8.2. ∎

Corollary 8.1 leads to

$$\|u - u_h^*\|_h \leq C h^k,$$

provided that we choose the optimal integer

$$L = L_{opt} = O\left(h^{-k/(\mu - \frac{1}{2})} \right).$$

In this case, the total number of unknown variables in (8.8) is

$$O(h^{-2}) + O(L_{opt}) = O(h^{-2}) + O\left(h^{-k/(\mu - \frac{1}{2})} \right).$$

Generally, $\mu \geq k + 1$, and then the number L of the unknown coefficients a_n and b_n is less than $O(h^{-1})$, which is much less than $O(h^{-2})$ as h is small. The latter is the number of element nodes in the FEM.

Obviously, the larger S_2 and μ are, the less the calculation and storage space in (8.8) are. Corollary 8.1 still holds for general elliptic equations if we choose the admissible functions according to the expansions of solutions in Bergman (1969) and Vekua (1967).

Remark 8.3 *According to the above analyses, the combined method (8.8) with the simplified hybrid trick should be used for singularity problems and unbounded problems. Also, we recommend that the combined method (8.8) be used for boundary value problems of elliptic equations if there exists a large subdomain where the solution is even sufficiently smooth.*

8.6 LAPLACE EQUATION

The basic algorithms and analysis of the simplified hybrid combinations are presented in Sections 8.1–8.5. However, from the point of view of practical application, further studies into the combinations have to be done. For instance, we have to approximate the simplified hybrid integral on Γ_0, to choose an empirical coupling relation between the RGM and FEM and in particular, to perform some numerical experiments that can verify the theoretical analyses made. All these are the purposes of the rest of this chapter.

Let us consider a typical Laplace equation with Dirichlet-Neumann boundary conditions (see Figure 4.1 in Chapter 4):

$$\Delta u = \frac{\partial^2 u}{\partial x^2} + \frac{\partial^2 u}{\partial y^2} = 0, \quad (x,y) \in S, \tag{8.41}$$

$$u = 0, \quad (x,y) \in \overline{AC}, \tag{8.42}$$

$$\frac{\partial u}{\partial n} = 0, \quad (x,y) \in \overline{AB}, \tag{8.43}$$

$$\frac{\partial u}{\partial n} = g, \quad (x,y) \in \overline{BF} \cup \overline{FG} \cup \overline{GC}, \tag{8.44}$$

where n is the outward normal of Γ and g is sufficiently smooth on the boundary $\overline{BF} \cup \overline{FG} \cup \overline{GC}$. Eqs. (8.41)-(8.44) can be written in the following weak form:

$$A(u,v) = g(v), \quad \forall v \in H_0^1(S), \ u \in H_0^1(S), \tag{8.45}$$

where the bilinear form is

$$A(u,v) = \iint_S (u_x v_x + u_y v_y)\,ds, \quad g(v) = \int_{\overline{BF \cup FG \cup GC}} gv\,d\ell,$$

$$H_0^1 = \{v | v,\ v_x,\ v_y \in L^2(S) \text{ and } v|_{\overline{AC}} = 0\}.$$

We note that there exists a singular point at the origin $(0,0)$, resulting from the intersection of the Dirichlet-Neumann boundary conditions when the angle $\Theta = \angle CAB > \frac{\pi}{2}$.

Expansions of solutions near the origin $(0,0)$:

$$u(r,\theta) = \sum_{l=0}^{\infty} D_l \left(\frac{r}{R}\right)^{\tau(l+\frac{1}{2})} \cos \tau(l+\frac{1}{2})\theta, \quad r \leq R, \qquad (8.46)$$

are given in Section 2.1 of Chapter 2, where D_l are expansion coefficients, (r,θ) are polar coordinates with the origin $(0,0)$, i.e., the singular point, and $\tau = \pi/\Theta$. Then

$$u(R,\theta) = \sum_{l=0}^{\infty} D_l \cos \tau(l+\frac{1}{2})\theta,$$

with the expansion coefficients

$$D_l = \frac{2}{\Theta} \int_0^{\Theta} u(R,\theta) \cos \tau(l+\frac{1}{2})\theta d\theta. \qquad (8.47)$$

Based on the expansion (8.46), it is convenient to choose the coupling boundary Γ_0 to be a circular arc l_{R^*} $(r = R^*,\ 0 \leq \theta \leq \Theta)$ with the condition

$$R^* < R. \qquad (8.48)$$

In this case, the subdomain S_2 is a sector $(r < R^*,\ 0 \leq \theta \leq \Theta)$, and S_1 is the rest of S.

Admissible functions in the combination are chosen as

$$v = \begin{cases} v^+ = \sum_{l=0}^{L} \widehat{D}_l \left(\frac{r}{R}\right)^{\tau(l+\frac{1}{2})} \cos \tau(l+\frac{1}{2})\theta & \text{in } S_2, \\ v^- = v_1 & \text{in } \widehat{S}_1^h, \end{cases} \qquad (8.49)$$

where $\widehat{D}_l$ are unknown coefficients to be solved, $\widehat{S}_1^h$ is the triangulation domain of S_1, (then $\widehat{S}_1^h \approx S_1$), and v_1 are piecewise linear interpolation functions on $\widehat{S}_1^h$. We note that there exists a small, overlapped region, $Area(S_2 \cap \widehat{S}_1^h) \neq 0$. This is a kind of the variational crimes, but it does not cause a reduced rate of convergence, see Section 7.2 of Chapter 7. Such a definition as (8.49) will lead to simplicity for both algorithms and error analysis below.

Let V_h^0 denote the space of the admissible function v, then $V_h^0 \notin H^1(S)$ owing to the noncontinuity of v on Γ_0, i.e., l_{R^*}. The simplified hybrid combination is designed to seek an approximate solution, $u_h \in V_h^0$, such that

$$A_h(u_h, v) = g(v), \quad \forall v \in V_h^0,$$

where the bilinear form is

$$A_h(u_h, v) = \iint_{\widehat{S}_1^h} \nabla u \, \nabla v \, ds + \iint_{S_2} \nabla u \, \nabla v \, ds + \int_{\Gamma_0} \left[\frac{\partial u^+}{\partial n} v^- - \frac{\partial v^+}{\partial n} u^- \right] d\ell,$$

where n is the outward normal to ∂S_2, $v^+ = v|_{\Gamma_0^+}$ and $v^- = v|_{\Gamma_0^-}$. It is remarkable that the simplified hybrid integral on Γ_0

$$I_S = \int_{\Gamma_0} \left[\frac{\partial u^+}{\partial n} v^- - \frac{\partial v^+}{\partial n} u^- \right] d\ell \tag{8.50}$$

plays a role of coupling the RGM and the FEM.

In practical calculation, the integral I_S cannot be computed exactly, and has to be approximated by some integration rules, e.g., by the trapezoidal rule

$$\int_\xi^{\xi + \Delta\xi} f(\xi) d\ell \approx \frac{1}{2} \Delta \xi [f(\xi) + f(\xi + \Delta\xi)].$$

Then we have

$$I_S \approx \widehat{I}_S = \widehat{\int_{\Gamma_0} \left[\frac{\partial u^+}{\partial n} v^- - \frac{\partial v^+}{\partial n} u^- \right] d\ell} \tag{8.51}$$

$$= \sum_{i=0}^N \left[\frac{\partial u^+(R^*, \theta_j)}{\partial r} v^-(R^*, \theta_j) - \frac{\partial v^+(R^*, \theta_j)}{\partial r} u^-(R^*, \theta_j) \right] \frac{1}{2} R^*(\Delta\theta_j + \Delta\theta_{j-1}),$$

where $\Delta\theta_j = \theta_{j+1} - \theta_j$ and $\Delta\theta_{N+1} = \Delta\theta_{-1} = 0$.

Let $\widehat{u}_h$ denote the numerical solution involving the integration approximation (8.51), and then

$$\widehat{A}_h(\widehat{u}_h, v) = g(v), \quad \forall v \in V_h^0, \tag{8.52}$$

where

$$\widehat{A}_h(u, v) = \iint_{\widehat{S}_1^h} \nabla u \, \nabla v \, ds + \iint_{S_2} \nabla u \, \nabla v \, ds + \widehat{\int_{\Gamma_0} \left[\frac{\partial u^+}{\partial n} v^- - \frac{\partial v^+}{\partial n} u^- \right] d\ell}. \tag{8.53}$$

Compared with the standard FEM, there exist two kinds of variational crimes:

(I) The integration approximation (8.51).

(II) The small overlap of $\widehat{S}_1^h$ and S_2; otherwise, the complicated, isoparametric elements must be employed (see Ciarlet (1978)).

Below, we shall estimate error bounds of the solution $\widehat{u}_h \in V_h^0$ by (8.52) with the variational crimes, (I) and (II).

8.7 ERROR BOUNDS AND COUPLING STRATEGY

Since $V_h^0 \notin H^1(S)$, define a norm over V_h^0:

$$\|v\|_1 = \left(\|v\|_{1,\widehat{S}_1^h}^2 + \|v\|_{1,S_2}^2 \right)^{\frac{1}{2}}.$$

We shall assess errors of the solutions, in the norm $\|\cdot\|_1$.

Lemma 8.2 *Let the family of all triangular elements in $\widehat{S}_1^h$ be quasiuniform with a maximal length h, and let the true solution u satisfy*

$$\frac{\partial u}{\partial n} \in H^2(\Gamma_0). \tag{8.54}$$

Also assume that the integration points (R^*, θ_j) in (8.51) are the element nodes on the common boundary Γ_0, (i.e., l_{R^*}) and the circular arcs $\Gamma_0^j (r = R^*, \theta_j \leq \theta \leq \theta_{j+1})$ fall into only one triangular element $\triangle_i$ in $\widehat{S}_1^h$, i.e., $\Gamma_0^j \in \triangle_i$. Then for the trapezoidal integration approximation (8.51), there exist the bounds

$$\left| \int_{\Gamma_0} \frac{\partial u}{\partial n} w^- d\ell - \widehat{\int_{\Gamma_0}} \frac{\partial u}{\partial n} w^- d\ell \right| \leq C \frac{H^2}{\sqrt{h}} \left\| \frac{\partial u}{\partial r} \right\|_{2,\Gamma_0} \|w\|_{1,\widehat{S}_1^h}, \quad \forall w \in V_h^0, \tag{8.55}$$

where H is the maximal mesh spacing of integration nodes, e.g., $H = R^* \max_j \triangle \theta_j$, and C is a bounded independent of h, H, L and u.

Proof. For the trapezoidal rule, we have

$$\left| \int_{\Gamma_0} \frac{\partial u}{\partial n} w^- d\ell - \widehat{\int_{\Gamma_0}} \frac{\partial u}{\partial n} w^- d\ell \right| \leq C H^2 \sum_j \left| \frac{\partial u}{\partial n} w^- \right|_{2,\Gamma_0^j} \tag{8.56}$$

$$\leq C H^2 \sum_j \left\| \frac{\partial u}{\partial n} \right\|_{2,\Gamma_0^j} \left(\|w^-\|_{0,\Gamma_0}^2 + |w^-|_{1,\Gamma_0}^2 + \sum_j |w^-|_{2,\Gamma_0^j}^2 \right)^{\frac{1}{2}},$$

where Γ_0^j denotes a small circular arc $(r = R^*, \theta_j \leq \theta \leq \theta_{j+1})$. We note that the norms $|w^-|_{2,\Gamma_0^j}$ exist since the admissible, piecewise linear interpolation functions w^- on Γ_0^j are sufficiently smooth by means of the assumption $\Gamma_0^j \in \triangle_i$.

Bounds can be found in Ciarlet (1978) for the piecewise linear interpolation function w^-

$$|w^-|_{2,\Gamma_0^j} \leq C\|w^-\|_{1,\Gamma_0^j}, \quad |w^-|_{1,\Gamma_0} \leq C\frac{1}{\sqrt{h}}\|w^-\|_{1,\widehat{S}_1^h},$$

to lead to

$$\|w^-\|_{0,\Gamma_0}^2 + |w^-|_{1,\Gamma_0}^2 + \sum_j |w^-|_{2,\Gamma_0^j}^2 \leq C\frac{1}{h}\|w\|_{1,\widehat{S}_1^h}^2. \tag{8.57}$$

The desired result (8.55) follows from (8.54), (8.56) and (8.57). ∎

Lemma 8.3 *Suppose that H and L satisfy*

$$\left[\tau\left(L + \frac{1}{2}\right)\right]^{\frac{5}{2}} H \leq C. \tag{8.58}$$

Then there exist the bounds

$$|\widehat{A}_h(u, v)| \leq C\|u\|_1\|v\|_1. \tag{8.59}$$

Proof. On the basis of (8.53), we obtain

$$
\begin{aligned}
|\widehat{A}_h(u_h, v)| \leq{}& \iint_{\widehat{S}_1^h} \nabla u \, \nabla v \, ds + \iint_{S_2} \nabla u \, \nabla v \, ds + \\
&+ \left|\int_{\Gamma_0}\left[\frac{\partial u^+}{\partial n}v^- - \frac{\partial v^+}{\partial n}u^-\right]d\ell\right| + \left|\int_{\Gamma_0}\frac{\partial u^+}{\partial n}v^-\,d\ell - \int_{\Gamma_0}\frac{\widehat{\partial u^+}}{\partial n}v^-\,d\ell\right| + \\
&+ \left|\int_{\Gamma_0}\frac{\partial v^+}{\partial n}u^-\,d\ell - \int_{\Gamma_0}\frac{\widehat{\partial v^+}}{\partial n}u^-\,d\ell\right|.
\end{aligned}
\tag{8.60}
$$

Since u^+ and v^+ in (8.49) are all particular solutions of the homogeneous Laplace equation, we can see from the trace theorem of Lions and Magenes (1968) and the Sobolev imbedding theorem (also see Babuška and Aziz (1972, p.32)):

$$\left|\int_{\Gamma_0}\frac{\partial u^+}{\partial n}v^-\,d\ell\right| \leq C\left\|\frac{\partial u^+}{\partial n}\right\|_{-\frac{1}{2},\Gamma_0}\|v^-\|_{\frac{1}{2},\Gamma_0} \leq C\|u\|_{1,S_2}\|v\|_{1,\widehat{S}_1^h}, \tag{8.61}$$

$$\left|\int_{\Gamma_0}\frac{\partial v^+}{\partial n}u^-\,d\ell\right| \leq C\left\|\frac{\partial v^+}{\partial n}\right\|_{-\frac{1}{2},\Gamma_0}\|u^-\|_{\frac{1}{2},\Gamma_0} \leq C\|v\|_{1,S_2}\|u\|_{1,\widehat{S}_1^h}. \tag{8.62}$$

On the other hand, the approximate integration $\widehat{\int_{\Gamma_0}}$ can be regarded as $\int_{\Gamma_0}$ but with the substitution of the integrand, such as $(\partial u^+/\partial n)v^-$ or $(\partial v^+/\partial n)u^-$, by their piecewise linear interpolation functions along Γ_0. We then have from Strang and Fix (1973),

$$\left| \int_{\Gamma_0} \frac{\partial u^+}{\partial n} v^- \, d\ell - \widehat{\int_{\Gamma_0} \frac{\partial u^+}{\partial n}} v^- \, d\ell \right| \le CH \left| \frac{\partial u^+}{\partial n} v^- \right|_{1,\Gamma_0}. \tag{8.63}$$

Also, using Babuska and Aziz (1972) and Sobolev (1963) yields

$$\left| \frac{\partial u^+}{\partial n} v^- \right|_{1,\Gamma_0} \le C \left\| \frac{\partial u^+}{\partial r} \right\|_{\frac{3}{2},\Gamma_0} \|v^-\|_{\frac{1}{2},\Gamma_0} \le C \left\| \frac{\partial u^+}{\partial r} \right\|_{2,\Gamma_0} \|v^-\|_{1,\widehat{S}_1^h}. \tag{8.64}$$

By noting the expansions (8.49) of the v^+ and the orthogonality of trigonometric functions, we obtain

$$\left\| \frac{\partial u^+}{\partial r} \right\|_{2,\Gamma_0}^2 = \int_0^\Theta \left[\left(\frac{1}{R^{*2}} \frac{\partial^3 u^+}{\partial \theta^2 \partial r} \right)^2 + \left(\frac{1}{R^*} \frac{\partial^2 u^+}{\partial \theta \partial r} \right)^2 + \left(\frac{\partial u^+}{\partial r} \right)^2 \right] R^* \, d\theta$$

$$= \frac{\Theta}{2} \sum_{l=0}^L \widehat{D}_l^2 \left(\frac{R^*}{R} \right)^{\tau(2l+1)} \left\{ \frac{1}{R^{*5}} \left[\tau \left(l + \frac{1}{2} \right) \right]^6 + \frac{1}{R^{*3}} \left[\tau \left(l + \frac{1}{2} \right) \right]^4 + \right.$$

$$\left. + \frac{1}{R^*} \left[\tau \left(l + \frac{1}{2} \right) \right]^2 \right\},$$

and by (4.35) in Chapter 4,

$$\iint_{S_2} \left(\frac{1}{r} \frac{\partial u^+}{\partial \theta} \right)^2 \, ds = \frac{\Theta}{4} \sum_{l=0}^L \widehat{D}_l^2 \tau \left(l + \frac{1}{2} \right) \left(\frac{R^*}{R} \right)^{\tau(2l+1)}.$$

It follows that

$$\left\| \frac{\partial u^+}{\partial r} \right\|_{2,\Gamma_0} \le C \left[\tau \left(L + \frac{1}{2} \right) \right]^{\frac{5}{2}} \left\{ \iint_{S_2} \left(\frac{1}{r} \frac{\partial u^+}{\partial \theta} \right)^2 \, dS \right\}^{\frac{1}{2}}$$

$$\le C \left[\tau \left(L + \frac{1}{2} \right) \right]^{\frac{5}{2}} \|u^+\|_{1,S_2}. \tag{8.65}$$

Finally from (8.63), (8.64) and (8.65) we obtain

$$\left| \int_{\Gamma_0} \frac{\partial u^+}{\partial n} v^- \, d\ell - \widehat{\int_{\Gamma_0} \frac{\partial u^+}{\partial n}} v^- \, d\ell \right| \le C \left[\tau \left(L + \frac{1}{2} \right) \right]^{\frac{5}{2}} H \|u\|_{1,S_2} \|v\|_{1,\widehat{S}_1^h}. \tag{8.66}$$

Similarly, we obtain

$$\left| \int_{\Gamma_0} \frac{\partial v^+}{\partial n} u^- d\ell - \widehat{\int_{\Gamma_0} \frac{\partial v^+}{\partial n}} u^- d\ell \right| \leq C \left[\tau \left(L + \frac{1}{2} \right) \right]^{\frac{5}{2}} H \|v\|_{1,S_2} \|u\|_{1,\widehat{S}_1^h}. \quad (8.67)$$

The desired result (8.59) follows from (8.58), (8.60), (8.61), (8.62), (8.66) and (8.67). ■

Theorem 8.4 *Let all conditions in Lemmas 8.2 and 8.3 hold, and* $Meas(\overline{AC} \cap \partial\widehat{S}_1^h) \neq 0$. *Then there exist the error bounds*

$$\|u - \widehat{u}_h\|_1 \leq C \left\{ \inf_{v \in V_h^0} \|u - v\|_1 + \sup_{w \in V_h^0} \frac{\left| \widehat{\int_{\Gamma_0} \frac{\partial u}{\partial r}} w^- d\ell - \int_{\Gamma_0} \frac{\partial u}{\partial r} w^- d\ell \right|}{\|w\|_1} \right.$$

$$\left. + \sup_{w \in V_h^0} \frac{\left| \int_{\Gamma_0} \frac{\partial u}{\partial r} w^- d\ell - \widehat{\int_{\Gamma_0} \frac{\partial u}{\partial n}} w^- d\ell \right|}{\|w\|_1} \right\}, \quad (8.68)$$

where $\widehat{\Gamma}_0$ *is the exterior boundary of* $\widehat{S}_1^h$, $\widehat{\Gamma}_0 \approx \Gamma_0$, *and n is the inward normal of* $\widehat{\Gamma}_0$ *to* $\partial\widehat{S}_1^h$.

Proof. In the simplified hybrid combinations (8.52), we note that the functions v in $\widehat{S}_1^h$ and v in S_2 are arbitrary and independent to each other. We then have

$$\iint_{\widehat{S}_1^h} \nabla u \, \nabla v ds + \widehat{\int_{\Gamma_0} \frac{\partial u^+}{\partial r}} v^- d\ell = g(v),$$

$$\iint_{S_2} \nabla u \, \nabla v ds - \widehat{\int_{\Gamma_0} \frac{\partial v^+}{\partial r}} u^- d\ell = 0. \quad (8.69)$$

Since the functions u^+ and v^+ both satisfy the Laplace equation in S_2, Eq. (8.69) is reduced to

$$\int_{\Gamma_0} \frac{\partial u^+}{\partial r} v^+ d\ell = \int_{\Gamma_0} \frac{\partial v^+}{\partial r} u^+ d\ell = \widehat{\int_{\Gamma_0} \frac{\partial v^+}{\partial r}} u^- d\ell. \quad (8.70)$$

Now let u be a true solution of (8.41)-(8.44), i.e., (8.45), we then obtain

$$\widehat{A}_h(u, v) = -\int_{\partial\widehat{S}_1^h} \frac{\partial u}{\partial n} v^- d\ell + \int_{\partial S_2} \frac{\partial u}{\partial n} v^+ d\ell +$$

$$+ \widehat{\int_{\Gamma_0}} \left[\frac{\partial u}{\partial r} v^- - \frac{\partial v^+}{\partial r} u \right] d\ell + g(v)$$

$$= - \int_{\widehat{\Gamma}_0} \frac{\partial u}{\partial n} v^- d\ell + \int_{\Gamma_0} \frac{\partial u}{\partial r} v^+ d\ell + \widehat{\int_{\Gamma_0}} \left[\frac{\partial u}{\partial r} v^- - \frac{\partial v^+}{\partial r} u \right] d\ell + g(v)$$

$$= - \int_{\widehat{\Gamma}_0} \frac{\partial u}{\partial n} v^- d\ell + \widehat{\int_{\Gamma_0}} \frac{\partial u}{\partial r} v^- d\ell + g(v). \tag{8.71}$$

In the last step in (8.71), we have applied (8.70).

Next, with the help of the assumptions of

$$Meas(\overline{AC} \cap \partial \widehat{S}_1^h) \neq 0 \quad \text{and} \quad Meas(\overline{AC} \cap \partial S_2) \neq 0,$$

which hold naturally, we can see from the Poincare-Friedrichs inequality

$$C_0 \|v\|_{1,\widehat{S}_1^h}^2 \leq |v|_{1,\widehat{S}_1^h}^2, \quad C_0 \|v\|_{1,S_2}^2 \leq |v|_{1,S_2}^2,$$

where C_0 is a positive constant independent of L, h, H and v. Therefore, the following uniformity V_h^0-elliptic inequality holds:

$$C_0 \|v\|_1^2 \leq |v|_{1,\widehat{S}_1^h}^2 + |v|_{1,S_2}^2 = \widehat{A}_h(v,v). \tag{8.72}$$

Below, let $v \in V_h^0$, and then by combining (8.72), (8.71) and (8.52) we have

$$C_0 \|v - \widehat{u}_h\|_1^2 \leq \widehat{A}_h(v - \widehat{u}_h, v - \widehat{u}_h)$$

$$= \widehat{A}_h(v - u, v - \widehat{u}_h) - \int_{\widehat{\Gamma}_0} \frac{\partial u}{\partial n} w^- d\ell + \widehat{\int_{\Gamma_0}} \frac{\partial u}{\partial r} w^- d\ell, \tag{8.73}$$

with $w = v - \widehat{u}_h \in V_h^0$. Moreover, we can see from Lemma 8.3

$$\widehat{A}_h(v - \widehat{u}_h, v - u) \leq C\|v - \widehat{u}_h\|_1 \|v - u\|_1.$$

Also it follows

$$\left| \int_{\widehat{\Gamma}_0} \frac{\partial u}{\partial n} w^- d\ell - \widehat{\int_{\Gamma_0}} \frac{\partial u}{\partial r} w^- d\ell \right| \tag{8.74}$$

$$\leq \left| \int_{\Gamma_0} \frac{\partial u}{\partial r} w^- d\ell - \int_{\widehat{\Gamma}_0} \frac{\partial u}{\partial n} w^- d\ell \right| + \left| \widehat{\int_{\Gamma_0}} \frac{\partial u}{\partial r} w^- d\ell - \int_{\Gamma_0} \frac{\partial u}{\partial r} w^- d\ell \right|.$$

Finally, we obtain from (8.73) and (8.74)

$$C_0 \|v - \widehat{u}_h\|_1 \leq C\|v - u\|_1 \tag{8.75}$$

$$+ \frac{1}{\|w\|_1} \left\{ \left| \widehat{\int_{\Gamma_0}} \frac{\partial u}{\partial r} w^- d\ell - \int_{\Gamma_0} \frac{\partial u}{\partial r} w^- d\ell \right| + \left| \int_{\Gamma_0} \frac{\partial u}{\partial r} w^- d\ell - \int_{\widehat{\Gamma}_0} \frac{\partial u}{\partial n} w^- d\ell \right| \right\},$$

and the desired bounds (8.68) result from (8.75) and the trigonometric inequality $\|u - \widehat{u}_h\|_1 \leq \|u - v\|_1 + \|v - \widehat{u}_h\|_1$. ∎

In Theorem 8.4, the second term on the right side of (8.68) is an error resulting from the approximation integration; and the third term is an error resulting from the noncoincidence of Γ_0 and $\widehat{\Gamma}_0$, i.e., of S_1 and $\widehat{S}_1^h$.

Theorem 8.5 *Let all conditions in Theorem 8.4 hold. Also suppose that $u \in H^2(\widehat{S}_1^h)$ and (8.48) are satisfied. Then there exist the bounds*

$$\|u - \widehat{u}_h\|_1 \leq C \left\{ h + \frac{H^2}{\sqrt{h}} + \left(\frac{R^*}{R} \right)^{\tau(L+\frac{3}{2})} \right\}. \tag{8.76}$$

Proof. Define an auxiliary function $\overline{w} \in V_h^0$ such that

$$\overline{w} = \begin{cases} \overline{w}^+ = \sum_{l=0}^{L} D_l \left(\frac{r}{R} \right)^{\tau(l+\frac{1}{2})} \cos \tau(l + \frac{1}{2})\theta, & \text{in } S_2, \\ u_1, & \text{in } S_1, \end{cases}$$

where D_l are the expansion coefficients defined by (8.47), and u_1 are piecewise linear interpolation function of $u \in \widehat{S}_1^h$. Also $u = \overline{w}^+ + R_L$ in S_2, with the remainder term

$$R_L = \sum_{l=L+1}^{\infty} D_l \left(\frac{r}{R} \right)^{\tau(l+\frac{1}{2})} \cos \tau(l + \frac{1}{2})\theta.$$

As a result, we have

$$\inf_{v \in V_h^0} \|u - v\|_1 \leq \|u - \overline{w}\|_1 \leq \|u - u_1\|_{1,\widehat{S}_1^h} + \|u - \overline{w}^+\|_{1,S_2} \leq Ch|u|_{2,\widehat{S}_1^h} + \|R_L\|_{1,S_2}.$$

Besides, we can see from Lemma 8.2 and (7.34) in Chapter 7

$$\left| \int_{\Gamma_0} \frac{\partial u}{\partial r} w^- d\ell - \int_{\widehat{\Gamma}_0} \frac{\partial u}{\partial r} w^- d\ell \right| \leq C \frac{H^2}{\sqrt{h}} \left\| \frac{\partial u}{\partial r} \right\|_{2,\Gamma_0} \|w\|_{1,\widehat{S}_1^h}$$

and

$$\left| \int_{\Gamma_0} \frac{\partial u}{\partial r} w^- d\ell - \int_{\widehat{\Gamma}_0} \frac{\partial u}{\partial n} w^- d\ell \right| \leq Ch^{\frac{3}{2}} \|w\|_1.$$

We therefore obtain from Theorem 8.4

$$\|u - \widehat{u}_h\|_1 \leq C \left\{ h|u|_{2,\widehat{S}_1^h} + h^{\frac{3}{2}} + \frac{H^2}{\sqrt{h}} \left\| \frac{\partial u}{\partial r} \right\|_{2,\Gamma_0} + \|R_L\|_{1,S_2} \right\}. \qquad (8.77)$$

Since $D_l = O(1/\tau(l + \frac{1}{2}))$, using (8.48) and the orthogonality of the trigonometric functions $\cos \tau(l + \frac{1}{2})\theta$ we obtain

$$\|R_L\|_{1,S_2}^2 = \frac{\Theta}{2} \sum_{l=L+1}^{\infty} D_l^2 \left(\frac{R^*}{R}\right)^{\tau(2l+1)} \left[\tau(l + \frac{1}{2}) + \frac{R^{*2}}{2 + \tau(2l + 1)}\right]$$

$$\leq C \left(\frac{R^*}{R}\right)^{\tau(2L+3)}. \qquad (8.78)$$

The bounds (8.76) are then obtained by combining (8.77) and (8.78). ∎

In (8.76), the bounds $O(H^2/\sqrt{h})$ are the errors that come from the integration approximation (8.51).

To obtain the optimal convergence rate from (8.76):

$$\|u - \widehat{u}_h\|_1 = O(h), \qquad (8.79)$$

the following two conditions should be satisfied:

$$\frac{H^2}{\sqrt{h}} \leq Ch, \qquad (8.80)$$

$$\left(\frac{R^*}{R}\right)^{\tau(l + \frac{3}{2})} = C_1 h. \qquad (8.81)$$

Eq. (8.81) leads to

$$L + 1 = \frac{\ln C_1 h}{\tau \ln \left(\frac{R^*}{R}\right)} - \frac{1}{2}, \qquad (8.82)$$

where C_1 is a bounded constant independent of L, h, H and u.

As usual, let

$$H = O(h). \qquad (8.83)$$

As the assumptions in Lemma 8.2, we always choose the element nodes (R^*, θ_j) on Γ_0 as the integration points in (8.51). Then Eq. (8.80) holds from (8.83), and condition (8.58) in Lemma 8.3 is automatically satisfied from (8.81) and (8.83). We have the following corollaries.

Corollary 8.2 *Let all conditions in Theorem 8.5 hold. Then there exists the optimal convergence rate (8.79) provided that (8.81)–(8.83) are satisfied.*

Corollary 8.3 *Let (8.83) and all conditions in Theorem 8.5 hold. Then there exists the optimal convergence rate (8.79) provided that the following coupling strategy between $L_h + 1$ and h is applied:*

$$L_h + 1 = (L_{h_0} + 1) + \frac{\left| \ln\left(\frac{h}{h_0} \right) \right|}{\left| \tau \ln\left(\frac{R^*}{R} \right) \right|}, \tag{8.84}$$

or

$$L_{\frac{1}{2}h} + 1 = (L_h + 1) + 1, \tag{8.85}$$

for a special case of $\Theta = \pi$, i.e., $\tau = \pi/\Theta = 1$, and $R^*/R = \frac{1}{2}$, where $L_h + 1$ is the total number of expansion functions in S_2, corresponding to a fixed, maximal boundary length h of quasiuniform elements in $\widehat{S}_1^h$.

Note that Corollaries 8.2 and 8.3 and condition (8.85) are analogous those in Chapters 4 and 7.

8.8 STABILITY ANALYSIS

From the combination (8.52), we should again study the numerical stability. We obtain linear algebraic equation systems

$$\mathbf{Bv} + \mathbf{Ed} = \mathbf{b}_1, \quad -\mathbf{E}^T \mathbf{v} + \mathbf{Dd} = 0, \tag{8.86}$$

where $\mathbf{v}$ is the unknown vector with the components v_{ij} on the elements nodes (i, j) in $\widehat{S}_1^h$, $\mathbf{d}$ is the unknown vector with coefficients $\widehat{D}_l$, $\mathbf{b}_1$ is known vector, and $\mathbf{B}$, $\mathbf{D}$, $\mathbf{E}$ and $\mathbf{E}^T$ are matrices.

The matrices $\mathbf{B}$ and $\mathbf{D}$ are obtained from the integrals $\iint_{\widehat{S}_1^h} \nabla u \nabla v \, ds$ and $\iint_{S_2} \nabla u \nabla v \, ds$ in (8.53), respectively. Let

$$Meas(\overline{AC} \cap \partial \widehat{S}_1^h) \neq 0, \tag{8.87}$$

the coefficient matrix $\mathbf{B}$ of the linear FEM is positive definite, symmetric and sparse. However, the matrix $\mathbf{D}$ is not only positive definite and symmetric, but also diagonal. In fact, since the circular arc l_{R^*} is chosen as the coupling boundary Γ_0, we let

$$u^+ = \sum_{l=0}^{L} \widehat{D}_l \left(\frac{r}{R}\right)^{\tau(l+\frac{1}{2})} \cos \tau(l+\frac{1}{2})\theta,$$

$$v^+ = \left(\frac{r}{R}\right)^{\tau(\ell+\frac{1}{2})} \cos \tau(l+\frac{1}{2})\theta. \tag{8.88}$$

Using the orthogonality of trigonometric functions yields

$$\iint_{S_2} \nabla u \nabla v ds = \int_{\partial S_2} \frac{\partial u}{\partial n} v d\ell = \int_{\Gamma_0} \frac{\partial u^+}{\partial r} v^+ d\ell$$

$$= \frac{1}{2}\Theta \sum_{l=0}^{L} \widehat{D}_l^2 \tau(l+\frac{1}{2}) \left(\frac{R^*}{R}\right)^{2\tau(l+\frac{1}{2})}.$$

Then, the matrix $\mathbf{D}$ has the components $\mathbf{D} = (d_{i,j})_{(L+1)\times(L+1)}$, where

$$d_{i,j} = 0 \text{ with } i \neq j, \quad d_{i,i} = \frac{1}{2}\Theta\tau(i-\frac{1}{2})\left(\frac{R^*}{R}\right)^{\tau(2i-1)}. \tag{8.89}$$

It is due to the diagonal matrix $\mathbf{D}$ that the solution procedure for (8.86) becomes much simpler. For instance, we solve (8.86) by

$$\mathbf{d} = \mathbf{D}^{-1}\mathbf{E}^T\mathbf{v}, \quad \mathbf{F}\mathbf{v} = \mathbf{b}_1, \tag{8.90}$$

where the coefficient matrix

$$\mathbf{F} = \mathbf{B} + \mathbf{E}\mathbf{D}^{-1}\mathbf{E}^T \tag{8.91}$$

is positive definite, symmetric and sparse because the matrix $\mathbf{E}\mathbf{D}^{-1}\mathbf{E}^T$ is symmetric and sparse. It is also due to the positive definite matrices $\mathbf{F}$ and $\mathbf{D}$ that the solutions $\mathbf{v}$ and $\mathbf{d}$ in the discrete problem (8.90) (i.e.,(8.86)) exist uniquely.

The stability can be measured by the condition numbers of the coefficient matrix $\mathbf{F}$ (and $\mathbf{D}$), defined by $Cond.(\mathbf{F}) = \lambda_{max}(\mathbf{F})\big/\lambda_{min}(\mathbf{F})$, where $\lambda_{max}(\mathbf{F})$ and $\lambda_{min}(\mathbf{F})$ are the maximal and minimal eigenvalues of the matrix $\mathbf{F}$, respectively. We have the next theorem.

Theorem 8.6 *Let the admissible functions (8.49) be given. Suppose (8.83) and (8.87) to be satisfied, and (R^*, θ_j) in (8.51) to be the element nodes on l_{R^*}. Then, there exist the bounds,*

$$Cond.(\mathbf{D}) \leq \frac{\left(\frac{R}{R^*}\right)^{2\tau L}}{2L+1}, \quad Cond.(\mathbf{F}) \leq C\left(\frac{1+hL^2}{h^2}\right), \tag{8.92}$$

where h is the maximal boundary length of regular elements in $\widehat{S}_1^h$.

Proof. The left estimate in (8.92) can be obtained directly from (8.89):

$$Cond.(\mathbf{D}) = \frac{d_{1,1}}{d_{L+1,L+1}} = \frac{\left(\frac{R}{R^*}\right)^{2\tau L}}{2L+1}.$$

From (8.91) we have

$$Cond.(\mathbf{F}) \leq \frac{\lambda_{max}(\mathbf{B}) + \lambda_{max}(\mathbf{ED}^{-1}\mathbf{E}^T)}{\lambda_{min}(\mathbf{B})}.$$

Since $\mathbf{B}$ is the coefficient matrix coming from the linear FEM, the assumption (8.87) gives (see Strang and Fix (1973))

$$\lambda_{min}(\mathbf{B}) \geq C_0 h^2, \quad \lambda_{max}(\mathbf{B}) \leq C,$$

where C_0 and C are positive constants independent of h. The estimate on the right side in (8.92) follows from the bounds

$$\lambda_{max}(\mathbf{ED}^{-1}\mathbf{E}^T) \leq ChL^2. \tag{8.93}$$

Let us prove (8.93) below.

In fact, the matrix $\mathbf{E}^T$ results from the integral $\widehat{\int_{l_{R^*}}}(\partial v^+ / \partial r)u^- d\ell$, whose values can be found by substituting v^+ with (8.88):

$$\widehat{\int_{l_{R^*}}} \frac{\partial v^+}{\partial r} u^- d\ell = \sum_{j=0}^{N} \frac{\partial v^+}{\partial r}(R^*, \theta_j)u_{0,j}\frac{1}{2}(R^*(\Delta\theta_j + \Delta\theta_{j-1}))$$

$$= \sum_{j=0}^{N} \left(\frac{R^*}{R}\right)^{\tau(l+\frac{1}{2})} \tau(l+\frac{1}{2})\cos\left(\tau(l+\frac{1}{2})\theta_j\right)\frac{1}{2}(\Delta\theta_j + \Delta\theta_{j-1})u_{0,j},$$

where $u_{0,j} = u_{0,j}^- = u(R^*, \theta_j)$, and subscripts $(0, j)$ denote the nodes on l_{R^*}. Therefore, the matrix $\mathbf{E}^T$ is

$$\mathbf{E}^T = \left(0, (\mathbf{E}^*)^T\right),$$

where $(\mathbf{E}^*)^T$ is an $(L+1) \times (N+1)$ matrix with the following components

$$e_{l,j}^* = \tau(l-\frac{1}{2})\cos\left(\tau(l-\frac{1}{2})\theta_{j-1}\right)\left(\frac{R^*}{R}\right)^{\tau(l-\frac{1}{2})} \times \frac{1}{2}(\Delta\theta_{j-1} + \Delta\theta_{j-2}),$$

$$1 \leq l \leq L+1 \quad \text{and} \quad 1 \leq j \leq N+1. \tag{8.94}$$

We now have

$$\mathbf{E}\mathbf{D}^{-1}\mathbf{E}^T = \begin{pmatrix} 0 & 0 \\ 0 & \mathbf{T} \end{pmatrix}, \tag{8.95}$$

with $\mathbf{T} = \mathbf{E}^*\mathbf{D}^{-1}(\mathbf{E}^*)^T$, and

$$\lambda_{max}(\mathbf{E}\mathbf{D}^{-1}\mathbf{E}^T) = \lambda_{max}(\mathbf{T}) = \lambda_{max}(\mathbf{E}^*\mathbf{D}^{-1}(\mathbf{E}^*)^T). \tag{8.96}$$

Moreover, the components T_{ij} of T can be obtained from (8.89), (8.94) and (8.95):

$$\mathbf{T}_{ij} = \sum_{l=1}^{L+1} \frac{e_{l,i}^* e_{l,j}^*}{d_{l,l}} = \frac{1}{2\Theta} \sum_{l=1}^{L+1} \tau(l - \frac{1}{2}) \cos\left(\tau(l - \frac{1}{2})\theta_{i-1}\right) \times$$

$$\times \cos\left(\tau(l - \frac{1}{2})\theta_{j-1}\right) (\Delta\theta_{i-1} + \Delta\theta_{i-2})(\Delta\theta_{j-1} + \Delta\theta_{j-2}).$$

Hence, we have from (8.83)

$$|\mathbf{T}_{ij}| \leq Ch^2 L^2, \ 1 \leq i,j \leq N+1.$$

The maximal eigenvalue of the $(N+1) \times (N+1)$ matrix T can be bounded by

$$\lambda_{max}(\mathbf{T}) \leq \max_j \sum_{i=1}^{N+1} |\mathbf{T}_{ij}| \leq ChL^2. \tag{8.97}$$

The desired result (8.93) follows from (8.96) and (8.97). This completes the proof of Theorem 8.6 ∎

Corollary 8.4 *Let all conditions in Theorem 8.6. hold. Also suppose that the coupling relations (8.82) and (8.84) are used, then there exist the bounds*

$$Cond.(\mathbf{D}) = O(h^{-2}), \quad Cond.(\mathbf{F}) = O(h^{-2}). \tag{8.98}$$

We note that the asymptotic relations in (8.98) are exactly the same as in the FEM (see Strang and Fix (1973)). With this coupling strategy, not only can the optimal convergence rates (8.79) be obtained, but also the optimal asymptotic rates (8.98) of condition numbers may be achieved. Now we can see how significant the coupling strategy of (8.81) and (8.84) is!

8.9   NUMERICAL EXPERIMENTS

The partition of S and the coupling between L and h are chosen as in Chapters 4 and 7. The results are given in Tables 8.1 and 8.2. It can be seen that there appear

Divisions	δ_1	δ_2	Max	$\|\varepsilon\|_{0,S}$	$\|\varepsilon\|_1$
$M = 2$ $L+1 = 4$	1.169	0.7191	4.221	2.414	35.53
$M = 3$ $L+1 = 5$	0.5617	0.3563	1.941	1.032	23.30
$M = 4$ $L+1 = 5$	0.3615	0.2197	1.102	0.5729	17.39
$M = 6$ $L+1 = 6$	0.1848	0.1116	0.4876	0.2526	11.56
$M = 8$ $L+1 = 6$	0.1114	0.06822	0.2744	0.1417	8.663

Table 8.1   The error norms of numerical solutions for $R^* = 0.5$. In this table $\delta_1 = \|\varepsilon^+ - \varepsilon^-\|_{\infty,l_{R^*}}$ and $\delta_2 = \|\varepsilon^+ - \varepsilon^-\|_{0,l_{R^*}}$.

the following asymptotic relations:

$$\|\varepsilon\|_1 = O(h), \tag{8.99}$$

$$\|\varepsilon\|_{0,S} = O(h^2), \quad \max = O(h^{2-\delta}), \quad 0 < \delta << 1, \tag{8.100}$$

$$\|\varepsilon^+ - \varepsilon^-\|_{0,l_{R^*}} = O(h^{\frac{5}{3}}), \quad \|\varepsilon^+ - \varepsilon^-\|_{\infty,l_{R^*}} = O(h^{\frac{5}{3}}), \tag{8.101}$$

$$\delta D_i = |D_i - \widehat{D}_i| = O(h^2), \quad i = 0, 1, 2. \tag{8.102}$$

Evidently, (8.99) and (8.100) are consistent with those in the standard FEM. In particular, the numerical result (8.99) agrees with the above theoretical analysis.

Let us compare (8.99)–(8.102) with those in Chapters 4 and 7. The error norms given in (8.101) on the common boundary Γ_0 have a little lower order $O(h^{\frac{5}{3}})$ of the convergence rates than $O(h^{2-\delta})$ in (7.57) of the penalty combination in Chapter 7. This can be explained by the fact that the simplified hybrid techniques (8.50) are not very restricted constraints between v^+ and v^- as the penalty coupling. However, the approximate coefficients $\widehat{D}_0$, $\widehat{D}_1$, $\widehat{D}_2$ obtained still have a good convergence rate (8.102). When $M = 8$ and $L + 1 = 6$, the value of $\widehat{D}_0$ is:

Divisions	$\tilde{D}_0$	$\tilde{D}_1$	$\tilde{D}_2$	$\tilde{D}_3$	$\tilde{D}_4$	$\tilde{D}_5$
$M = 2$ $L+1 = 4$	399.172	88.2237	18.7449	-11.8927		
$M = 3$ $L+1 = 5$	400.285	87.8612	17.9537	-10.3167	1.36658	
$M = 4$ $L+1 = 5$	400.666	87.7659	17.6624	-9.52901	1.75661	
$M = 6$ $L+1 = 6$	400.940	87.7044	17.4375	-8.81086	1.75810	-0.069337
$M = 8$ $L+1 = 6$	401.037	87.6834	17.3532	-8.51183	1.66907	0.057042
True Coeffs.	401.162	87.6559	17.2379	-8.07122	1.44027	0.331055

Table 8.2 The calculated coefficients for $R^* = 0.5$.

$$\hat{D}_0 = 401.037 \qquad\qquad (8.103)$$

with a small relative error 0.0003. It is spectacular that the digits of $\hat{D}_0$ in (8.103) are exactly the same as those in Table 4.4 of Chapter 4 by the nonconforming combinations!

Remark 8.4 *We may consider the general, hybrid techniques with $P_c = 0$ and $\alpha + \beta = 1$ in (6.43) in Chapter 6. The numerical experiments for Motz's problem are provided in Li and Bui (1988a). The simplified hybrid combination, i.e., $\alpha = 1$ and $\beta = 0$, is significant to singularity problems of homogeneous elliptic equations. The case of $\alpha = 1$ and $\beta = 0$ is best, having optimal convergence rates and good stability of numerical solutions. Moreover, the numerical solutions therein for the cases of $\alpha = 0 \wedge \beta = 1$ and $\alpha = \beta = \frac{1}{2}$ encounter some troubles in either inaccuracy or instability so that we should employ the coupling techniques in this chapter, or in the next chapter by means of the mixed type of penalty plus hybrid techniques.*

9

PENALTY PLUS HYBRID TECHNIQUES

For matching different methods, an effective coupling strategy is essential, in order to yield optimal convergence rates and good stability of numerical solutions. In this chapter, we peruse again additional integrals with penalty plus hybrid terms along the common boundary of different methods, since these techniques have been applied in the standard FEM in Arnold (1982), Baker (1977), Barrett and Elliott (1986), Nitsche (1971) and Fairweather (1978). However, in those papers, it is assumed that smooth solutions exist throughout the entire solution domain S. Therefore, the new coupling techniques here are developed for the problems with singularities. Also, two (or more) different methods are used simultaneously, rather than only the FEM is used. Moreover, since the coupling strategy is employed only on the boundary of different methods, not on *all* the element boundary as in the nonconforming FEM in Ciarlet (1978), more interesting conclusions can be reached.

In the first section, a general strategy using additional integrals is described for the combination of the Ritz-Galerkin and finite element methods (RGM-FEM). Error bounds for numerical solutions are derived in the Sobolev norms in the next section, and stability analysis is given in Section 9.3. All these analyses form a theoretical basis for a number of combinations of RGM-FEM. In Section 9.4, a summary of six coupling techniques in Chapters 7-9 is given, and in Section 9.5, the detailed computational algorithms are provided for users.

The important theoretical results are given in Theorems 9.1–9.3. Theorem 9.2 indicates that the optimal convergence rates

$$\|\varepsilon\|_h = O(h^k)$$

can be achieved for coupling the k−order FEM with the RGM, contrasted to only the linear FEM used in the penalty combination of Chapter 7. Also Theorem 9.3 indicates that the condition number of the associated matrix is

$$Cond. = O(h^{-2} + P_c h^{-(2+\sigma)}), \quad \sigma \geq 1 \text{ (or } > 0),$$

which is fairly good, although not optimal.

Another important result is given in Section 9.4, to overview, link and compare all the six coupling techniques in both theory and numerical experiments which are described in Part II, Chapters 7, 8 and this chapter, separately. These coupling techniques will be applied to other combinations and other problems, to be stated in the rest of this book. Various effective coupling techniques are one of the significant developments from the previous book in Li (1990), where only the nonconforming coupling was introduced.

9.1 DESCRIPTION OF METHODS

The general coupling strategy is developed for the combinations of the RGM-FDM, accompanied with error bounds of the solutions. Consider a general elliptic equation

$$-\frac{\partial}{\partial x}\left(p\frac{\partial u}{\partial x}\right) - \frac{\partial}{\partial y}\left(p\frac{\partial u}{\partial y}\right) + cu = f, \ (x, y) \in S, \tag{9.1}$$

and general boundary conditions

$$u = t \ \text{ on } \ \Gamma_D, \ \ p\frac{\partial u}{\partial n} + qu = g \ \text{ on } \ \Gamma_N, \tag{9.2}$$

where S is a bounded, simply connected, and polygonal domain, Γ is the exterior boundary of S such that $\Gamma = \Gamma_D \cup \Gamma_N$, $Meas(\Gamma_D) > 0$ if $q \equiv 0$, and the functions p, c, f, q, t and g are sufficiently smooth. Moreover, the functions in (9.1) satisfy

$$c = c(x, y) \geq 0, \ \ q = q(x, y) \geq 0, \ \ p = p(x, y) \geq p_0 > 0.$$

Let us define

$$H^1_*(S) = \left\{v | v, \ v_x, \ v_y \in L^2(S), \ \text{and } v|_{\Gamma_D} = t\right\}.$$

Then, the solution of (9.1) and (9.2) belongs to $H^1_*(S)$, i.e., $u \in H^1_*(S)$, and u can be represented in the weak form:

$$a(u, v) = f(v), \ \ \forall v \in H^1_0(S),$$

where the notations are:

$$a(u, v) = \iint_{S_1} [p \bigtriangledown u \bigtriangledown v + cuv]\, ds + \iint_{S_2} [p \bigtriangledown u \bigtriangledown v + cuv]\, ds + \int_{\Gamma_N} quvd\ell,$$

$$f(v) = \iint_{S} fvds + \int_{\Gamma_N} gvd\ell,$$

$$H^1_0(S) = \left\{v | v, \ v_x, \ v_y \in L^2(S), \ \text{and } v|_{\Gamma_D} = 0\right\}.$$

Let the solution domain S be divided by a common boundary Γ_0, which is a piece-wise straight line, into two subdomains S_1 and S_2. The RGM is used in S_2 containing a singular point, and the $k(\geq 1)-$ order Lagrange FEM is used in S_1 where $u \in H^{k+1}(S_1)$. Therefore, we may choose the following admissible functions:

$$v = \begin{cases} v^- = V_1^k & \text{in } S_1, \\ v^+ = \sum_{i=1}^{L} a_i \Psi_i & \text{in } S_2, \end{cases} \tag{9.3}$$

where V_1^k are piecewise k-order Lagrange interpolation polynomials on a quasi-uniform triangulation of S_1 with a maximal boundary length h of elements, $\{\Psi_i\}$ in S_2 are known, complete and linearly independent, and a_i are unknown coefficients to be sought.

Since the admissible functions (9.3) are not continuous on the common boundary Γ_0, i.e., $v^+ \neq v^-$ on Γ_0, we shall employ some additional integrals along Γ_0 to match v^+ and v^- in (9.3). Define a new space H such that

$$H = \left\{ v | v \in L^2(S), \ v \in H^1(S_1), \ v \in H^1(S_2) \right\},$$

and let $V_h^* \in H$ be a finite-dimensional collection of the admissible functions (9.3). Consequently, the general combinations using penalty plus hybrid functions are designed to seek a solution $\bar{u}_h \in V_h^*$ such that

$$A_h(\bar{u}_h, v) = F(v), \quad v \in V_h^*, \tag{9.4}$$

where

$$A_h(u, v) = B_h(u, v) + D_h(u, v), \tag{9.5}$$

$$B_h(u, v) = \iint_{S_1} [p \, \nabla u \, \nabla v + cuv] \, ds + \iint_{S_2} [p \, \nabla u \, \nabla v + cuv] \, ds +$$

$$+ \int_{\Gamma_N} quv d\ell, \tag{9.6}$$

$$D_h(u, v) = \frac{P_c}{h^\sigma} \int_{\Gamma_0} (u^+ - u^-)(v^+ - v^-) d\ell -$$

$$- \int_{\Gamma_0} \left[\left(\alpha p^+ \frac{\partial u^+}{\partial n} + \beta p^- \frac{\partial u^-}{\partial n} \right) (v^+ - v^-) + \right.$$

$$\left. + \left(\alpha p^+ \frac{\partial v^+}{\partial n} + \beta p^- \frac{\partial v^-}{\partial n} \right) (u^+ - u^-) \right] d\ell +$$

$$+ \frac{P_c}{h^\sigma} \int_{\Gamma_D} uv d\ell - \int_{\Gamma_D} \bar{\alpha} p \left(\frac{\partial u}{\partial n} v + \frac{\partial v}{\partial n} u \right) d\ell, \tag{9.7}$$

and

$$F(v) = F_S(v) + F_B(v), \tag{9.8}$$

$$F_S(v) = \iint_S fv \, ds + \int_{\Gamma_N} gv \, d\ell, \tag{9.9}$$

$$F_B(v) = \frac{P_c}{h^\sigma} \int_{\Gamma_D} vt \, d\ell - \int_{\Gamma_D} \overline{\alpha} p \frac{\partial v}{\partial n} t \, d\ell. \tag{9.10}$$

In (9.4)-(9.10), the additional integrals $B_h(u, v)$ and $F_B(v)$ are employed along the common boundary Γ_0 and the Dirichlet boundary Γ_D, where $P_c(> 0)$ is the penalty constant, σ is the penalty power, h is the maximal boundary length of quasi-uniform triangular elements in S_1, n is the outward normal to ∂S_2 (or ∂S), and α, β and $\overline{\alpha}$ are bounded constants to be discussed.

It should be noted that the bounded constants, α, β and $\overline{\alpha}$, are arbitrary. Based on the analysis in the next section, it is better to choose

$$\alpha + \beta = 1, \quad \overline{\alpha} = 1.$$

9.2   ERROR ANALYSIS

For simplicity, we let $\Gamma_N = \Gamma$, i.e., there exists only the Robin condition (9.2)

$$p \frac{\partial u}{\partial n} + qu = g \quad \text{on } \Gamma. \tag{9.11}$$

In this case, the integral terms in (9.4)-(9.10) involving the Dirichlet boundary Γ_D disappear. In order to ensure a unique solution of (9.1) and (9.11), we assume that there exists a positive constant q_0 such that

$$Meas(\Gamma|_{q \geq q_0 > 0}) > 0. \tag{9.12}$$

Define a new norm over the Sobolev space H,

$$\|v\|_h = \left\{ \|v\|_{1,S_1}^2 + \|v\|_{1,S_2}^2 + \frac{P_c}{h^\sigma} \|v^+ - v^-\|_{0,\Gamma_0}^2 \right\}^{\frac{1}{2}},$$

where the Sobolev norms are (see Sobolev 1963):

$$\|v\|_{m,\Omega} = \left\{ \sum_{|\alpha| \leq m} \iint_\Omega |D^\alpha v|^2 ds \right\}^{\frac{1}{2}}, \quad |v|_{m,\Omega} = \left\{ \sum_{|\alpha| = m} \iint_\Omega |D^\alpha v|^2 ds \right\}^{\frac{1}{2}}.$$

We also suppose that for $v \in V_h^*$, there exists a bounded constant C independent of h, L, u and v such that

$$\left\|\frac{\partial v^+}{\partial n}\right\|_{0,\Gamma_0} \leq CL^\mu \|v\|_{1,S_2}, \tag{9.13}$$

where μ is another positive constant independent of h, L, u and v. Below, the values of C may be different in different contexts. We can easily obtain the following lemma from the norms of piecewise polynomials.

Lemma 9.1 *Let the k-order Lagrange triangular elements be quasiuniform, then*

$$\left\|\frac{\partial v^-}{\partial n}\right\|_{0,\Gamma_0} \leq \frac{C}{\sqrt{h}}|v|_{1,S_1}, \quad \forall v \in V_h^*,$$

where h is the maximal boundary length of triangular elements in S_1.

Also, we need another lemma.

Lemma 9.2 *Let the k-order Lagrange triangular elements be quasiuniform, and suppose (9.13) and the following conditions hold:*

$$\sigma \geq 1 \quad when \quad \beta \neq 0, \tag{9.14}$$

$$\sigma > 0 \quad and \quad h^\sigma L^{2\mu} \leq C \quad when \quad \alpha \neq 0. \tag{9.15}$$

Then when the bounded penalty constant P_c is suitably large but independent of h, L and u, there always exist the bounds:

$$\alpha_0 \|v\|_1^2 \leq A_h(v,v), \tag{9.16}$$

$$A_h(u,v) \leq C \|u\|_h \|v\|_h, \tag{9.17}$$

where α_0 is a positive constant independent of h, L, u and v.

Proof. Since either $\Gamma \cap S_1$ or $\Gamma \cap S_2$ has to satisfy (9.12), we here let

$$Meas(\Gamma \cap S_1|_{q \geq q_0 > 0}) > 0,$$

without loss of generality. We then obtain from the theorem of Sobolev (1963)

$$\alpha_0 \|v\|_{1,S_1} \leq |v|_{1,S_1} + |v^-|_{0,\Gamma^*}, \tag{9.18}$$

where Γ^* denotes the part of $\Gamma \cap S_1$ and $q \geq q_0 > 0$.

Similarly, we have

$$\alpha_0\|v\|_{1,S_2} \leq |v|_{1,S_2} + |v^+|_{0,\Gamma_0}, \leq |v|_{1,S_2} + |v^-|_{0,\Gamma_0} + |v^+ - v^-|_{0,\Gamma_0}.$$

Use of the Sobolev embedding theorem and (9.18) gives

$$|v^-|_{0,\Gamma_0} \leq C\|v\|_{1,S_1} \leq \frac{C}{\alpha_0}\left(|v|_{1,S_1} + |v^-|_{0,\Gamma_\ast}\right). \tag{9.19}$$

Therefore, we can see from (9.18) and (9.19)

$$\alpha_0\left(\|v\|_{1,S_1}^2 + \|v\|_{1,S_2}^2\right) \leq C\left(|v|_{1,S_1}^2 + |v|_{1,S_2}^2 + |v^-|_{0,\Gamma_\ast}^2 + \|v^+ - v^-\|_{0,\Gamma_0}^2\right)$$

$$\leq \frac{C}{\min(p_0,q_0)}\left\{\iint_{S_1}\left[p(\nabla v)^2 + cv^2\right]ds + \iint_{S_2}\left[p(\nabla v)^2 + cv^2\right]ds\right.$$

$$\left.+ \int_\Gamma qv^2 d\ell\right\} + C\|v^+ - v^-\|_{0,\Gamma_0}^2. \tag{9.20}$$

Using (9.5) and (9.6) we obtain

$$A_h(v,v) \geq \alpha_0\left(\|v\|_{1,S_1}^2 + \|v\|_{1,S_2}^2\right) + \left(\frac{P_c}{h^\sigma} - \alpha_1\right)\|v^+ - v^-\|_{0,\Gamma_0}^2$$

$$-2\left|\int_{\Gamma_0}\left(\alpha p^+\frac{\partial v^+}{\partial n} + \beta p^-\frac{\partial v^-}{\partial n}\right)(v^+ - v^-)d\ell\right|, \tag{9.21}$$

where α_1 is also a positive constant independent of h, L, u and v.

Next, by means of the assumption (9.13) and Lemma 9.1, we have

$$\left|\int_{\Gamma_0} p^+\frac{\partial v^+}{\partial n}(v^+ - v^-)d\ell\right| \leq \max_{\Gamma_0} p^+\left\|\frac{\partial v^+}{\partial n}\right\|_{0,\Gamma_0}\|v^+ - v^-\|_{0,\Gamma_0}$$

$$\leq CL^\mu\|v\|_{1,S_2}\|v^+ - v^-\|_{0,\Gamma_0},$$

and

$$\left|\int_{\Gamma_0} p^-\frac{\partial v^-}{\partial n}(v^+ - v^-)d\ell\right| \leq \max_{\Gamma_0} p^-\left\|\frac{\partial v^-}{\partial n}\right\|_{0,\Gamma_0}\|v^+ - v^-\|_{0,\Gamma_0}$$

$$\leq \frac{C}{\sqrt{h}}\|v\|_{1,S_1}\|v^+ - v^-\|_{0,\Gamma_0}.$$

Hence with the help of $2ab \leq a^2 + b^2$, Eq. (9.21) will lead to

$$A_h(v,v) \geq \alpha_0\left(\|v\|_{1,S_1}^2 + \|v\|_{1,S_2}^2\right) + \left(\frac{P_c}{h^\sigma} - \alpha_1\right)\|v^+ - v^-\|_{0,\Gamma_0}^2 -$$

$$-C\left(|\alpha|L^\mu\|v\|_{1,S_2} + \frac{|\beta|}{\sqrt{h}}\|v\|_{1,S_1}\right)\|v^+ - v^-\|_{0,\Gamma_0}$$

$$\geq \frac{\alpha_0}{2}\left(\|v\|_{1,S_1}^2 + \|v\|_{1,S_2}^2\right) +$$

$$+ \left\{\frac{P_c}{h^\sigma} - \alpha_1 - \frac{C^2}{2\alpha_0}\left(\alpha^2 L^{2\mu} + \frac{\beta^2}{h}\right)\right\} \times \|v^+ - v^-\|_{0,\Gamma_0}^2$$

$$\geq \frac{\alpha_0}{2}\left(\|v\|_{1,S_1}^2 + \|v\|_{1,S_2}^2\right) + \frac{P_c}{2h^\sigma}\|v^+ - v^-\|_{0,\Gamma_0}^2. \tag{9.22}$$

In (9.22), the penalty constant P_c can be found such that

$$\frac{P_c}{h^\sigma} - \alpha_1 - \frac{C^2}{2\alpha_0}\left(\alpha^2 L^{2\mu} + \frac{\beta^2}{h}\right) \geq \frac{P_c}{2h^\sigma},$$

owing to the assumptions (9.14) and (9.15). Consequently, the first desired result (9.16) is obtained. The proof of (9.17) is similar. ∎

Now we give the main theorem for error bounds:

Theorem 9.1 *Let all conditions in Lemma 9.2 be given, then there exist the error bounds*

$$\|u - \bar{u}_h\|_h \leq C\left\{\inf_{v \in V_h^*}\|u - v\|_h + [1 - (\alpha + \beta)]h^{\frac{\sigma}{2}}\left[\int_{\Gamma_0}\left(\frac{\partial u}{\partial n}\right)^2 d\ell\right]^{\frac{1}{2}}\right\}. \tag{9.23}$$

Proof. Let u be the exact solution of (9.1) and (9.11). We then have

$$A_h(u, v) = \iint_{S_1}(p\nabla u\,\nabla v + cuv)\,ds + \iint_{S_2}(p\nabla u\,\nabla v + cuv)\,ds + \int_{\Gamma_N}quvd\ell$$

$$- \int_{\Gamma_0}(\alpha + \beta)p\frac{\partial u}{\partial n}(v^+ - v^-)d\ell$$

$$= F(v) + [1 - (\alpha + \beta)]\int_{\Gamma_0}p\frac{\partial u}{\partial n}(v^+ - v^-)d\ell. \tag{9.24}$$

Therefore, the following bounds can be found from Lemma 9.2, (9.24) and the Schwarz inequality

$$\alpha_0\|u_h - v\|_h^2 \leq A_h(u_h - v, u_h - v)$$

$$= A_h(u - v, u_h - v) - [1 - (\alpha + \beta)]\int_{\Gamma_0}p\frac{\partial u}{\partial n}(w_h^+ - w_h^-)d\ell \tag{9.25}$$

$$\leq \ C\left\{\|u-v\|_h + |1-(\alpha+\beta)|h^{\frac{\sigma}{2}}\left[\int_{\Gamma_0}\left(\frac{\partial u}{\partial n}\right)^2 d\ell\right]^{\frac{1}{2}}\right\}\|u_h-v\|_h,$$

where $w_h = u_h - v$ for any $v \in V_h^*$. The desired bounds (9.23) are obtained by dividing two sides of (9.25) with $\|u_h - v\|_h$ and applying the triangular inequality. ∎

Since the basis functions Ψ_i in (9.3) are complete in S_2, the exact solution u can be expanded into

$$u = \sum_{i=1}^{\infty} a_i^* \Psi_i = u_L + R_L \text{ in } S_2,$$

where a_i^* are the exact expansion coefficients, and

$$u_L = \sum_{i=1}^{L} a_i^* \Psi_i, \quad R_L = \sum_{i=L+1}^{\infty} a_i^* \Psi_i.$$

Theorem 9.2 *Let all conditions in Theorem 9.1 be given, and suppose*

$$u \in H^{k+1}(S_1), \quad u \in H^{k+1}(\Gamma_0).$$

Then there always exist the bounds

$$\|u-\overline{u}_h\|_h \ \leq \ C\left\{h^k + h^{k+1-\frac{\sigma}{2}} + \|R_L\|_{1,S_2} + h^{-\frac{\sigma}{2}}\|R_L\|_{0,\Gamma_0}\right.$$
$$\left. +|1-(\alpha+\beta)|h^{\frac{\sigma}{2}}\right\}. \tag{9.26}$$

In (9.26), the terms $O(h^k)$ and $O(\|R_L\|_{1,S_2})$ are the errors resulting from the k-order Lagrange FEM in S_1 and the RGM in S_2 respectively, and the remaining terms are the errors resulting from the coupling techniques of the penalty plus hybrid functions on Γ_0.

Since the k-order Lagrange FEM is used in S_1, optimal convergence rates of solutions by the RGM-FEM, (9.4), will be: $\|u - \overline{u}_h\|_h = O(h^k)$. Now we have the following corollary

Corollary 9.1 *Let all the conditions in Theorem 9.2 be given. Also suppose*

$$\alpha + \beta = 1, \tag{9.27}$$
$$\sigma \leq 2. \tag{9.28}$$

Then there exist the error bounds

$$\|u - \bar{u}_h\|_h \leq C \left\{ h^k + \|R_L\|_{1,S_2} + h^{-\frac{\sigma}{2}} \|R_L\|_{0,\Gamma_0} \right\}.$$

We have from (9.14), (9.15) and (9.28) the following conditions for RGM-FEM:

$$1 \leq \sigma \leq 2 \quad \text{when} \quad \beta \neq 0, \qquad (9.29)$$

$$0 < \sigma \leq 2 \quad \text{and} \quad h^\sigma L^{2\mu} \leq C \quad \text{when} \quad \alpha \neq 0. \qquad (9.30)$$

Let us now apply Corollary 9.1 to the three typical combinations. As a result, the following limitations of the constants σ, h and L have to be fulfilled:

1. $1 \leq \sigma \leq 2$ for Combination $I(\alpha = 0$ and $\beta = 1)$;

2. $0 < \sigma \leq 2$ and $h^\sigma L^{2\mu} \leq C$ for Combination $II(\alpha = 1$ and $\beta = 0)$;

3. $1 \leq \sigma \leq 2$ and $h^\sigma L^{2\mu} \leq C$ for Symmetric Combination $(\alpha = \beta = \frac{1}{2})$.

We note that the condition (9.27) does not have to hold, i.e.,

$$\alpha + \beta \neq 1. \qquad (9.31)$$

For instance, when $\alpha = \beta = 0$, the RGM-FEM (9.4) leads to the penalty combined method in Chapter 7. In addition, a necessary condition for the optimal convergence rate is also derived from Theorem 9.2:

$$\sigma = 2, \quad k = 1. \qquad (9.32)$$

It is interesting to point out that any bounded constants α and β (even < 0) satisfying (9.31) and (9.32) can be chosen in RGM-FEM (9.4), causing the same optimal convergence rates of the penalty combined method in Chapter 7.

It is remarkable that the above error bounds only require $u \in H^1(S_2)$. Consequently, the combinations in this chapter can be applied to solving the problems with singular points in S_2, such as angular singularity problems in Birkhoff (1972), interface singularity problems in Kellogg (1975), and some unbounded domain problems. For Laplace's equation with the angular singularity, we can also derive a coupling relation $L = O(|\ln h|)$ between L and h, see Section 7.3 of Chapter 7 and Section 8.7 of Chapter 8.

9.3   STABILITY ANALYSIS

Linear algebraic equations can be reduced from (9.4)

$$\mathbf{Ax} = \mathbf{b},$$

where $\mathbf{x}$ is an unknown vector with the components a_i and $v_{ij} = v^-(Q_{ij})$, Q_{ij} are finite element nodes, $\mathbf{b}$ is a known vector, and the coefficient matrix $\mathbf{A}$ is symmetric because $A_h(u, v) = A_h(v, u)$.

It is important that the uniformly elliptic inequality (9.16) ensure the positive definite property of $\mathbf{A}$. Hence the solution $\mathbf{x}$ (i.e., $\overline{u}_h$ in (9.4)) can easily be obtained.

Now we will discuss the stability of numerical solutions. The stability is measured by the condition number $Cond.(\mathbf{A})$ defined by

$$Cond.(\mathbf{A}) = \frac{\lambda_{max}(\mathbf{A})}{\lambda_{min}(\mathbf{A})}, \tag{9.33}$$

where $\lambda_{max}(\mathbf{A})$ and $\lambda_{min}(\mathbf{A})$ are the maximal and minimal eigenvalues of $\mathbf{A}$ respectively. For simplicity, we still consider the case (9.11). Then we have the following theorem.

Theorem 9.3 *Let all conditions in Lemma 9.2 be given, then there exist the bounds*

$$Cond.(\mathbf{A}) \le C \frac{1 + P_c h^{1-\sigma} + \left(1 + \frac{P_c}{h^\sigma}\right) \lambda_{max}(\mathbf{D})}{Min\left[h^2, \lambda_{min}(\mathbf{D})\right]}, \tag{9.34}$$

where the matrix $\mathbf{D}$ is defined by

$$\|v\|_{1,S_2}^2 = \mathbf{d}^T \mathbf{D} \mathbf{d}, \quad v \in V_h^*,$$

$\mathbf{d} = (a_1, \ a_2, \ ..., \ a_L)^T$, *and matrix $\mathbf{D} = (d_{ij})_{L \times L}$ with the matrix components*

$$d_{ij} = \iint_{S_2} \left(\nabla \Psi_i \, \nabla \Psi_j + \Psi_i \Psi_j\right) ds.$$

Proof. We have from Lemma 9.2

$$\alpha_0 \left(\|v\|_{1,S_1}^2 + \|v\|_{1,S_2}^2\right) \le \alpha_0 \|v\|_h^2 \le A_h(v, v), \tag{9.35}$$

$$A_h(v, v) \le C\|v\|_h^2 \le C \left\{\|v\|_{1,S_1}^2 + \|v\|_{1,S_2}^2 + \frac{P_c}{h^\sigma} \left(\|v^+\|_{0,\Gamma_0}^2 + \|v^-\|_{0,\Gamma_0}^2\right)\right\}. \tag{9.36}$$

By means of the Sobolev embedding theorem:

$$\|v^+\|_{0,\Gamma_0} \le C\|v\|_{1,S_2},\tag{9.37}$$

Eq. (9.36) leads to

$$A_h(v,v) \le C\left\{\|v\|_{1,S_1}^2 + \frac{P_c}{h^\sigma}\|v^-\|_{0,\Gamma_0}^2 + \left(1+\frac{P_c}{h^\sigma}\right)\|v\|_{1,S_2}^2\right\}.$$

Denote

$$A_h(v,v) = \mathbf{x}^T\mathbf{A}\mathbf{x}, \quad \mathbf{x} = \begin{pmatrix} \mathbf{v} \\ \mathbf{d} \end{pmatrix}, \quad \mathbf{v} = (...v_{ij}...)^T.$$

Then we obtain from (9.35) and (9.36)

$$\lambda_{min}(\mathbf{A}) \ge \alpha_0 \min\left[\lambda_{min}(\mathbf{F}), \lambda_{min}(\mathbf{D})\right],\tag{9.38}$$

$$\lambda_{max}(\mathbf{A}) \le C\left\{\lambda_{max}(\mathbf{F}) + \frac{P_c}{h^\sigma}\lambda_{max}(\mathbf{E}) + \left(1+\frac{P_c}{h^\sigma}\right)\lambda_{max}(\mathbf{D})\right\},\tag{9.39}$$

where the matrices $\mathbf{F}$ and $\mathbf{E}$ are defined by

$$\|v\|_{1,S_1}^2 = \mathbf{v}^T\mathbf{F}\mathbf{v}, \quad \|v^-\|_{0,\Gamma_0}^2 = \mathbf{v}^T\mathbf{E}\mathbf{v}.\tag{9.40}$$

Since F and E are the coefficient matrices coming from FEM used in S_1, we can find the eigenvalue bounds from Strang and Fix (1973)

$$\lambda_{min}(\mathbf{F}) \ge \alpha_0 h^2, \ \lambda_{max}(\mathbf{F}) \le C, \ \lambda_{max}(\mathbf{E}) \le Ch.$$

Consequently, Eq. (9.38) and (9.39) yield

$$\lambda_{min}(\mathbf{A}) \ge \alpha_0 \min\left[h^2, \lambda_{min}(\mathbf{D})\right],\tag{9.41}$$

$$\lambda_{max}(\mathbf{A}) \le C\left[1 + P_c h^{1-\sigma} + \left(1+\frac{P_c}{h^\sigma}\right)\lambda_{max}(\mathbf{D})\right].\tag{9.42}$$

The desired results (9.34) follow from (9.33), (9.41) and (9.42). ■

Moreover, when choose

$$\lambda_{max}(\mathbf{D}) = O(1),\tag{9.43}$$

by using some normalized basis functions Ψ_i, we then obtain the following corollary.

Corollary 9.2 *Let all conditions in Lemma 9.2 and (9.43) be given, then*

$$Cond.(\mathbf{A}) \leq C \left(1 + \frac{P_c}{h^\sigma}\right)(h^{-2} + Cond.(\mathbf{D})). \tag{9.44}$$

In (9.44), the terms h^{-2} and $Cond.(\mathbf{D})$ are the condition numbers resulting from FEM in S_1 and RGM in S_2 respectively. However, an additional factor $\frac{P_c}{h^\sigma}$ comes from the penalty integral along Γ_0, and the assumptions (9.14) and (9.15) are caused by the hybrid integral along Γ_0. Let us draw some interesting conclusions from Theorem 9.3 and Corollary 9.2.

First, the smaller the penalty power σ is, the smaller the bounds of $Cond.(\mathbf{A})$ are. Therefore, $\sigma = 1$ is best for stability of Combination I and Symmetric Combination where $1 \leq \sigma \leq 2$ is a necessary condition for optimal convergence rates. Let $\sigma = 1$, Corollary 9.2 gives

$$Cond.(\mathbf{A}) \leq C \left(1 + \frac{P_c}{h}\right)(h^{-2} + Cond.(\mathbf{D})),$$

which shows a good stability of numerical solutions by our combinations. Therefore, $\sigma = 1$ is a good choice for Combination I and Symmetric Combination. For Combination II where (9.30) is necessary conditions for optimal convergence rates, the values of σ in Combination II should be chosen as small as possible. Since $L = O(|\ln h|)$ given in Chapters 4, 7, and 8, we may choose $\sigma << 1$ as $h << 1$, to reach the nearly optimal stability from (9.44).

Second, the solutions of Combinations I, II and Symmetric Combination will approach the solution of the nonconforming combination provided that $P_c \to \infty$, based on Proposition 9.1 stated in Section 9.4 later. Thus the penalty constant P_c is only allowed to be suitably large (but not infinite), from the point of view of stability. This also explains why $P_c = 10^6$ is employed for the numerical experiments in Li and Bui (1988c).

Third, it should also be noted that the above stability analysis is valid for the penalty combined method where $\alpha = \beta = 0$ in Chapter 7 and the simplified hybrid combination $\alpha = 1$ and $P_c = \beta = 0$ for the combination of RGM-FEM in Chapter 8.

9.4 SUMMARY OF COUPLING TECHNIQUES

In this section, overall comparisons are made for six efficient couplings for combinations of the RGM-FEM. The comparisons are done by using theoretical analysis and numerical experiments. Significant relations among the six combinations described in Chapters 4 and 7–9 are also found. A survey of these combinations and their coupling strategies are given. These combinations are important not only for matching the RGM-FEM but also for matching other numerical methods, such as various FEMs in Chapter 10 and the RGM-FDM in Chapter 11.

9.4.1 Six Combinations

Consider a general elliptic equation

$$-\frac{\partial}{\partial x}(p\frac{\partial u}{\partial x}) - \frac{\partial}{\partial x}(p\frac{\partial u}{\partial x}) + cu = f, (x,y) \in S, \qquad (9.45)$$

with the boundary conditions

$$u = g_1 \text{ in } \Gamma_D, \qquad (9.46)$$

$$p\frac{\partial u}{\partial n} + qu = g_2 \text{ in } \Gamma_N, \qquad (9.47)$$

where the domain S is a bounded, simply connected polygon with the exterior boundary Γ, $\Gamma = \Gamma_D \cup \Gamma_N$, $\text{Meas}(\Gamma_D) > 0$, the functions p, c, f, q, g_1 and g_2 are sufficiently smooth, and

$$c = c(x,y) \geq 0, q = q(x,y) \geq 0, p = p(x,y) \geq p_0 > 0,$$

here p_0 is a constant.

The solution of the problem (9.45)—(9.47) can be equivalently expressed by minimizing a quadratic functional $I(v)$:

$$I(u) = \min_{v \in H^1_*(v)} I(v),$$

where the quadratic functional is

$$I(v) = \frac{1}{2}\iint_S [p(\nabla v)^2 + cv^2]ds + \frac{1}{2}\int_{\Gamma_N} qv^2 d\ell - \iint_S fvds - \int_{\Gamma_N} g_2 vd\ell,$$

and $H^1_*(S)$ is the Sobolev space defined by

$$H^1_*(S) = \{v \mid v, v_x, v_y \in L^2(S), \text{ and } v \mid_{\Gamma_D} = g_1\}.$$

Let S be divided by Γ_0, the piecewise straight line into S_1 and S_2. The Ritz-Galerkin method is used in S_2, and the $k(\geq 1)-$ order Lagrange finite element method is still used in S_1 where the true solution is supposed to be smooth enough such that $u \in H^{k+1}(S_1)$. Therefore, the admissible functions can be written as follows:

$$v = \begin{cases} v^- = V_1^k & \text{in } S_1, \\ v^+ = \sum_{i=1}^N a_i \Psi_i & \text{in } S_2, \end{cases} \tag{9.48}$$

where V_1^k are piecewise k-order Lagrange interpolation polynomials on the quasiuniform triangulation of S_1 with the maximal boundary length h, $\{\Psi_i\}$ are analytic, complete and linear independent basis functions on S_2, and a_i are unknown coefficients to be solved. Note that the admissible functions v^+ in (9.48) may not satisfy the elliptic equation (9.45) exactly.

Define a space

$$H = \{v \mid v \in L^2(S), v \in H^1(S_1), v \in H^1(S_2) \text{ and } v \mid_{\Gamma_D \cap S_1} = g_1\}.$$

Let $V_h^* \in H$ be a finite-dimensional collection of the functions (9.48). For simplicity, we assume that the functions $v \in V_h^*$ will strictly satisfy the boundary condition (9.46) on $\Gamma_D \cap S_1$ where the finite element method is used (otherwise, see Strang and Fix (1973)).

In this subsection, the combination using the hybrid plus penalty techniques is designed to seek an approximate solution $u_h \in V_h^*$ such that

$$I_h^{(\alpha,\beta)}(u_h) = \min_{v \in V_h^*} I_h^{(\alpha,\beta)}(v), \tag{9.49}$$

where

$$
\begin{aligned}
I_h^{(\alpha,\beta)}(v) =\ & \frac{1}{2}\iint_{S_1}(p(\nabla v)^2 + cv^2)ds + \frac{1}{2}\iint_{S_2}(p(\nabla v)^2 + cv^2)ds \\
& + \frac{1}{2}\int_{\Gamma_N} qv^2 d\ell + \frac{P_c}{2h^{*\sigma}}\int_{\Gamma_D \cap S_2}(v^+ - g_1)^2 d\ell \\
& - \int_{\Gamma_D \cap S_2} p\frac{\partial v^+}{\partial n}(v^+ - g_1)d\ell + \frac{P_c}{2h^{*\sigma}}\int_{\Gamma_0}(v^+ - v^-)^2 d\ell \\
& - \int_{\Gamma_0} p(\alpha\frac{\partial v^+}{\partial n} + \beta\frac{\partial v^-}{\partial n})(v^+ - v^-)d\ell - \iint_S fvds - \int_{\Gamma_N} g_2 v d\ell.
\end{aligned}
\tag{9.50}
$$

In fact, the combinations as in (9.50) are general approaches using additional integrals, which contain the following six important combined methods:

1. **Combination I:** $P_c > 0$, $\alpha = 0$ and $\beta = 1$.

2. **Combination II:** $P_c > 0$, $\alpha = 1$ and $\beta = 0$.

3. **Symmetric Combination:** $P_c > 0$ and $\alpha = \beta = \frac{1}{2}$.

4. **The Simplified Hybrid Combined Method:** $P_c = 0$, $\alpha = 1$ and $\beta = 0$.

5. **The Penalty Combined Method:** $P_c > 0$ and $\alpha = \beta = 0$.

6. **The Nonconforming Combination:** as $P_c \longrightarrow \infty$, or $\sigma \longrightarrow \infty$.

In the last combination, a direct, nonconforming constraint of the admissible functions such that

$$v^+(Q_i) = v^-(Q_i), \quad \forall Q_i \in \Gamma_0 \tag{9.51}$$

can also be employed, without using any additional integrals on Γ_0. In (9.51), Q_i denote the element nodes on Γ_0. Define by $\overline{V}_h \in H$ a dimensional collection of the admissible functions (9.48) that satisfy the constraint conditions (9.51) and the Dirichlet boundary condition (9.46). We obtain from (9.49) as $P_c = \alpha = \beta = 0$

$$I_h^{(0,0)}(u_h) = \min_{v \in \overline{V}_h} I_h^{(0,0)}(v).$$

Also, the Lagrange multiplier method is given in Chapter 6. The bilinear form (9.4) is modified as

$$B(\lambda, \mu; u, v) = \iint_{S_1} (p \nabla u \nabla v + cu)ds + \iint_{S_2} (p \nabla u \nabla v + cu)ds + \\ +D(\lambda, \mu; u, v),$$

where the boundary integrals are

$$D(\lambda, \mu; u, v) = \int_{\Gamma_N} quvd\ell - \int_{\Gamma_D \cap S_2} \lambda vd\ell - \int_{\Gamma_D \cap S_2} \mu ud\ell \\ - \int_{\Gamma_0} \lambda(v^+ - v^-)d\ell - \int_{\Gamma_0} \mu(u^+ - u^-)d\ell, \tag{9.52}$$

and the Lagrangian multiplier λ has the true solution, $\lambda = p\frac{\partial u}{\partial n}\big|_{\Gamma_0}$. In (9.52), the λ is treated as an extra variable. A relation of these coupling techniques is illustrated in Figure 9.1.

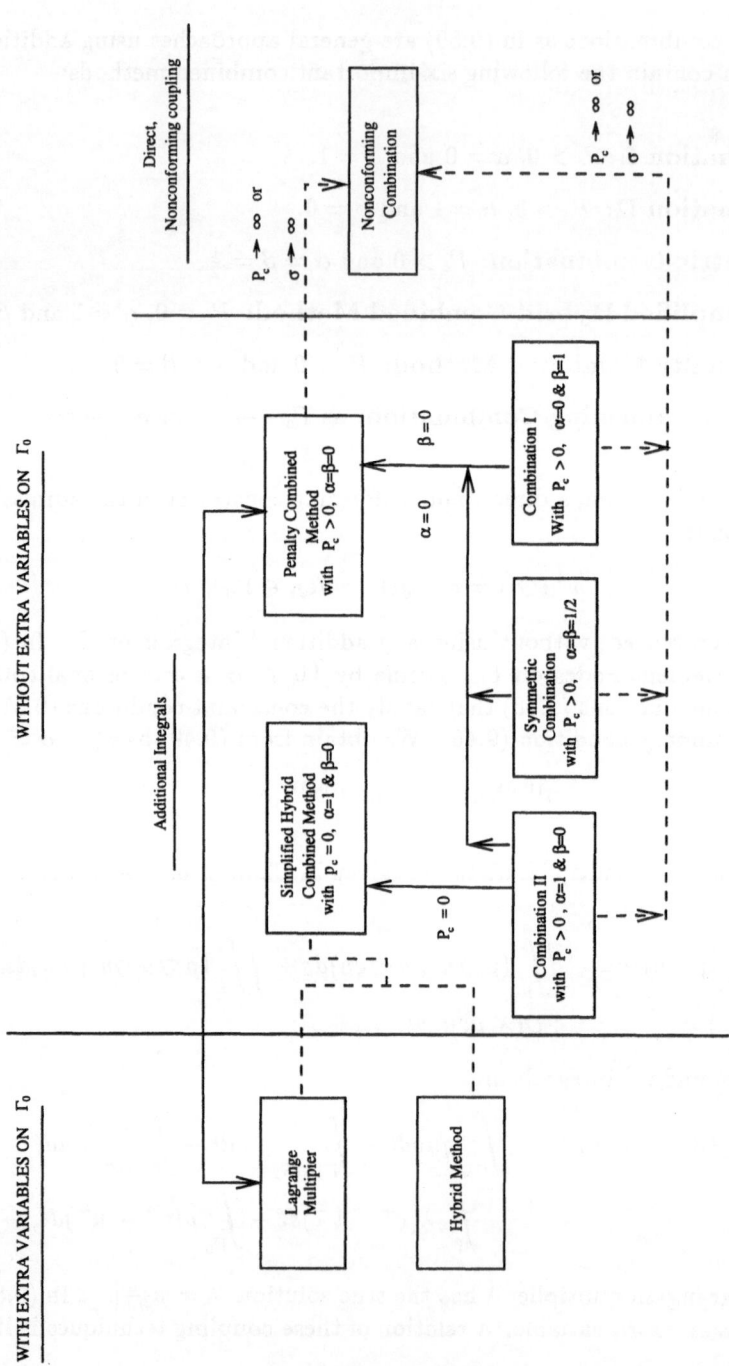

Figure 9.1 Comparison of different couplings in combinations.

9.4.2 Linkage of Combinations

Below we apply six combinations to Laplace's equation with the model problem in Chapter 4 (see Figure 4.2).

$$\Delta u = \frac{\partial^2 u}{\partial x^2} + \frac{\partial^2 u}{\partial y^2} = 0, \tag{9.53}$$

$$u = 0, \quad (x,y) \in \overline{AC}, \tag{9.54}$$

$$\frac{\partial u}{\partial n} = 0, \quad (x,y) \in \overline{AB}, \tag{9.55}$$

$$\frac{\partial u}{\partial n} = g, \quad (x,y) \in \overline{BF} \cup \overline{FG} \cup \overline{GC}. \tag{9.56}$$

The admissible function can be chosen as

$$v = \begin{cases} v^- = v_1^1, & \text{in } \widehat{S}_1^h, \\ v^+ = \sum_{\ell=0}^{L} \tilde{D}_\ell \left(\frac{r}{R}\right)^{\tau(\ell+1/2)} \cos\left[\tau\left(\ell+\frac{1}{2}\right)\theta\right], & \text{in } S_2, \end{cases} \tag{9.57}$$

where $\tau = \pi/\Theta$, (r,θ) are the polar coordinates with the origin $(0,0)$, $\tilde{D}_\ell$ are unknown coefficients, and v_1^1 are piecewise linear interpolation functions on the triangulation $\widetilde{S}_1^h$ of S_1.

Denote by $v_{i,j}$ the unknowns of V_1^1 on the element nodes (i,j) in S_1. Then the constraint condition (9.51) is reduced to

$$v_{0,j} = \sum_{\ell=0}^{L} \tilde{D}_\ell \left(\frac{R^*}{R}\right)^{\tau(\ell+\frac{1}{2})} \cos\left[\tau\left(\ell+\frac{1}{2}\right)\theta_j\right], \quad \forall j, \tag{9.58}$$

where $v_{0,j} = v_1^1(R^*, \theta_j)$, and (R^*, θ_j) are the element nodes $(0,j)$ on ℓ_{R^*}.

The overall relations are schematically illustrated in Figure 9.1, which shows six combinations linked by the relation lines.

In Chapter 7, while $P_c \longrightarrow \infty$ and under certain integration rules, the solution of the penalty combined method will, in theory, approach that of the nonconforming combination. The same approaching property also happens to Combinations I and II and Symmetric Combination. Let us explain this below.

The additional integrals along Γ_0 in (9.50) can be approximated by the simplest trapezoidal rule:

$$\int_x^{x+\Delta x} f(x)dx \approx \frac{\Delta x}{2}\left[f(x+\Delta x) + f(x)\right], \tag{9.59}$$

where the element nodes (R^*, θ_j) on Γ_0 are chosen as the integration nodes for the integrals in (9.50) along Γ_0. Therefore, a difference equation for $\tilde{D}_\ell$ is obtained from (9.49),

$$
\begin{aligned}
0 = & \ L_\ell(\tilde{D}_\ell, v_{ij}, \alpha, \beta) + \\
& + \frac{P_c}{h^{*\sigma}} \sum_{j=0}^{N} \left[\sum_{\ell=0}^{L} \tilde{D}_\ell \left(\frac{R^*}{R} \right)^{\tau(\ell + \frac{1}{2})} \cos\left\{ \tau \left(\ell + \frac{1}{2} \right) \theta_j \right\} - v_{0,j} \right] \times \\
& \times \left(\frac{R^*}{R} \right)^{\tau(\ell + 1/2)} \cos\left(\tau \left(\ell + \frac{1}{2} \right) \theta_j \right) \frac{R^*(\triangle\theta_j + \triangle\theta_{j-1})}{2}, \qquad (9.60)
\end{aligned}
$$

where the linear functional $L_\ell(\tilde{D}_\ell, v_{ij}, \alpha, \beta)$ of $\tilde{D}_\ell$ and v_{ij}, which results from (9.49), is independent of P_c. Consequently, while $P_c \longrightarrow \infty$ or $\sigma \longrightarrow \infty$, Eq. (9.49) yields the constraint condition (9.58) of the nonconforming method. No matter what the values of the bounded constants α and β are, the ultimate method of combinations using the penalty integrals with $P_c \longrightarrow \infty$ or $\sigma \longrightarrow \infty$ is the very nonconforming combinations. Such a conclusion has been confirmed by the numerical experiments.

Proposition 9.1 *When the penalty plus hybrid integrals are approximated by the trapezoidal rule, then when P_c (or σ) is chosen properly large, the solutions of numerical Combinations I, II and Symmetric Combinations approach those of the nonconforming combination.*

We should point out that the solution method for the nonconforming combination using high order FEMs is troublesome, owing to complicated elimination procedure. However, the combinations using the additional integrals along Γ_0 are considerably simple to carry out. Therefore, it is better to use one of Combinations I and II and Symmetric Combination with a suitably large value of P_c, than to use the nonconforming combination in Chapters 4 and 5 when high order FEMs are used.

9.4.3   Comparisons

ubsectionComparisons The above analyses also imply that the nonconforming combination is the most fundamental approach in matching the RGM-FEM. Research into the coupling strategy of additional integrals is also important because it may provide much more flexibility when using the combined methods for solving general elliptic boundary value problems with singularities.

Next, Tables 9.1 and 9.2 summarize the comparisons for six combinations. It is

Methods	Limitations for Applications	Values of k for optimal convergence rates	Values of σ for optimal convergence rates	Properties of **A** in **Ax** = **b**
Nonconf.	No	Any*	NA	Symm. and posit. def.
Penalty	No	$k = 1$	$\sigma = 2$	Symm. and posit. def.
Simplified Hybrid	$f \equiv 0$ in S_2	Any*	NA	Symm. only or posit. def. only
Comb. I	No	Any*	$1 \leq \sigma \leq 2$	Symm. and posit. def.
Comb. II	No	Any*	$0 < \sigma \leq 2$	Symm. and posit. def.
Symm.	No	Any*	$1 \leq \sigma \leq 2$	Symm and posit. def.

Table 9.1 Comparisons of six combinations for general elliptic equations $L(u) = -\frac{\partial}{\partial x}(p\frac{\partial u}{\partial x}) - \frac{\partial}{\partial y}(p\frac{\partial u}{\partial y}) + cu = f$. *The order k of the FEM's is usually less than 4, and "NA" means "not available".

Methods	Necessary constraints between L and h	Optimal matching between L and h	Calculated error norms
Nonconf.	$h[\tau(L + \frac{1}{2})]^{3/2} \leq C$	Six Combs. share:	Six Combs. have:
Penalty	$h[\tau(L + \frac{1}{2})]^{3/2} \leq C$		
Simplified hybrid	$h[\tau(L + \frac{1}{2})^{\frac{5}{2}}] \leq C$	$(\frac{R^*}{R})^{\tau(L+3/2)} = O(h)$,	$\|\varepsilon\|_1 = O(h)$,
Comb. I	No		$\|\varepsilon\|_{0,s} = O(h^2)$, $\|\varepsilon\|_{\infty,s} = O(h^{2-\delta})$,
Comb. II	$h[\tau(L + \frac{1}{2})]^{\frac{3}{2}} \leq C$	$L + 1 = O(|\ell n h|)$	$\delta D_0 = O(h^2)$,
Symm.	$h[\tau(L + \frac{1}{2})]^{\frac{3}{2}} \leq C$		as $h \to 0$

Table 9.2 Comparisons of six combinations for Laplace's equation with singularities when $k = 1$.

noted that the homogeneous condition

$$f \equiv 0, \quad \text{in } S_2 \tag{9.61}$$

is required by the simplified hybrid combination which may also be formally re-moved.

Define a norm over H for the piecewise straight line Γ_0

$$\|v\|_1 = (\|v\|_{1,S_1}^2 + \|v\|_{1,S_2}^2)^{1/2}, \tag{9.62}$$

or for the circular arc Γ_0

$$\|v\|_1 = (\|v\|_{1,\widehat{S}_1^h}^2 + \|v\|_{1,S_2}^2)^{1/2}. \tag{9.63}$$

The norm (9.63) is valid for Laplace's equation with singularities in which the linear FEM is used in $\widehat{S}_1^h$, i.e., $k = 1$. A criterion for efficient combined methods is the requirement of optimal convergence rates for the numerical solutions, i.e.,

$$\|\varepsilon\|_1 = O(h^k), \tag{9.64}$$

where $\varepsilon = u - u_h$, u_h are the numerical solutions and h is the maximal boundary length of k-order elements on S_1.

To obtain (9.64), the constants k and σ should, from theoretical analysis, satisfy certain limitations. For instance, the penalty combined method grants (9.64) if and only if $k = 1$ and $\sigma = 2$. However, for Combination I and the Symmetric Combination, any $k \geq 1$ but $1 \leq \sigma \leq 2$ can ensure the optimal convergence rates. Moreover, the analysis in Section 9.3 shows that the smaller σ is, the better the stability of numerical solutions is. Hence, the best value of $\sigma = 1$ is found for Combination I and Symmetric Combination.

Other interesting results are that there may exist some necessary constraints be-tween L and h for the optimal convergence rates (9.64). Take the case of Laplace's equations (9.53)– (9.56) with singularities as an example. Since $k = 1$, Eq. (9.64) becomes

$$\|\varepsilon\|_1 = O(h), \tag{9.65}$$

where $\|\cdot\|_1$ is defined by (9.63). The necessary constraints between L and h are given in Table 9.2 for the combinations, except Combination I (also see Li and Bui (1988b)). It is remarkable that all six combinations share the same significant coupling relation as

$$L + 1 = O(|\ln h|) \tag{9.66}$$

that grants to six combinations high efficiency for solving Laplace's equation with singularities.

In addition, a linear algebraic equation system

$$\mathbf{A}\mathbf{x} = \mathbf{b} \tag{9.67}$$

is obtained from the six combinations, where $\mathbf{x}$ is an unknown vector with the components a_i (or $\tilde{D}_\ell$) and $V_{i,j}$, $\mathbf{b}$ is a known vector, and $\mathbf{A}$ is the coefficient matrix.

For combinations given in this chapter, the matrix $\mathbf{A}$ in (9.67) is symmetric and sparse. The matrix $\mathbf{A}$ is also positive definite for the nonconforming combination with elimination precess, the penalty combination, and Combinations I and II and the Symmetric Combination when the penalty constant P_c is not very small.

As for the simplified hybrid combination, defined in (9.49) with $P_c = \beta = 0$ and $\alpha = 1$, the matrix $\mathbf{A}$ is symmetric but not positive definite. However, for the alternative presentation of (8.30) and (8.31) of Chapter 8, the corresponding matrix $\mathbf{A}$ can be positive definite but not symmetric.

Numerical experiments have shown to be even more promising. For the typical Motz problem in the next subsection, only a few particular solutions in S_2 are required for practical purposes. Moreover, the following asymptotic formulae have been observed:

$$\|\varepsilon\|_1 = O(h), \quad \|\varepsilon\|_{0,S} = O(h^2), \quad \|\varepsilon\|_{\infty,S} = O(h^{2-\delta}), \quad 0 < \delta << 1, \tag{9.68}$$
$$\delta D_0 = |D_0 - \tilde{D}_0| = O(h^2). \tag{9.69}$$

We note that the asymptotic formulae (9.68) and (9.69) agree with those in the standard FEM.

For numerical stability, the optimal rates of condition number, $O(h^{-2})$, can be obtained by using the nonconforming combinations and the simplified hybrid combinations; but only the fair rates of $O(h^{-2} + P_c h^{-(2+\sigma)})$ with small $\sigma (\geq 1)$ by using penalty combinations, Combination I and Symmetric combination, and with $\sigma > 0$ by using Combination II.

Methods	δ_1	δ_2	Max	$\|\varepsilon\|_{0,S}$	$\|\varepsilon\|_h$	δD_0	δD_1
Nonconf. Penalty	.1412*	.06932*	.2690	.1397	8.690	.1434	.0176
$\sigma = 2, P_c = 14$ Simplified	.1646	.1048	.2653*	.1392	8.690	.1906	.0198
hybrid Combination I	.3305	.1771	.2692	.1420	8.687*	.1604	.0592
$\sigma = 1, P_c = 8$ Symmetric	.2037	.1255	.2687	.1377*	8.692	.1493*	.0129*
$\sigma = 1, P_c = 5$	.1937	.1128	.2689	.1417	8.690	.1815	.0332

Table 9.3   Comparisons of error norms by five combined methods for $M = 8$ and $L + 1 = 6$. In this table, asterisks indicate the best results among all combinations, $\delta_1 = \|\varepsilon^+ - \varepsilon^-\|_{\infty,\ell_{R*}}$ and $\delta_2 = \|\varepsilon^+ - \varepsilon^-\|_{0,\ell_{R*}}$.

Divisions	$\|\varepsilon^+ - \varepsilon^-\|_{\infty,l_{R*}}$	$\|\varepsilon^+ - \varepsilon^-\|_{0,l_{R*}}$	Max	$\|\varepsilon\|_{0,S}$	$\|\varepsilon\|_h$
$M = 2$ $L + 1 = 4$	1.683	0.9467	4.451	2.438	35.75
$M = 3$ $L + 1 = 5$	1.006	0.4897	1.941	1.013	23.49
$M = 4$ $L + 1 = 5$	0.5675	0.2731	1.085	0.5625	17.50
$M = 6$ $L + 1 = 6$	0.2504	0.1219	0.4802	0.2491	11.61
$M = 8$ $L + 1 = 6$	0.1412	0.0693	0.2690	0.1397	8.690

Table 9.4   The error norms of calculated solutions with $P_c = 10^6$ and $\sigma = 1$ for Combinations I and II and Symmetric Combination.

9.4.4   Numerical Experiments

For comparisons, we collect error norms and the approximate coefficients $\tilde{D}_0$ and $\tilde{D}_1$ of $M = 8$ and $L + 1 = 6$ in Table 9.3. The values are slightly different among combinations, with the asterisk signalling the best results. It is worth noting that the values of the error norms, $\|\varepsilon^+ - \varepsilon^-\|_{0,\ell_{R*}}$ and $\|\varepsilon^+ - \varepsilon^-\|_{\infty,\ell_{R*}}$, are smallest for the nonconforming combinations, but largest for the simplified hybrid combination.

Divisions	$\tilde{D}_0$	$\tilde{D}_1$	$\tilde{D}_2$	$\tilde{D}_3$	$\tilde{D}_4$	$\tilde{D}_5$
$M = 2$						
$L + 1 = 4$	398.900	88.1605	18.0945	-16.2052		
$M = 3$						
$L + 1 = 5$	400.136	87.8196	17.6125	-12.3557	-10.8763	
		7 |	|4			
$M = 4$						
$L + 1 = 5$	400.587	87.7370	17.4545	-10.6984	-5.78267	
					| 6 |	
$M = 6$						
$L + 1 = 6$	409.074	87.6875	17.3377	-9.33121	-1.87735	-2.0978
			| 9 |		3 | 4 |	
$M = 8$						
$L + 1 = 6$	409.019	87.6735	17.2951	-8.80235	-0.447959	-1.12089
				| 7 | 6	56 | 63 |	88 | 91 |
True coefficients	401.162	87.6559	17.2379	-8.07122	1.440273	0.331055

Table 9.5 The calculated coefficients by the Nonconforming Combination, as well as Combinations I and II and Symmertric Combination with $P_c = 10^6$ and $\sigma = 1$.

This can be explained by the facts that the condition (9.58) is strongest in constraining v^+ and v^- on Γ_0, and that there do not exist straightforward constraints between v^+ and v^- in the simplified hybrid combination.

Next, letting $P_c = 10^6$ and $\sigma = 1$ in Combination I and II and the Symmetric Combination, we provide the calculated error norms in Table 9.4. It is spectacular that all digits appearing in Table 9.4 are exactly the same as those by the nonconforming combination!

Also, Table 9.5 lists the approximate coefficients $\tilde{D}_i$ by the nonconforming combination and by the three combinations: I, II, and Symmetric Combination. The complete digits of $\tilde{D}_i$ in Table 9.5 denote the results obtained by the nonconforming combination. We provide the different digits obtained by the other three combinations, shown in the left, middle and right positions on the next lines if having, as

Left:	Middle:	Right:
different digits	different digits	different digits
from	from	from
Combination II	Symmetric Combination	Combination I

For example, when $M = 8$ and $L + 1 = 6$, $\tilde{D}_4 = -0.447959$ for the nonconforming combination and Combination I, $\tilde{D}_4 = -0.447956$ for Combination II, and $\tilde{D}_4 = -0.447963$ for Symmetric Combination.

9.5 APPENDIX: IMPLEMENTS IN COMPUTATIONS FOR MOTZ PROBLEM

In this appendix, we provide how to conduct the explicit algebraic equations in the combinations of RGM-FEM, using the penalty plus hybrid techniques.

Consider Motz's problem. The admissible functions can be chosen as in Chapter 4:

$$v = \begin{cases} v^- = V_1^1 & \text{in } \hat{S}_1, \\ v^+ = \sum_{\ell=0}^{L} \tilde{D}_\ell \, r^{\ell+\frac{1}{2}} \cos(\ell + \tfrac{1}{2})\theta & \text{in } S_2, \end{cases} \tag{9.70}$$

where $\tilde{D}_\ell$ are unknown coefficients to be calculated, and V_1^1 are piecewise linear interpolation functions on the triangulation of $\hat{S}_1$.

In this case, the combinations (9.49) may be written as

$$I_h^{(\alpha,\beta)}(u_h) = \min_{v \in V_h^*} I_h^{(\alpha,\beta)}(v),$$

where

$$
\begin{aligned}
I_h^{(\alpha,\beta)}(v) &= \frac{1}{2}\iint_{S_1}(\nabla v)^2 ds + \frac{1}{2}\iint_{S_2}(\nabla v)^2 ds + \frac{P_c}{2h^{*\sigma}}\int_{\ell_{R^*}}(v^+ - v^-)^2 d\ell - \\
&\quad - \int_{\ell_{R^*}}\left(\alpha\frac{\partial v^+}{\partial r} + \beta\frac{\partial v^-}{\partial r}\right)(v^+ - v^-)d\ell.
\end{aligned}
\tag{9.71}
$$

We note that the admissible functions v^+ in (9.70) satisfy the Dirichlet-Neumann boundary conditions at $y = 0$. Here the space $V_h^* \in H$ is a finite-dimensional collection of (9.70) satisfying the homogeneous Dirichlet condition. In (9.71), the integral, $\iint_{S_1}(\nabla v)^2 ds$, can be substituted by an approximation $\iint_{\hat{S}_1}(\nabla v)^2 d\ell$, in Chapter 4. Also, the additional integrals along ℓ_{R^*} need to be approximated by integrations rules. For simplicity, we can use the trapezoidal rule:

$$\int_x^{x+\Delta x} f(x)dx \approx \int_x^{\widetilde{x+\Delta x}} f(x)dx = \frac{\Delta x}{2}(f(x) + f(x + \Delta x)). \tag{9.72}$$

Then the solution $\tilde{u}_h$ is obtained by

$$\tilde{I}_h^{(\alpha,\beta)}(\tilde{u}_h) = \min_{v \in V_h^*} \tilde{I}_h^{(\alpha,\beta)}(v), \tag{9.73}$$

where

$$\tilde{I}_h^{(\alpha,\beta)}(v) = \frac{1}{2} \iint_{\widehat{S}_1} (\nabla v)^2 ds + \frac{1}{2} \iint_{S_2} (\nabla v)^2 ds + \frac{P_c}{2h^{*\sigma}} \widetilde{\int_{\ell_{R_*}}} (v^+ - v^-)^2 d\ell -$$
$$- \widetilde{\int_{\ell_{R_*}}} (\alpha \frac{\partial v^+}{\partial r} + \beta \frac{\partial v^-}{\partial r})(v^+ - v^-)^2 d\ell, \tag{9.74}$$

where $V_h^* \in H$, and the definition of H is revised by

$$H = \{v \mid v \in L^2(S), v \in H^1(\widehat{S}_1), v \in H^1(S_2)\}.$$

As a result, Combinations I, II and III lead to:

Combination I. ($\alpha = 0$ and $\beta = 1$):

$$\tilde{I}_h^{(0,1)}(\tilde{u}_h^I) = \min_{v \in V_h^*} \tilde{I}_h^{(0,1)}(v). \tag{9.75}$$

Combination II. ($\alpha = 1$ and $\beta = 0$):

$$\tilde{I}_h^{(1,0)}(\tilde{u}_h^{II}) = \min_{v \in V_h^*} \tilde{I}_h^{(1,0)}(v). \tag{9.76}$$

Combination III. ($\alpha = \beta = \frac{1}{2}$):

$$\tilde{I}_h^{(\frac{1}{2},\frac{1}{2})}(\tilde{u}_h^{III}) = \min_{v \in V_h^*} \tilde{I}_h^{(\frac{1}{2},\frac{1}{2})}(v). \tag{9.77}$$

Here $\tilde{u}_h^I$ is the solutions of u_h^I resulting from the integration approximation and the non-coincidence between $\widehat{S}_1 = \widehat{S}_1^h$ and S_1.

Now, let us establish the concrete difference equations

$$\mathbf{A}\mathbf{x} = \mathbf{b}, \tag{9.78}$$

where $\mathbf{x}$ is the unknown vector with the components a_i and $V_1^k(Q_{ij})$ on the linear element nodes Q_{ij}, $\mathbf{b}$ is a known vector, and $\mathbf{A}$ is a symmetric matrix with respect to the unknowns $\widehat{D}_\ell$ and $v_{i,j}$. Note that $v_{i,j} = v^-(r_j, \theta_j)$ are the linear element nodes in $\widehat{S}_1$.

First by means of (9.70) and the orthogonality of trigonometric functions, we obtain

$$
\iint_{S_2} (\nabla v)^2 ds = \int_{\partial S_2} v \frac{\partial v}{\partial n} d\ell = \int_{\ell_{R^*}} v \frac{\partial v}{\partial r}\Big|_{r=R^*} d\ell
$$

$$
= \int_0^\pi \left[\sum_{\ell=0}^{L} \tilde{D}_\ell (R^*)^{\ell+\frac{1}{2}} \cos(\ell+\frac{1}{2})\theta \right] \times \left[\sum_{\ell=0}^{L} \tilde{D}_\ell (\ell+\frac{1}{2})(R^*)^{\ell-\frac{1}{2}} \cos(\ell+\frac{1}{2})\theta \right] R^* d\theta
$$

$$
= \frac{\pi}{2} \sum_{\ell=0}^{L} (\ell+\frac{1}{2}) \tilde{D}_\ell^2 (R^*)^{2\ell+1}. \tag{9.79}
$$

Next, the derivatives $\frac{\partial v^+}{\partial r}$ in (9.74) are evaluated by

$$
\frac{\partial v^+}{\partial r} = \sum_{\ell=0}^{L} \tilde{D}_\ell (\ell+\frac{1}{2})(R^*)^{\ell-\frac{1}{2}} \cos(\ell+\frac{1}{2})\theta.
$$

For the derivatives $\frac{\partial v^-}{\partial r}$, we give a simple approximation by

$$
\frac{\partial v^-}{\partial r}\Big|_{(R^*,\theta_j)} \approx \frac{1}{\Delta r_j} (v_{1,j} - v_{0,j}),
$$

provided that both $v_{0,j}$ and $v_{1,j}$ are the solutions v^- at the element nodes $(0,j)$ and $(1,j)$ along the same coordinate line $\theta = \theta_j$, respectively (see Figure 4.4 in Chapter 4), i.e.,

$$
v_{0,j} = v^- (R^*, \theta_j), \quad v_{1,j} = v^- (r_j^1, \theta_j),
$$

with $\Delta r_j = r_j^1 - R^*$. Denote as M the number of elements in $\overline{DB}$ and $\overline{BE}$ combined in an equidistribution, and M=2 for Figure 4.4. Then 4M is the element number along the semi-circle ℓ_{R^*} but in a non-equidistribution. In this case, the trapezoidal rule (9.72) yields the following approximate integration:

$$
\widetilde{\int_{\Gamma_0}} \left(\alpha \frac{\partial v^+}{\partial r} + \beta \frac{\partial v^-}{\partial r} \right) (v^+ - v^-) d\ell \tag{9.80}
$$

$$
= \sum_{j=0}^{4M} \frac{R^*(\Delta\theta_j + \Delta\theta_{j-1})}{2} \left[\alpha \sum_{\ell=0}^{L} \tilde{D}_\ell (\ell+\frac{1}{2})(R^*)^{\ell-\frac{1}{2}} \cos(\ell+\frac{1}{2})\theta_j \right.
$$

$$
\left. + \beta \frac{1}{\Delta r_j} (v_{1,j} - v_{0,j}) \right] \times \left[\sum_{\ell=0}^{L} \tilde{D}_\ell (R^*)^{\ell+\frac{1}{2}} \cos(\ell+\frac{1}{2})\theta_j - v_{0,j} \right],
$$

where $\Delta\theta_j = \theta_j - \theta_{j-1}$ and $\Delta\theta_{4M} = \Delta\theta_{-1} = 0$. Similarly, we have

$$\int_{\ell_{R^*}} (v^+ - v^-)^2 d\ell \qquad (9.81)$$

$$= \sum_{j=0}^{4M} \frac{R^*(\Delta\theta_j + \Delta\theta_{j-1})}{2} \left[\sum_{\ell=0}^{L} \tilde{D}_\ell (R^*)^{\ell+\frac{1}{2}} \cos(\ell + \frac{1}{2})\theta_j - v_{0,j} \right]^2.$$

Therefore, we obtain from (9.73), (9.79), (9.80) and (9.81)

$$\tilde{I}_h^{(\alpha,\beta)}(v) = \frac{1}{2} \iint_{\hat{S}_1} (\nabla v)^2 ds + \frac{\pi}{4} \sum_{\ell=0}^{L} (\ell + \frac{1}{2}) \tilde{D}_\ell^2 (R^*)^{2\ell+1}$$

$$+ \frac{P_c}{2h^{*\sigma}} \sum_{j=0}^{4M} \frac{R^*(\Delta\theta_j + \Delta\theta_{j-1})}{2} \left[\sum_{\ell=0}^{L} \tilde{D}_\ell (R^*)^{\ell+\frac{1}{2}} \cos(\ell + \frac{1}{2})\theta_j - v_{0,j} \right]^2$$

$$- \sum_{j=0}^{4M} \frac{R^*(\Delta\theta_j + \Delta\theta_{j-1})}{2} \left[\alpha \sum_{\ell=0}^{L} \tilde{D}_\ell (\ell + \frac{1}{2})(R^*)^{\ell-\frac{1}{2}} \cos(\ell + \frac{1}{2})\theta_j \right.$$

$$\left. + \beta \frac{1}{\Delta r_j}(v_{1,j} - v_{0,j}) \right] \left[\sum_{\ell=0}^{L} \tilde{D}_\ell (R^*)^{\ell+\frac{1}{2}} \cos(\ell + \frac{1}{2})\theta_j - v_{0,j} \right],$$

where the average boundary length of the element along ℓ_{R^*} is $h^* = \frac{\pi}{4M} R^*$.

Finally, the linear difference equations of (9.78) can be obtained by differentiating $\tilde{I}_h^{(\alpha,\beta)}(v)$ with respect to $\tilde{D}_\ell$ and $v_{i,j}$ respectively:

$$0 = \frac{\partial \tilde{I}^{(\alpha,\beta)}(v)}{\partial \tilde{D}_\ell}$$

$$= \frac{\pi}{2}(\ell + \frac{1}{2})(R^*)^{2\ell+1} \tilde{D}_\ell + \frac{P_c}{h^{*\sigma}} \sum_{j=0}^{4M} \frac{R^*(\Delta\theta_j + \Delta\theta_{j-1})}{2} \times$$

$$\times \left[\sum_{\ell=0}^{L} \tilde{D}_\ell (R^*)^{\ell+\frac{1}{2}} \cos(\ell + \frac{1}{2})\theta_j - v_{0,j} \right] (R^*)^{\ell+\frac{1}{2}} \cos(\ell + \frac{1}{2})\theta_j -$$

$$- \sum_{j=0}^{4M} \frac{R^*(\Delta\theta_j + \Delta\theta_{j-1})}{2} \left\{ \alpha \sum_{k=0}^{L} (k + \ell + 1)(R^*)^{\ell+k} \cos((k + \frac{1}{2})\theta_j) \times \right.$$

$$\times \cos((\ell + \frac{1}{2})\theta_j) \tilde{D}_k + (R^*)^{\ell+\frac{1}{2}} \cos(\ell + \frac{1}{2})\theta_j \left[\frac{\beta}{\Delta r_j} v_{1,j} - \right.$$

$$\left. \left. - \left(\frac{\beta}{\Delta r_j} + \frac{\alpha(\ell + \frac{1}{2})}{R^*} \right) v_{0,j} \right] \right\}, \qquad (9.82)$$

$$0 = \frac{\partial \widetilde{I}^{(\alpha,\beta)}(v)}{\partial v_{0,j}} = \frac{\partial}{\partial v_{0,j}} \left\{ \frac{1}{2} \iint_{\widehat{S}_1} \nabla v^2 ds \right\} +$$

$$+ \frac{P_c}{h^{*\sigma}} \sum_{j=0}^{4M} \frac{R^*(\Delta\theta_j + \Delta\theta_{j-1})}{2} \left[v_{0,j} - \sum_{\ell=0}^{L} \widetilde{D}_\ell (R^*)^{\ell+\frac{1}{2}} \cos(\ell + \frac{1}{2})\theta_j \right] +$$

$$+ \sum_{j=0}^{4M} \frac{R^*(\Delta\theta_j + \Delta\theta_{j-1})}{2} \left[\sum_{\ell=0}^{L} \left(\frac{\alpha(\ell+\frac{1}{2})}{R^*} + \frac{\beta}{\Delta r_j} \right) (R^*)^{\ell+\frac{1}{2}} \times \right.$$

$$\left. \times \cos((\ell+\frac{1}{2})\theta_j)\widetilde{D}_\ell + \beta \frac{1}{\Delta r_j}(v_{1,j} - 2v_{0,j}) \right], \tag{9.83}$$

$$0 = \frac{\partial \widetilde{I}^{(\alpha,\beta)}(v)}{\partial v_{1,j}}$$

$$= \frac{\partial}{\partial v_{1,j}} \left\{ \frac{1}{2} \iint_{\widehat{S}_1} \nabla v^2 ds \right\} - \sum_{j=0}^{4M} \frac{R^*(\Delta\theta_j + \Delta\theta_{j-1})}{2} \frac{\beta}{\Delta r_j} \times$$

$$\times \left[\sum_{\ell=0}^{L} (R^*)^{\ell+\frac{1}{2}} \cos((\ell + \frac{1}{2})\theta_j)\widetilde{D}_\ell - v_{0,j} \right], \tag{9.84}$$

$$0 = \frac{\partial}{\partial v_{i,j}} \left\{ \frac{1}{2} \iint_{\widehat{S}_1} (\nabla v)^2 ds \right\} \quad \text{with } i > 0. \tag{9.85}$$

In (9.83)–(9.85), the part of the part of the difference equations from $\frac{\partial}{\partial v_{i,j}} \left\{ \frac{1}{2} \iint_{\widehat{S}_1} (\nabla v)^2 ds \right\}$ can be found in the standard FEM. Moreover, the above straightforward derivations for (9.78) can be easily rewritten in the matrix-vector forms as in Zienkiewciz (1977).

OPTIMAL COMBINATIONS
FOR VARIOUS FEMS

Let us study again the questions arising in Chapter 5. Suppose that the true solution is smooth such that $u \in H^{k+1}(S)$ where $k \geq 1$, and S is the solution domain. Then the k-order Lagrange FEM can be used, and the resulting solutions and their generalized derivatives have the optimal convergence rate $O(H^k)$, where H is the maximal length of the k-order quasiuniform elements. Quite often in practical problems, the smoothness of the true solution u is not uniform throughout the entire solution domain S. For instance, the solution u in one part, i.e., S_2, of S is smoother than u in the rest, S_1, of S:

$$u \in H^{m+1}(S), \quad u \in H^{k+1}(S_2), \quad 1 \leq m < k,$$

where $S = S_1 \cup S_2, S_1 \cap S_2 = \varnothing$, and $H^m(S)$ is the Sobolev space. As a consequence, we can use the lower m-order Lagrange FEM in the entire solution domain S. By means of the assumption $u \in H^{k+1}(S_2)$, we may also use the higher k-order Lagrange FEM in the subdomain S_2 to gain a reduction of CPU time and computer storage, in particular when S_2 is much larger than S_1. Therefore, the use of FEM combinations is important in both theoretical and practical aspects. Unfortunately, the reduced rates of convergence are often caused by the nonconforming constraints in Chapter 5. Naturally a question arises: Can we find other coupling techniques in Chapters 7–9 to match different FEMs such that the optimal convergence rates can *always* be achieved? This chapter attempts to give a positive answer to this question.

In the next section, in order to combine different FEMs we present the coupling technique of the penalty integral plus the simplified hybrid integrals only on the side of $\partial S_2 \cap \Gamma_0$, where Γ_0 is the common boundary: $\Gamma_0 = \partial S_1 \cap \partial S_2$. This combination is, indeed, similar to Combination II in Chapter 9. In Sections 10.2 and 10.3, we derive error bounds of the numerical solutions in the Sobolev norms, and bounds of the condition numbers of the associated coefficient matrix. Based on these bounds, both the optimal rates of convergence and optimal asymptotic condition numbers can be obtained for coupling *any* two different Lagrange FEMs. In Section 10.4, we prove that the above coupling technique is *unique* among all penalty plus hybrid techniques, which always maintains the optimal convergence rates. In the last section, we use the same coupling technique to combine the finite difference method (FDM) with different FEMs, also to yield the optimal convergence rates.

In summary, this chapter again displays the significance of the generalized coupling techniques given in the previous chapter, which may lead to the optimal convergence and stability of combinations of various FEMs. The important results are given in Theorems 10.2 – 10.4. Moreover, in Section 10.5, the combinations of FEM-FDM may follow the combinations of RGM-FEM described in the previous chapters, to achieve the optimal convergence and stability, based on an interpretation of FDM as FEM in Section 1.5 of Chapter 1. The combinations of RGM-FDM will be studied again in the next chapter, to gain the superconvergence $O(h^{2-\delta})$, $0 < \delta << 1$.

10.1   PENALTY PLUS SIMPLIFIED HYBRID TECHNIQUES

Consider a general elliptic equation

$$Lu = -\frac{\partial}{\partial x}\left(p\frac{\partial u}{\partial x}\right) - \frac{\partial}{\partial y}\left(p\frac{\partial u}{\partial y}\right) + cu = f, \quad (x,y) \in S, \tag{10.1}$$

with the Dirichlet boundary condition

$$u = g \quad in \quad \Gamma, \tag{10.2}$$

where S is a bounded, simply connected, and convex polygon domain, Γ is the exterior boundary ∂S of S, and the functions p, c, f and g are sufficiently smooth. Moreover, the functions in (10.1) and (10.2) satisfy

$$c = c(x,y) \geq 0, \quad p = p(x,y) \geq p_0 > 0.$$

A functional space is defined by

$$H_*^1(S) = \left\{v | v, v_x, v_y \in L^2(S), \text{ and } v|_\Gamma = g\right\}.$$

Then, the solution u of (10.1) and (10.2) belongs to $H_*^1(S)$, i.e., $u \in H_*^1(S)$, and can be defined in a weak form:

$$a(u,v) = f(v), \quad \forall v \in H_0^1(S),$$

where the notations are:

$$a(u,v) = \iint_S (p \nabla u \nabla v + cuv)\, ds,$$

$$f(v) = \iint_S fv\, ds, \tag{10.3}$$

$$H_0^1(S) = \left\{v | v, v_x, v_y \in L^2(S), \text{ and } v|_\Gamma = 0\right\}.$$

We assume that the true solution u of (10.1) and (10.2) and the flux $p(\partial u/\partial n)$ are continuous across any interior boundary. Let the solution domain S be divided by an interior boundary Γ_0, a piecewise straight line, into two subdomains S_1 and S_2. In this chapter, we shall discuss the combinations of all Lagrange finite element methods (simply written Lagrange FEMs). Without loss of generality, we combine two different methods: the $m(\geq 1)-$ and $k(> m)-$ order Lagrange FEMs used in S_1 and S_2 respectively, where $u \in H^{m+1}(S_1)$ and $u \in H^{k+1}(S_2)$. The admissible functions are chosen as

$$
v_h = \begin{cases} v^- = V_m, & (x,y) \in S_1, \\ v^+ = V_k, & (x,y) \in S_2, \end{cases} \tag{10.4}
$$

where V_m are the piecewise $m-$ order Lagrange interpolation polynomials.

Define two spaces H and H^* such that

$$
\begin{aligned}
H &= \left\{ v | v \in L^2(S),\ v \in H^1(S_1),\ v \in H^1(S_2),\ v|_\Gamma = g \right\}, \\
H^* &= \left\{ v | v \in L^2(S),\ v \in H^1(S_1),\ v \in H^1(S_2),\ v|_\Gamma = 0 \right\},
\end{aligned}
$$

and let $V_h \in H$ and $V_h^* \in H^*$ be finite-dimensional collections of the admissible functions (10.4). Consequently, the combination using penalty plus simplified hybrid integrals is designed to seek a solution $\tilde{u}_h \in V_h$ such that

$$
B_h(\tilde{u}_h, v) = f(v),\ v \in V_h^*, \tag{10.5}
$$

where

$$
\begin{aligned}
B_h(u,v) &= \iint_{S_1} (p\,\nabla u\,\nabla v + cuv)ds + \iint_{S_2} (p\,\nabla u\,\nabla v + cuv)ds + \\
&\quad + \frac{P_c}{H^\sigma} \int_{\Gamma_0} (u^+ - u^-)(v^+ - v^-)d\ell - \\
&\quad - \int_{\Gamma_0} \left[p^+ \frac{\partial u^+}{\partial n}(v^+ - v^-) + p^+ \frac{\partial v^+}{\partial n}(u^+ - u^-) \right] d\ell. \tag{10.6}
\end{aligned}
$$

In (10.6), $P_c(> 0)$ is the penalty constant, σ the penalty power, H the maximal boundary length of the k-order triangular elements in S_2, and n the outward normal of $\partial S_2 \cap \Gamma_0$. We note that in (10.6) the simplified hybrid technique $p^+ \frac{\partial u^+}{\partial n}$ is employed on the side of $\Gamma_0 \in \partial S_2$ only, for coupling the FEM-FEM, similar to Combination II in Chapter 9.

10.2   ERROR BOUNDS AND
OPTIMAL CONVERGENCE RATES

Define a new norm over the space H^*:

$$\|v\|_h = \left\{ \|v\|_{1,S_1}^2 + \|v\|_{1,S_2}^2 + \frac{P_c}{H^\sigma} \|v^+ - v^-\|_{0,\Gamma_0}^2 \right\}^{\frac{1}{2}}. \qquad (10.7)$$

Then we can easily prove the following lemmas by a similar argument in Lemma 5.2:

Lemma 10.1 *Let the k-order Lagrange triangular elements be quasiuniform, then there exists a bounded constant $C(> 0)$ independent of h, H, and v such that*

$$\left\| \frac{\partial v^+}{\partial n} \right\|_{0,\Gamma_0} \leq \frac{C}{H^{\frac{1}{2}}} |v|_{1,S_2}, \quad \forall v \in V_h^*,$$

where H is the maximal boundary length of the k-order quasiuniform triangular elements in S_2.

In the following the values of C may be different in different contexts but independent of h, H and v, where h is the maximal boundary length of the m-order quasiuniform triangular elements in S_1. Also we need another lemma.

Lemma 10.2 *Let the k- and m-order Lagrange triangular elements in S_1 and S_2 be quasiuniform, and suppose*

$$\sigma \geq 1. \qquad (10.8)$$

Then the bounded penalty constant P_c can be chosen suitably large but independent of h, H and v such that:

$$\alpha_0 \|v\|_h^2 \leq B_h(v, v), \quad \forall v \in V_h^*, \qquad (10.9)$$

and

$$B_h(u, v) \leq C\|u\|_h \|v\|_h, \quad \forall v \in V_h^*, \qquad (10.10)$$

where α_0 is a positive constant independent of h, H and v.

Proof. Without loss of generality, let $\Gamma \cap \partial S_1 \neq \emptyset$. We obtain from the Poincare-Friedrichs inequality

$$\alpha_0 \|v\|_{1,S_1} \leq |v|_{1,S_1}, \quad \forall v \in H^*, \tag{10.11}$$

and

$$\alpha_0 \|v\|_{1,S_2} \leq |v|_{1,S_2} + \|v^+\|_{0,\Gamma_0} \leq |v|_{1,S_2} + \|v^-\|_{0,\Gamma_0} + \|v^+ - v^-\|_{0,\Gamma_0}.$$

Using (10.11) and the Sobolev imbedding theorem gives

$$\|v^-\|_{0,\Gamma_0} \leq C\|v\|_{1,S_1} \leq \frac{C}{\alpha_0}|v|_{1,S_1}. \tag{10.12}$$

We can see from (10.11) and (10.12)

$$\alpha_0 \left(\|v\|_{1,S_1}^2 + \|v\|_{1,S_2}^2 \right) \leq C \left(|v|_{1,S_1}^2 + |v|_{1,S_2}^2 + \|v^+ - v^-\|_{0,\Gamma_0}^2 \right)$$

$$\leq \frac{C}{p_0} \left\{ \iint_{S_1} \left[p\,(\nabla v)^2 + cv^2 \right] ds + \iint_{S_2} \left[p\,(\nabla v)^2 + cv^2 \right] ds \right\} +$$

$$+ C\|v^+ - v^-\|_{0,\Gamma_0}^2. \tag{10.13}$$

Therefore, we obtain from (10.6) and (10.13)

$$B_h(v,v) \geq \alpha_0 \left[\|v\|_{1,S_1}^2 + \|v\|_{1,S_2}^2 \right] + \left(\frac{P_c}{H^\sigma} - \alpha_1 \right) \|v^+ - v^-\|_{0,\Gamma_0}^2 -$$

$$-2 \left| \int_{\Gamma_0} p^+ \frac{\partial v^+}{\partial n} (v^+ - v^-) d\ell \right|, \tag{10.14}$$

where α_1 is also a positive constant independent of h, H and v.

Next, based on Lemma 10.1 we have

$$\left| \int_{\Gamma_0} p^+ \left(\frac{\partial v^+}{\partial n} \right) (v^+ - v^-) d\ell \right| \leq \max_{\Gamma_0} p^+ \left\| \frac{\partial v^+}{\partial n} \right\|_{0,\Gamma_0} \|v^+ - v^-\|_{0,\Gamma_0}$$

$$\leq \frac{C}{H^{\frac{1}{2}}} \|v\|_{1,S_2} \|v^+ - v^-\|_{0,\Gamma_0}. \tag{10.15}$$

Hence, with the help of $2ab \leq a^2 + b^2$, Eq. (10.14) will lead to

$$B_h(v,v) \geq \alpha_0 \left(\|v\|_{1,S_1}^2 + \|v\|_{1,S_2}^2 \right) + \left(\frac{P_c}{H^\sigma} - \alpha_1 \right) \|v^+ - v^-\|_{0,\Gamma_0}^2 -$$

$$- \frac{C}{H^{\frac{1}{2}}} \|v\|_{1,S_2} \|v^+ - v^-\|_{0,\Gamma_0}$$

$$\geq \frac{\alpha_0}{2} \left(\|v\|_{1,S_1}^2 + \|v\|_{1,S_2}^2 \right) + \left(\frac{P_c}{H^\sigma} - \alpha_1 - \frac{C^2}{2\alpha_0} \frac{1}{H} \right) \|v^+ - v^-\|_{0,\Gamma_0}^2$$

$$\geq \frac{\alpha_0}{2} \left(\|v\|_{1,S_1}^2 + \|v\|_{1,S_2}^2 \right) + \frac{P_c}{2H^\sigma} \|v^+ - v^-\|_{0,\Gamma_0}^2. \tag{10.16}$$

The last step in (10.16) follows from the estimates (10.17) and (10.18) below. It is due to the assumptions $\sigma \geq 1$ that the penalty constant P_c can be chosen suitably large but independent of h, H and v such that

$$\frac{P_c}{H^\sigma} - \alpha_1 - \frac{C^2}{2\alpha_0}\frac{1}{H} \geq \frac{P_c}{2H^\sigma}, \tag{10.17}$$

i.e.,

$$\frac{P_c}{H^\sigma} \geq 2\alpha_1 + \frac{C^2}{\alpha_0}\frac{1}{H}. \tag{10.18}$$

Consequently, the first desired result (10.9) is obtained from the definition (10.7).

Next let us prove (10.10). From the Schwarz inequality and the Sobolev imbedding theorem, we obtain

$$\iint_{S_1} (p\,\nabla u\,\nabla v + cuv)\,ds + \iint_{S_2} (p\,\nabla u\,\nabla v + cuv)\,ds$$
$$\leq \quad C\,(\|u\|_{1,S_1} + \|u\|_{1,S_2})\,(\|v\|_{1,S_1} + \|v\|_{1,S_2})\,. \tag{10.19}$$

By combining (10.15) and (10.19), it is easy to see from $\sigma \geq 1$ that

$$B_h(u,v) \leq C\left[1 + H^{\left(\frac{\sigma-1}{2}\right)}\right]\|u\|_h\|v\|_h \leq C\|u\|_h\|v\|_h. \quad \blacksquare$$

Now we give a main theorem of error bounds.

Theorem 10.1 *Let all conditions in Lemma 10.2 be given, then there exist the error bounds,*

$$\|u - \tilde{u}_h\|_h \leq C \inf_{v \in V_h} \|u - v\|_h. \tag{10.20}$$

Proof. Let u be the exact solution of u. Since u and $p\frac{\partial u}{\partial n}$ are continuous on Γ_0, we have

$$B_h(u,v) = f(v), \quad \forall v \in V_h^*.$$

It follows from (10.5) that

$$B_h(u - \tilde{u}_h, v) = 0, \quad \forall v \in V_h^*. \tag{10.21}$$

Based on Lemma 10.2 and (10.21), we obtain

$$
\begin{aligned}
\alpha_0 \left\| \tilde{u}_h - v \right\|_h^2 &\leq B_h(\tilde{u}_h - v, \tilde{u}_h - v) = B_h(u - v, \tilde{u}_h - v) \\
&\leq C \left\| u - v \right\|_h \left\| \tilde{u}_h - v \right\|_h .
\end{aligned}
\tag{10.22}
$$

The desired result (10.20) is obtained by dividing two sides of (10.22) by $\left\| \tilde{u}_h - v \right\|_h$, and by applying the triangular inequality. ■

Theorem 10.2 *Let all conditions in Lemma 10.2 be given, and suppose*

$$
u \in H^{m+1}(S_1), u \in H^{k+1}(S_2), u \in H^{k+1}(\Gamma_0), 1 \leq m < k.
\tag{10.23}
$$

Then there always exist the bounds

$$
\begin{aligned}
\left\| u - \tilde{u}_h \right\|_h &\leq C \Big\{ h^m |u|_{m+1,S_1} + H^k |u|_{k+1,S_2} + \frac{h^{m+1}}{H^{\sigma/2}} |u|_{m+1,\Gamma_0} + \\
&\quad + H^{k+1-\frac{\sigma}{2}} |u|_{k+1,\Gamma_0} \Big\}.
\end{aligned}
$$

Proof. Construct an auxiliary function $\hat{u}_h \in V_h$ such that

$$
\hat{u}_h = \left\{ \begin{array}{ll} \hat{u}_h^m & \text{in } S_1, \\ \hat{u}_H^k & \text{in } S_2, \end{array} \right.
$$

where $\hat{u}_h^m$ and $\hat{u}_H^k$ are the piecewise m- and k-order Lagrange interpolation polynomials on the triangulation of S_1 and S_2, respectively. We then obtain from Theorem 10.1 and the assumption (10.23)

$$
\left\| u - \tilde{u}_h \right\|_h \leq C \inf_{v \in V^h} \left\| u - v \right\|_h \leq C \left\| u - \hat{u}_h \right\|_h
\tag{10.24}
$$

$$
\leq C \left\{ \left\| u - \hat{u}_h^m \right\|_{1,S_1} + \left\| u - \hat{u}_H^k \right\|_{1,S_2} + \frac{P_c^{1/2}}{H^{\sigma/2}} \left(\left\| u - \hat{u}_h^m \right\|_{0,\Gamma_0} + \left\| u - \hat{u}_H^k \right\|_{0,\Gamma_0} \right) \right\}
$$

$$
\leq C \left\{ h^m |u|_{m+1,S_1} + H^k |u|_{k+1,S_2} + \frac{h^{m+1}}{H^{\sigma/2}} |u|_{m+1,\Gamma_0} + H^{k+1-\sigma/2} |u|_{k+1,\Gamma_0} \right\}. \blacksquare
$$

In (10.24), the terms $O(h^m)$ and $O(H^k)$ are the errors resulting from the m- and k-order Lagrange FEMs in S_1 and S_2 respectively, and the remaining terms are the errors resulting from the coupling techniques of the penalty plus simplified hybrid integrals on Γ_0.

We can easily obtain the following corollary from Theorem 10.2.

Corollary 10.1 *Let all the conditions in Theorem 10.2 be given, and suppose*

$$\sigma \leq 2 \tag{10.25}$$

and

$$h \leq CH, \tag{10.26}$$

then

$$\|u - \tilde{u}_h\|_h = O(h^m) + O(H^k). \tag{10.27}$$

Also let the following optimal balance between h and H be given:

$$h^m = O(H^k), \tag{10.28}$$

then the optimal rates of convergence are obtained:

$$\|u - \tilde{u}_h\|_h = O(H^k) = O(h^m). \tag{10.29}$$

Combining (10.8) and (10.25) yields a necessary condition of σ for optimal convergence rates:

$$1 \leq \sigma \leq 2. \tag{10.30}$$

Since $k > m$, Eq. (10.26) can always be true. Also the assumption (10.23) is the same as in Chapter 5. Therefore, all the conditions in Corollary 10.1 can be easily satisfied. Contrasted to Chapter 5, conclusion (10.29) is significant because optimal convergence rates can *always* be achieved for matching any different Lagrange FEMs. Furthermore, we have the following corollary from Corollary 10.1.

Corollary 10.2 *Let (10.23) and (10.30) and all conditions in Lemma 10.1 hold. Then when $H = O(h)$, there exist the bounds, $\|u - \tilde{u}_h\|_h = O(h^m)$.*

As an example, $m = 1$, i.e., the combination of the linear and $k(> 1)$-order Lagrange FEMs will yield the convergence rate $O(h)$, which is better than $O(h^{\frac{1}{2}})$ given in Section 5.6 of Chapter 5. Consequently, the coupling technique of penalty plus simplified integrals in (10.5) and (10.6) is promising among the combined methods.

Below, we use the Nitsche trick of Nitsche (1971) to provide error bounds in the L_2 norm:

$$\|e_h\|_{0,S} = \|u - \tilde{u}_h\|_{0,S} = \left(\iint_S e_h^2 \, ds \right)^{\frac{1}{2}},$$

where $\tilde{u}_h$ is the approximate solution from (10.5) and (10.6).

Theorem 10.3 *Let (10.26) and all conditions in Lemma 10.2 hold, then*

$$\|e_h\|_{0,S} \leq Ca_0 H \|e_h\|_h, \tag{10.31}$$

where $e_h = u - \tilde{u}_h$, *and the coefficient*

$$a_0 = \left(1 + H^{\frac{1-\sigma}{2}} + \frac{h^{\frac{1}{2}}}{H^{\sigma/2}}\right). \tag{10.32}$$

Proof. Let $<u,v>_S = \iint_S uv\,ds$, then

$$\|e_h\|_{0,S} = \sup_{\eta \in L^2(S)} \frac{<e_h, \eta>_S}{\|\eta\|_{0,S}}. \tag{10.33}$$

Define the function Z by satisfying

$$LZ = \eta \quad \text{in} \quad S, \quad Z = 0 \quad \text{on} \quad \Gamma, \tag{10.34}$$

where the operator L is defined in (10.1). Using the Green theorem and noting $Z^+ = Z^-$ on Γ_0, we obtain

$$
\begin{aligned}
<e_h, \eta>_S &= <e_h, LZ>_{S_1} + <e_h, LZ>_{S_2} \\
&= \iint_{S_1} (p\,\nabla e_h\,\nabla Z + ce_h Z)\,ds + \iint_{S_2} (p\,\nabla e_h\,\nabla Z + ce_h Z)\,ds \\
&\quad - \int_{\Gamma_0} p^+ \frac{\partial Z^+}{\partial n}(\tilde{u}_h^+ - \tilde{u}_h^-)\,d\ell \\
&= B_h(Z, e_h).
\end{aligned}
\tag{10.35}
$$

From (10.21) and (10.10), Eq. (10.35) is reduced to

$$<e_h, \eta>_S = B_h(Z - v_h, e_h) \leq C\|Z - v_h\|_h\|e_h\|_h, \quad \forall v_h \in V_h^*. \tag{10.36}$$

Also we obtain

$$
\begin{aligned}
\|v^+ - v^-\|_{0,\Gamma_0} &\leq \|Z - v^+\|_{0,\Gamma_0} + \|Z - v^-\|_{0,\Gamma_0} \\
&\leq C\left(\|Z - v^+\|_{\frac{1}{2},S_2} + \|Z - v^-\|_{\frac{1}{2},S_1}\right),
\end{aligned}
\tag{10.37}
$$

where the norm with half derivatives is defined by

$$\|v\|_{\frac{1}{2},\Omega} = \left\{\|v\|_{0,\Omega}^2 + \int_\Omega \int_\Omega \frac{[v(t) - v(T)]^2}{\|t - T\|^3}\,dt\,dT\right\}^{\frac{1}{2}}.$$

Choose an auxiliary function

$$\widehat{Z}_h = \left\{ \begin{array}{l} \widehat{Z}_h^m \quad \text{in } S_1, \\[2mm] \widehat{Z}_H^k \quad \text{in } S_2, \end{array} \right.$$

where $\widehat{Z}_h^m$ and $\widehat{Z}_H^k$ are the m- and k-order Lagrange interpolation functions of Z on S_1 and S_2, respectively. Then we have from (10.37) and Ciarlet (1978)

$$\|Z - \widehat{Z}_h\|_h \leq C\Big\{ h\|Z\|_{2,S_1} + H\|Z\|_{2,S_2} + \frac{1}{H^{\sigma/2}}\Big(h^{3/2}\|Z\|_{2,S_1} + \\ + H^{3/2}\|Z\|_{2,S_2} \Big) \Big\}. \tag{10.38}$$

There also exist the bounds of Z in (10.34) from Grisvard (1985):

$$\|Z\|_{2,S_i} \leq C\|LZ\|_{0,S_i} \leq C\|\eta\|_{0,S}, \quad i = 1, 2. \tag{10.39}$$

Consequently, the desired result (10.31) is obtained from (10.33), (10.36), (10.38) and (10.39). ∎

When $\sigma = 1$, the coefficient (10.32) satisfies $a_0 \leq C$. Then we have the following corollary.

Corollary 10.3 *Let all the conditions in Theorem 10.2 hold, and suppose $\sigma = 1$ and (10.28), then*

$$\|e_h\|_{0,S} = O(H^{k+1}). \tag{10.40}$$

Eq. (10.40) is also the optimal convergence rates for the k-order Lagrange FEM. When $\sigma = 1$, the optimal rates of k-order elements can be obtained for both $\|\cdot\|_h$ and $\|\cdot\|_{0,s}$ norms. Note that the convergence rate $O(h^{m+1})$ can not be obtained due to (10.29) and $m < k$.

10.3   STABILITY ANALYSIS

Linear algebraic equations can be reduced from (10.5)

$$\mathbf{Ax = b},$$

where $\mathbf{x}$ is an unknown vector with the components $v_{ij} = v^{\pm}(Q_{ij})$, Q_{ij} are finite element nodes, $\mathbf{b}$ is a known vector, and the coefficient matrix A is symmetric and positive definite.

Theorem 10.4 *Let all conditions in Lemma 10.2 be given, also suppose $\partial S_1 \cap \Gamma \neq \varnothing$ and $\partial S_2 \cap \Gamma \neq \varnothing$ then there exist the bounds*

$$\text{Cond.}(\mathbf{A}) \leq Ch^{-2} \left[1 + P_c H^{-\sigma}(H + h) \right].$$

Proof. We have from Lemma 10.2

$$\alpha_0 \left(\|v\|_{1,S_1}^2 + \|v\|_{1,S_2}^2 \right) \leq \alpha_0 \|v\|_h^2 \leq B^h(v,v), \tag{10.41}$$

and

$$\begin{aligned} B_h(v,v) &\leq C\|v\|_h^2 \tag{10.42} \\ &\leq C \left\{ \|v\|_{1,S_1}^2 + \|v\|_{1,S_2}^2 + \frac{2P_c}{H^\sigma} \left(\|v^+\|_{0,\Gamma_0}^2 + \|v^-\|_{0,\Gamma_0}^2 \right) \right\}. \end{aligned}$$

Let

$$B_h(v,v) = \frac{1}{2}\mathbf{x}^T \mathbf{A}\mathbf{x}, \qquad \|v\|_{1,S_1}^2 + \|v\|_{1,S_2}^2 = \frac{1}{2}\mathbf{x}^T \mathbf{F}\mathbf{x},$$

$$\|v^-\|_{0,\Gamma_0}^2 = \frac{1}{2}\mathbf{x}_1^T \mathbf{E}_1 \mathbf{x}_1, \qquad \|v^+\|_{0,\Gamma_0}^2 = \frac{1}{2}\mathbf{x}_2^T \mathbf{E}_2 \mathbf{x}_2,$$

where $\mathbf{x}_1$ and $\mathbf{x}_2$ are the vectors consisting of v_{ij}^+ and v_{ij}^- near Γ_0^+ and Γ_0^- respectively, and $\mathbf{F}$, $\mathbf{E}_1$ and $\mathbf{E}_2$ are the coefficient matrices from FEMs. We can find the eigenvalue bounds from Strang and Fix (1973):

$$\lambda_{min}(\mathbf{F}) \geq \alpha_0 h^2, \qquad \lambda_{max}(\mathbf{F}) \leq C,$$
$$\lambda_{max}(\mathbf{E}_1) \leq Ch, \qquad \lambda_{max}(\mathbf{E}_2) \leq CH.$$

Consequently, we have from (10.41) and (10.42)

$$\lambda_{min}(\mathbf{A}) \geq \lambda_{min}(\mathbf{F}) \geq \alpha_0 h^2,$$

and

$$\begin{aligned} \lambda_{max}(\mathbf{A}) &\leq C \left[\lambda_{max}(\mathbf{F}) + \frac{P_c}{H^\sigma} \left(\lambda_{max}(\mathbf{E}_1) + \lambda_{max}(\mathbf{E}_2) \right) \right] \\ &\leq C \left[1 + \frac{P_c}{H^\sigma}(h + H) \right]. \end{aligned}$$

Therefore, it follows that

$$\frac{\lambda_{max}(\mathbf{A})}{\lambda_{min}(\mathbf{A})} \leq Ch^{-2} \left[1 + \frac{P_c}{H^\sigma}(h + H) \right]. \qquad \blacksquare$$

Corollary 10.4 *Let $\sigma = 1$, (10.26) and all conditions in Theorem 4 hold, then*

$$\frac{\lambda_{max}(\mathbf{A})}{\lambda_{min}(\mathbf{A})} = O(h^{-2}).$$

It is noted that $\sigma = 1$ is the best choice that grants not only the optimal convergence rates in $\|e_h\|_1$ and $\|e_h\|_{0,S}$, but also the optimal asymptotic condition number of FEM.

10.4   OTHER COUPLING TECHNIQUES

In the combination (10.5) and (10.6), the normal derivative (i.e., the hybrid term) $\frac{\partial u^+}{\partial n}$ is used only from one side of Γ_0. Can we use $\frac{\partial u^-}{\partial n}$ from the opposite side to maintain optimal convergence rates? For example, let us consider a general form of penalty plus hybrid techniques on Γ_0 described in Chapter 9, to seek a solution $\widetilde{u}_h^* \in V_h$ such that

$$\bar{B}_h(\widetilde{u}_h^*, v) = f(v), \quad v \in V_h^*, \tag{10.43}$$

where

$$
\begin{aligned}
\bar{B}_h(u, v) = & \iint_{S_1} (p \, \nabla u \, \nabla v + cuv) \, ds + \iint_{S_2} (p \, \nabla u \, \nabla v + cuv) \, ds \\
& + P_c \left(\frac{\alpha}{H^{\sigma_2}} + \frac{\beta}{h^{\sigma_1}} \right) \int_{\Gamma_0} (u^+ - u^-)(v^+ - v^-) d\ell \\
& - \int_{\Gamma_0} \left[\left(\alpha p^+ \frac{\partial u^+}{\partial n} + \beta p^- \frac{\partial u^-}{\partial n} \right) (v^+ - v^-) \right. \\
& \left. + \left(\alpha p^+ \frac{\partial v^+}{\partial n} + \beta p^- \frac{\partial v^-}{\partial n} \right) (u^+ - u^-) \right] d\ell,
\end{aligned}
$$

where the real constants P_c, α, β and σ_i satisfy

$$P_c > 0, \quad \alpha + \beta \neq 0, \quad \sigma_i \geq 0, \quad i = 1, 2.$$

A necessary condition of α and β for optimal convergence rates will be proven later as

$$\alpha + \beta = 1. \tag{10.44}$$

Then, there are three typical choices of α and β.

1. $\alpha = 1$ and $\beta = 0$, i.e., the combination (10.5) and (10.6), which is similar to Combination II of the RGM and FEM in Chapter 9;

2. $\alpha = 0$ and $\beta = 1$, similar to Combination I in Chapter 9;

3. $\alpha = \beta = \frac{1}{2}$, similar to Symmetric Combination in Chapter 9.

For the combination (10.43), we define another norm over H^* in this section:

$$\|v\|_h = \left[\|v\|_{1,S_1}^2 + \|v\|_{1,S_2}^2 + P_c \left(\frac{\alpha}{H^{\sigma_2}} + \frac{\beta}{h^{\sigma_1}} \right) \|v^+ - v^-\|_{0,\Gamma_0}^2 \right]^{\frac{1}{2}}.$$

We will give a lemma and a theorem below without proofs.

Lemma 10.3 *Let m- and $k(> m)$-order Lagrange triangular elements in S_1 and S_2 be quasiuniform, and suppose*

$$\sigma_2 \geq 1 \ \ if \ \alpha \neq 0; \quad \sigma_1 \geq 1 \ \ if \ \beta \neq 0.$$

Then the bounded penalty constant P_c can be chosen suitably large but independent of h, H and v such that

$$\alpha_0 \|v\|_h^2 \leq \bar{B}_h(v,v), \quad \bar{B}_h(u,v) \leq C \|u\|_h \|v\|_h, \quad \forall v \in V_h^*,$$

where α_0 is a positive constant independent of h, H and v.

Theorem 10.5 *Let all the conditions in Lemma 10.3 hold, then*

$$\|u - \tilde{u}_h^*\|_h \leq C \left\{ \inf_{v \in V_h} \|u - v\|_h + \frac{1}{b_0} |1 - (\alpha + \beta)| \left\| \frac{\partial u}{\partial n} \right\|_{0,\Gamma_0} \right\},$$

where the constant

$$b_0 = P_c^{\frac{1}{2}} \left(\frac{\alpha^{\frac{1}{2}}}{H^{\sigma_2/2}} + \frac{\beta^{\frac{1}{2}}}{h^{\sigma_1/2}} \right).$$

Also let (10.23) be given, then

$$\|u - \tilde{u}_h^*\|_h \ \leq \ C \left\{ h^m |u|_{m+1,S_1} + H^k |u|_{k+1,S_2} + \right.$$

$$\left. + b_0 \left(h^{m+1} |u|_{m+1,\Gamma_0} + H^{k+1} |u|_{k+1,\Gamma_0} \right) + \frac{|1 - (\alpha + \beta)|}{b_0} \left\| \frac{\partial u}{\partial n} \right\|_{0,\Gamma_0} \right\}.$$

In order to guarantee the optimal rates (10.29), Eq. (10.44) and the following conditions have to be fulfilled:

$$1 \le \sigma_2 \le 2 \quad \text{if} \quad \alpha \ne 0, \tag{10.45}$$
$$1 \le \sigma_1 \le 2 \quad \text{if} \quad \beta \ne 0, \tag{10.46}$$

and when $H \to 0$,

$$h^{-\sigma_1} \le CH^{-\sigma_2} \quad \text{if} \quad \beta \ne 0 \text{ and } \alpha \ne 0,$$
$$h^{-\sigma_1} \le CH^{-r}, \ r \le 2 \quad \text{if} \quad \beta \ne 0 \text{ and } \alpha = 0. \tag{10.47}$$

Under the optimal balance (10.28), Eqs. (10.47) are reduced to

$$H^{-\frac{k}{m}\sigma_1} \le CH^{-r}, \quad \beta \ne 0,$$

where

$$r = \sigma_2 \in [1, 2] \quad \text{if} \quad \alpha \ne 0;$$
$$r \le 2 \quad \text{if} \quad \alpha = 0.$$

This leads to

$$\frac{k}{m}\sigma_1 \le r. \tag{10.48}$$

Suppose $\beta \ne 0$. First choose $r = \sigma_2 = 1$, which is the necessary condition for the optimal rates in both the zero norm $\|e_h\|_{0,S}$ and the optimal asymptotic condition number for (10.5) and (10.6) made in Sections 10.2 and 10.3. However, the bounds (10.48) as

$$\frac{k}{m}\sigma_1 \le 1$$

never hold because of $k > m$ and (10.46). This indicates that $\beta \ne 0$ is not a good choice since it is impossible to reach the optimal asymptotic behavior in $\|e_h\|_{0,S}$ and $Cond.(\mathbf{A})$.

Next, consider the norms $\|e_h\|_h$ under

$$\sigma_1 = 1 \quad \text{and} \quad \sigma_2 = 2 \text{ (or } r = 2 \text{ if } \alpha = 0). \tag{10.49}$$

In this case Eq. (10.48) is reduced to

$$\frac{k}{m} \le 2, \tag{10.50}$$

where $1 \leq m < k$. There do exist some multiple solutions m and k of (10.49). When $\beta \neq 0$, a necessary condition as (10.49) is found for certain combinations of FEMs to maintain the optimal rates in $\|\epsilon_h\|_h$ only. It should be noted that the multiple combinations with optimal convergence rates in $\|\epsilon_h\|_h$ are also better than a unique combination found in penalty combinations, see Chapter 7.

In order to obtain always the optimal convergence rates, we must choose the constant, $\beta = 0$, and then from (10.44)

$$\alpha = 1 \quad \text{and} \quad \beta = 0.$$

This is the very coupling technique of combinations discussed in Sections 10.1–10.3. Let us summarize this result as a corollary.

Corollary 10.5 *Among the general forms of penalty plus hybrid techniques in (10.43), the coupling strategy with $\sigma = 1$ in (10.5) and (10.6) is unique for the combinations of different Lagrange FEMs to obtain the optimal rates of k-order elements in $\|e_h\|_1$ and $\|e_h\|_{0,s}$, as well as the optimal asymptotic condition number.*

Based on Corollary 10.5 the combination (10.5) and (10.6) is highly recommended for combining different FEMs, rather than (10.43) with $\alpha = 0$ and $\beta = 1$ or $\alpha = \beta = \frac{1}{2}$. The latter is the traditional symmetric treatment in FEM.

In addition, suppose $\alpha + \beta \neq 1, \alpha \neq 0$ and $\beta \neq 0$, then we obtain from Theorem 10.5

$$\|u - \widetilde{u}_h^*\|_h = O(H^{\sigma_2/2}) + O(h^{\sigma_1/2}).$$

On the basis of (10.45) and (10.46), we can choose $\sigma_1 = \sigma_2 = 2$, then

$$\|u - \widetilde{u}_h^*\|_h = O(H).$$

Since $k > 1$, this is also a reduced rate of convergence, compared with (10.27).

Furthermore, Eq. (10.50) is still valid for a special choice $\alpha = \beta = 0$. Then the pure penalty technique on Γ_0 will always cause the reduced convergence rates for the combinations. The above analysis suggests that the condition (10.44) be chosen for combining different FEMs. We remind again that this conclusion is drawn under the norm definition $\|v\|_h$ given in (10.7).

10.5   COMBINATIONS OF FDM AND FEM

Although the FEM today is the most popular approach for solving partial differential equations, the finite difference method (FDM) is still important because the discrete algebraic equations can be obtained in a straightforward manner. The remarkable advantage of FDM over FEM is that much less work is required for programming and forming the discrete algebraic equations. In fact, using FDM is a better choice for the problems where the solution domain can be, at least approximately, partitioned by quasiuniform difference grids. When the solution u may be more smooth in S_2:

$$u \in H^2(S_1), \quad u \in H^{k+1}(S_2),$$

where $k \geq 1$, we may use the $k(\geq 1)$-order Lagrange FEM in S_2. This leads to a combination of FDM with various FEMs. In this section, we also allow $k = 1$, i.e., the combination of FDM with the linear Lagrange FEM. Such a special combination is reported in Boulbrachene, Cortey–Dumont and Miellou (1988).

Based on Section 1.5.3 in Chapter 1, the traditional FDM used in S_1 can be regarded as a special kind of FEM by the following view:

1. The subdomain S_1 is divided by the horizontal and vertical lines into small rectangles $\square_{ij}$ and right triangles $\triangle_{ij}$. Then $S = \cup_{ij}\square_{ij} \cup_{ij} \triangle_{ij}$.

2. The piecewise bilinear and linear functions v_h^0 are chosen to be the admissible functions in $\square_{ij}$ and $\triangle_{ij}$, and different integration rules are used.

Let $S = S_1 \cup S_2$, and the FDM and the $k(\geq 1)$-order Lagrange FEM be used in S_1 and S_2, respectively. Choose the admissible functions

$$v_h = \left\{ \begin{array}{ll} v_h^0, & \text{if } (x,y) \in S_1, \\ V_k, & \text{if } (x,y) \in S_2, \end{array} \right. \qquad (10.51)$$

where v_h^0 are piecewise bilinear or linear functions in S_1, and V_k are the k-order Lagrange interpolation functions in S_2. Let $V_h^D(\in H)$ and $V_h^0(\in H^*)$ be finite dimensional collections of (10.51). Therefore, the combination of FDM with FEM is defined as

$$B_h^d(\widetilde{u}_h^d, v) = f^d(v), \qquad \forall \, v \in V_h^0, \qquad (10.52)$$

where the approximate solution $\widetilde{u}_h^d \in V_h^D$, and the notations are

$$B_h^d(u,v) \;\; = \;\; \widetilde{\iint_{S_1}} (p \, \nabla u \, \nabla v + cuv)ds + \iint_{S_2} (p \, \nabla u \, \nabla v + cuv)ds +$$

$$+\frac{P_c}{H^\sigma}\int_{\Gamma_0}(u^+ - v^-)(v^+ - v^-)d\ell -$$

$$-\int_{\Gamma_0}\left[p^+\frac{\partial u^+}{\partial n}(v^+ - v^-) + p^+\frac{\partial v^+}{\partial n}(u^+ - u^-)\right]d\ell, \qquad (10.53)$$

and

$$f^d(v) = \widetilde{\iint_{S_1}} fv\,ds + \iint_{S_2} fv\,ds, \qquad (10.54)$$

where $\sigma > 0$, $P_c > 0$ and H is the maximal boundary of the k-order triangular elements in S_2. In (10.52) and (10.53), the approximate integrations $\widetilde{\iint_{S_1}}$ will follow the special rules given in Section 1.5.3. It is noted that the simplified hybrid term $\frac{\partial u^+}{\partial n}$ is employed only from the side of $\Gamma_0 \cap \partial S_2$ using the FEM.

We can obtain a lemma (see Lemma 11.1 of the next chapter):

Lemma 10.4 *Let the difference grids be quasiuniform, then*

$$\widetilde{\iint_{S_1}}(p\,\nabla u\,\nabla v + cuv)\,ds \geq \alpha_0|v|^2_{1,S_1}.$$

We have from (10.52)–(10.54)

$$B_h^d(u,v) = B_h(u,v) + \left(\widetilde{\iint_{S_1}} - \iint_{S_1}\right)[p\,\nabla u\,\nabla v + cuv]\,ds,$$

and

$$f^d(v) = f(v) + \left(\widetilde{\iint_{S_1}} - \iint_{S_1}\right)fv\,ds,$$

where $B_h(u,v)$ and $f(v)$ are given in (10.6) and (10.3) respectively. Following the proof in Section 10.2, we can obtain the following theorem.

Theorem 10.6 *Let the difference grids in S_1 and the k-order Lagrange triangular elements in S_2 be quasiuniform, and suppose that $\sigma \geq 1$ and the constant P_c is bounded, and suitably large but independent of h, H and v. Then there exist the following error bounds of the solution $\tilde{u}_h^d \in V_h^D$ from the combination (10.52)–(10.54) of FDM and FEM:*

$$\|u - \tilde{u}_h^d\|_h \leq C \inf_{v \in V_h^D}(\|u - v\|_h + T),$$

where

$$T = \sup_{w_h \in V_h^0} \left\{ \frac{1}{\|w_h\|_h} \left| \left(\widetilde{\iint}_{S_1} - \iint_{S_1} \right) p \, \nabla w_h \, \nabla v ds \right| + \left| \left(\widetilde{\iint}_{S_1} - \iint_{S_1} \right) f w_h ds \right| \right\}.$$

Since the integration formulas used in Section 1.5.3 of Chapter 1 hold without errors for the linear functions u and v on $\square_{ij}$ and $\triangle_{ij}$, we can see from Ciarlet (1978) that $T = O(h)$, provided that

$$u \in H^2(S_1), \quad f \in H^1(S_1), \quad p \in C^1(S_1),$$

where h is the maximal mesh spacings of difference grids in S_1, and $C^1(S_1)$ denotes the space over S_1 of the functions with continuous derivatives. Therefore, we have the following corollary.

Corollary 10.6 *Let all conditions in Theorem 10.6 hold, and suppose that*

$$u \in H^2(S), \quad u \in H^{k+1}(S_2), \quad u \in H^{k+1}(\Gamma_0), \quad k \geq 1,$$
$$f \in H^1(S), \quad p \in C^1(S_1),$$

and $1 \leq \sigma \leq 2$. Then there exist the error bounds

$$\|u - \widetilde{u}_h^d\|_h = O(h) + O(H^k),$$

where $h \leq CH$, h and H are the maximal boundary lengths of the difference grids in S_1 and the triangular elements in S_2 respectively. Moreover, when $H^k = O(h)$, then $\|u - \widetilde{u}_h^d\|_h = O(h)$.

Corollary 10.6 implies that the FDM can be combined with any $k(\geq 1)$-order Lagrange FEM as in (10.52)–(10.54), to reach the optimal convergence rates. Therefore, to save CPU time we may employ both the simplest difference schemes in S_1 and the $k(\geq 1)$- order elements in S_2. The rest of analysis on $\|e_h\|_{0,s}$ and stability of the solutions from (10.52)–(10.54) can be done similarly. Also when the integration in FEM used in S_2 is involved in approximation, error bounds can be similarly obtained from Ciarlet (1978).

In this section, we have completed an analysis on the combinations of FDM and FEM, two most important methods for solving elliptic equations. If the FDM is regarded as a kind of linear FEMs, the combinations of RGM-FDM follow immediately from Chapters 4–10 (see the next chapter). This section also demonstrates an easy way to apply the stated combinations for creating new kinds of combinations. For instance, we can easily develop the combinations of the finite volume method (FVM) with FEM, FDM, RGM etc., based on the view of FVM as the Galerkin FEM described in Section 1.6.6 of Chapter 1.

11

COMBINATIONS OF RGM AND FDM

We follow Section 10.5 and study again the Combinations of RGM-FDM. This chapter illustrates an example how to design other kinds of combinations, based on the combinations that have been described in the previous chapters for Combinations of RGM-FEM. Moreover, our further attempts are pointed to pursue the average nodal superconvergence in this chapter; the global superconvergence rates will be explored in Chapter 14. The important results are given Theorems 11.2, 11.3 and 11.4. Under the new norms defined in (11.14) and (11.15), involving discrete summations of solutions and derivatives in S_1, the average nodal superconvergence rates, $O(h^{2-\delta})$, $0 < \delta << 1$, can be achieved by all six combinations, where the quasiuniform rectangular elements are used. The convergence rates $O(h^{\frac{3}{2}})$ can be obtained if the triangular elements $\triangle_{ij}$ are also used close to the slant exterior and interior boundaries, Γ and Γ_0. Below, we focus on the superconvergence rate $O(h^{2-\delta})$; the lower case $O(h^{\frac{3}{2}})$ is discussed in Li (1997a).

Effects of different norms on convergence are demonstrated in Section 11.5, to display significance of the new norms $\overline{\|\varepsilon\|}_h$, instead of $\|\varepsilon\|_h$ defined in Chapters 7 and 9. As a consequence, the limitation of penalty combinations drawn in Chapter 7 may be removed. Moreover, Lemma 11.9 reveals a deep insight between the nonconforming combinations and other combinations with penalty integrals. Note that the approaching algorithms while $P_c \longrightarrow \infty$ or $\sigma \longrightarrow \infty$ as $h < 1$ are really mimic to the popular techniques in dealing with the Dirichlet conditions (see Zienkiewicz (1977)). In this case, however, the condition number $O(P_c h^{-(2+\sigma)}) \longrightarrow \infty$, based on the stability analysis in Chapter 9. It is due to optimal convergence rates or superconvergence, and optimal or fair good numerical stability that the six combinations that are described in this book stand as individual computational algorithms in combinations. Let us briefly speak of these combinations resulting from different coupling techniques: the nonconforming combination is basic; the penalty combination simplest; the simplified hybrid combination analogous to the Lagrange multiplier method and the hybrid FEM method; and Combinations I, II and Symmetric combination more flexible in use.

11.1   ALGORITHMS

Consider the Poisson equation with the Dirichlet boundary condition

$$-\triangle u = -(\frac{\partial^2 u}{\partial x^2} + \frac{\partial^2 u}{\partial y^2}) = f(x,y), \qquad (x,y) \in S, \qquad (11.1)$$

$$u = 0, \qquad (x,y) \in \Gamma, \qquad (11.2)$$

where S is a polygonal domain, Γ the exterior boundary ∂S of S, and f smooth enough. Let the solution domain S be divided by a piecewise straight line Γ_0 into two subdomains S_1 and S_2. The RGM is used in S_2 where there may exist a singular point, and the FDM is used in S_1. For simplicity, the subdomain S_1 is again split by difference grids into small quasiuniform rectangles $\square_{ij}$ only, where $\square_{ij} = \{(x,y), x_i \leq x \leq x_{i+1}, y_j \leq y \leq y_{j+1}\}$. This confines the subdomain S_1 to be rectangles or the "L" shape, and those consisting of rectangles. Denote $u_{i,j} = u(x_i, y_j)$, $h_i = x_{i+1} - x_i$, $k_j = y_{j+1} - y_j$, and the maximal mesh spacing $h = \max_{i,j}(h_i, k_j)$. The quasiuniform difference grids imply that there exists a bounded constant C independent of h_i and k_j such that $h/\min_{i,j}(h_i, k_j) \leq C$. We assume that the boundary difference nodes (i,j) are always placed on ∂S_1.

From Section 1.5.3 in Chapter 1, the FDM can be regarded as a special kind of FEM, by using piecewise bilinear interpolatory functions $v_1(x,y)$ on $\square_{ij}$ as

$$v_1(x,y) = \frac{1}{h_i k_j} \Big\{ (x_{i+1} - x)(y_{j+1} - y)v_{ij} + (x - x_i)(y_{j+1} - y)v_{i+1,j} \quad (11.3)$$

$$+ (x_{i+1} - x)(y - y_j)v_{i,j+1} + (x - x_i)(y - y_j)v_{i+1,j+1} \Big\}, \text{ for } (x,y) \in \square_{ij},$$

and by the approximate integrals evaluated by the following specific rules:

$$\widehat{\iint_{S_1}} \nabla u \nabla v ds = \sum_{ij} \iint_{\square_{ij}} \nabla u \nabla v ds$$

$$= \sum_{ij} \Big\{ \widehat{\iint_{\square_{ij}}} u_x v_x ds + \widehat{\iint_{\square_{ij}}} u_y v_y ds \Big\}, \qquad (11.4)$$

$$\widehat{\iint_{\square_{ij}}} u_x v_x ds = \frac{h_i k_j}{2} [u_x(i + \tfrac{1}{2}, j)v_x(i + \tfrac{1}{2}, j) +$$

$$u_x(i + \tfrac{1}{2}, j + 1)v_x(i + \tfrac{1}{2}, j + 1)], \qquad (11.5)$$

$$\sum_{ij} \widehat{\iint_{\square_{ij}}} f v ds = \sum_{ij} \Big\{ \frac{h_i k_j}{4} [f_{ij} v_{ij} + f_{i+1,j} v_{i+1,j} + f_{i,j+1} v_{i,j+1} +$$

$$+f_{i+1,j+1}v_{i+1,j+1}]\Big\},\qquad\qquad(11.6)$$

where $u_x(i+\frac{1}{2},j) = u_x(x_{i+\frac{1}{2}},y_j)$, and $x_{i+\frac{1}{2}} = \frac{1}{2}(x_i + x_{i+1})$.

In S_2, we assume that the solution u can be spanned by $u = \sum_{i=1}^{\infty} a_i\psi_i$, where a_i are the expansion coefficients, and $\psi_i(i = 1, 2, \cdots, \infty)$ are complete and linearly independent basis functions. $\{\psi_i\}$ may be the chosen as analytical and singular functions. Then the admissible functions of combinations of the RGM-FDMs are written as:

$$v = \begin{cases} v^- = v_1, & \text{in } S_1, \\ v^+ = f_L(\tilde{a}_i), & \text{in } S_2, \end{cases}\qquad\qquad(11.7)$$

where $\tilde{a}_i$ are unknown coefficients to be sought, and $f_L(a_i) = \sum_{i=1}^{L} a_i\psi_i$. If the particular solutions of (11.1) and (11.2) are chosen as ψ_i, the total number of ψ_i used will greatly decrease for a given accuracy of the approximate solutions. Considering the discontinuity of solutions on Γ_0, i.e.,

$$v^+ \neq v^- \quad \text{on } \Gamma_0 ,$$

we define another space $H = \{v|v \in L^2(S),\ v \in H^1(S_1),\ \text{and}\ v \in H^1(S_2)\}$, where $H^1(S_1)$ is the Sobolev space.

Let $V_h(\subseteq H)$ denote a finite dimensional collection of the function v in (11.7) satisfying (11.2). The combinations of the RGM-FDMs involving integral approximation on Γ_0 can be expressed by (see Chapters 6–9)

$$\hat{a}_h(u_h, v) = \hat{f}_h(v), \quad \forall v \in V_h,\qquad\qquad(11.8)$$

where

$$\hat{a}_h(u,v) = \widehat{\iint_{S_1}} \nabla u\, \nabla v ds + \iint_{S_2} \nabla u\, \nabla v ds + \hat{D}(u,v),$$

$$\hat{f}_h(v) = \widehat{\iint_{S_1}} f v ds + \iint_{S_2} f v ds,\quad \widehat{\iint_{S_1}} f v ds = \sum_{ij} \widehat{\iint_{\square_{ij}}} f v ds,$$

$$\hat{D}(u,v) = \frac{P_c}{h^\sigma}\widehat{\int_{\Gamma_0}} (u^+ - u^-)(v^+ - v^-)d\ell - \int_{\Gamma_0}\Big(\alpha\frac{\partial u^+}{\partial n} + \beta\frac{\Delta u^-}{\Delta n}\Big)(v^+ - v^-)d\ell$$

$$- \int_{\Gamma_0}\Big(\alpha\frac{\partial v^+}{\partial n} + \beta\frac{\Delta v^-}{\Delta n}\Big)(u^+ - u^-)d\ell,\qquad\qquad(11.9)$$

where $\left(\triangle u^- / \triangle x\right)\big|_{(x_i,y_j)\in\Gamma_0} = (u^-(x_i + h_i, y_j) - u^-(x_i, y_j))/h_i$. In the coupling integral (11.9), $P_c(\geq 0)$ is the penalty constant, σ is the penalty power, and $\alpha(\geq 0)$ and $\beta(\geq 0)$ satisfy $\alpha + \beta = 1$ or 0. The first term on the right side of (11.9) is called the penalty integral, and the second and third terms the hybrid integrals. Different combinations of (11.8) are obtained from different parameters in (11.9).

- (I)   Combination I: $\alpha = 0$ and $\beta = 1$,
- (II)   Combination II: $\alpha = 1$ and $\beta = 0$,
- (III) Symmetric Combination: $\alpha = \beta = \frac{1}{2}$,
- (IV)   Penalty Combination: $\alpha = \beta = 0$,
- (V)   Simplified Hybrid Combinations: $P_c = 0$, $\alpha = 1$ and $\beta = 0$.
- (VI)   Besides, a direct continuity constraint at the difference nodes Z_k on Γ_0 is given in Chapter 4 as

$$v^+(Z_k) = v^-(Z_k), \quad \forall Z_k \in \Gamma_0, \tag{11.10}$$

where the interface difference nodes Z_k are located just on Γ_0. We then obtain the nonconforming combination

$$\widehat{I}_h(u_N, v) = \widehat{f}_h(v), \quad \forall v \in \overline{V}_h, \text{ where} \tag{11.11}$$

$$\widehat{I}_h(u, v) = \widehat{\iint_{S_1}} \nabla u \, \nabla v \, ds + \iint_{S_2} \nabla u \, \nabla v \, ds,$$

and $\overline{V}_h(\subset H)$ is a finite dimensional collection of (11.7) satisfying both (11.10) and (11.2). Eq. (11.11) is regarded as (11.8) as $P_c = \alpha = \beta = 0$ but with (11.10).

The approximate integrals $\widehat{D}(u, v)$ on Γ_0 using the integration rules are

$$\int_{\Gamma_0} \xi\eta d\ell \approx \widehat{\int_{\Gamma_0}} \xi\eta d\ell = \int_{\Gamma_0} \widehat{\xi}\widehat{\eta} d\ell$$

$$= \sum_{k=1}^{N_1} \frac{\overline{Z_{k-1}Z_k}}{6} \{2\xi(Z_{k-1})\eta(Z_{k-1}) + \xi(Z_{k-1})\eta(Z_k) +$$

$$+ \xi(Z_k)\eta(Z_{k-1}) + 2\xi(Z_k)\eta(Z_k)\}, \tag{11.12}$$

where $\Gamma_0 = \cup_{k=1}^{N_1}\Gamma_0^{(k)}$, $\Gamma_0^{(k)} = Z_{k-1}Z_k$, $\overline{Z_{k-1}Z_k}$ denotes the length of $Z_{k-1}Z_k$, and $\widehat{\xi}$ and $\widehat{\eta}$ are the piecewise linear interpolatory functions on Γ_0. For the interior boundary Γ_0 we have

$$\int_{\Gamma_0} \frac{\partial u^-}{\partial n}(v^+ - v^-)d\ell = \int_{\Gamma_0} \frac{\partial u^-}{\partial x}(v^+ - v^-)dy + \int_{\Gamma_0} \frac{\partial u^-}{\partial y}(v^+ - v^-)dx$$

$$\approx \widehat{\int_{\Gamma_0}} \frac{\triangle u^-}{\triangle x}(v^+ - v^-)dy + \widehat{\int_{\Gamma_0}} \frac{\triangle u^-}{\triangle y}(v^+ - v^-)dx = \widehat{\int_{\Gamma_0}} \frac{\triangle u^-}{\triangle n}(v^+ - v^-)d\ell.$$

Note that the above integration is also suited for the slant boundary Γ_0 when using triangular elements. Suitable norms are defined for evaluating error bounds of the solutions by combinations. We thus define

$$\|v\|_h = \left(\|v\|_{1,S_1}^2 + \|v\|_{1,S_2}^2 + \frac{P_c}{h^\sigma}\|v^+ - v^-\|_{0,\Gamma_0}^2\right)^{1/2}, \qquad (11.13)$$

$$\|v\|_1 = \left(\|v\|_{1,S_1}^2 + \|v\|_{1,S_2}^2\right)^{1/2}.$$

Optimal convergence rates, $\|\epsilon\|_h = \|u - u_h\|_h = O(h)$, of numerical solutions, have been obtained in previous chapters. In this chapter, we pursue superconvergence based on the new norms involving discrete summation:

$$\overline{\|v\|}_h = \left(\overline{\|v\|}_{1,S_1}^2 + \|v\|_{1,S_2}^2 + \frac{P_c}{h^\sigma}\overline{\|v^+ - v^-\|}_{0,\Gamma_0}^2\right)^{1/2}, \qquad (11.14)$$

$$\overline{\|v\|}_1 = \left(\overline{\|v\|}_{1,S_1}^2 + \|v\|_{1,S_2}^2\right)^{1/2}, \qquad (11.15)$$

$$\overline{|v|}_1 = \left(\overline{|v|}_{1,S_1}^2 + |v|_{1,S_2}^2\right)^{\frac{1}{2}}, \qquad (11.16)$$

where the norms with discrete summation are defined by

$$\overline{\|v\|}_{1,S_1}^2 = \overline{|v|}_{1,S_1}^2 + \overline{\|v\|}_{0,S_1}^2,$$

$$\overline{|v|}_{1,S_1}^2 = \sum_{ij} \iint_{\Box_{ij}} (\nabla v)^2 ds, \quad \overline{\|v\|}_{0,S_1}^2 = \sum_{ij} \iint_{\Box_{ij}} v^2 ds,$$

$$\overline{\|v^+ - v^-\|}_{0,\Gamma_0}^2 = \int_{\Gamma_0} (v^+ - v^-)^2 d\ell. \qquad (11.17)$$

11.2 NONCONFORMING COMBINATION

For the norms in (11.16) over V_h^0:

$$\overline{|v|}_{1,S_1}^2 = \sum_{i,j} \frac{h_i k_j}{2} \left[v_x^2(A_{ij}) + v_x^2(C_{ij}) + v_y^2(B_{ij}) + v_y^2(D_{ij})\right],$$

$$\overline{\|v\|}_{1,S_1}^2 = \overline{|v|}_{1,S_1}^2 + \sum_{i,j} \frac{h_i k_j}{4} \sum_{k=1}^4 v^2(k_{ij}),$$

where $S_1 = \cup_{ij}\Box_{ij}$, A_{ij}, B_{ij}, C_{ij} and D_{ij} denote the boundary middle points of $\Box_{ij}$, and k_{ij} the four corner points of $\Box_{ij}$ (see Figure 1.3 of Chapter 1).

We note that the norm definitions (11.15) and (11.16) agree with the integral approximations (11.4)–(11.6). In fact, the errors ε in the norm $\overline{\|\varepsilon\|}_1$ are taken into account only at particular points on S_1. For example, the derivative errors ε_x are evaluated only at the middle points $(x_{i+\frac{1}{2}}, y_j)$, ε_y only at the middle points $(x_i, y_{j+\frac{1}{2}})$, and the errors ε only at the corner points (x_i, y_j). Therefore, the values of $\overline{\|\varepsilon\|}_1$ also provide a measure of average errors and average derivative errors of numerical solutions in S_2 and on all these specific kinds of points in S_1.

The goal in this section is to pursue superconvergence under the quasiuniform assumption. First, we state a lemma:

Lemma 11.1 *For all $v \in \overline{V}_h$ there exist two bounded constants α and C independent of L, h and v such that*

$$\widehat{I}_h(u, v) \le \overline{\|u\|}_1 \overline{\|v\|}_1, \tag{11.18}$$

$$\|v\|_1 \le \overline{\|v\|}_1 \le 3\|v\|_1, \tag{11.19}$$

$$|v|_1 \le \overline{|v|}_1 \le \sqrt{3}|v|_1, \tag{11.20}$$

and

$$\alpha \overline{\|v\|}_1^2 \le \widehat{I}_h(v, v). \tag{11.21}$$

Proof: Based on the Schwarz inequality, we have

$$\widehat{\iint}_{S_1} \nabla u \, \nabla v \, ds, \quad = \quad \sum_{i,j} \frac{h_i k_j}{2} [u_x v_x(A_{ij}) + u_x v_x(C_{ij}) + u_y v_y(B_{ij}) + u_y v_y(D_{ij})]$$

$$\le \quad \overline{|v|}_{1,S_1} \overline{|v|}_{1,S_1}.$$

Then,

$$\widehat{I}_h(u, v) \le \widehat{I}_h(u, u)\widehat{I}_h(v, v) = \overline{|u|}_1 \overline{|v|}_1 \le \overline{\|u\|}_1 \overline{\|v\|}_1.$$

This is (11.18).

Also, Eqs. (11.19) and (11.20) can be found by directly computing the values of $|v|_1$, $\overline{|v|}_1$, $\|v\|_1$ and $\overline{\|v\|}_1$, or by using (7.66) in Chapter 7.

Moreover, we obtain from (11.19), (11.20) and the Poincare-Friedrichs inequality:

$$\widehat{I}_h(v, v) = \overline{|v|}_1^2 \ge |v|_1^2 \ge \alpha\|v\|_1^2 \ge \frac{\alpha}{9}\overline{\|v\|}_1^2,$$

with a positive constant α independent of v. ∎

Eqs. (11.19) and (11.20) show that over the space $\overline{V}_h$, the norms $\overline{\|v\|}_1$ and $\overline{|v|}_1$ are equivalent to $\|v\|_1$ and $|v|_1$, respectively. Now, we will prove the main theorem for error bounds.

Theorem 11.1 *The solution from the nonconforming combination (11.11) has the bounds:*

$$\overline{\|u_N - u\|}_1 \leq C\left\{ \inf_{v \in \overline{V}_h} \overline{\|u - v\|}_1 + \|\frac{\partial u}{\partial n}\|_{0,\Gamma_0} \sup_{w \in \overline{V}_h} \frac{\|w^+ - w^-\|_{0,\Gamma_0}}{\|w\|_1} \qquad (11.22)$$

$$+ \sup_{w \in \overline{V}_h} \frac{\left| \left(\iint_{S_1} - \widehat{\iint_{S_1}}\right) \nabla u \, \nabla w \, ds \right|}{\|w\|_1} + \sup_{w \in \overline{V}_h} \frac{\left| \left(\iint_{S_1} - \widehat{\iint_{S_1}}\right) f w \, ds \right|}{\|w\|_1} \right\},$$

where C is a bounded constant independent of L, h v and u.

Proof: We notice that the solution, $u \in H_0^1(S)$, satisfy

$$a^*(u, v) = \iint_{S_1} \nabla u \, \nabla v \, ds + \iint_{S_2} \nabla u \, \nabla v \, ds$$

$$= \oint_{\Gamma_0} \frac{\partial u}{\partial n}(v^+ - v^-) d\ell + \iint_S f v \, ds, \quad \forall v \in H_0^1(S).$$

We then have from (11.21) that for $v \in \overline{V}_h$ and $w_h = u_N - v$:

$$\alpha \overline{\|w_h\|}_1^2 \leq \widehat{I}_h(u_N - v, w_h)$$

$$= \widehat{I}_h(u - v, w_h) + \iint_{S_1} f w_h \, ds + \iint_{S_2} f w_h \, ds$$

$$+ [a^*(u, w) - \widehat{I}_h(u, w_h)] - \int_{\Gamma_0} \frac{\partial u}{\partial \nu}(w_h^+ - w_h^-) d\ell - \iint_S f w_h \, ds$$

$$= \widehat{I}_h(u - v, w_h) - \int_{\Gamma_0} \frac{\partial u}{\partial \nu}(w_h^+ - w_h^-) d\ell$$

$$+ \left(\iint_{S_1} - \widehat{\iint_{S_1}}\right) \nabla u \, \nabla w_h \, ds - \left(\iint_{S_1} - \widehat{\iint_{S_1}}\right) f w_h \, ds. \qquad (11.23)$$

Also from (11.18) we can obtain

$$\widehat{I}_h(u - v, u_N - v) \leq \overline{\|u - v\|}_1 \overline{\|u_N - v\|}_1, \qquad (11.24)$$

and

$$\left| \int_{\Gamma_0} \frac{\partial u}{\partial \nu} (w_h^+ - w_h^-) d\ell \right| \leq \left[\int_{\Gamma_0} \left(\frac{\partial u}{\partial \nu} \right)^2 d\ell \right]^{\frac{1}{2}} \left[\int_{\Gamma_0} (w_h^+ - w_h^-)^2 d\ell \right]^{\frac{1}{2}}. \qquad (11.25)$$

It then follows from (11.23), (11.24) and (11.25):

$$\overline{\|u_N - v\|}_1 \leq \left\{ \overline{\|u - v\|}_1 + \frac{\left[\int_{\Gamma_0} \left(\frac{\partial u}{\partial \nu} \right)^2 d\ell \right]^{\frac{1}{2}} \left[\int_{\Gamma_0} (w_h^+ - w_h^-)^2 d\ell \right]^{\frac{1}{2}}}{\|w_h\|_1} \right. \qquad (11.26)$$

$$\left. + \frac{|(\iint_{S_1} - \widehat{\iint}_{S_1}) \nabla u \nabla w_h ds|}{\|w_h\|_1} + \frac{|(\iint_{S_1} - \widehat{\iint}_{S_1}) f w_h ds|}{\|w_h\|_1} \right\}.$$

Consequently, the desired result (11.22) is derived from (11.26) and the triangular inequality:

$$\overline{\|u - u_N\|}_1 \leq \overline{\|u - v\|}_1 + \overline{\|v - u_N\|}_1. \qquad \blacksquare$$

We can easily prove the following lemma.

Lemma 11.2 *Assume*

$$|v^+|_{\ell,\Gamma_0} \leq C L^{\ell\mu} \|v^+\|_{0,\Gamma_0}, \quad \ell = 1, 2, \quad \forall v \in \overline{V}_h, \qquad (11.27)$$

where $\mu(> 0)$ is a bounded power independent of L, h and v. Then

$$\|w^+ - w^-\|_{0,\Gamma_0} \leq C h^2 L^{2\mu} \overline{\|w\|}_1, \quad \forall w \in \overline{V}_h. \qquad (11.28)$$

Let $\Gamma_0^* (\subset S_1)$ consist of the difference coordinate lines, and S_1 be divided by Γ_0 into S_1^* and S_0^* $(S_1 = S_1^* \cup S_0^*)$ such that the middle region S_0^* is located between S_1^* and S_2. Denoting $S_2^* = S_2 \cup S_0^*$, we can also prove the following lemma.

Lemma 11.3 *Let*

$$u \in C^3(S_1) \qquad (11.29)$$

hold, where $C^k(S_1)$ denotes the space of functions having k-order continuous derivatives. Then

$$\inf_{v \in \overline{V}_h} \overline{\|u - v\|}_1 \leq C h^2 + \|R_L\|_{1, S_2^*},$$

where the remainder $R_L = \sum_{\ell=L+1}^{\infty} a_\ell \psi_\ell$.

Define $M_n(u) = \max_{\substack{i+j=k \le n \\ (x,y) \in S_1}} \left| \frac{\partial^k u}{\partial x^i \partial y^j} \right|$. Below we prove the following important lemma.

Lemma 11.4 *Let (11.27), (11.29) and*

$$f \in C^2(S_1) \tag{11.30}$$

hold. Then there exist the bounds

$$\left| \left(\iint_{S_1} - \widehat{\iint_{S_1}} \right) \nabla u \, \nabla w ds \right| \le Ch^2 L^\mu M_3(u) \overline{\|w\|}_1, \quad \forall w \in \overline{V}_h, \tag{11.31}$$

$$\left| \left(\iint_{S_1} - \widehat{\iint_{S_1}} \right) f w ds \right| \le Ch^2 M_2(f) \overline{\|w\|}_1, \quad \forall w \in \overline{V}_h. \tag{11.32}$$

Proof: Since

$$\left| \left(\iint_{S_1} - \widehat{\iint_{S_1}} \right) \nabla u \, \nabla w ds \right|$$

$$\le \left| \left(\iint_{S_1} - \widehat{\iint_{S_1}} \right) u_x w_x ds \right| + \left| \left(\iint_{S_1} - \widehat{\iint_{S_1}} \right) u_y w_y ds \right|, \tag{11.33}$$

we only prove bounds of one term in the right side of (11.33), for example,

$$\left| \left(\iint_{S_1} - \widehat{\iint_{S_1}} \right) u_x w_x ds \right| \le Ch^2 L^\mu M_3(u) \overline{\|w\|}_1, \quad \forall w \in \overline{V}_h. \tag{11.34}$$

The proof for bounds of the other term is the same. By Taylor's formula, we obtain

$$\iint_{\square_{ij}} g ds = \frac{h_i k_j}{2} \left[g(i + \tfrac{1}{2}, j) + g(i + \tfrac{1}{2}, j + 1) \right] + R_{i,j}^{(1)},$$

where the truncation errors

$$R_{i,j}^{(1)} = h_i k_j \left\{ \frac{1}{24} \left(h_i^2 \frac{\partial^2 \widetilde{g}_{ij}^{(1)}}{\partial x^2} - 2k_j^2 \frac{\partial^2 \widetilde{g}_{ij}^{(2)}}{\partial y^2} \right) + \frac{1}{32} h_i k_j \left[\frac{\partial^2 \widetilde{g}_{ij}^{(3)}}{\partial x \partial y} - \frac{\partial^2 \widetilde{g}_{ij}^{(4)}}{\partial x \partial y} \right] \right\}, \tag{11.35}$$

$$\widetilde{g}_{ij}^{(k)} = g(\xi_{ij}^{(k)}), \quad \xi_{ij}^{(k)} \in \square_{ij}, \quad k = 1, 2, 3, 4.$$

Since $w(\in V_h)$ is a bilinear function on $\square_{ij}$, then we have from (11.3)

$$w_{xx} = w_{yy} = 0 \quad \text{in} \quad \square_{ij},$$

$$w_{xy} = \frac{1}{h_i k_j} [w_{ij} - w_{i+1,j} - w_{i,j+1} + w_{i+1,j+1}] \text{ in } \square_{ij}. \qquad (11.36)$$

Let $g = u_x w_x$, we have

$$g_{xx} = u_{xxx}w_x, \quad g_{yy} = u_{xyy}w_x + 2u_{xy}w_{xy}, \quad g_{xy} = u_{xxy}w_x + u_{xx}w_{xy}.$$

We can apply (11.35) to the integration (11.5), to yield the following bounds through some operations.

$$\left| \left(\iint_{S_1} - \widehat{\iint_{S_1}} \right) u_x w_x ds \right| = \left| \sum_{ij} \left(\iint_{\square_{ij}} - \widehat{\iint_{\square_{ij}}} \right) u_x w_x ds \right|$$

$$= \sum_{ij} R_{ij}^{(1)} \le C \left\{ h^2 M_3(u) \sum_{ij} h_i k_j |w_x(\eta_{ij})| + \left| \sum_{ij} h_i k_j^3 \frac{\partial^2 \tilde{u}_{ij}^{(2)}}{\partial x \partial y} w_{xy} \right| \right.$$

$$\left. + \left| \sum_{ij} h_i^2 k_j^2 \left(\frac{\partial^2 \tilde{u}_{ij}^{(3)}}{\partial x^2} - \frac{\partial^2 \tilde{u}_{ij}^{(4)}}{\partial x^2} \right) w_{xy} \right| \right\}, \qquad (11.37)$$

where $\eta_{ij} \in \square_{ij}$, $u_{ij}^{(k)} = u(\xi_{ij}^{(k)})$, and $\xi_{ij}^{(k)} \in \square_{ij}, k = 1, 2, 3, 4$. Bounds of the first term of the rightmost side in (11.37) can be obtained from the Schwarz inequality:

$$T_{\mathrm{I}} = h^2 M_3(u) \sum_{ij} h_i k_j |w_x(\eta_{ij})|$$

$$\le C h^2 M_3(u) \sum_{ij} h_i k_j [|w_x(i + \tfrac{1}{2}, j)| + |w_x(i + \tfrac{1}{2}, j + 1)|]$$

$$\le C h^2 M_3(u) \overline{|w|}_1 \le C h^2 M_3(u) \, \overline{\|w\|}_1. \qquad (11.38)$$

For the third term in (11.37), we can see from (11.29), (11.36) and the Schwarz inequality,

$$T_{\mathrm{III}} = |\sum_{ij} h_i^2 k_j^2 \left(\frac{\partial^2 \tilde{u}_{ij}^{(3)}}{\partial x^2} - \frac{\partial^2 \tilde{u}_{ij}^{(4)}}{\partial x^2} \right) w_{xy}| \le C M_3(u) h |\sum_{ij} h_i^2 k_j^2 w_{xy}|$$

$$\le C M_3(u) h \sum_{ij} h_i k_j [|w_{ij} - w_{i+1,j} - w_{i,j+1} + w_{i+1,j+1}|]$$

$$\le C M_3(u) h \sum_{ij} h_i^2 k_j \left(\frac{|w_{i+1,j} - w_{i,j}|}{h_i} + \frac{|w_{i+1,j+1} - w_{i,j+1}|}{h_i} \right)$$

$$\le C M_3(u) h^2 \overline{|w|}_{1,S_1} \le C M_3(u) h^2 \, \overline{\|w\|}_1. \qquad (11.39)$$

By applying the boundary conditions and coupling relations, more efforts are paid to estimate carefully the second term of the rightmost side in (11.37),

$$
T_{\mathrm{II}} = \left| \sum_{ij} h_i k_j^3 \frac{\partial^2 \tilde{u}_{ij}^{(2)}}{\partial x \partial y} w_{xy} \right|
$$

$$
= \left| \sum_{ij} k_j^2 \frac{\partial^2 \tilde{u}_{ij}^{(2)}}{\partial x \partial y} [w_{ij} - w_{i+1,j} - w_{i,j+1} + w_{i+1,j+1}] \right|
$$

$$
= \left| \sum_j k_j^2 \sum_i \frac{\partial^2 \tilde{u}_{ij}^{(2)}}{\partial x \partial y} [w_{ij} - w_{i+1,j} - w_{i,j+1} + w_{i+1,j+1}] \right|.
$$

Denote by $\overline{P_{ij} P_{i,j+1}}$ a vertical segment of $\partial \square_{ij}$, between the difference vertices (i,j) and $(i, j+1)$. From the assumption, we may locate the vertical segments $\{\overline{P_{ij} P_{i,j+1}}\}$ either inside of S_1 or just on the boundary ∂S_1: **Case I**: $\overline{P_{ij} P_{i,j+1}} \in S_1$, or **Case II**: $\overline{P_{ij} P_{i,j+1}} \in \partial S_1$. Since $\partial S_1 = (\partial S_1 \cap \Gamma) \cup (\partial S_1 \cap \Gamma_0)$, Case II can also be split into two following sub-cases. **Sub-Case IIa**: $\overline{P_{ij} P_{ij+1}} \in \partial S_1 \cap \Gamma$. It is due to $w \in \overline{V}_h$ and the Dirichlet boundary condition (11.2) that

$$
\sum_j \sum_{\substack{i \\ \text{Case IIa}}} |w_{i,j+1} - w_{ij}| = 0, \tag{11.40}
$$

where $w_{i,j} = w_{i,j+1} = 0$.

Sub-Case IIb: $\overline{P_{ij} P_{ij+1}} \in \partial S_1 \cap \Gamma_0$. In this case we obtain from the constraint coupling condition (11.10) as well as assumption (11.27)

$$
\sum_j \sum_{\substack{i \\ \text{Case IIb}}} |w_{i,j+1}^- - w_{ij}^-| = \sum_j \sum_{\substack{i \\ \text{Case IIb}}} |w_{i,j+1}^+ - w_{ij}^+|
$$

$$
= \sum_j \sum_{\substack{i \\ \text{Case IIb}}} k_j \frac{|w_{i,j+1}^+ - w_{i,j}^+|}{k_j}
$$

$$
\leq C|w^+|_{1,\Gamma_0^y} \leq C|w^+|_{1,\Gamma_0} \leq CL^\mu \|w^+\|_{0,\Gamma_0}
$$

$$
\leq CL^\mu \|w^+\|_{1,S_2} \leq CL^\mu \|\overline{w}\|_1, \tag{11.41}
$$

where Γ_0^y is the vertical segment of Γ_0. Therefore, the second term in (11.37) can be reduced to

$$
T_{\mathrm{II}} \leq \sum_j k_j^2 \left\{ \sum_{\substack{i \\ \text{Case I}}} \left| (w_{i,j+1} - w_{ij}) \left\{ \frac{\partial^2 \tilde{u}_{ij}^{(2)}}{\partial x \partial y} - \frac{\partial^2 \tilde{u}_{i-1,j}^{(2)}}{\partial x \partial y} \right\} \right| \right\} \tag{11.42}
$$

$$+ \sum_i \frac{\partial^2 \widetilde{u}_{ij}^{(2)}}{\partial x \partial y} |w_{i,j+1} - w_{ij}| + \sum_i \frac{\partial^2 \widetilde{u}_{ij}^{(2)}}{\partial x \partial y} |w_{i,j+1} - w_{ij}| \Bigg\}.$$
$$\text{Case IIa} \qquad\qquad\qquad\qquad \text{Case IIb}$$

For the first term in the right side of (11.42), we obtain from the Schwarz inequality

$$\sum_j k_j^2 \left| \sum_i (w_{i,j+1} - w_{ij}) \left\{ \frac{\partial^2 \widetilde{u}_{ij}^{(2)}}{\partial x \partial y} - \frac{\partial^2 \widetilde{u}_{i-1,j}^{(2)}}{\partial x \partial y} \right\} \right|$$
$$\text{Case I}$$

$$\leq \sum_j k_j^2 \sum_i |w_{i,j+1} - w_{ij}| \left| \frac{\partial^2 \widetilde{u}_{ij}^{(2)}}{\partial x \partial y} - \frac{\partial^2 \widetilde{u}_{i-1,j}^{(2)}}{\partial x \partial y} \right|$$
$$\text{Case I}$$

$$\leq CM_3(u)h \sum_j k_j^2 \sum_i |w_{i,j+1} - w_{ij}| = CM_3(u)h \sum_j k_j^3 \sum_i \frac{|w_{i,j+1} - w_{ij}|}{k_j}$$
$$\text{Case I} \qquad\qquad\qquad\qquad\qquad \text{Case I}$$

$$\leq Ch^2 M_3(u)\overline{|w|}_1 \leq Ch^2 M_3(u)\overline{\|w\|}_1. \tag{11.43}$$

Combining (11.40)—(11.43) yields

$$T_{\text{II}} = \left| \sum_{ij} h_i k_j^3 \frac{\partial^2 \widetilde{u}_{ij}^{(2)}}{\partial x \partial y} w_{xy} \right|$$
$$\leq Ch^2 (M_3(u) + M_2(u)L^\mu) \overline{\|w\|}_1 \leq Ch^2 L^\mu M_3(u)\overline{\|w\|}_1 . \tag{11.44}$$

The desired result (11.34) is obtained from (11.37)—(11.39) and (11.44); this completes the proof of (11.31).

For proving (11.32), we have similarly from Taylor's formula

$$\iint_{\Box_{ij}} g\,ds = \frac{h_i k_j}{4}(g_{ij} + g_{i+1,j} + g_{i,j+1} + g_{i+1,j+1}) + R_{ij}^{(2)},$$

where the truncation errors

$$R_{ij}^{(2)} = -\frac{1}{12}h_i k_j \left(h_i^2 \frac{\partial^2 \widetilde{g}_{ij}^{(1)}}{\partial x^2} + k_j^2 \frac{\partial^2 \widetilde{g}_{ij}^{(2)}}{\partial y^2} \right) - \frac{3}{32}h_i^2 k_j^2 \left(\frac{\partial^2 \widetilde{g}_{ij}^{(3)}}{\partial x \partial y} - \frac{\partial^2 \widetilde{g}_{ij}^{(4)}}{\partial x \partial y} \right).$$

Letting $g = fw$, then

$$g_{xx} = f_{xx}w + 2f_x w_x, \quad g_{yy} = f_{yy}w + 2f_y w_y, \quad g_{xy} = f_{xy}w + 2f_x w_y + f w_{xy}.$$

Hence, we can obtain similarly

$$\left|(\iint_{S_1} - \widehat{\iint_{S_1}})fwds\right| = \left|\sum_{ij}(\iint_{\Box_{ij}} - \widehat{\iint_{\Box_{ij}}})fwds\right|$$

$$= \sum_{ij} R_{ij}^{(2)} \le Ch^2 M_2(f) \sum_{ij} h_i k_j \left[|w(\xi_{ij}^{(1)})| + |w_x(\xi_{ij}^{(2)})| + |w_y(\xi_{ij}^{(3)})|\right]$$

$$+ C\left|\sum_{ij} h_i^2 k_j^2 w_{xy}(f(\xi_{ij}^{(4)}) - f(\xi_{ij}^{(5)}))\right| \le Ch^2 M_2(f)\overline{\overline{\|w\|}}_1. \quad \blacksquare$$

Based on Theorem 11.1 and Lemmas 11.2 and 11.4, we have the following theorem.

Theorem 11.2 *Let all the conditions in Lemmas 11.3 and 11.4 hold. Then the solution u_N from the nonconforming combination (11.11) has the error bounds*

$$\overline{\overline{\|u_N - u\|}}_1 \le C(h^2 + \|R_L\|_{1,S_2^-} + h^2 L^{2\mu}).$$

Also suppose that the number L of the basis functions used for u^+ is chosen such that

$$\|R_L\|_{1,S_2^-} = O(h^2) \quad and \quad L = O(|\ell n h|). \tag{11.45}$$

Then there exist the superconvergence rates, $\overline{\overline{\|u_N - u\|}}_1 = O(h^{2-\delta})$, of the solution u_N, where $\delta(> 0)$ is arbitrarily small.

Theorem 11.2 is a development of superconvergence in Li (1989a, 1990) to quasiuniform rectangles used.

11.3   COMBINATIONS I, II AND SYMMETRIC COMBINATION

First we give the following theorem and lemma without proofs.

Theorem 11.3 *Suppose that there exist two bounded constants $C_0(> 0)$ and C_1 independent of L, h and u and v such that*

$$C_0 \overline{\overline{\|v\|}}_h^2 \le \hat{a}_h(v,v), \quad \forall v \in V_h, \tag{11.46}$$

$$|\hat{a}_h(u,v)| \le C_1 \overline{\overline{\|u\|}}_h \overline{\overline{\|v\|}}_h, \quad \forall v \in V_h. \tag{11.47}$$

Then the solution u_h from the combinations (11.8) has the error bounds

$$\overline{\|u - u_h\|}_h \leq C \left\{ \inf_{v \in V_h} \overline{\|u - v\|}_h + \right.$$

$$+ \sup_{w \in V_h} \frac{|(\iint_{S_1} - \widehat{\iint_{S_1}}) \bigtriangledown u \bigtriangledown w \, ds|}{\|w\|_h} + \sup_{w \in V_h} \frac{|(\iint_{S_1} - \widehat{\iint_{S_1}}) f w \, ds|}{\|w\|_h} +$$

$$+ |1 - (\alpha + \beta)| \sup_{w \in V_h} \frac{|\int_{\Gamma_0} \frac{\partial u}{\partial n}(w^+ - w^-) d\ell|}{\|w\|_h} +$$

$$+ (\alpha + \beta) \sup_{w \in V_h} \frac{|(\int_{\Gamma_0} - \widehat{\int_{\Gamma_0}}) \frac{\partial u}{\partial n}(w^+ - w^-) d\ell|}{\|w\|_h} +$$

$$+ \beta \sup_{w \in V_h} \left. \frac{|\widehat{\int_{\Gamma_0}}(\frac{\partial u}{\partial n} - \frac{\Delta u^-}{\Delta n})(w^+ - w^-) d\ell|}{\|w\|_h} \right\}, \tag{11.48}$$

where u is the true solution of (11.1) and (11.2).

Lemma 11.5 *Let*

$$\|\frac{\partial v^+}{\partial n}\|_{0,\Gamma_0} \leq C L^\mu \|v^+\|_{0,\Gamma_0}, \ \forall v \in V_h,$$

and suppose that $\sigma \geq 1$ when $\beta > 0$, or $\sigma > 0$ and $h^\sigma L^{2\mu} \leq C$ when $\alpha > 0$. Then, when $P_c > 0$ is chosen suitably large but independent of L, h and v, the inequalities (11.46) and (11.47) hold.

The proof of Lemma 11.5 is deferred in Chapter 15, see Lemma 15.2. Below we prove the following lemma involving the coupling relations along Γ_0.

Lemma 11.6 *Let (11.27) for all $v \in V_h$, (11.29) and $\partial u/\partial n \in H^2(\Gamma_0)$ be given, then for all $w \in V_h$ such that*

$$\left|\left(\int_{\Gamma_0} - \widehat{\int_{\Gamma_0}}\right) \frac{\partial u}{\partial n}(w^+ - w^-) d\ell\right| \leq C \left\{ (hL^\mu)^2 \left\|\frac{\partial u}{\partial n}\right\|_{0,\Gamma_0} + h^{1+\frac{\sigma}{2}} \left|\frac{\partial u}{\partial n}\right|_{2,\Gamma_0} \right\} \overline{\|w\|}_h,$$

$$\left|\widehat{\int_{\Gamma_0}}(\frac{\partial u}{\partial n} - \frac{\Delta u^-}{\Delta n})(w^+ - w^-) d\ell\right| \leq C M_2(u) h^{1+\frac{\sigma}{2}} \overline{\|w\|}_h. \tag{11.49}$$

Proof: We first have

$$|(\int_{\Gamma_0} - \widehat{\int_{\Gamma_0}})\xi\eta d\ell| = |\int_{\Gamma_0}(\xi\eta - \widehat{\xi\eta})d\ell| \leq |\int_{\Gamma_0}\xi(\eta - \widehat{\eta})d\ell| + |\int_{\Gamma_0}(\xi - \widehat{\xi})\widehat{\eta}d\ell|$$

$$\leq \|\xi\|_{0,\Gamma_0}\|\eta - \widehat{\eta}\|_{0,\Gamma_0} + \|\xi - \widehat{\xi}\|_{0,\Gamma_0}\|\widehat{\eta}\|_{0,\Gamma_0}.$$

Let $\xi = \frac{\partial u}{\partial n}$, and $\eta = w^+ - w^-$. Since w^- is the piecewise linear function, then $\eta - \widehat{\eta} = w^+ - \widehat{w}^+$. Therefore we obtain from (11.27) for all $v \in V_h$,

$$|(\int_{\Gamma_0} - \widehat{\int_{\Gamma_0}})\frac{\partial u}{\partial n}(w^+ - w^-)d\ell|$$

$$\leq \|\frac{\partial u}{\partial n}\|_{0,\Gamma_0}\|w^+ - \widehat{w}^+\|_{\Gamma_0} + \|\frac{\partial u}{\partial n} - \frac{\partial \widehat{u}}{\partial n}\|_{0,\Gamma_0}\overline{\|w^+ - w^-\|}_{0,\Gamma_0}$$

$$\leq C\left\{h^2\|\frac{\partial u}{\partial n}\|_{0,\Gamma_0}|w^+|_{2,\Gamma_0} + h|\frac{\partial u}{\partial n}|_{2,\Gamma_0}\overline{\|w^+ - w^-\|}_{0,\Gamma_0}\right\}$$

$$\leq C\left\{h^2 L^{2\mu}\|\frac{\partial u}{\partial n}\|_{0,\Gamma_0} + h^{1+\frac{\varepsilon}{2}}|\frac{\partial u}{\partial n}|_{2,\Gamma_0}\right\}\overline{\|w\|}_h.$$

This is the first result of Lemma 11.6. Next, we obtain

$$|\int_{\Gamma_0}(\frac{\partial u}{\partial n} - \frac{\Delta u^-}{\Delta n})(w^+ - w^-)d\ell| \leq \overline{\|\frac{\partial u}{\partial n} - \frac{\Delta u^-}{\Delta n}\|}_{0,\Gamma_0}\overline{\|w^+ - w^-\|}_{0,\Gamma_0}$$

$$\leq ChM_2(u)\overline{\|w^+ - w^-\|}_{0,\Gamma_0} \leq Ch^{1+\frac{\varepsilon}{2}}M_2(u)\overline{\|w\|}_h. \quad\blacksquare$$

Since the norm $\overline{\|v^+ - v^-\|}_{0,\Gamma_0}$ is defined by the values $(v^+ - v^-)$ only on the mesh nodes $Z_k \in \Gamma_0$, the solution u_N $(\in V_h)$ from the nonconforming combination (11.11) has $\overline{\|u_N^+ - u_N^-\|}_{0,\Gamma_0} = 0$. Then we have from Lemma 11.3

$$\inf_{v \in V_h}\overline{\|u - v\|}_h \leq \overline{\|u - u_N\|}_h = \overline{\|u - u_N\|}_1 \leq Ch^2 + \|R_L\|_{1,S_2^*}. \quad (11.50)$$

Similarly to Lemma 11.4, we have the following lemma.

Lemma 11.7 *Let (11.27) for $\forall v \in V_h$, (11.29) and (11.30) hold. Then*

$$\left|(\iint_{S_1} - \widehat{\iint_{S_1}})\nabla u \nabla w ds\right| \leq C\{h^2 L^\mu + h^{1+\frac{\varepsilon}{2}}\}M_3(u)\overline{\|w\|}_h, \ \forall w \in V_h,$$

$$(11.51)$$

$$\left|(\iint_{S_1} - \widehat{\iint_{S_1}})fwds\right| \leq Ch^2 M_2(f)\overline{\|w\|}_h, \ \forall w \in V_h. \quad (11.52)$$

Proof: To compare the proofs in (11.31), the crucial different step of the proofs for (11.51) is that the estimates on (11.41) should be modified, due to lack of the constraint condition (11.10) used in the nonconforming combination. Since

$$|w_{i,j+1}^- - w_{ij}^-| \leq |w_{i,j+1}^+ - w_{ij}^+| + |w_{i,j+1}^+ - w_{i,j+1}^-| + |w_{i,j}^+ - w_{ij}^-|,$$

we can obtain the estimates for (11.41) by referring the discrete penalty integral on Γ_0:

$$\sum_j \underset{\text{Case IIb}}{\sum_i} |w_{i,j+1}^- - w_{ij}^-|$$

$$\leq \sum_j \underset{\text{Case IIb}}{\sum_i} \{|w_{i,j+1}^+ - w_{ij}^+| + |w_{i,j+1}^+ - w_{i,j+1}^-| + |w_{ij}^+ - w_{ij}^-|\}$$

$$\leq C \left\{ L^\mu \overline{\|w\|}_1 + \frac{1}{h} \overline{\|w^+ - w^-\|}_{0,\Gamma} \right\}.$$

Hence the bounds of (11.44) should be modified as follows

$$T_{\text{II}} \leq C \left[h^2 + h^2 L^\mu + h^{1+\frac{\varsigma}{2}} \right] M_3(u) \overline{\|w\|}_1 \leq C \left[h^2 L^\mu + h^{1+\frac{\varsigma}{2}} \right] M_3(u) \overline{\|w\|}_h.$$

Since the bounds of T_{I} and T_{III} are the same as in (11.38) and (11.39), then (11.51) is obtained by noting $\overline{\|w\|}_1 \leq \overline{\|w\|}_h$. ∎

The constants satisfy $\alpha + \beta = 1$ for Combinations I, II and the Symmetric Combination. Hence the fourth term of the right side of (11.48) is zero. So far, we have provided all bounds of other terms in (11.48) from Lemmas 11.6, 11.7 and (11.50). This leads to the following theorem.

Theorem 11.4 *Let all the conditions in Lemmas 11.5–11.7 hold. Then the solution u_h from Combinations I and II and the Symmetric Combination has the error bounds*

$$\overline{\|u_h - u\|}_h \leq C(h^2 + h^{1+\frac{\varsigma}{2}} + \|R_L\|_{1,S_2^*} + h^2 L^{2\mu}).$$

Moreover, when $\sigma \geq 2$ and (11.45) are given, then $\overline{\|u - u_h\|}_h = O(h^{2-\delta})$.

11.4   PENALTY COMBINATION

For the penalty combination, the constants $\alpha = \beta = 0$ are chosen. We then have from (11.8)

$$\widehat{a}_h(u_h, v) = \widehat{f}_h(v), \quad \forall v \in V_h, \tag{11.53}$$

where

$$\hat{a}_h(u,v) = \iint_{S_1} \nabla u \, \nabla v ds + \iint_{S_2} \widehat{\nabla u \, \nabla w} ds + \frac{P_c}{h^\sigma} \int_{\Gamma_0} \widehat{(u^+ - u^-)(v^+ - v^-)} d\ell \; .$$

It is easy to prove that for any $P_c(>0)$ and $\sigma(\geq 0)$, both (11.46) and (11.47) hold true. Also, the error bounds of solution u_h are obtained from (11.48) by letting $\alpha = \beta = 0$:

$$\overline{\|u - u_h\|}_h$$

$$\leq C \left\{ \inf_{v \in V_h} \overline{\|u - u_h\|}_h + \sup_{w \in V_h} \frac{|(\iint_{S_1} - \widehat{\iint_{S_1}}) \nabla u \, \nabla w ds|}{\|w\|_h} + \right. \qquad (11.54)$$

$$\left. + \sup_{w \in V_h} \frac{|(\iint_{S_1} - \widehat{\iint_{S_1}}) f w ds|}{\|w\|_h} + \sup_{w \in V_h} \frac{|(\int_{\Gamma_0} \frac{\partial u}{\partial n}(w^+ - w^-) d\ell|}{\overline{\|w\|}_h} \right\} .$$

The bounds of the first three terms in the right sides of (11.54) have been provided in Section 11.3 already; only the last terms need to be estimated.

Lemma 11.8 *Let (11.27) be given* $\forall v \in V_h$ *, then*

$$\|v^+ - v^-\|_{0,\Gamma_0} \leq \overline{\|v^+ - v^-\|}_{0,\Gamma_0} + C(hL^\mu)^2 \|v\|_{1,S_2} \, , \quad \forall v \in V_h.$$

Proof: Since $\overline{\|w\|}_{0,\Gamma_0} = \|\hat{w}\|_{0,\Gamma_0}$, we have from the triangular inequality

$$\|w\|_{0,\Gamma_0} - \overline{\|w\|}_{0,\Gamma_0} = \|w\|_{0,\Gamma_0} - \|\hat{w}\|_{0,\Gamma_0} \leq \|w - \hat{w}\|_{0,\Gamma_0}. \qquad (11.55)$$

Letting, $w = v^+ - v^-$ we obtain from (11.55)

$$\|v^+ - v^-\|_{0,\Gamma_0} - \overline{\|v^+ - v^-\|}_{0,\Gamma_0} \leq \|v^+ - \hat{v}^+\|_{0,\Gamma_0}$$

$$\leq Ch^2|v^+|_{2,\Gamma_0} \leq Ch^2L^{2\mu}|v^+|_{0,\Gamma_0} \leq Ch^2L^{2\mu}|v|_{1,S_2}. \qquad \blacksquare$$

From Lemma 11.8 we can see the bounds:

$$\left| \int_{\Gamma_0} \frac{\partial u}{\partial n}(w^+ - w^-) d\ell \right| \leq \|\frac{\partial u}{\partial n}\|_{0,\Gamma_0} \|w^+ - w^-\|_{0,\Gamma_0}$$

$$\leq \left\| \frac{\partial u}{\partial n} \right\|_{0,\Gamma_0} \left[\overline{\|w^+ - w^-\|}_{0,\Gamma_0} + C(hL^\mu)^2 \|w\|_{1,S_2} \right]$$

$$\leq C \left\| \frac{\partial u}{\partial n} \right\|_{0,\Gamma_0} [h^{\frac{\varsigma}{2}} + (hL^\mu)^2] \overline{\|w\|}_h. \qquad (11.56)$$

Based on (11.54), (11.56), and (11.50)—(11.52), we provide the following theorem.

Theorem 11.5 *Let (11.29), (11.30) and the condition in Lemma 11.8 hold. Then the solution u_h from the penalty combination (11.53) has the error bounds*

$$\overline{\|u_h - u\|}_h \leq C \left\{ h^2 + h^{\sigma/2} + \|R_L\|_{1,S_2^*} + (L^\mu h)^2 \right\}.$$

Moreover, when $\sigma \geq 4$ and (11.45) are given, then $\overline{\|u_h - u\|}_h = O(h^{2-\delta})$.

Remark 11.1 *The superconvergence rates in the norm $\overline{\|\epsilon\|}_h$ given in (11.14) are significant to penalty combinations. Note that the limitation of $\sigma = 2$ is derived for optimal convergence rates, based on the norm $\|\cdot\|_h$ given in (11.13) and in Chapter 9. This comparison shows that a suitable choice of error norms is important to evaluation of the proposed algorithms. From the error analysis and the numerical experiments in the last section, the penalty combination seems to be promising among all combined methods.*

11.5   RELATIONS OF COUPLINGS

Let us discover the relation with different couplings to the nonconforming constraints (11.10), to implement the discussions in Section 7.5 of Chapter 7 and Section 9.4.2 of Chapter 9.

Lemma 11.9 *Let all conditions in Theorem 11.5 hold. Then when $\sigma \geq 4$ the average jumps of the solution $v = u_h^p$ of the penalty combinations at the element nodes $Z_i \in \Gamma_0$ have the convergence rates*

$$E(v^+ - v^-) = \frac{1}{N_1 + 1} \sum_{i=0}^{N_1} |v^+(Z_i) - v^-(Z_i)| = O(h^{2+\frac{\sigma}{2}-\delta}), \qquad (11.57)$$

where N_1 is the number of element nodes on Γ_0.

Moreover, let all conditions in Theorem 11.4 hold. Then when $\sigma \geq 2$, the average jumps of the solution $v = u_h$ of Combinations I, II and Symmetric Combination also have bounds (11.57).

Proof: Denote the trapezoidal rule

$$\widetilde{\int_{\Gamma_0} v^2 d\ell} = \sum_{k=1}^{N_1} \frac{\overline{Z_{k-1} Z_k}}{2} \left[v^2(Z_{k-1}) + v^2(Z_k) \right].$$

Since $x^2 + y^2 \leq 2(x^2 + xy + y^2)$, we have

$$\widetilde{\int_{\Gamma_0}} v^2 d\ell \;\leq\; 3\sum_{k=1}^{N_1} \frac{\overline{Z_{k-1}Z_k}}{3}\left[v^2(Z_{k-1}) + v(Z_{k-1})v(Z_k) + v^2(Z_k)\right]$$

$$= \; 3\widetilde{\int_{\Gamma_0}} v^2 d\ell. \tag{11.58}$$

Since the rectangles are quasiuniform, we obtain from the Schwarz inequality

$$\left(\sum_{i=0}^{N_1} \frac{v(Z_i)}{N_1+1}\right)^2 \;\leq\; \sum_{i=0}^{N_1} \frac{1}{N_1+1} v^2(Z_i) \leq C\sum_{k=1}^{N_1} \frac{\overline{Z_{k-1}Z_k}}{2}\left[v^2(Z_{k-1}) + v^2(Z_k)\right]$$

$$= \; C\widetilde{\int_{\Gamma_0}} v^2 d\ell \leq 3C\widetilde{\int_{\Gamma_0}} v^2 d\ell = 3C\overline{\|v\|}_{0,\Gamma_0}^2.$$

Letting $v = (u_h^p)$ and $w = u - u_h^p$ we have

$$E(v^+ - v^-) \;=\; \sqrt{\frac{1}{N_1+1}\sum_{i=0}^{N_1}|(u_h^p)_{Z_i}^+ - (u_h^p)_{Z_i}^-|} \leq C\overline{\|(u_h^p)^+ - (u_h^p)^-\|}_{0,\Gamma_0}$$

$$= \; C\overline{\|w^+ - w^-\|}_{0,\Gamma_0} \leq Ch^{\frac{\sigma}{2}}\|w^-\|_h.$$

When $\sigma \geq 4$ we obtain from Theorem 11.5

$$E(v^+ - v^-) \leq Ch^{2+\frac{\sigma}{2}-\delta}.$$

The proof for rest of Lemma 11.9 is similar. ■

Remark 11.2 *Lemma 11.9 displays an interesting fact that the average jumps between $(u_h^p)^+$ and $(u_h^p)^-$ are $O(h^{2+\frac{\sigma}{2}-\delta})$. When $\sigma \longrightarrow \infty$, the penalty combination, Combinations I, II and Symmetric Combination all approach the same nonconforming combinations. Note that the large values of $\sigma(\geq 4)$ do not incur the reduced convergence rates.*

On the common boundary Γ_0, the norm, $\frac{P_c^{1/2}}{h^{\sigma/2}}\overline{\|v^+ - v^-\|}_{0,\Gamma_0}$, is a summation of discrete solutions over all the interface mesh nodes Z_k (see (11.17) and (11.12)). This discrete penalty technique plays a coupling role in ensuring that v^+ and v^- will be close to each other only at the interface mesh nodes. The larger the ratio P_c/h^σ is, the smaller the differences $|v^+(Z_k) - v^-(Z_k)|$ at all Z_k are. Therefore, this norm will never cause deterioration of the global error norm $\|\epsilon\|_h$ even when $P_c/h^\sigma \to \infty$ (i.e., $P_c \to \infty$ or $\sigma \to \infty$ while $h < 1$). In this case the discrete penalty technique leads to the nodal continuity constraints (11.10) in the nonconforming combination.

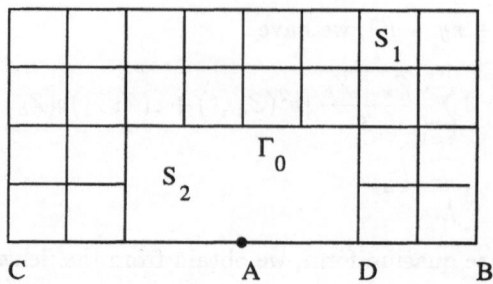

Figure 11.1   A partition of Motz's problem with $M_S = 2$.

By contrast, a continuous penalty integral, $\frac{P_c^{1/2}}{h^{\sigma/2}} \|v^+ - v^-\|_{0,\Gamma_0}$, also plays a role in coupling v^+ and v^- on the *entire* interface boundary Γ_0. Since the admissible functions v^+ and v^- in (11.7) are different, the norm $\|v^+ - v^-\|_{0,\Gamma_0}$ can never diminish except for the null solution. Therefore, when P_c/h^σ is large or infinite, so are the above norm given and the norm $\|\cdot\|_h$ in (11.13). This leads to a unboundedness of $\|\varepsilon\|_h$ in Chapters 7 and 9.

11.6   NUMERICAL EXPERIMENTS

Numerical solutions for Motz's problem with the partition in Figure 11.1 have been obtained from Combinations I and II, the Symmetric Combination and the Penalty Combination, and their error norms are listed in Tables 11.1 and 11.2. Since results from Combinations I and II and the Symmetric Combination are close to each other, we provide only those from the Symmetric Combination. From Table 11.1 we can see the following empirical, asymptotic relations,

$$\overline{\|\epsilon\|}_1 = \overline{\|u_h - u\|}_1 = O(h^{2-\delta}), \tag{11.59}$$

$$\|\epsilon\|_h = O(h), \quad \|\epsilon\|_1 = O(h), \quad \overline{\|\epsilon\|}_{0,S} = O(h^2), \tag{11.60}$$

$$\max = \|\epsilon\|_{\infty,S} = O(h^{2-\delta}),$$

$$\|\epsilon^+ - \epsilon^-\|_{0,\Gamma_0} = O(h^2), \quad \|\epsilon^+ - \epsilon^-\|_{\infty,\Gamma_0} = O(h^2),$$

$$|D_i - \tilde{D}_i| = O(h^2) \text{ as } i = 0, 1. \tag{11.61}$$

Eq. (11.59) is consistent with the superconvergence rates; Eqs. (11.60)–(11.61) are all optimal convergence rates.

We are more interested in the penalty combination. We will investigate convergence rates in different norm definitions and the influence of σ upon convergence rates.

Divis.	δ_1	δ_2	Max	$\|\epsilon\|_{0,S}$	$\|\epsilon\|_1$	$\|\epsilon\|_h$	$\overline{\|\epsilon\|}_1$	$\bar{D}_0$	$\bar{D}_1$
$M_S = 2$ L+1=4	1.963	3.037	3.370	1.1671	21.66	32.95	6.295	399.4392	86.6969
M_S=4 L+1=5	.4680	.7310	.9451	.2930	10.49	15.82	1.696	400.8742	87.3944
M_S=6 L+1=5	.2043	.3196	.4531	.1315	.6941	10.40	.7691	401.0323	87.5239
M_S=8 L+1=6	.1139	.1803	.2823	.0572	5.192	7.758	.4505	401.0887	87.6455
M_S=10 L+1=6	.0727	.1147	.1859	.0484	4.148	6.192	.2944	401.1152	87.6493

Table 11.1 Error norms and approximate coefficients from Symmetric Combination of RGM-FDMs with $P_c = 10$ and $\sigma = 2$, where $\delta_1 = \|\varepsilon^+ - \varepsilon^-\|_{0,\Gamma_0}$ and $\delta_2 = \|\varepsilon^+ - \varepsilon^-\|_{\infty,\Gamma_0}$.

Divis.	$\|\epsilon\|_1$					$\overline{\|\epsilon\|}_h$				
	$\sigma=1$	$\sigma=2$	$\sigma=3$	$\sigma=4$	$\sigma=5$	$\sigma=1$	$\sigma=2$	$\sigma=3$	$\sigma=4$	$\sigma=5$
M_S=2 L+1=4	39.07	24.07	21.87	21.67	21.65	69.59	35.75	18.70	10.79	7.631
M_S=4 L+1=5	22.56	11.09	10.50	10.49	10.49	50.65	18.31	6.662	2.819	1.842
M_S=6 L+1=5	16.38	7.204	6.941	6.939	6.939	41.76	12.32	3.627	1.264	.7976
M_S=8 L+1=6	13.10	5.337	5.191	5.191	5.191	36.35	9.284	2.358	.7216	.4535
M_S=10 L+1=6	11.04	4.239	4.147	4.147	4.147	32.61	7.450	1.688	.4645	.2899

Table 11.2 The error norms $\|\epsilon\|_1$ and $\overline{\|\epsilon\|}_h$.

From Table 11.2 and Figure 11.2 we find that

$$\overline{\|\epsilon\|}_h = O(h^{\sigma/2}) \ as \ \sigma = 1,2,3, \quad \overline{\|\epsilon\|}_h = O(h^{2-\delta}) \ as \ \sigma = 4,5. \quad (11.62)$$

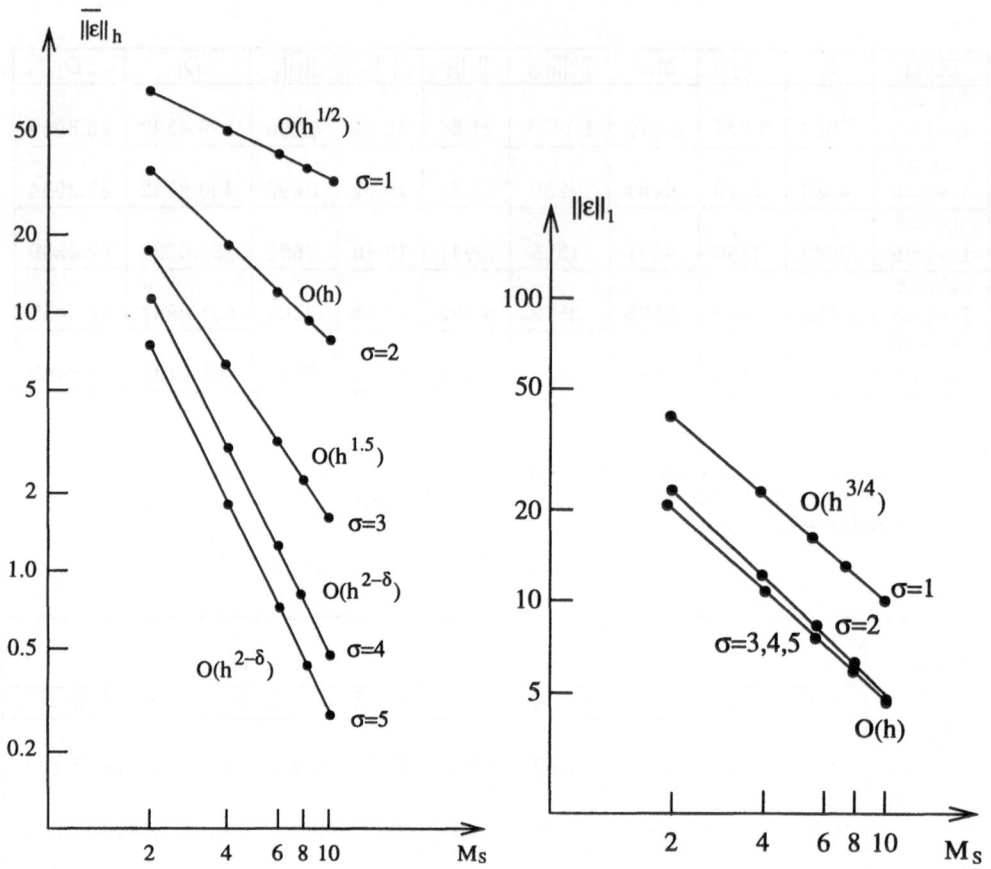

Figure 11.2 The error norm $\overline{\|\epsilon\|}_h$.   **Figure 11.3** The error norm $\|\epsilon\|_1$.

Eqs. (11.62) perfectly verify Theorem 11.5. From Table 11.2 and Figure 11.3, we can also see that

$$\|\epsilon\|_1 = O(h^{\frac{3}{4}}) \ as \ \sigma = 1, \quad \|\epsilon\|_1 = O(h) \ as \ \sigma = 2, 3, 4, 5.$$

Surprisingly, the optimal convergence rates $O(h)$ in $\|\epsilon\|_1$ can always be obtained when $\sigma \geq 2$. This comparison underscores the importance of the norms chosen. The norms, $\|\epsilon\|_1$, in particular $\overline{\|\epsilon\|}_1$ and $\overline{\|\epsilon\|}_h$, should be used to replace $\|\epsilon\|_h$ in both analysis and computation. Overall, the penalty combination is highly recommended because of its simplicity, and because of superconvergence of the solutions produced.

There are no whole truths;
all truths are half-truths.
It is trying to treat them as whole truths
that plays the devil.
—— *Dialoges (1953)* ——

Alfred North Whitehead
(1861 – 1947)

PART IV

APPLICATIONS AND ADVANCED TOPICS

The last part of this book consists of five chapters, to display a wide range of applications of the combinations, and various advanced topics the combinations may follow.

Chapter 12: Crack-Infinity Problem, using the nonconforming combinations of RGM-FEM in Chapter 4.

Chapter 13: Wind Flow over Buildings, using the penalty combinations of RGM-FEM in Chapter 7.

Chapter 14: Global Superconvergence in Combinations, for six coupling techniques in Parts II and III for the combinations of RGM-FEM.

Chapter 15: Iterative Substructing Methods, applied to the combinations of RGM-FDM in Chapter 11.

Chapter 16: Schwarz Alternating Method, applied to the combination in Chapter 12.

In Chapters 12, 13 and 16, the combined methods are applied to unbounded domain problems. In Chapter 14, the global superconvergence of combinations is discovered in the combinations, based on Lin's techniques (see Lin and Yan (1996)). In Chapters 15 and 16, the current domain decomposition methods are well suited to implement the combinations in this book into parallel computation. The contents of this part are adapted mainly from Li (1989a, 1990, 1994, 1996b, c, 1997c), Li and Bui (1992b), and Li, Lin and Yan (1997).

In this part, we focus on application to the unbounded domain problems only, although the combinations can be applied to interface problems easily. Surely, the combinations in this book may be extended to other elliptic equations, such as the Helmholtz equation, the biharmonic equation, elastic plate problems, Stoke equations, etc. Moreover, the eigenvalue and parabolic problems may also be solved effectively by combinations, whenever involving singularity or not. Studies on some above topics have been completed, to show that the combinations are still promising; details appear elsewhere.

In philosophy, the combined ideas are suited to solving hyperbolic equations. The coupling techniques, however, must be re-studied cautiously. It seems that the least squares method may also be chosen as a framework to link different numerical methods.

From our analysis, the advantages in single numerical methods such as superconvergence will remain in the combinations; the significance of the combinations will also hold in the domain decomposition methods. By the new imbedding techniques introduced in Chapter 15, the existing iterative substructing methods may also fit for all the combinations in this book. In the Schwarz alternative methods, different numerical methods should be naturally used in different subproblems, which are described in the last chapter of this book.

For solving elliptic equations we recommend the harmonious family of combined methods, in which all methods not only contribute their best but also collaborate with others well. These are the ideal numerical methods, the methods that can solve the challenging problems we may encounter in the future.

12

CRACK-INFINITY PROBLEM

In this and next chapters, the combinations of RGM-FEM in Parts II and III are applied to unbounded domain problems, which also involve angular singularities. The basic error analysis can follow the outlines given in Parts II and III; but more exploitation may still be needed.

In this chapter, the typical crack-infinity problem in Section 3.6.3 is chosen as a computational example, by using the nonconforming combination in Part II. The unbounded solution domain is divided into several bounded subdomains and an exterior, infinite subdomain in the combined method. The linear FEM is still used in the bounded subdomains S_i where $u \in H^2(S_i)$; but the RGM with particular solutions is used on the singular and exterior subdomains. In addition, the admissible functions chosen here are imposed to be continuous only at the element nodes on the common boundary of these subdomains called *the nonconforming combination* in Chapter 4. We notice that a linear algebraic equation system can be finally reduced, with the coefficient matrix being positive definite, symmetric and sparse.

In this chapter, the Debye-Huckel equation on an unbounded domain is solved by the nonconforming combination. Error norms in the Sobolev norm sense can easily be derived, based on the outlines in Chapter 4. A significant coupling relation

$$N + 1 = O(|\ln h|) \tag{12.1}$$

which originated in Chapter 4 has also been proved to be efficient for unbounded domain problems, where $N + 1$ is the total number of particular solutions used in the unbounded domain, and h is the maximal boundary length used in the FEM.

The important results are given in Theorem 12.1, to lead to the optimal convergence rate $O(h)$. The numerical experiments in Section 12.4 display outstanding advantages in using the coupling relation (12.1), in which only two or three terms of asymptotic particular solutions near the infinity are, in fact, needed for engineering application.

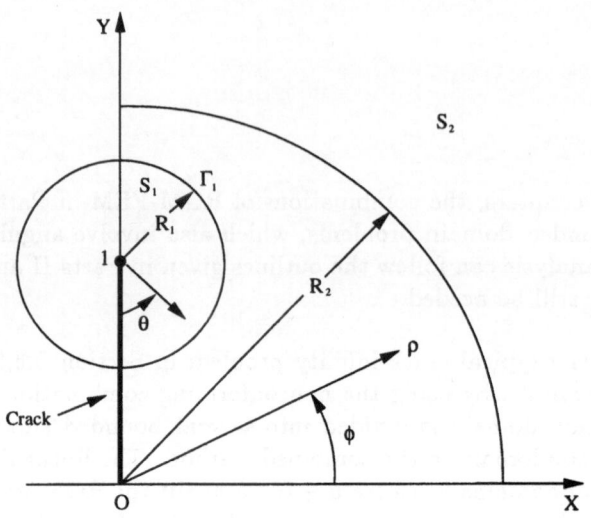

Figure 12.1 A division of S.

12.1   DESCRIPTION OF METHOD

Consider the sample problem on unbounded domains of two dimensions discussed
in Section 3.6.3 of Chapter 3:

$$-\Delta u + u = 0, \quad \text{in} \quad S^*, \tag{12.2}$$
$$u = 1, \quad \text{on} \quad y = 0,$$
$$u = 1, \quad \text{on} \quad x = 0 \cap 0 < y < 1,$$

where $\Delta = \partial^2/\partial x^2 + \partial^2/\partial y^2$, and S^* is the upper semi-plane but excluding the
crack section

$$x = 0 \cap 0 < y < 1.$$

Because of the model's symmetry, we can solve the problem on the first quadrant
only (Figure 12.1):

$$-\Delta u + u = 0, \quad \text{in} \quad S, \tag{12.3}$$
$$u = 1, \quad \text{on} \quad y = 0 \cap x > 0, \tag{12.4}$$
$$u = 1, \quad \text{on} \quad x = 0 \cap 0 < y < 1, \tag{12.5}$$
$$\frac{\partial u}{\partial x} = 0, \quad \text{on} \quad x = 0 \cap y > 1, \tag{12.6}$$

where S is the unbounded domain: $x > 0$ and $y > 0$.

The sample problem (12.3)–(12.6) can also be described in a weak form

$$B(u, v) = 0, \quad \forall v \in H_0^1(S),$$

where the solution $u \in H_*^1(S)$, and the bilinear form is

$$B(u, v) = \iint_S (u_x v_x + u_y v_y + uv) \, ds.$$

Here, $H_*^1(S)$ and $H_0^1(S)$ are the first-order Sobolev spaces over the unbounded domain S satisfying nonhomogeneous and homogeneous boundary conditions of (12.4)–(12.5), respectively.

We recall from Section 3.6.3 that there exist three singularities (the crack singularity, the infinite point and the corner singularity at the origin O) for the model problem (12.3)-(12.6) on S. Since the corner singularity at the origin $(0, 0)$ is mild with an asymptotic behavior:

$$u = 1 + O(\rho^2 \ln \rho) \quad \text{as } \rho \to 0,$$

the expansion solution $v^{(3)}$ of Li and Mathon (1990a) in a subdomain S_3 including the origin will satisfy $v^{(3)} \in H^2(S_3)$. Therefore, the linear FEM can still be applied to the subdomain including the origin. However, we have to use the RGM for the subdomains including the crack singularity $(0, 1)$ and the infinity. With the help of the method of separation of variables, we can find the particular solutions near the crack singular point

$$u = \cosh(r \sin \theta) + \sum_{n=0}^{\infty} a_n I_{n+\frac{1}{2}}(r) \sin((n + \frac{1}{2})\theta), \tag{12.7}$$

and the asymptotic solutions near the infinity

$$u = e^{-\rho \sin \phi} + \sum_{n=0}^{\infty} c_n K_{2n+1}(\rho) \sin((2n + 1)\phi), \tag{12.8}$$

where a_n and c_n are real expansion coefficients, (r, θ) and (ρ, ϕ) are the polar

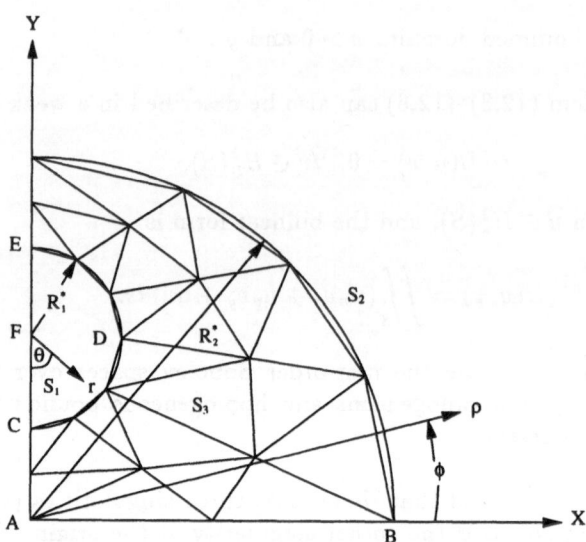

Figure 12.2   A division of S for the combination.

coordinates with the origins $(0,1)$ and $(0,0)$ respectively, and $I_\mu(r)$ and $K_n(\rho)$ are
the Bessel and Hankel functions for a purely imaginary argument defined by Watson
(1944):

$$I_\mu(r) = \sum_{k=0}^{\infty} \frac{1}{\Gamma(k+1)\Gamma(k+\mu+1)} \left(\frac{r}{2}\right)^{2k+\mu},$$

$$K_n(\rho) = \frac{1}{2}\int_{-\infty}^{\infty} e^{-\rho\cosh\eta - n\eta}\,d\eta. \tag{12.9}$$

In order to consider these two singular points, we divide the solution domain S into
three subdomains (see Figure 12.2): $S = S_1 \cap S_2 \cap S_3$, where S_1 is a semicircular
disc ($r < R_1^* < 1$ and $x \ge 0$) containing the crack singular point, S_2 is an exterior
domain ($\rho > R_2^* > 1$ and $x \ge 0$), and S_3 is the subdomain between S_1 and S_2.
Therefore, the radii R_1^* and R_2^* have to satisfy the relation

$$R_2^* > 1 + R_1^*.$$

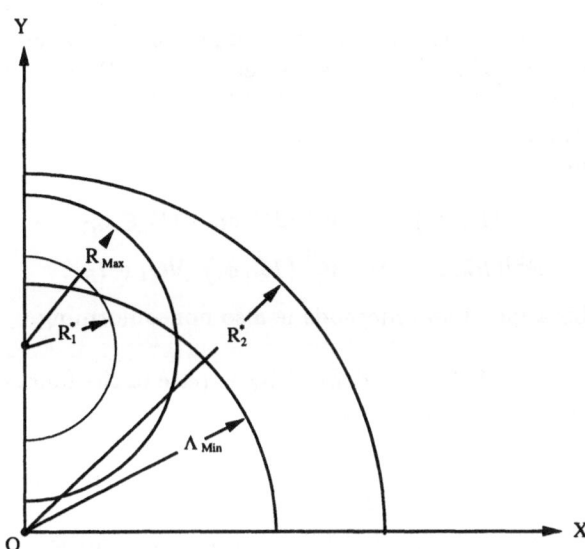

Figure 12.3 Radius relations : $R_1^* < R_{Max} \leq 1$,

Since $u \in H^2(S_3)$, we use the linear FEM in S_3 and use the RGM in S_1 and S_2 of Figure 12.3. This is, indeed, an approach combining the RGM-FEM.

Based on the solution expansions (12.7) and (12.8), we will choose the following admissible functions:

$$
v = \begin{cases}
v^{(1)} = \cosh(r\sin\theta) + \sum_{l=0}^{L} \widetilde{a}_l \dfrac{I_{l+\frac{1}{2}}(r)}{I_{l+\frac{1}{2}}(R_{Max})} \sin(l+\tfrac{1}{2})\theta, & \text{in } S_1, \\[3mm]
v^{(2)} = e^{-\rho\sin\phi} + \sum_{l=0}^{N} \widetilde{c}_l \dfrac{K_{2l+1}(\rho)}{K_{2l+1}(\Lambda_{Min})} \sin(2l+1)\phi, & \text{in } S_2, \\[3mm]
v_1^{(3)}, & \text{in } \widehat{S}_3^h,
\end{cases} \qquad (12.10)
$$

where $\widetilde{a}_l$ and $\widetilde{c}_l$ are unknown coefficients, $\widehat{S}_3^h$ is the triangulation domain for S_3 ($\widehat{S}_3^h \approx S_3$), and $v_1^{(3)}$ are the piecewise linear interpolation polynomials on the triangulation of $\widehat{S}_3^h$ satisfying the Dirichlet boundary conditions (12.4) and (12.5).

In (12.10), the functions $I_{l+\frac{1}{2}}(R_{Max})$ and $K_{2l+1}(\Lambda_{Min})$ are employed for scale constants, where the radii R_{Max} and Λ_{Min} are chosen as (Figure 12.3)

$$
R_{Max} \leq 1 \quad \text{and} \quad \Lambda_{Min} \geq 1.
$$

Now, we couple two numerical methods as follows:

Let $l_{R_1^*}(r = R_1^*, 0 \le \theta \le \pi)$ denote the common boundary between S_1 and S_3, and $P_i(R_1^*, \theta_i)$ the element nodes on $l_{R_1^*}$. Also, let $l_{R_2^*}(\rho = R_2^*, 0 \le \phi \le \pi/2)$ denote the common boundary of S_2 and S_3, and $Q_i(R_2^*, \phi_i)$ the element nodes on $l_{R_2^*}$. We constrain the continuity conditions only at the element nodes on $l_{R_1^*}$ and $l_{R_2^*}$ for the admissible functions v of (12.10):

$$v^{(1)}(R_1^*, \theta_i) = v_1^{(3)}(R_1^*, \theta_i), \quad \forall P_i \in l_{R_1^*}, \tag{12.11}$$

$$v^{(2)}(R_2^*, \phi_i) = v_1^{(3)}(R_2^*, \phi_i), \quad \forall Q_i \in l_{R_2^*}. \tag{12.12}$$

Clearly, this combination of two methods is also nonconforming.

We need some notations. Let V_h be defined by a space of the functions (12.10), and V_h^0 by another space of the following functions w,

$$w = \begin{cases} w^{(1)} = \sum_{l=0}^{L} \tilde{a}_l \dfrac{I_{l+\frac{1}{2}}(r)}{I_{l+\frac{1}{2}}(R_{Max})} \sin(l + \tfrac{1}{2})\theta, & \text{in } S_1, \\[2ex] w^{(2)} = \sum_{l=0}^{N} \tilde{c}_l \dfrac{K_{2l+1}(\rho)}{K_{2l+1}(\Lambda_{Min})} \sin(2l + 1)\phi, & \text{in } S_2, \\[2ex] w_1^{(3)}, & \text{in } \widehat{S}_3^h, \end{cases} \tag{12.13}$$

accompanied by satisfying the continuity conditions (12.11) and (12.12), where $w_1^{(3)}$ are piecewise linear interpolating polynomials on $\widehat{S}_3^h$, satisfying the homogeneous Dirichlet boundary conditions:

$$u = 0, \quad \text{for } y = 0 \cap x > 0,$$

$$u = 0, \quad \text{for } x = 0 \cap 0 < y < 1.$$

Therefore, the nonconforming combination of the RGM-FEM is designed to find an approximate solution, $u_h^* \in V_h$, such that

$$B_h(u_h^*, v_h) = 0, \quad \forall v_h \in V_h^0, \tag{12.14}$$

where the bilinear form is

$$B_h(u, v) = \sum_{i=1}^{2} \iint_{S_i} (u_x v_x + u_y v_y + uv) ds + \iint_{\widehat{S}_3^h} (u_x v_x + u_y v_y + uv) ds.$$

12.2   ERROR BOUNDS AND COUPLING STRATEGY

Define a norm

$$\|u\|_1 = (\|u\|_{1,S_1}^2 + \|u\|_{1,S_2}^2 + \|u\|_{1,\widehat{S}_3^h}^2)^{\frac{1}{2}}. \tag{12.15}$$

We notice that S_2 is the exterior domain $(\rho > R_2^* \cap x > 0)$. Because of (12.2) the norm definition (12.15) is applicable.

Below, we shall assess errors of the solution u_h^* by the nonconforming combination (12.14), in the norm $\| \cdot \|_1$. Error bounds are provided by Theorem 12.1, whose proof is deferred until Section 12.3.

Theorem 12.1 *Let the bounded radii satisfy (Figure 12.3):*

$$R_1^* < R_{Max}, \quad R_2^* > \Lambda_{Min}, \quad 1 + R_1^* < R_2^*. \tag{12.16}$$

Also, suppose that $u \in H^2(S_3)$ and that the family of triangular elements in $\widehat{S}_3^h$ is quasiuniform. Then there exists a bounded constant C independent of L, N, h and u such that

$$\|u - u_h^*\|_1 \leq C \left\{ h + \left(\frac{R_2}{R_{Max}} \right)^{L + \frac{3}{2}} + \left(\frac{\Lambda_{Min}}{\Lambda_2} \right)^{2N+3} + \right.$$

$$\left. + \left(L + \frac{1}{2} \right)^2 h^2 + (2N + 1)^2 h^2 \right\}, \tag{12.17}$$

where h is the maximal boundary length of quasiuniform triangular elements in $\widehat{S}_3^h$, and the radii R_2 and Λ_2 satisfy

$$R_1^* < R_2 < R_{Max}, \quad \Lambda_{Min} < \Lambda_2 < R_2^*, \quad 1 + R_2 < \Lambda_2.$$

From Theorem 12.1 we have the following corollary:

Corollary 12.1 *Suppose that all the conditions in Theorem 12.1 hold. Then,*

$$\|u - u_h^*\|_1 = O(h),$$

provided that

$$\left(\frac{R_2}{R_{Max}} \right)^{L + \frac{3}{2}} \leq \overline{C} h \text{ and } \left(\frac{\Lambda_{Min}}{\Lambda_2} \right)^{2N+3} \leq \overline{C}^* h,$$

where $\overline{C}$ and $\overline{C}^*$ are the bounded constants independent of h, L, N and u.

On the basis of Corollary 12.1, we can choose the total numbers of particular solutions as

$$L + 1 = \frac{\ln h + \ln \overline{C}}{\ln(R_2/R_{Max})} - \frac{1}{2}, \tag{12.18}$$

$$N + 1 = \frac{\ln h + \ln \overline{C}^*}{2\ln(\Lambda_{Min}/\Lambda_2)} - \frac{1}{2}. \tag{12.19}$$

Then when $h \to 0$,

$$L + 1 = O(|\ln h|) \quad \text{and} \quad N + 1 = O(|\ln h|).$$

This shows high efficiency by using the combination (12.14) with the coupling relations (12.18) and (12.19) because only a few terms of particular solutions are required for matching a very small h, where h is the maximal boundary length of quasiuniform elements in $\widehat{S}_3^h$.

Since the constants $\overline{C}$ and $\overline{C}^*$ in (12.18) and (12.19) are unknown, we shall develop an empirical choice for the total numbers, $L + 1$ and $N + 1$. Denote $L_h + 1$ and $N_h + 1$ as (12.18) and (12.19) with respect to a fixed h. We can obtain from (12.18) and (12.19) by following Chapter 4:

$$L_{h_1} + 1 = (L_h + 1) + \frac{\ln(h_1/h)}{\ln(R_2/R_{Max})}, \tag{12.20}$$

$$N_{h_1} + 1 = (N_h + 1) + \frac{\ln(h_1/h)}{2\ln(\Lambda_{Min}/\Lambda_2)}. \tag{12.21}$$

Consequently, if one group of $L_h + 1, N_h + 1$ and h is found to be good matching for the combination (12.14) by numerical experiments, other good groups of $L_{h_1} + 1, N_{h_1} + 1$ and $h_1(< h)$ can be, directly from (12.20) and (12.21), anticipated to be also good matching.

For example, the radii R_{Max} and Λ_{Min} in Figure 12.3 are chosen as

$$R_{Max} = \Lambda_{Min} = 1,$$

when $R_1^* = 1/2$ and $R_2^* = 2$, i.e., $R_1^*/R_{Max} = 1/2$ and $R_2^*/\Lambda_{Min} = 2$, we can obtain the following useful relations from (12.20) and (12.21):

$$L_{h/2} + 1 \approx (L_h + 1) + 1, \tag{12.22}$$

$$N_{h/4} + 1 \approx (N_h + 1) + 1, \tag{12.23}$$

by $R_2 \to R_1^*$ and $\Lambda_2 \to R_2^*$. In other words, when h decreases to its half, almost one more term of particular solutions is added to the expansions (12.13) on S_1. More significantly, when h decreases to its fourth, almost one more term of particular solutions is added in those on S_2. In fact, we only need two or three terms in S_2 for engineering application (see Section 12.4). This clearly shows how little calculation work and storage space are required for problems of the unbounded domains if the nonconforming combination and the coupling strategy are used!

12.3 PROOF OF THEOREM 12.1

First, let us give several auxiliary theorems and lemmas by following the proofs in Chapter 4.

Theorem 12.2 *There exists a bounded constant C independent of L, N, h and u such that*

$$\|u - u_h^*\|_1 \leq C \left\{ \inf_{v \in V_h} \|u - v\|_1 + \right.$$

$$+ \sum_{i=1}^{2} \left\{ \sup_{w_h \in V_h^0} \left| \int_{l_{R_i^*}} \frac{\partial u}{\partial \nu} w_h^- \, dl - \int_{\hat{l}_{R_i^*}} \frac{\partial u}{\partial \nu} w_h^- \, dl \right| \Big/ \|w_h\|_1 \right.$$

$$+ \sup_{w_h \in V_h^0} \left[\int_{l_{R^*}} \left(\frac{\partial u}{\partial \nu} \right)^2 dl \right]^{\frac{1}{2}} \left[\int_{l_{R^*}} (w_h^+ - w_h^-)^2 dl \right]^{\frac{1}{2}} \Big/ \|w_h\|_1 \right\} \right\},$$

where ν is the outward normal on $\hat{l}_{R_i^*}$ (or $l_{R_i^*}$), h is the maximal boundary length of triangular elements on $\hat{S}_3^h$, and $\hat{l}_{R_i^*}$ $(i = 1, 2)$ are the circular arcs in Figure 12.2:

$$l_{R_1^*} \ : \ r = R_1^* \ \text{and} \ 0 \leq \theta \leq \pi, \ \hat{l}_{R_1^*} \approx l_{R_1^*},$$

$$l_{R_2^*} \ : \ \rho = R_2^* \ \text{and} \ 0 \leq \phi \leq \pi/2, \ \hat{l}_{R_2^*} \approx l_{R_2^*},$$

and the notations are

$$w_h^- \ = \ w_h \big|_{\hat{S}_3^h} = v_1^{(3)} \ \text{in} \ \hat{S}_3^h,$$

$$w_h^+ \ = \ w_h \big|_{S_1 \cup S_2} = \begin{cases} v^{(1)} \ \text{in} \ S_1, \\ v^{(2)} \ \text{in} \ S_2. \end{cases}$$

Lemma 12.1 *For the admissible function (12.10), there exist the bounds*

$$\left[\int_{l_{R_1^*}} (w_h^+ - w_h^-)^2 dl \right]^{\frac{1}{2}} \leq C \left\{ h^{\frac{3}{2}} + \left(L + \frac{1}{2} \right)^2 h^2 \right\} \|w_h\|_1,$$

$$\left[\int_{l_{R_2^*}} (w_h^+ - w_h^-)^2 dl \right]^{\frac{1}{2}} \leq C \left\{ h^{\frac{3}{2}} + (2N+1)^2 h^2 \right\} \|w_h\|_1.$$

Lemma 12.2 *Let (12.16) and $u \in H^2(S_3)$ be given. Also, suppose that the triangular elements in $\widehat{S}_3^h$ are quasiuniform. Then an auxiliary function $\overline{W} \in V_h$ can be found from Chapter 4 such that*

$$\|u - \overline{W}_h\|_1 \leq C \left(h + \|\Sigma_1\|_{1,S_{R_2}} + \|\Sigma_2\|_{1,S_\Lambda^\infty} \right),$$

where S_{R_2} is the domain $(r \leq R_2, 0 \leq \theta \leq \pi)$, $S_{\Lambda_2}^\infty$ is the exterior domain $(\rho \geq \Lambda_2, 0 \leq \theta \leq \pi/2)$, and Σ_1 and Σ_2 are the remainders

$$\Sigma_1 = \sum_{l=L+1}^{\infty} a_l \frac{I_{l+\frac{1}{2}}(r)}{I_{l+\frac{1}{2}}(R_{Max})} \sin\left(l + \frac{1}{2} \right) \theta,$$

$$\Sigma_2 = \sum_{l=N+1}^{\infty} c_l \frac{K_{2l+1}(\rho)}{K_{2l+1}(\Lambda_{Min})} \sin(2l+1)\phi.$$

Theorem 12.3 *Let all the conditions in Lemma 12.2 be given, then*

$$\|u - u_h^*\|_1 \leq C \left\{ h + \|\Sigma_1\|_{1,S_{R_2}} + \|\Sigma_2\|_{1,S_\Lambda^\infty} + \left(L + \frac{1}{2} \right)^2 h^2 + (2N+1)^2 h^2 \right\}.$$

It has been found (see Li (1989c, 1990)) that

$$\|\Sigma_1\|_{1,S_{R_2}} \leq C \left(\frac{R_2}{R_{Max}} \right)^{L+\frac{3}{2}}. \tag{12.24}$$

So, our attention is paid to obtaining similar bounds for $\|\Sigma_2\|_{1,S_{\Lambda_2}^\infty}$. First, we prove the following:

Lemma 12.3 *The Hankel functions (12.9) for a purely imaginary argument have the bounds*

$$\frac{K_{2l+1}(\Lambda_2)}{K_{2l+1}(\Lambda_{Min})} \leq C \left(\frac{\Lambda_{Min}}{\Lambda_2} \right)^{2l+1}, \tag{12.25}$$

where $l \geq 1$ and $\Lambda_{Min} < \Lambda_2 < R_2^*$.

Proof. The Hankel functions can also be defined by Watson (1944, p. 185):

$$K_\mu(xz) = \frac{\Gamma(\mu + \frac{1}{2})(2z)^\mu}{x^\mu \Gamma\left(\frac{1}{2}\right)} \int_0^\infty \frac{\cos x\xi d\xi}{(x^2 + z^2)^{\mu + \frac{1}{2}}}.$$

We then obtain

$$\frac{K_{2l+1}(\Lambda_2)}{K_{2l+1}(\Lambda_{Min})} = \left(\frac{\Lambda_{Min}}{\Lambda_2}\right)^{2l+1} \delta_{2l+1},$$

where the notations are

$$\delta_{2l+1} = F_{2l+\frac{3}{2}}(\Lambda_{Min}, x) \Big/ F_{2l+\frac{3}{2}}(\Lambda_{Min}, 1),$$

with $x = \Lambda_2/\Lambda_{Min} > 1$, and

$$F_\mu(z, x) = \int_0^\infty \frac{\cos x\xi}{(x^2 + z^2)^\mu} d\xi.$$

Clearly, the desired results (12.25) are obtained if we can show that the ratio δ_{2l+1} is a bounded constant independent of ℓ.

Since $F_\mu(z, x) > 0$ and $\partial F\mu(z, x)/\partial\mu < 0$ for $\mu \geq 1$, we have

$$\delta_{2l+1} \leq F_{2l+1}(\Lambda_{Min}, x) \Big/ F_{2l+2}(\Lambda_{Min}, 1). \tag{12.26}$$

For integers n, the values of $F_n(\Lambda_{Min}, x)$ are found in Watson (1944, p. 188)

$$F_n(\Lambda_{Min}, x) = \frac{\pi e^{-\Lambda_{Min}x}}{(2\Lambda_{Min})^{2n-1}} \frac{1}{(n-1)!} \sum_{m=0}^{n-1} \frac{(2\Lambda_{Min}x)^m(2n - m - 1)}{(n - m - 1)!}. \tag{12.27}$$

On the basis of (12.27) and the Stirling formula (see Courant and Hilbert (1953)):

$$n! = \left(\frac{n}{e}\right)^n \sqrt{2\pi n}\left(1 + o\left(\frac{1}{n^{1/5}}\right)\right),$$

we can prove the following inequalities:

$$F_{2l+1}(\Lambda_{min}, x) \leq C\left(\frac{2l+1}{\pi}\right)^{\frac{1}{2}} \Big/ \Lambda_{Min}^{2l+1}, \tag{12.28}$$

$$F_{2l+2}(\Lambda_{min}, 1) \leq C^*(2l+2)^{\frac{1}{2}} \Big/ \Lambda_{Min}^{2l+3}, \tag{12.29}$$

where C^* is a positive constant independent of l. Therefore, the use of (12.26) and (12.28) and (12.29) yields

$$\delta_{2l+1} \leq C,$$

where C is also a bounded constant independent of l. This completes the proof of Lemma 12.3. ∎

Lemma 12.4 *Let*

$$\Lambda_{Min} < \Lambda_2 < R_2^*, \tag{12.30}$$

then

$$\|\Sigma_2\|_{1,S_{\Lambda_2}^\infty} \leq C \left(\frac{\Lambda_{Min}}{\Lambda_2}\right)^{2N+3}. \tag{12.31}$$

Proof. By using the derivative formulae (Watson (1944)),

$$\frac{d}{dz}K_\mu(z) = -\frac{\mu}{z}K_\mu(z) - K_{\mu-1}(z),$$

$$\frac{dK_\mu(z)}{d\mu} = \int_0^\infty \eta e^{-z\cosh\eta}\sinh(\mu\eta)d\eta > 0,$$

we obtain

$$\left|\frac{d}{dz}K_\mu(z)\right| \leq \left(\frac{\mu}{z}+1\right)|K_\mu(z)|, \tag{12.32}$$

for $\mu \geq 1$ and $z > 0$.

Also, we find from (12.3)–(12.6) and Green's formula:

$$\|\Sigma_2\|_{1,S_{\Lambda_2}^\infty}^2 = \iint_{S_{\Lambda_2}^\infty}[\Sigma_2^2 + (\nabla\Sigma_2)^2]ds$$

$$= \int_{\partial S_{\Lambda_2}^\infty}\Sigma_2\frac{\partial\Sigma_2}{\partial\nu}dl = -\int_0^{\pi/2}\Sigma_2\frac{\partial\Sigma_2}{\partial\rho}\bigg|_{\rho=\Lambda_2}\Lambda_2 d\phi.$$

Consequently, we have from (12.32), Lemma 12.3, and $c_l = O(1/l)$,

$$\|\Sigma_2\|_{1,S_{\Lambda_2}^\infty}^2 \leq \frac{\pi\Lambda_2}{4}\sum_{l=N+1}^\infty c_l^2\left[1+\frac{(2l+1)}{\Lambda_2}\right]\left[\frac{K_{2l+1}(\Lambda_2)}{K_{2l+1}(\Lambda_{Min})}\right]^2 \leq C\sum_{l=N+1}^\infty\left(\frac{\Lambda_{Min}}{\Lambda_2}\right)^{2(2l+1)}.$$

By noting the assumption (12.30), the desired result (12.31) follows immediately.
∎

Finally, the error bounds (12.17) can be obtained from Theorem 12.3, Lemma 12.4 and (12.24). This completes all proofs of Theorem 12.1. ∎

12.4 NUMERICAL EXPERIMENTS

In Figure 12.2, let M denote the number of elements along the section $\overline{AC}$ (e.g., $M = 2$ in Figure 12.2). Also, let M and $3M$ be the numbers of elements along $\overline{AB}$ and the semicircle $C\widehat{D}E(r = R_1^*)$, respectively. Then $h = O(1/M)$.

The radii R_{Max} and Λ_{Min} can be chosen as $R_{Max} = \Lambda_{Min} = 1$. In calculation, we also choose the radii $R_1^* = \frac{1}{2}$ and $R_2^* = 2$. First, let $M = 3$ in a finer division than that in Figure 12.2. Error norms, condition numbers and approximate coefficients have been provided for different matches of the numbers $(L + 1)$ and $(N + 1)$. It has been observed from the trial combination given in Li (1989b, 1990) that the match of $L + 1 = 4$ and $N + 1 = 2$ is good for obtaining small error norms and small condition numbers. We also notice that when the total numbers, $(L + 1)$ and $(N + 1)$, of particular solutions are rather large, the condition numbers will be tremendously large. Therefore, useful total numbers should be small, such as

$$L + 1 \leq 8 \quad \text{and} \quad N + 1 \leq 4.$$

Next, we increase the number M, i.e., decrease h. Since

$$R_1^* \big/ R_{Max} = \frac{1}{2}, \quad R_2^* \big/ \Lambda_{Min} = 2,$$

the formulae (12.22) and (12.23) can be applied to find good matches of M, $L + 1$ and $N + 1$.

A good match, $L + 1 = 4$ and $N + 1 = 2$ for $M = 3$, has been found by numerical trials (see Li (1989b)). Therefore, $L + 1 = 5$ for $M = 6$, $L + 1 = 6$ for $M = 12$, and $N + 1 = 3$ for $M = 12$ should be also good matches, on the basis of (12.22) and (12.23). By using such good matches, calculated results have been provided in Tables 12.1 and 12.2, where

$$Cond.(\mathbf{T}) = \frac{\lambda_{Max}(\mathbf{T})}{\lambda_{Min}(\mathbf{T})},$$

M	$L+1$	$N+1$	$\|\varepsilon\|_{\infty,S}$	$\|\varepsilon\|_{0,S}$	$\|\varepsilon\|_1$	Cond.
2	3	2	0.4160×10^{-1}	0.2517×10^{-1}	0.1849	131
3	4	2	0.2165×10^{-1}	0.1151×10^{-1}	0.1258	157
4	4	2	0.1312×10^{-1}	0.6663×10^{-2}	0.9677×10^{-1}	177
6	5	3	0.6469×10^{-2}	0.3065×10^{-2}	0.6607×10^{-1}	1748
8	5	3	0.3823×10^{-2}	0.1668×10^{-2}	0.4850×10^{-1}	1943
12	6	3	0.1929×10^{-2}	0.8040×10^{-3}	0.3341×10^{-1}	2184

Table 12.1   The error norms and condition numbers by the nonconforming combination when changing M.

M	$L+1$	$N+1$	Coefficients a_l in $v^{(1)}$ of (12.10)					
			a_0	a_1	a_2	a_3	a_4	a_5
2	3	2	−1.1329928	−0.2214788	0.1753092			
3	4	2	−1.1327693	−0.2105333	0.1763861	0.0640194		
4	4	2	−1.1332042	−0.2072779	0.1772435	0.0578327		
6	5	3	−1.1337013	−0.2051583	0.1787996	0.0528620	0.0285947	
8	5	3	−1.1339437	−0.2044463	0.1792901	0.0514486	0.0226290	
12	6	3	−1.1341469	−0.2039440	0.1797691	0.0505960	0.0181092	0.0127454
True Coefs.			−1.1343218	−0.2035664	0.1802645	0.0498517	0.0144863	0.0073170

M	$L+1$	$N+1$	Coefficients c_l in $v^{(2)}$ of (12.10)		
			c_0	c_1	c_2
2	3	2	0.2786721	−0.1682026	
3	4	2	0.2779603	−0.1258769	
4	4	2	0.2776018	−0.1107171	
6	5	3	0.2772499	−0.0996663	0.0307501
8	5	3	0.2771073	−0.0957463	0.0372170
12	6	3	0.2769958	−0.0929280	0.0418685
True Coefs.			0.2769005	−0.090665	0.0456200

Table 12.2   Calculated coefficients by the nonconforming combination for $M=3$, $N+1=2$, and different $L+1$, where true coefficients are from Section 3.6.3.

$\lambda_{Max}(\mathbf{T})$ and $\lambda_{Min}(\mathbf{T})$ are the maximal and minimal eigenvalues of the coefficient matrix resulting from (12.14). When the approximate solutions u_h^* have been obtained, the errors, $\varepsilon = u_h^* - u$, can be computed by means of the "true solution" u in Chapter 3. It is seen from Table 12.1 that the following asymptotic relations holds:

$$\|\varepsilon\|_1 = O(h), \qquad \|\varepsilon\|_{0,S} = O(h^2), \qquad (12.33)$$
$$\|\varepsilon\|_{\infty,S} = O(h^{2-\delta}), \qquad (12.34)$$

where δ is an arbitrary small, positive constant. Obviously, Eqs. (12.33) and (12.34) coincide with those in the linear FEM. In particular, the left side of (12.33) is consistent with the theoretical analyses in Sections 12.2 and 12.3. It is significant that only six basic functions used in S_1 and three basic functions used in S_2 are used for matching the finest division in S_3 (i.e. $M = 12$). From Table 12.2, the best coefficient values are obtained:

$$a_0 = -1.1341469, \qquad c_0 = 0.2769958,$$

with relative errors of 0.00015 and 0.00034, respectively. It is also remarkable to find a good stability of numerical solutions owing to small condition numbers in Table 12.1 while using the coupling strategies (12.22) and (12.23), i.e., (12.20) and (12.21).

It is worth pointing out that the norm (12.15) is valid for the infinite domain of $-\triangle u + u = 0$. For general unbounded domain problems, the norms should be modified by a weighted Sobolev norm since $\|v\|_{0,S}$ may be infinite (see Babuška and Aziz (1972) and Fufner (1985)).

In summary, we recommend from the above theoretical analysis and numerical experiments that the nonconforming combination and the coupling strategies in this chapter be used for solving problems on unbounded domains, provided that asymptotic solutions near the infinite point can be found.

13

WIND FLOW OVER BUILDINGS

In this chapter, we apply the penalty combination of RGM-FEM in Chapter 7 to a unbounded domain problem, which results from a real engineering problem: wind flow over building on the vast ground. Although the optimal convergence rates follow easily from Chapter 7, new, interesting results are provided in Lemma 13.1 and Theorem 13.2, to find the answers to two *puzzles* (13.17) and (13.29), the unboundedness of the used energy and the infinity of velocity of wind flow over the roof corners. The latter can be explained by the fact the real stress effects in practice rely on average velocity along a certain length; the finite number of isolate corners with unbounded velocity is insignificant.

Since the building contains multiple corners and the infinity, there exist multiple singularities in potential flow (see Habashi (1979)). Using the penalty techniques is more advantageous for coupling different methods along multiple interfaces than using the nonconforming coupling (see Chapter 7). Moreover, the computation of potential flow is the first step of the random vortex method of Chorin (1973) and Bui and Oppenheim (1987), so that the method in this chapter can also be applied to study of turbulent wind flow.

13.1   POTENTIAL FLOW

Consider the potential flow over a building:

$$\triangle \psi = 0, \tag{13.1}$$

$$\frac{\partial \psi}{\partial y} = u, \qquad \frac{\partial \psi}{\partial x} = v, \tag{13.2}$$

where ψ is the stream function, (u, v) are the components of the wind velocity u along the coordinates x and y. For the model building in two dimensions (see Figure 13.1), Eqs. (13.1) and (13.2) govern ψ on an unbounded domain S. On the ground $y = 0$ and on the surface of the building, we have

$$\mathbf{u} \cdot \mathbf{n} = -\frac{\partial \psi}{\partial \ell} = 0, \tag{13.3}$$

369

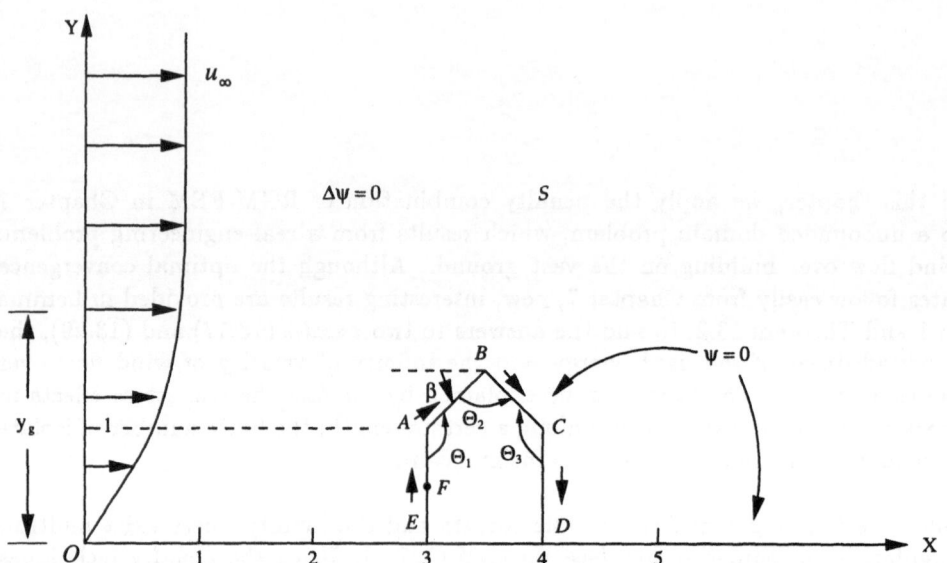

Figure 13.1   A model of buildings on the earth with an inlet wind.

where $\mathbf{u} = (u, v)$, $\mathbf{n}$ is the unit normal to ∂S, and ℓ is the tangential direction of ∂S. Then we can simply let

$$\psi = 0, \tag{13.4}$$

on $y = 0$ and on the building surface, because adding a constant to ψ does not change u and v. Also, the wind profile is supposed to be known (see Summers, Hanson, and Wilson (1985)):

$$u(y) = \begin{cases} u_\infty \left(\dfrac{y}{y_g}\right)^\alpha, & \text{for } y \leq y_g, \\ u_\infty, & \text{for } y \geq y_g, \end{cases} \tag{13.5}$$

where u_∞ is the velocity at the infinity, and α and y_g are positive constants (see Figure 13.1). We then obtain the boundary condition at $x = 0$ (a point of the upstream far from the building):

$$\psi = \begin{cases} \dfrac{u_\infty}{\alpha+1} y \left(\dfrac{y}{y_g}\right)^\alpha, & \text{for } y \leq y_g, \\ u_\infty y - C_0, & \text{for } y \geq y_g, \end{cases} \tag{13.6}$$

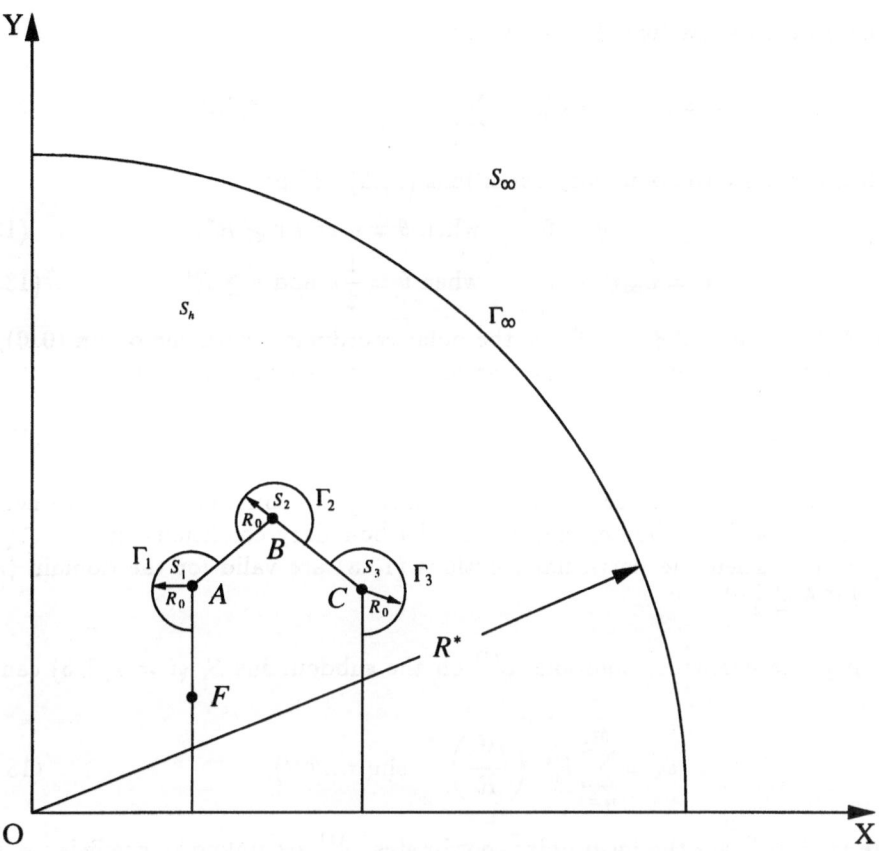

Figure 13.2 The division of the solution domain.

where the constant $C_0 = [\alpha/(\alpha+1)]y_g u_\infty$.

We note that in this problem, there exist three angular points and an infinity. Naturally, we divided the solution domain S into five subdomains (Figure 13.2)

$$S = S_\infty \cup S_h \cup S_1 \cup S_2 \cup S_3, \tag{13.7}$$

where S_1, S_2 and S_3 are the sectors with the same radius R_0, including the angular points A, B and C, respectively, S_∞ is an unbounded subdomain ($r \geq R^*$, $0 \leq \theta \leq \frac{1}{2}\pi$), and S_h is the rest of S.

In order to apply the combined methods, we have to find asymptotic expansions near the angular points and the infinity. By the method of separation of variables, we can obtain the particular solutions in S_∞:

$$\psi_\infty = u_\infty y - \frac{2}{\pi} C_0 \theta + \sum_{n=1}^{L} a_n \left(\frac{r}{R_\infty} \right)^{-2n} \sin 2n\theta, \qquad (13.8)$$

which also satisfy the boundary conditions (13.3)–(13.5):

$$\psi = 0, \qquad \text{when } \theta = 0 \text{ and } r \geq R^*, \qquad (13.9)$$

$$\psi = u_\infty y - C_0, \qquad \text{when } \theta = \frac{1}{2}\pi \text{ and } r \geq R^*, \qquad (13.10)$$

with $R^* \geq y_g$. In (13.8), (r, θ) are the polar coordinates with the origin $(0,0)$, a_n are unknown coefficients, and R_∞ a scale factor which can be chosen as

$$R_\infty = \text{Max} \left\{ y_g, \max_{\text{on the building}} r \right\} < R^*. \qquad (13.11)$$

Evidently, the particular solutions $\{r^{-2n} \sin 2n\theta\}$ are complete for ψ governed by (13.1) in S_∞ with the homogeneous Dirichlet boundary conditions on $\partial S_\infty \cap (x = 0 \cup y = 0)$. Then the particular solutions (13.8) are valid for the domain $(r \geq R_\infty, \ 0 \leq \theta \leq \frac{1}{2}\pi)$.

Similarly, the particular solutions $\psi^{(i)}$ on the subdomains S_i $(i = 1, 2, 3)$ can be obtained:

$$\psi_i = \sum_{n=1}^{M_i} b_n^{(i)} \left(\frac{r^{(i)}}{R} \right)^{\sigma_i n} \sin(\sigma_i n \theta^{(i)}), \qquad (13.12)$$

where $(r^{(i)}, \theta^{(i)})$ are the local polar coordinates, $b_n^{(i)}$ are unknown coefficients, and the constants

$$\sigma_i = \frac{\pi}{\theta_M^{(i)}}, \quad \theta_M^{(i)} = 2\pi - \Theta_i. \qquad (13.13)$$

Here Θ_i are the corner angles of the building. In the denominator of (13.12), a suitable radius $R \ (> R_0)$ also serves as a scale factor.

Once the asymptotic solutions on S_1, S_2, S_3 and S_∞ are found, the penalty combination below can be easily developed.

13.2   PENALTY COMBINATION

In this section, we give the algorithm of the penalty combination. We also derive a quasi-optimal convergence rates of stream functions and error bounds for approximate wind velocity and approximate average shear stress on the building roof.

Based on the analysis of asymptotic solutions, the RGM is used in $S_i (i = 1, 2, 3)$ and S_∞, and the traditional FEM is used in the subdomain S_h, where $\psi \in H^2(S_h)$. We choose the admissible functions as

$$
\hat{\psi} = \begin{cases}
\psi_\infty = u_\infty y - \frac{2}{\pi} C_0 \theta + \sum\limits_{n=1}^{L} a_n \left(\frac{r}{R_\infty} \right)^{-2n} \sin 2n\theta, & \text{in } S_\infty, \\
\psi_i = \sum\limits_{n=1}^{M_i} b_n^{(i)} \left(\frac{r^{(i)}}{R} \right)^{\sigma_i n} \sin(\sigma_i n \theta^{(i)}), & \text{in } S_i, \ i = 1, 2, 3, \\
\psi_h^1, & \text{in } S_h,
\end{cases} \tag{13.14}
$$

where ψ_h^1 are piecewise linear interpolations functions on S_h. Denote the common boundaries, $\Gamma_i = S_i \cap S_h$ and $\Gamma_\infty = S_\infty \cap S_h$, with the element nodes $(r_j^{(i)}, \theta_k^{(i)})$ and (r_j, θ_k), respectively.

Since there exist four coupling boundaries, we will choose the penalty combination in Chapter 7. Denote by V_h the space of all functions $\hat{\psi}$ in (13.14) satisfying (13.4) and (13.6), and by V_h^0 another space of the function

$$
\hat{\psi} = \begin{cases}
\psi_\infty = \sum\limits_{n=1}^{L} a_n \left(\frac{r}{R_\infty} \right)^{-2n} \sin 2n\theta, & \text{in } S_\infty, \\
\psi_i = \sum\limits_{n=1}^{M_i} b_n^{(i)} \left(\frac{r^{(i)}}{R} \right)^{\sigma_i n} \sin(\sigma_i n \theta^{(i)}), & \text{in } S_i, \ i = 1, 2, 3, \\
\psi_h^1, & \text{in } S_h,
\end{cases}
$$

where ψ_h^1 are piecewise linear interpolations satisfying (13.4) and $\psi|_{x=0} = 0$. Therefore, the penalty combination is to seek an approximate solution $\tilde{\psi}_h \in V_h$ such that

$$
B(\tilde{\psi}_h, \psi) = 0, \quad \forall \psi \in V_h^0, \tag{13.15}
$$

where the bilinear form is

$$
B(\psi, \phi) = \iint_{S_\infty} \nabla \psi \, \nabla \phi \, ds + \iint_{\hat{S}_h} \nabla \psi \, \nabla \phi \, ds + \sum_{i=1}^{3} \iint_{S_i} \nabla \psi \, \nabla \phi \, ds + \tag{13.16}
$$

$$
+ \sum_{i=1}^{3} \frac{P_c}{h^2} \int_{\Gamma_i} (\psi_i - \psi_h^1)(\phi_i - \phi_h^1) d\ell + \frac{P_c}{h^2} \int_{\Gamma_\infty} (\psi_\infty - \psi_h^1)(\phi_\infty - \phi_h^1) d\ell.
$$

In (13.16), $\hat{S}_h$ is the triangulation domain of S_h so that $\hat{S}_h \approx S_h$.

A difficulty here is how to express the infinity condition for the stream function $\hat{\psi}$, because the expressions in (13.14) are unbound:

$$\psi\Big|_{\substack{r \longrightarrow \infty \\ \theta = \frac{\pi}{2}}} = \infty. \tag{13.17}$$

However, we will prove the following proposition.

Proposition 13.1 *For $\psi \in V_h$ and $\psi \in V_h^0$, the bilinear forms in (13.15) and (13.16) are still bounded.*

Proof: It suffices to show for $\psi \in V_h$ and $\phi \in V_h^0$,

$$\iint_{S_\infty} \nabla\psi \, \nabla\phi \, ds < C, \tag{13.18}$$

where C is a bounded constant, since only S_∞ is the unbounded domain. In fact, we have from the Green Theorem

$$\iint_{S_\infty} \nabla\psi \, \nabla\phi \, ds = \int_{\partial S_\infty} \frac{\partial\psi}{\partial\nu}\phi \, d\ell \tag{13.19}$$

$$= -\int_0^{\pi/2} \left(\frac{\partial\psi}{\partial r}\phi r\right)\Big|_{r=R^*} d\theta + \lim_{r \longrightarrow \infty} \int_0^{\pi/2} \frac{\partial\psi}{\partial r}\phi r \, d\theta,$$

by noting that the function, $\phi \in V_h^0$, satisfies the homogeneous conditions $\phi|_{\theta=0} = \phi|_{\theta=\pi/2} = 0$ on ∂S_∞.

Denote $\phi = (r/R_\infty)^{-2n}\sin 2n\theta$ and substitute $\psi = \psi_\infty$ of (13.14) into the following integral:

$$I = \int_0^{\pi/2} \frac{\partial\psi_\infty}{\partial r}\phi r \, d\theta \tag{13.20}$$

$$= \int_0^{\pi/2} \left[u_\infty \sin\theta + \sum_{n=1}^{L} a_n(-2n)\left(\frac{r}{R_\infty}\right)^{-2n}\frac{\sin 2n\theta}{r}\right] \times \left(\frac{r}{R_\infty}\right)^{-2n}\sin 2n\theta r \, d\theta$$

$$= -\frac{1}{2}\pi n a_n\left(\frac{r}{R_\infty}\right)^{-4n} + u_\infty r\left(\frac{r}{R_\infty}\right)^{-2n}\int_0^{\pi/2} \sin\theta \sin 2n\theta \, d\theta.$$

Since $n \geq 1$, $\lim\limits_{r \longrightarrow \infty} I = 0$. The inequality (13.18) holds from (13.19) and (13.20). ∎

The difficulty (13.17) can be bypassed since the useful velocities (13.2) are computed from the derivatives of ψ, and since the numerical magnitude involving in the combination, such as (13.15) and (13.16) as well as (13.18), is still bounded.

To give error bounds of u and v, we define a norm

$$\|\psi\|_1 = \left\{ \sum_{i=1}^{3} \|\psi\|_{1,S_i}^3 + \|\psi\|_{1,\widehat{S}_h}^2 + |\psi|_{1,S_\infty}^2 \right\}^{1/2}, \tag{13.21}$$

where $\|\psi\|_{1,S_i}$, are the Sobolev norms. It should be noted that on the unbounded domain S_∞, the semi-norm in (13.21) is defined only with derivatives

$$|\psi|_{1,S_\infty} = \left\{ \iint_{S_\infty} (\psi_x^2 + \psi_y^2) ds \right\}^{1/2}. \tag{13.22}$$

The norm definitions, (13.21) and (13.22), are plausible since only the velocities u and v (i.e., the derivatives of ψ) are of our concern in practical application.

13.3 ERROR ANALYSIS

We obtain the following theorem from Chapter 7.

Theorem 13.1 *Let the radii be*

$$R_\infty < R^*, \quad R_0 < R, \tag{13.23}$$

and the coupling relations satisfy

$$\left(\frac{R_\infty}{R^*}\right)^{2L} \le Ch, \quad \left(\frac{R_0}{R}\right)^{\sigma_k M_k} \le Ch, \ k = 1, 2, 3, \tag{13.24}$$

where C is a bounded constant, independent of L, M_k, h and ψ. Then, a quasi-optimal convergence rate of numerical solutions by the penalty combination can be achieved:

$$\|\widehat{\psi}_h - \psi\|_1 = O(h), \tag{13.25}$$

where $\widehat{\psi}_h$ and ψ are the approximate and exact solutions, and h is the maximal boundary length of quasiuniform triangular elements in S_h.

Denote by $\widehat{u}_h$ and $\widehat{v}_h$ the approximate velocities from $\widetilde{\psi}_h$. We have the following corollary from (13.2) and Theorem 13.1.

Corollary 13.1 *Let all conditions in Theorem 13.1 hold. Then there exist optimal convergence rates for $\tilde{u}_h$ and $\tilde{v}_h$ on the unbounded solution domain S:*

$$\|\tilde{u}_h - u\|_{0,S} = O(h), \quad \|\tilde{v}_h - v\|_{0,S} = O(h), \tag{13.26}$$

where the norm $\|w\|_{0,S} = \left(\iint_S w^2 ds \right)^{1/2}$.

Below we consider the singularities of solutions. Suppose that the angles Θ_i of building corners with *convex figuration* are

$$0 < \Theta_i < \pi. \tag{13.27}$$

Note that the building corners with *concave figuration* are not singular points because near them the derivatives of ψ are bounded. Then from (13.13), the constants σ_k satisfy

$$\frac{1}{2} < \sigma_k < 1. \tag{13.28}$$

Consequently, the point velocity of wind flow at the corners A, B and C in Figure 13.2 will be infinite:

$$|\mathbf{v}_k|_{A,B \text{ or } C} = \left| \frac{\partial \psi_k}{r^{(k)} \partial \theta^{(k)}} \right|_{A,B \text{ or } C} = O([r^{(k)}]^{\sigma_k - 1}) \longrightarrow \infty. \tag{13.29}$$

This paradox seems to conflict with our intuition. However, the ideal point velocity can never be measured. The velocity (or stress) unboundedness at only a few solitary points may not be serious. In fact, the average wind velocity $\bar{v}$, which is bounded, will play an important role. Hence we define an average velocity of wind along the building edge $\overline{AF}$ (see Figure 13.1), containing the singular point A:

$$\bar{v}_{\overline{AF}} = \left\{ \frac{1}{\overline{AF}} \int_{\overline{AF}} v_\ell^2 d\ell \right\}^{1/2}, \tag{13.30}$$

where v_ℓ is the wind velocity along $\overline{AF}$. Then it follows that

$$\bar{v}_{\overline{AF}} = \frac{1}{\sqrt{\overline{AF}}} |v_\ell|_{\overline{AF}}, \quad |v_\ell|_{\overline{AF}} = \left\{ \int_{\overline{AF}} v_\ell^2 d\ell \right\}^{1/2}. \tag{13.31}$$

The value of $\bar{v}_{\overline{AF}}$ is, indeed, an upper bound

$$v_{\overline{AF}} \leq \bar{v}_{\overline{AF}} \tag{13.32}$$

of the absolute average velocity $v_{\overline{AF}}$ defined by

$$v_{\overline{AF}} = \frac{1}{\overline{AF}} \int_{\overline{AF}} |v_\ell| d\ell. \tag{13.33}$$

Eq. (13.32) can be verified by the Schwarz inequality. Then we have the following lemma.

Lemma 13.1 *Let (13.23), (13.27) and (13.28) hold. Then for $\overline{AF} > 0$, the value of $|v|_{\overline{AF}}$ is bounded. Also suppose*

$$0 < \varepsilon_0 \leq \overline{AF} \leq R_0, \tag{13.34}$$

where ε_0 is a constant independent of L, M_k and h. Then the average velocity $\bar{v}_{\overline{AF}}$ is also bounded.

Proof: Since $|v_\ell| = |\mathbf{u}| = (u^2 + v^2)^{1/2}$ from (13.3), we have from (13.14) and the bounded coefficients obtained from (13.15) and (13.16) that

$$
\begin{aligned}
|v_\ell|^2_{\overline{AF}} &= |\mathbf{u}|^2_{\overline{AF}} = \int_0^{r^{(i)}} \left[\frac{\partial \psi_i}{r^{(i)} \partial \theta^{(i)}} \bigg|_{\theta=0} \right]^2 dr^{(i)} \\
&= \int_0^{r^{(i)}} \left[\sum_{n=1}^{M_i} b_n^{(i)} \left(\frac{r^{(i)}}{R} \right)^{\sigma_i n} \frac{\sigma_i n}{r^{(i)}} \right]^2 dr^{(i)} \\
&= \sum_{m=1}^{M_i} \sum_{n=1}^{M_i} b_m^{(i)} b_n^{(i)} \frac{\sigma_i^2 mn}{\sigma_i(m+n) - 1} \left(\frac{r^{(i)}}{R} \right)^{\sigma_i(m+n)} \frac{1}{r^{(i)}}. \tag{13.35}
\end{aligned}
$$

Even when $r^{(i)} \longrightarrow 0$, (13.35) is reduced to

$$|v_\ell|^2_{\overline{AF}} \longrightarrow O\left(\frac{1}{2\sigma_i - 1} \left[r^{(i)} \right]^{2\sigma_i - 1} \right). \tag{13.36}$$

The assumptions (13.27) and (13.28) guarantee the boundedness of $|v_\ell|_{\overline{AF}}$. Therefore, the bounds of $\bar{v}_{\overline{AF}}$ with $\overline{AF} = r^{(i)}$ are obtained from (13.30) and (13.34)–(13.36). ∎

Based on Lemma 13.1, we can evaluate the shear stress along building edges (see Summers, Hanson and Wilson (1985)):

$$\bar{\tau}_{\overline{AF}} = 0.03 \mathrm{Re}^{-0.2} \rho \bar{v}^2_{\overline{AF}}, \tag{13.37}$$

where Re is the Reynold number and ρ is the density of wind flow.

The approximate shear stress is computed by

$$\tilde{\tau}_{\overline{AF}} = 0.03 \mathrm{Re}^{-0.2} \rho \tilde{v}^2_{\overline{AF}}, \tag{13.38}$$

where $\tilde{v}_{\overline{AF}}$ is given by

$$\tilde{v}_{\overline{AF}} = \left\{ \frac{1}{\overline{AF}} \int_{\overline{AF}} \tilde{v}_\ell^2 d\ell \right\}^{1/2}, \tag{13.39}$$

$\tilde{v}_\ell = (\partial \tilde{\psi}_i / (r^{(i)} \partial \theta))|_{\theta=0}$, and $\tilde{\psi}_i$ are obtained from the penalty combination (13.15) and (13.16). Then we have the following theorem.

Theorem 13.2 *Let (13.27), (13.28) and (13.34) and the conditions in Theorem 13.1 hold. Then there exist the error bounds*

$$|\tilde{\bar{\tau}}_{\overline{AF}} - \bar{\tau}_{\overline{AF}}| = O(h). \tag{13.40}$$

Proof: Let $\tilde{b}_n^{(i)}$ and $b_n^{(i)}$ denote the approximate and exact coefficients. We have from Theorem 13.1,

$$\sum_{n=1}^{\infty} (\tilde{b}_n^{(i)} - b_n^{(i)})^2 \sigma_i n \left(\frac{r^{(i)}}{R} \right)^{2\sigma_i n} \frac{1}{2} \theta_M^{(i)} = \iint_{S_i} (\varepsilon_x^2 + \varepsilon_y^2) ds \leq C h^2, \tag{13.41}$$

where $\varepsilon = \tilde{\psi}_h - \psi$, $r^{(i)} \leq R_0$ and $\tilde{b}_n^{(i)} \equiv 0$ for $n > M_i$. Now, we obtain

$$|\tilde{\bar{\tau}}_{\overline{AF}} - \bar{\tau}_{\overline{AF}}| \leq 0.03 \mathrm{Re}^{-0.2} \rho \left(\frac{1}{\overline{AF}} \int_{\overline{AF}} (\tilde{v}_\ell + v_\ell)^2 d\ell \right)^{1/2} \times$$

$$\times \left(\frac{1}{\overline{AF}} \int_{\overline{AF}} (\tilde{v}_\ell - v_\ell)^2 d\ell \right)^{1/2}. \tag{13.42}$$

It follows from the boundedness of Lemma 13.1 that

$$|\tilde{\bar{\tau}}_{\overline{AF}} - \bar{\tau}_{\overline{AF}}| \leq C \left(\frac{1}{\overline{AF}} \int_{\overline{AF}} (\tilde{v}_\ell - v_\ell)^2 d\ell \right)^{1/2}, \tag{13.43}$$

where C is a bounded constant independent of L, M_k and h.

Next, we can also see from Lemma 13.1 that

$$\frac{1}{\overline{AF}} \int_{\overline{AF}} (\tilde{v}_\ell - v_\ell)^2 d\ell = \sum_{m,n=1}^{\infty} (\tilde{b}_m^{(i)} - b_m^{(i)})(\tilde{b}_n^{(i)} - b_n^{(i)}) \times$$

$$\times (-1)^{k(m+n)} \frac{\sigma_i^2 mn}{\sigma_i(m+n) - 1} \left(\frac{r^{(i)}}{R} \right)^{\sigma_i(m+n)} \frac{1}{(r^{(i)})^2}, \tag{13.44}$$

where $k = 0$ or 1 for $\theta = 0$ or $\theta_M^{(i)}$. Since

$$\frac{\sigma_i mn}{\sigma_i(m+n)-1} \leq C\sqrt{mn}, \tag{13.45}$$

we can obtain from (13.34), (13.41) and the Schwarz inequality

$$\frac{1}{\overline{AF}} \int_{\overline{AF}} (\tilde{v}_\ell - v_\ell)^2 d\ell \leq C \left\{ \sum_{n=1}^{\infty} (\tilde{b}_n^{(i)} - b_n^{(i)})^2 \sigma_i n \left(\frac{r^{(i)}}{R} \right)^{2\sigma_i n} \frac{1}{2} \theta_M^{(i)} \right\}$$

$$= C \iint_{S_i} (\varepsilon_x^2 + \varepsilon_y^2) ds \leq Ch^2. \tag{13.46}$$

The desired result (13.40) is obtained from (13.43) and (13.46). ∎

The above analysis on the average velocity and stress grants a good precision of approximate values near the building corners by the penalty combination, bypassing the difficulty (13.17) and the paradox (13.29). Above all, the combined algorithms and the theoretical analysis in this chapter are significant for solving singularity problems.

13.4 NUMERICAL EXPERIMENTS

We consider a model building with different roof configurations by the pitch angle β (see Figure 13.1). Let $R = 1$, $R_0 = 0.33$, $R^* = 7$ and $R_\infty = \sqrt{17}$. For the boundary condition (13.5), the constants are given by $\alpha = 0.3$, $u_\infty = 1$ and $y_g = 3$.

Two triangulations in $\widehat{S}_h$ are shown in Figures 13.3 and 13.4. The stream functions $\tilde{\psi}_h$ are computed by the penalty combination (13.15) and (13.16) with $P_c = 10^6$, and the velocity field of wind flow is depicted in $\widehat{S}_h$ only. In the subdomains S_i and S_∞, the coefficients are listed in Table 13.1, from which the wind stress on the building edges can be obtained, see Li and Bui (1992b). Note that from the analysis in Chapters 7 and 9, the solutions of the penalty combinations as $P_c = 10^6$ approximate actually the solutions of the nonconforming combinations. However, the algorithms in this chapter is much simpler owing to four interfaces involved in.

For the coarse partition in Figure 13.3, a good match of

$$M_1 = M_2 = M_3 = 4, \quad L = 3,$$

Coefficients	l	The rough partition in Figure 13.3	The fine partition in Figure 13.4
$b_l^{(1)}$	1	0.4108213	0.4170790
	2	0.1529620	0.1576584
	3	0.0204069	0.0166925
	4	−0.0182768	−0.0028578
	5	−	−0.0424585
$b_l^{(2)}$	1	0.7822667	0.7718139
	2	0.0018099	0.0072472
	3	−0.0575886	−0.0313106
	4	0.0263113	0.0078526
	5	−	−0.0484961
$b_l^{(3)}$	1	0.4373974	0.4444181
	2	−0.1517123	−0.1582070
	3	0.0241213	0.0138468
	4	−0.0023637	−0.0025032
	5	−	−0.0426734
a_l	1	−1.2292486	−1.2436789
	2	0.8130880	0.9679558
	3	−1.3929712	−1.8918433
	4	−	2.8204021

Table 13.1 The calculated coefficients from the penalty combinations for $\beta = 40°$.

is found after trial calculations. Therefore, we can expect a good match of

$$M_1 = M_2 = M_3 = 5, \quad L = 4,$$

for the fine partition in Figure 13.4, where the maximal boundary length of elements in S_h decreases to its half, based on the coupling relation in Chapter 7. Obviously, all the results in the coarse and fine partitions are close to each other.

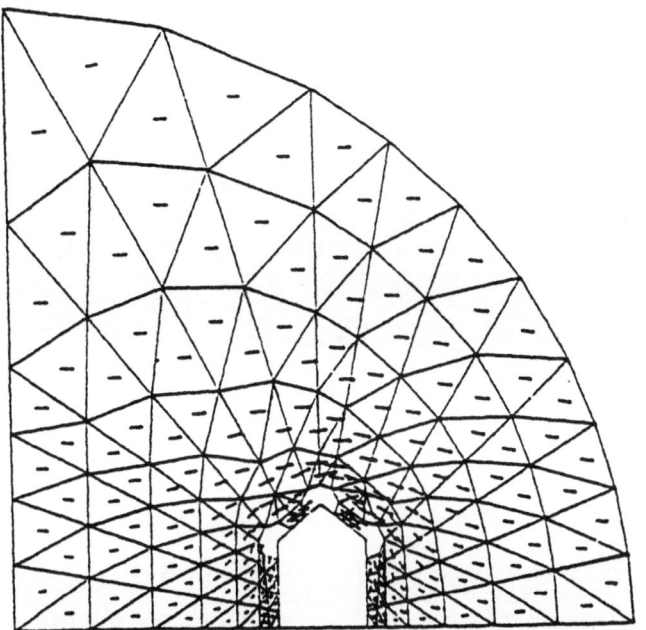

Figure 13.3 A coarse partition with velocity field for $\beta = 40°$.

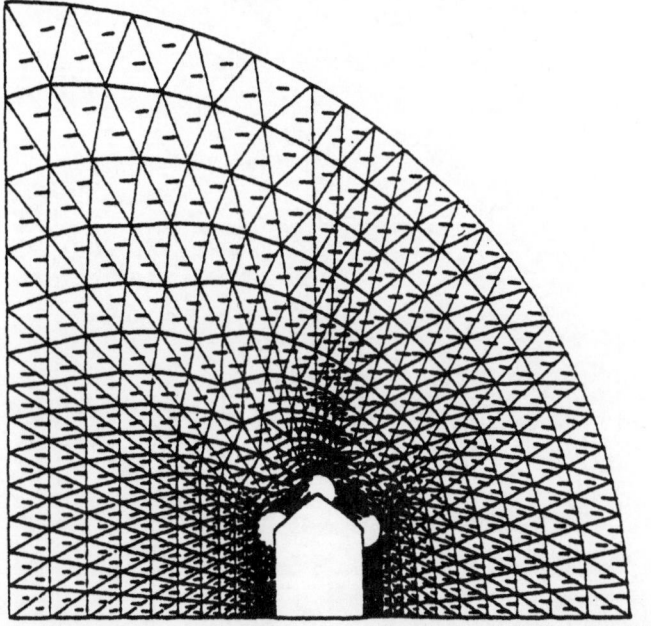

Figure 13.4 A fine partition with velocity field for $\beta = 40°$.

14

GLOBAL SUPERCONVERGENCE IN COMBINATIONS

This chapter combines the piecewise bilinear elements on quasiuniform rectangular elements with the singular functions, to seek the solutions of elliptic boundary value problems of corner singularity. The *global* superconvergence rates $O(h^{2-\delta})$ for the combinations of RGM-FEM can be achieved for different coupling strategies in Parts II and III. A little effort in computation is paid to conduct posterior interpolation of the numerical solutions u_h, only on the FEM subregion. Now, the global superconvergence rates of solution derivatives on the entire solution domain are pursued by combinations of RGM-FEMs. There exist many reports on superconvergence at specific points, see Krizek and Neittaanmaki (1984, 1987), MacKinnon and Carey (1990), Nakao (1987), Pekhlivanov, Lazarov et. al. (1992), Lin (1991a, b), Lin and Whiteman (1990), Lin and Zhu (1994), Zhu and Lin (1989), Wheeler and Whiteman (1987), and in particular in the monographs of Wahlbin (1991) and Lin and Yan (1996). The traditional superconvergence is devoted to specific nodal solutions; this chapter is devoted to the global superconvergence in the entire subdomains in combinations of RGM-FEM, based on Lin and Yan (1996).

Optimal convergence $O(h)$ of RGM-FEM is given in Chapters 4 and 7–9, superconvergence $O(h^{2-\delta})$, $0 < \delta << 1$, of combinations of the RGM-FDM in Chapter 11, to average nodal derivatives, where h is the maximal boundary length of finite elements or the maximal mesh length of difference grids. An equivalence is also explored in the last section between the global superconvergence and the superconvergence in Chapter 11.

Let the solution domain S be divided into the singular and regular subdomains S_1 and S_2. Suppose that the regular domain S_1 can be partitioned into quasiuniform rectangles. Then the bilinear elements are chosen as in the FEM in S_1. Six different combinations are also discussed in this chapter. Under the assumption $u \in H^3(S_1)$, the global superconvergence rates can be achieved as:

$$\|u_I - \widetilde{u}_h\|_{1,S_1} + \|u_L - \widetilde{u}_L\|_{1,S_2} = O(h^{2-\delta}),$$
$$\|u - \Pi_{2h}^2 \widetilde{u}_h\|_{1,S_1} + \|u_L - \widetilde{u}_L\|_{1,S_2} = O(h^{2-\delta}), \quad 0 < \delta << 1,$$

where u_I, $\widetilde{u}_h$ and $\Pi_{2h}^2 \widetilde{u}_h$ are the solution interpolant, the numerical solution and a posterior interpolant respectively. Here u_L and $\widetilde{u}_L$ are the true and approximate expansions of L singular functions.

14.1 COMBINATIONS OF RGM AND FEM

Consider the Poisson equation with the Dirichlet boundary condition

$$-\Delta u = -\left(\frac{\partial^2 u}{\partial x^2} + \frac{\partial^2 u}{\partial y^2}\right) = f(x,y), \qquad (x,y) \text{ in } S,$$

$$u|_\Gamma = 0, \qquad (x,y) \text{ on } \Gamma, \qquad (14.1)$$

where S is a polygonal domain, Γ the exterior boundary ∂S of S, and f smooth enough. Let the solution domain S be divided by a piecewise straight line Γ_0 into S_1 and S_2. The RGM is used in S_2, where a singular point may exist, and the FEM is used in S_1. For simplicity, the subdomain S_1 is again split into small quasiuniform rectangles $\square_{ij}$ only, where $\square_{ij} = \{(x,y), x_i \leq x \leq x_{i+1}, y_j \leq y \leq y_{j+1}\}$.

In S_2, we assume that the solution u can be spanned by $u = \sum_{i=1}^{\infty} a_i \psi_i$, where a_i are the expansion coefficients, and $\psi_i (i = 1, 2, \cdots, \infty)$ are complete and linearly independent basis functions. $\{\psi_i\}$ may be the chosen as analytical and singular functions. Then the admissible functions of combinations of the RGM-FEM are written as:

$$v = \begin{cases} v^- = v_1, & \text{in } S_1, \\ v^+ = f_L(\tilde{a}_i), & \text{in } S_2, \end{cases} \qquad (14.2)$$

where $\tilde{a}_i$ are unknown coefficients to be sought, and $f_L(a_i) = \sum_{i=1}^{L} a_i \psi_i$. Considering the discontinuity of solutions on Γ_0, i.e., $v^+ \neq v^-$ on Γ_0, we define another space

$$H = \{v | v \in L^2(S), v \in H^1(S_1), \text{ and } v \in H^1(S_2)\},$$

where $H^1(S_1)$ is the Sobolev space. Let $V_h(\subseteq H)$ denote a finite dimensional collection of the function v in (14.2) satisfying (14.1). The combinations of the RGM-FEM can be expressed by

$$a_h(u_h, v) = f(v), \ \forall v \in V_h, \qquad (14.3)$$

where

$$a_h(u,v) = \iint_{S_1} \nabla u \, \nabla v ds + \iint_{S_2} \nabla u \, \nabla v ds + \widehat{D}(u,v),$$

$$f(v) = \iint_S fv ds,$$

$$\widehat{D}(u,v) = \frac{P_c}{h^\sigma} \widehat{\int_{\Gamma_0}} (u^+ - u^-)(v^+ - v^-)d\ell - \widehat{\int_{\Gamma_0}} (\alpha\frac{\partial u^+}{\partial n} + \beta\frac{\Delta u^-}{\Delta n})(v^+ - v^-)d\ell$$

$$- \widehat{\int_{\Gamma_0}} (\alpha\frac{\partial v^+}{\partial n} + \beta\frac{\Delta v^-}{\Delta n})(u^+ - u^-)d\ell, \qquad (14.4)$$

where $\frac{\partial}{\partial n}$ on Γ_0 is the outside normal of ∂S_2.

In the coupling integrals (14.4), $P_c(\geq 0)$ is the penalty constant, σ is the penalty power, and $\alpha(\geq 0)$ and $\beta(\geq 0)$ satisfy $\alpha + \beta = 1$ or 0. The approximate integrals $\hat{D}(u,v)$ on Γ_0 are obtained using the integration rules in (11.12) in Chapter 11. Five combinations of (14.3) are obtained from different parameters:

(I)	Combination I:	$P_c > 0$, $\alpha = 0$ and $\beta = 1$,
(II)	Combination II:	$P_c > 0$, $\alpha = 1$ and $\beta = 0$,
(III)	Symmetric Combination:	$P_c > 0$, $\alpha = \beta = \frac{1}{2}$,
(IV)	Penalty Combination:	$P_c > 0$, $\alpha = \beta = 0$,
(V)	Simplified Hybrid Combination:	$P_c = 0$, $\alpha = 1$ and $\beta = 0$,

(VI)   Besides, a direct continuity constraint at the difference nodes Z_k on Γ_0 is given in Chapter 4 as

$$v^+(Z_k) = v^-(Z_k), \quad \forall Z_k \in \Gamma_0, \tag{14.5}$$

where the interface difference nodes Z_k are located just on Γ_0. We then obtain the nonconforming combination,

$$I(u_h^N, v) = f_h(v), \quad \forall v \in \overline{V}_h, \tag{14.6}$$

where

$$I(u,v) = \iint_{S_1} \nabla u \, \nabla v \, ds + \iint_{S_2} \nabla u \, \nabla v \, ds,$$

and $\overline{V}_h(\subset H)$ is a finite dimensional collection of (14.2) satisfying both (14.5) and (14.1).

The stability analysis is given in Parts II and III for six combinations. The global superconvergence rates of $\left\| u - \prod_{2h}^2 u_h \right\|_h = O(h^{2-\delta})$ can be achieved by all six combinations for the quasiuniform rectangles.

14.2   NONCONFORMING COMBINATION

Let S_1 be partitioned into quasiuniform rectangles in 2×2 fashion. S_Δ is a boundary layer consisting of $\square_{ij}$, to separate S_2 and S^*, where (see Figure 14.1)

$$S_1 = S^* \cup S_\Delta, \quad S^* \cap S_\Delta = \emptyset,$$

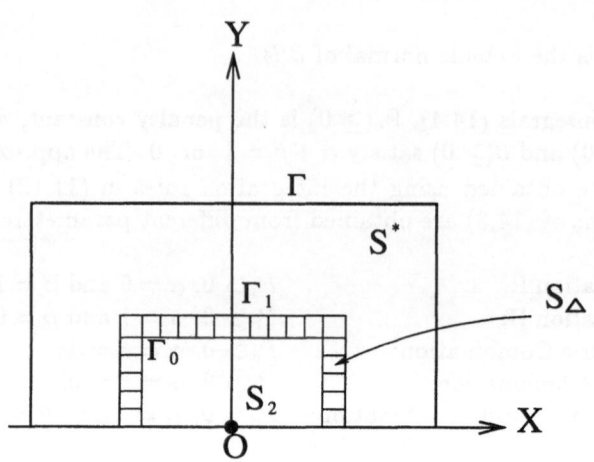

Figure 14.1   Partition of $S = S_1 \cup S_2$.

such that $S^* \subset\subset S_1$. To construct the following interpolant

$$\hat{u}_{I,L} = \begin{cases} u_I^- = u_I, & \text{in } S^*, \\ \hat{u}_I, & \text{in } S_\Delta, \\ u_L^+ = f_L(a_i), & \text{in } S_2, \end{cases} \tag{14.7}$$

where u_I is the piecewise bilinear interpolant of the true solution on rectangulation of S_1 and a_i are the true coefficients. We choose S_Δ with width of just one rectangle $\square_{ij}$, and the function $\hat{u}_I$ in S_Δ as the piecewise bilinear function with the corner values

$$\hat{u}_I(P_i) = \begin{cases} u(P_i), & \text{if } P_i \in \Gamma_1 = S^* \cap S_\Delta, \\ u_L^+(P_i), & \text{if } P_i \in \Gamma_0, \\ 0, & \text{if } P_i \in \Gamma. \end{cases} \tag{14.8}$$

Hence $\hat{u}_{I,L} \subset \overline{V}_h$. Denote

$$\|v\|_1 = (\|v\|_{1,S_1}^2 + \|v\|_{1,S_2}^2)^{1/2}.$$

We have the following basic theorem.

Theorem 14.1 *There exist the error bounds between the numerical and interpolant solutions of (14.6) and (14.7)*

$$\|u_h^N - \hat{u}_{I,L}\|_1 \leq C \sup_{w \in \overline{V}_h} \frac{1}{\|w\|_1} \left\{ \left| \iint_{S_1} \nabla(u - u_I)\, \nabla w\, ds \right| + \right. \tag{14.9}$$

$$+ \left. \left| \iint_{S_\Delta} \nabla(u_I - \hat{u}_I)\, \nabla w\, ds \right| + \left| \iint_{S_2} \nabla(u - u_L)\, \nabla w\, ds \right| + \left| \int_{\Gamma_0} \frac{\partial u}{\partial n}(w^+ - w^-)\, d\ell \right| \right\},$$

where C is a bounded constant independent of h, L, u and w.

Proof :   For the true solution u, we have

$$I(u - u_h^N, v) = \int_{\Gamma_0} \frac{\partial u}{\partial n}(v^+ - v^-)d\ell, \quad \forall v \in \overline{V}_h. \tag{14.10}$$

Let $w = u_h^N - \widehat{u}_{I,L} \in \overline{V}_h$. Then from the definition (14.7),

$$
\begin{aligned}
C_0\|w\|_1^2 \;\leq\; |w|_1^2 &= I(u - \widehat{u}_{I,L}, w) - \int \frac{\partial w}{\partial n}(w^+ - w^-)d\ell \\
&= \iint_{S^*} \nabla(u - u_I)\,\nabla w\,ds + \iint_{S_\Delta} \nabla(u - \widehat{u}_I)\,\nabla w\,ds + \\
&\quad + \iint_{S_2} \nabla(u - u_L^+)\,\nabla w\,ds - \int_{\Gamma_0} \frac{\partial u}{\partial n}(w^+ - w^-)d\ell. \tag{14.11}
\end{aligned}
$$

Since $u - \widehat{u}_I = u - u_I + u_I - \widehat{u}_I$ in S_Δ, we have

$$
\begin{aligned}
&\iint_{S^*} \nabla(u - u_I)\,\nabla w\,ds + \iint_{S_\Delta} \nabla(u - \widehat{u}_I)\,\nabla w\,ds \\
&= \iint_{S_1} \nabla(u - u_I)\,\nabla w\,ds + \iint_{S_\Delta} \nabla(u_I - \widehat{u}_I)\,\nabla w\,ds. \tag{14.12}
\end{aligned}
$$

The desired results (14.9) are obtained from (14.10)-(14.12). ∎

Below, we will derive the bounds of all terms in the right side of (14.9).

Lemma 14.1 *Let the rectangles $\Box_{ij}$ be quasiuniform, then for $w \in \overline{V}_h$*

$$\left| \iint_{S_\Delta} \nabla(u_I - \widehat{u}_I)\,\nabla w\,ds \right| \leq Ch^{-\frac{1}{2}}\|R_L\|_{0,\Gamma_0}\|w\|_1. \tag{14.13}$$

Proof :   From the Schwarz inequality,

$$\left| \iint_{S_\Delta} \nabla(u_I - \widehat{u}_I)\nabla\,wds \right| \leq |u_I - \widehat{u}_I|_{1,S_\Delta}|w|_1. \tag{14.14}$$

Denoting $\phi = u_I - \widehat{u}_I$, we obtain from the inverse estimates

$$|u_I - \widehat{u}_I|_{1,S_\Delta} = |\phi|_{1,S_\Delta} \leq Ch^{-1}\|\phi\|_{0,S_\Delta}. \tag{14.15}$$

Note that ϕ is the piecewise bilinear function on S_Δ and the fact that $\phi(Z_i) = 0$ for all element nodes $Z_i \notin \Gamma_0$ in S_Δ, the maximal values of $|\phi|$ along any vertical and horizontal lines in S_Δ are just located on Γ_0, i.e.,

$$\|\phi\|_{0,S_\Delta}^2 = \int_{S_\Delta} \phi^2 ds \leq Ch \int_{\Gamma_0} \phi^2 dl = Ch\|u_I - \widehat{u}_I\|_{0,\Gamma_0}^2. \qquad (14.16)$$

Moreover, since

$$u_I - \widehat{u}_I = u_I - \widehat{u}_L = \widehat{R}_L, \quad on \quad \Gamma_0.$$

and since $\widehat{R}_L$ is the piecewise linear interpolant of the remainder R_L on Γ_0^h, we obtain from the triangular inequality

$$\begin{aligned} \|u_I - \widehat{u}_I\|_{0,\Gamma_0} &= \|\widehat{R}_L\|_{0,\Gamma_0} \leq \|R_L\|_{0,\Gamma_0} + Ch^\varepsilon\|R_L - \widehat{R}_L\|_{\varepsilon,\Gamma_0} \\ &\leq \|R_L\|_{0,\Gamma_0} + Ch^\varepsilon\|R_L\|_{\varepsilon,\Gamma_0} \leq C\|R_L\|_{0,\Gamma_0} \end{aligned} \qquad (14.17)$$

as $\varepsilon \longrightarrow 0$. Hence, we have from (14.15)–(14.17)

$$|u_I - \widehat{u}_I|_{1,S_\Delta} = |\phi|_{1,S_\Delta} \leq Ch^{-1}h^{\frac{1}{2}}\|R_L\|_{0,\Gamma_0} = Ch^{-\frac{1}{2}}\|R_L\|_{0,\Gamma_0}. \qquad (14.18)$$

Combining (14.14) and (14.18) yields (14.13). ∎

Lemma 14.2 *Let $u \in H^3(S_1)$, then*

$$\left|\iint_{S_1} \nabla(u - u_I)\,\nabla w\,ds\right| \leq Ch^2|u|_{3,S_1}\|w\|_{1,S_1}, \quad \forall w \in \overline{V}_h. \qquad (14.19)$$

Proof : We may follow Lin and Yan (1996), p. 5, to prove (14.19), but rather provide the straightforward argument below. Denote the second order interpolant of u by

$$u_I^{(2)} = \sum_{i+j=0}^{2} \alpha_{ij}x^i y^j = u_I + \alpha_{20}x^2 + \alpha_{02}y^2.$$

By the help of the following equality shown later,

$$\iint_{\square_{ij}} \nabla(u - u_I)\,\nabla w\,ds = \iint_{\square_{ij}} \nabla(u - u_I^{(2)})\,\nabla w\,ds, \qquad (14.20)$$

we obtain

$$\left| \iint_{S_1} \nabla(u - u_I) \nabla w \, ds \right| = \left| \sum_{ij} \iint_{\square_{ij}} \nabla(u - u_I) \nabla w \, ds \right|$$

$$= \left| \sum_{ij} \iint_{\square_{ij}} \nabla(u - u_I^{(2)}) \nabla w \, ds \right| = C \sum_{ij} \left| u - u_I^{(2)} \right|_{1,\square_{ij}} |w|_{1,\square_{ij}}$$

$$\leq Ch^2 \sum_{ij} |u|_{3,\square_{ij}} |w|_{1,\square_{ij}} \leq Ch^2 |u|_{3,S_1} |w|_1.$$

This is the desired result (14.19).

Now we show (14.20). Denote

$$\square = \left\{ (\hat{x}, \hat{y}), -\frac{1}{2} \leq \hat{x} \leq \frac{1}{2}, \quad -\frac{1}{2} \leq \hat{y} \leq \frac{1}{2} \right\}.$$

Then by the linear transformations

$$\hat{x} = -\frac{1}{2} + (x - x_i)/h_i, \quad \hat{y} = -\frac{1}{2} + (y - y_j)/k_j,$$

the integrals on $\square_{ij}$ are

$$\iint_{\square_{ij}} (u - u_I^{(2)})_x w_x \, ds = \frac{h_i k_j}{h_i^2} \iint_{\square} (\hat{u} - \hat{u}_I^{(2)})_{\hat{x}} \hat{w}_{\hat{x}} \, d\hat{s}$$

$$= \frac{k_j}{h_i} \iint_{\square} (\hat{u} - \hat{u}_I - \hat{\alpha}_{20}\hat{x}^2 + \hat{\alpha}_{02}\hat{y}^2)_{\hat{x}} \hat{w}_{\hat{x}} \, d\hat{s}$$

$$= \frac{k_j}{h_i} \iint_{\square} (\hat{u} - \hat{u}_I)_{\hat{x}} \hat{w}_{\hat{x}} \, d\hat{s} - 2\hat{\alpha}_{20} \iint_{\square} \hat{x}\hat{w}_{\hat{x}} \, d\hat{s}$$

$$= \frac{k_j}{h_i} \iint_{\square} (\hat{u} - \hat{u}_I)_{\hat{x}} \hat{w}_{\hat{x}} \, d\hat{s}, \tag{14.21}$$

where $\hat{\alpha}_{20}$ and $\hat{\alpha}_{02}$ are also constant. In the last equality of (14.21), we have used the following equation for any bilinear functions $\hat{w}$ on $\square$,

$$\iint_{\square} \hat{x}\hat{w}_{\hat{x}} \, d\hat{s} = \left(\int_{-\frac{1}{2}}^{\frac{1}{2}} \hat{x} \, d\hat{x} \right) \left(\int_{-\frac{1}{2}}^{\frac{1}{2}} \hat{w}_{\hat{x}} \, d\hat{y} \right) = 0.$$

By the inverse transformation,

$$\iint_{\square} (\hat{u} - \hat{u}_I)_{\hat{x}} \hat{w}_{\hat{x}} \, d\hat{s} = \frac{h_i^2}{h_i k_j} \iint_{\square_{ij}} (u - u_I)_x w_x \, ds. \tag{14.22}$$

Combining (14.21) and (14.22) yields

$$\iint_{\square_{ij}} (u - u_I^{(2)})_x w_x ds = \iint_{\square_{ij}} (u - u_I)_x w_x ds. \tag{14.23}$$

Similarly, we have

$$\iint_{\square_{ij}} (u - u_I^{(2)})_y w_y ds = \iint_{\square_{ij}} (u - u_I)_y w_y ds. \tag{14.24}$$

Adding (14.23) and (14.24) leads to (14.20). This completes the proof of Lemma 14.2. ■

We can easily prove the following lemma.

Lemma 14.3 *Assume that there exists a bounded power $\mu(> 0)$ independent of L, h and v such that*

$$|v^+|_{\ell,\Gamma_0} \le CL^{\ell\mu}\|v^+\|_{0,\Gamma_0}, \quad \ell = 1, 2, \ \forall v \in \overline{V}_h. \tag{14.25}$$

Then

$$\left|\int_{\Gamma_0} \frac{\partial u}{\partial n}(w^+ - w^-)d\ell\right| \le C(hL^\mu)^2 \left\|\frac{\partial u}{\partial n}\right\|_{0,\Gamma_0} \|w\|_{1,S_2}.$$

Now we prove an important theorem.

Theorem 14.2 *Let $u \in H^3(S_1)$ and (14.25) be given. Then there exist the error bounds between u_h^N and $\hat{u}_{I,L}$,*

$$\|u_h^N - \hat{u}_{I,L}\|_1 \le \varepsilon_1$$
$$= C\left\{h^2|u|_{3,S_1} + |R_L|_{1,S_2} + (hL^\mu)^2 \left\|\frac{\partial u}{\partial n}\right\|_{0,\Gamma_0} + h^{-\frac{1}{2}}\|R_L\|_{0,\Gamma_0}\right\}. \tag{14.26}$$

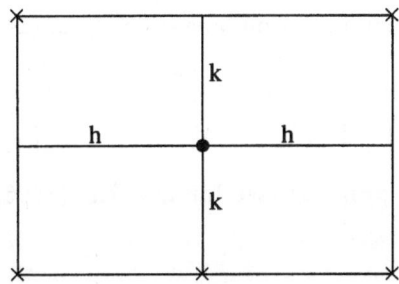

Figure 14.2 Rectangles in 2 × 2 fashion.

Proof :   We have

$$\left| \iint_{S_2} \nabla (u - u_L) \, \nabla w \, ds \right| \le \| u - u_L \|_{1, S_2} \| w \|_{1, S_2} \le \| R_L \|_{1, S_2} \| w \|_1.$$

The bounds of other terms in the right side of (14.9) are given in Lemmas 14.3–14.3. This leads to (14.26) directly. ∎

Theorem 14.2 provides the global errors between u_h^N and $\widehat{u}_{h,L}$ only, but not the true errors between u_h^N and u. The key idea in raising convergence rates is to form the posterior interpolant, $\prod_p u_h^N$, of u_h^N because the errors $\left\| u - \prod_p u_h^N \right\|_1$ can reach the global superconvergence. The solution of (14.6)

$$u_h^N = \left\{ \begin{array}{ll} (u_h^N)^-, & \text{in } S_1, \\ (u_L^N)^+, & \text{in } S_2. \end{array} \right.$$

Based on Lin's techniques in Lin and Yan (1996) for the FEM solutions $(u_h^N)^-$ in S_1, the bi-quadratic interpolant, $\Pi_{2h}^2 (u_h^N)^-$, can be formed in 2 × 2 neighbouring rectangles shown in Figure 14.2. The posterior process is made only in S_1. Denote

$$\Pi_p v = \left\{ \begin{array}{ll} \Pi_{2h}^2 v^-, & \text{in } S_1, \ \forall v \in \overline{V}_h, \\ v^+, & \text{in } S_2, \ \forall v \in \overline{V}_h. \end{array} \right. \tag{14.27}$$

We obtain the following theorem.

Theorem 14.3 *Let all conditions in Theorem 14.2 hold. Then there exist the error bounds*

$$\left\| u - \Pi_p u_h^N \right\|_1 \le C \varepsilon_1, \tag{14.28}$$

where ε_1 is given in Theorem 14.2.

Proof :   By noting the interpolation operation Π_p in (14.27) we have

$$\left\| u - \Pi_p u_h^N \right\|_1^2 \leq \left\| u - \Pi_p \widehat{u}_{h,L} \right\|_1^2 + \left\| \Pi_p (\widehat{u}_{h,L} - u_h^N) \right\|_1^2 \tag{14.29}$$

$$= \left\| u - \Pi_{2h}^2 u_I \right\|_{1,S_1}^2 + \left\| \Pi_{2h}^2 (\widehat{u}_I - u_I) \right\|_{1,S_\Delta}^2 + \left\| R_L \right\|_{1,S_2}^2 + \left\| \Pi_p (\widehat{u}_{h,L} - u_h^N) \right\|_1^2.$$

We cite the interpolation property (see Lin and Yan (1996, p. 176)).

$$\left\| \Pi_{2h}^2 v \right\|_{\ell,S_1} \leq C \|v\|_{\ell,S_1}, \qquad \forall v \in \overline{V}_h.$$

Letting $v = \widehat{u}_I - u_I$ we then obtain from (4.18)

$$\left\| \Pi_{2h}^2 (\widehat{u}_I - u_I) \right\|_{1,S_\Delta} \leq C \| \widehat{u}_I - u_I \|_{1,S_\Delta} \leq C h^{-\frac{1}{2}} \| R_L \|_{0,\Gamma_0}. \tag{14.30}$$

Since $\Pi_{2h}^2 u_I$ is the bi-quadratic interpolant of u, then we have

$$\left\| u - \Pi_{2h}^2 u_I \right\|_{1,S_1} \leq C h^2 \|u\|_{3,S_1}.$$

Also from Theorem 14.2,

$$\left\| \Pi_p (\widehat{u}_{h,L} - u_h^N) \right\|_{1,S_1} + \| R_L \|_{1,S_2} \leq C(\| \widehat{u}_{h,L} - u_h^N \|_{1,S_1} + \| R_L \|_{1,S_2})$$

$$\leq C \| \widehat{u}_{h,L} - u_h^N \|_1 \leq C \varepsilon_1. \tag{14.31}$$

Combining (14.29)–(14.31) yields the desired result (14.28).   ∎

Based on Theorems 14.2 and 14.3 we obtain the following corollary.

Corollary 14.1 *Let the conditions in Theorem 14.2 hold. Assume the expansion term L is chosen such that*

$$|R_L|_{1,S_2} = O(h^2), \quad \| R_L \|_{0,\Gamma_0} = O(h^{\frac{5}{2}}). \tag{14.32}$$

Then

$$\| u_h^N - \widehat{u}_{I,L} \|_1 = O(h^2) + O(h^2 L^{2\mu}), \quad \left\| u - \Pi_p u_h^N \right\|_1 = O(h^2) + O(h^2 L^{2\mu}).$$

Also, if such an L is so small to satisfy

$$L = O(|\ln h|), \tag{14.33}$$

then

$$\| u_h^N - \widehat{u}_{I,L} \|_1 = O(h^{2-\delta}), \quad \left\| u - \Pi_p u_h^N \right\|_1 = O(h^{2-\delta}),$$

where $0 < \delta << 1$ as $h \longrightarrow 0$.

The superconvergence in Corollary 14.1 is *global* in the entire solution domain S_1. In particular,

$$\left\| u - \Pi_{2h}^2 u_h^N \right\|_{1,S_1} = O(h^{2-\delta}).$$

In fact, the posterior quadratic interpolation Π_{2h}^2 on the solution u_h^N costs a little more computation. In Chapter 11, the superconvergence in the discrete norm $\overline{\|\cdot\|}_h$ implies that the average nodal derivatives are $O(h^{2-\delta})$; so are the maximal nodal derivatives in majority. On the other hand, the assumption $u \in H^3(S_1)$ in Theorem 14.1 is weaker than $u \in C^3(S_1)$ given in Chapter 11.

14.3   OTHER COMBINATIONS

In this section, we list the important conclusions for other combinations; their proofs may follow Section 14.2 and Chapter 11. Define the new norms:

$$\|v\|_h^* = \left(\|v\|_{1,S_1}^2 + \|v\|_{1,S_2}^2 + \frac{P_c}{h^\sigma} \overline{\|v^+ - v^-\|}_{0,\Gamma_0}^2 \right),$$

where the norms with discrete summation are assigned on Γ_0 only:

$$\overline{\|v^+ - v^-\|}_{0,\Gamma_0}^2 = \widehat{\int_{\Gamma_0}} (v^+ - v^-)^2 d\ell.$$

For the solution u_h^P by the penalty combinations, (14.3) as $\alpha = \beta = 0$, we have the following theorems without proofs (see Li, Lin and Yan (1997)).

Theorem 14.4 *Let all conditions in Theorem 14.2 hold. Then there exist the error bounds of u_h^P*

$$\|u_h^P - \widehat{u}_{I,L}\|_h^* = \varepsilon_2$$

$$\leq C \left\{ h^2 |u|_{3,S_1} + \|R_L\|_{1,S_2} + \left\| \frac{\partial u}{\partial n} \right\|_{0,\Gamma_0} \left(h^{\sigma/2} + (hL^\mu)^2 \right) + h^{-\frac{1}{2}} \|R_L\|_{0,\Gamma_0} \right\}.$$

Moreover for the posterior interpolant $\Pi_p u_h^P$,

$$\left\| u - \Pi_p u_h^P \right\|_h^* \leq C \varepsilon_2.$$

Corollary 14.2 *Let (14.32) and all condition in Theorem 14.2 hold. Then*

$$\|u - \Pi_p u_h^p\|_h^* = O(h^2) + O(h^2 L^{2\mu}) + O(h^{\frac{\sigma}{2}}).$$

Moreover, let $\sigma \geq 4$ and (14.33) be given, then

$$\|u - \Pi_p u_h^p\|_h^* = O(h^{2-\delta}), \qquad 0 < \delta << 1.$$

The penalty combination conducts the same global convergence rates as the non-conforming combination because

$$\|u - \Pi_p u_h^p\|_1 \leq \|u - \Pi_p u_h^p\|_h^* = O(h^{2-\delta}).$$

Moreover, we have

$$\overline{\|u - \Pi_p u_h^p\|}_{0,\Gamma_0} \leq C h^{\frac{\sigma}{2}} \|u - \Pi_p u_h^p\|_h^* = O(h^{2+\frac{\sigma}{2}-\delta}).$$

When $\alpha + \beta = 1$, we denote by u_h^c the three combinations. We have the following theorem without proofs (see Li, Lin and Yan (1987)).

Theorem 14.5 *Let (14.25) be given, and suppose that $\sigma \geq 1$ when $\beta > 0$, or $\sigma > 0$ and $h^\sigma L^{2\mu} \leq C$ when $\alpha > 0$. Assume $u \in H^3(S_1)$, $u \in C^2(S_1)$, and $\frac{\partial u}{\partial n} \in H^2(\Gamma_0)$. Then there exist the bounds*

$$\|u_h^c - \widehat{u}_{I,L}\|_h^* \leq \varepsilon_3$$
$$= C\left\{ h^2|u|_{3,S_1} + \|R_L\|_{1,S_2} + \left(h^{-\frac{1}{2}} + h^{\frac{\sigma}{2}-1}\right) \|R_L\|_{0,\Gamma_0} + \right.$$
$$\left. + h^{\frac{\sigma}{2}} \left\|\frac{\partial R_L}{\partial n}\right\|_{0,\Gamma_0} + (hL^\mu)^2 \left\|\frac{\partial u}{\partial n}\right\|_{0,\Gamma_0} + h^{1+\frac{\sigma}{2}} \left\|\frac{\partial u}{\partial n}\right\|_{2,\Gamma_0} + h^{1+\frac{\sigma}{2}} \|u\|_{2,\infty,S_\Delta} \right\}.$$

Moreover

$$\|u - \Pi_p u_h^c\|_h^* \leq C\varepsilon_3.$$

Corollary 14.3 *Let all conditions in Theorem 14.5 be given. Assume $\sigma \geq 4$ and (14.32) hold. Then*

$$\|u - \Pi_p u_h^c\|_h^* = O(h^2) + O(h^2 L^{2\mu}).$$

Moreover if L satisfies (14.33), then

$$\|u - \Pi_p u_h^c\|_h^* = O(h^{2-\delta}), \qquad 0 < \delta << 1.$$

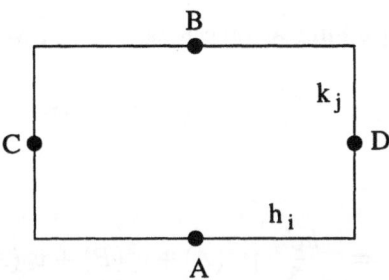

Figure 14.3 The rectangle $\square_{ij}$.

14.4 COMPARISONS

Let us explore the relation between the superconvergence in Chapter 11 and in this chapter.

In Chapter 11, the semi-norm in (11.13) is defined as

$$\overline{|v|}_{1,S_1} = \left\{ \sum_{ij} \widehat{\iint}_{\square_{ij}} \left(v_x^2 + v_y^2 \right) ds \right\}^{1/2},$$

where

$$\widehat{\iint}_{\square_{ij}} v_x^2 ds = \frac{h_i k_j}{2} \left(v_x^2(A) + v_x^2(B) \right), \qquad \widehat{\iint}_{\square_{ij}} v_y^2 ds = \frac{h_i k_j}{2} \left(v_y^2(C) + v_y^2(D) \right),$$

and A, B, C and D are the mid-points of the edges of $\square_{ij}$, shown in Figure 14.3. Then we have the following lemma.

Lemma 14.4 *Let*

$$u \in C^3(S_1), \tag{14.34}$$

and assume that the solutions $u_h \in V_h$ (or V_h^0) have the errors

$$\overline{|u - u_h|}_{1,S_1} = O(h^{2-\delta}), \qquad 0 \le \delta < 1.$$

Then the global errors between u_h and the solution interpolant u_I have the error bounds

$$|u_I - u_h|_{1,S_1} = O(h^{2-\delta}). \tag{14.35}$$

Proof : For the piecewise bilinear functions, $v \in V_h^0$, we have

$$|v|_{1,S_1}^2 = \sum_{ij} \iint_{\square_{ij}} \left(v_x^2 + v_y^2 \right) ds,$$

where

$$\iint_{\square_{ij}} v_x^2 ds = \frac{h_i k_j}{3} \left[v_x^2(A) + v_x^2(B) + v_x(A)v_x(B) \right],$$

$$\iint_{\square_{ij}} v_y^2 ds = \frac{h_i k_j}{3} \left[v_y^2(C) + v_y^2(D) + v_y(C)v_y(D) \right].$$

Since $x^2 + y^2 + xy \leq \frac{3}{2}(x^2 + y^2)$, we obtain for $v \in V_h^0$

$$\iint_{\square_{ij}} v_x^2 ds \leq \frac{h_i k_j}{2} \left[v_x^2(A) + v_x^2(B) \right] = \widehat{\iint_{\square_{ij}} v_x^2 ds}.$$

Hence

$$|v|_{1,S_1} \leq \overline{|v|}_{1,S_1}, \quad \forall v \in V_h \text{ (or } V_h^0).$$

We then have

$$|u_I - u_h|_{1,S} \leq \overline{|u_I - u_h|}_{1,S_1} = \overline{|u - u_h|}_{1,S_1} + \overline{|u - u_I|}_{1,S_1}. \tag{14.36}$$

Also,

$$\overline{|u - u_I|}_{1,S_1} = Ch^2 |u|_{3,\infty,S_1} = O(h^2). \tag{14.37}$$

The desired results (14.35) follow (14.36) and (14.37). ∎

Lemma 14.5 *Let (14.34) and*

$$|u_I - u_h|_{1,S_1} = O(h^{2-\delta}), \quad 0 \leq \delta < 1$$

be given. Then

$$\overline{|u - u_h|}_{1,S_1} = O(h^{2-\delta}).$$

Proof : Similarly, since $x^2 + y^2 \leq 2(x^2 + xy + y^2)$ we have

$$\overline{|u_I - u_h|}_{1,S_1} \leq \sqrt{3} |u_I - u_h|_{1,S_1}.$$

Then

$$\overline{|u - u_h|}_{1,S_1} \leq \overline{|u - u_I|}_{1,S_1} + \overline{|u_I - u_h|}_{1,S_1} \leq O(h^2) + \sqrt{3}|u_I - u_h|_{1,S_1}$$
$$= O(h^2) + O(h^{2-\delta}) = O(h^{2-\delta}). \qquad \blacksquare$$

Lemmas 14.4 and 14.5 imply that under the assumption (14.34), the superconvergences in this chapter and in Chapter 11 are equivalent to each other. We then conclude that the order $O(h^{2-\delta})$ also holds for the solutions of RGM-FEM in this chapter, to the average nodal derivatives at the edge mid-points of $\Box_{ij}$, so it does for the maximal nodal derivatives in the majority.

On the other hand, based on Lemma 14.4, the solutions in (11.8) by the combinations of RGM-FDM in Chapter 11 also have the *global* superconvergence as Corollaries 14.1–14.3. However, the proofs in this chapter using Lin's techniques are simpler so as to be extend to other kinds of singularity problems, i.e., biharmonic equations and elastic plates with cracks. Details appear elsewhere. Also note that the assumption $u \in H^3(S_1)$ in Theorem 14.2 is weaker than (14.34).

In summary, five combinations of RGM-FEM have been developed above to provide the global superconvergence rates $O(h^{2-\delta})$, $0 < \delta << 1$ on the entire subdomains S_1 and S_2. The theoretical results have been verified by the numerical experiments in Li, Lin and Yan (1997). The superconvergence proof for the simplified hybrid method of RGM-FEM appears elsewhere. In summary, this global superconvergence is also equivalent to the superconvergence in Chapter 11, where the average nodal derivative errors have order $O(h^{2-\delta})$.

ITERATIVE SUBSTRUCTING
METHODS

In the last two chapters, we focus on parallel computation for the combinations given in this book. The combined algorithms can be easily implemented into parallel by the iterative substructing methods, based on the new imbedding techniques provided in this chapter. Note that although the parallel algorithms are developed for only the combinations of RGM-FDM, they may also applied to other kinds of combinations, such as combinations of RGM-FEM.

Iterative substructing methods are applied to implement the combined methods in parallel computing, in which the RGM is used in the subdomain where singularities exist and the FDM is used in the rest of the solution domain. The singular subdomains can be regarded not only as S_I where the mixed Neumann-Dirichlet problem is solved in the inverse preconditioning matrix, but also as the interfaces of the domain decomposition methods because only a few singular functions are needed to cope with the FDM. Both analysis and numerical experiments show that the existing domain decomposition methods can be extended to combinations of the RGM-FDM.

The main theoretical results are given in Theorems 15.1 and 15.2; the preconditioning condition number in parallel combinations will not deteriorate, compared to the existing domain decomposition methods.

15.1   INTRODUCTION

The domain decomposition methods (DDM) have been studied very actively and extensively since the pioneer work of Dryja (1982). Of a number of reports on this subject, only a few, e.g., Anderson (1989) and Mroz (1989), are concerned with solution singularities. In many distinguished works on the subject, such as in Bjorstad and Wildlund (1986), Bramble, Pasciak and Schatz (1986), Chan (1989) and Dryja (1982), the problems with the concave domains and material interfaces are chosen as examples for illustration, but unfortunately, the angular and interface singularities are still being ignored. Furthermore, the unbounded domain problems are also important to practical application. Therefore, it is necessary to deal fully with singularities in parallel computing for solving elliptic equations. Although the local refinement of Bramble, Ewing, et. al. (1992) and the p−version of the finite element methods of Babuska and Osborn (1991) can handle singularity problems, we will follow the combined methods in Parts II and III and explore new techniques to implement them in parallel.

In the singular subdomains, singular basis functions are chosen as admissible functions, and the Ritz-Galerkin method (RGM) is employed. In other subdomains, the finite element method (FEM) or the finite difference method (FDM) is employed as usual. In Chapters 4, 7 and 11, we may choose the total term of singular functions by the matching formula

$$L = O(|\ln h|), \tag{15.1}$$

where h is the maximal boundary length of quasiuniform finite elements or quasi-uniform difference grids.

Since the FEM and the FDM are already taken into account in the DDM, our attention is turned to the RGM and particularly to the coupling strategies. Since the coupling constraints and the coupling integrals are involved only along the common boundary Γ_0 between the singular subdomains and the other subdomains, Γ_0 is also chosen as the interface Γ in the capacity matrix method. The singular subdomains may be chosen as S_I where the mixed Neumann-Dirichlet problem is solved. However, since a few numbers of unknown coefficients D_ℓ are needed to cope with the discrete solutions $v_{ij} \in \Gamma_0$, we may simply choose $\{D_\ell\}$ to be included in the vector x_3 corresponding to the Schur matrix, and regard the singular subdomains as the interface *zone*.

Through a new approach in this chapter, the DDM can be easily applied to combinations; an analysis also shows that this new approach will not cause a deterioration of preconditioning numbers for the associated matrix obtained.

In Section 15.2 we apply the iterative substructing methods to implement the RGM-FDM of Chapter 11 in parallel. In Section 15.3, an analysis is given to derive bounds of the preconditioned condition numbers based on the existing bounds or proof techniques. In the last section, numerical experiments are carried out for Motz's problem to support the analysis. It is noted that the sequential errors can be reduced by a factor of $\frac{1}{2} \times 10^{-6}$ through only 7–9 iterations.

15.2 COMBINATIONS OF RGM AND FDM

Consider the Poisson equation with the Dirichlet boundary condition

$$-\triangle u = -\left(\frac{\partial^2 u}{\partial x^2} + \frac{\partial^2 u}{\partial y^2}\right) = f(x, y), \qquad (x, y) \in S, \qquad (15.2)$$

$$u = 0, \qquad (x, y) \in \partial S, \qquad (15.3)$$

where S is a polygonal domain, and the function f is smooth enough. For simplicity, we assume that only one singular point of the solution $u(x, y)$ exists in $\bar{S} = (S \cup \partial S)$. Let S be divided by a piecewise straight line Γ_0 into two subdomains S^+ and S^-. The Ritz-Galerkin method is used in $S^+ \cup \partial S$ including the singular point, and the finite difference method is used in S^-. The subdomain S^- is again split by quasiuniform difference grids into small rectangles $\square_{ij}$ and triangles $\triangle_{ij}$. Denote $u_{i,j} = u(x_i, y_i)$, as the solutions on the difference nodes (i, j). The finite difference method can be regarded as a special kind of finite element method, if piecewise bilinear and linear interpolatory functions are used, and if special integration rules are used to approximate the integrals involved.

Assume that the solution u in S can be spanned by $u = \sum_{i=1}^{\infty} D_i \Psi_i$, where D_i are the expansion coefficients, and $\Psi_i (i = 1, 2, \cdots, \infty)$ are complete and linearly independent bases in $L^2(S)$ that are known. If the problem, (15.2) and (15.3), has a unique solution for any $f \in L^2(S)$, the particular solutions Ψ_i of (15.2) can be employed. Then the admissible functions in combinations of the RGM-FDM are written as:

$$v = \begin{cases} v^- & = & v_1, \text{ in } S^-, \\ v^+ & = & \sum_{i=1}^{L} \tilde{D}_i \Psi_i, \text{ in } S^+, \end{cases} \qquad (15.4)$$

where $\tilde{D}_i$ are unknown coefficients to be sought. It is due to the particular solutions, of (15.2) and (15.3) chosen as Ψ_i, that their total number as (15.1) will greatly decrease for a given accuracy of approximate solutions.

Since there occurs discontinuity of the admissible functions on Γ_0, i.e.,

$$v^+ \neq v^- \text{ on } \Gamma_0, \qquad (15.5)$$

we define another space

$$H = \{v \in L^2(S), v \in H^1(S^-), \text{ and } v \in H^1(S^+)\},$$

where $H^1(S)$ is the Sobolev space. Let $V_h(\subseteq H)$ denote a finite dimensional collection of the functions v in (15.4) satisfying (15.3). We shall employ the following integrals on Γ_0 to couple v^+ and v^-.

$$\widehat{D}(u,v) = \frac{P_c}{h^\sigma}\widehat{\int_{\Gamma_0}}(u^+ - u^-)(v^+ - v^-)d\ell \qquad (15.6)$$

$$- \widehat{\int_{\Gamma_0}}(\alpha\frac{\partial u^+}{\partial n} + \beta\frac{\partial u^-}{\partial n})(v^+ - v^-)d\ell - \widehat{\int_{\Gamma_0}}(\alpha\frac{\partial v^+}{\partial n} + \beta\frac{\partial v^-}{\partial n})(u^+ - u^-)d\ell,$$

where h is the maximal mesh spacing of difference grids, $P_c(> 0)$ the penalty constant, $\sigma(\geq 0)$ the penalty power, and the parameters $\alpha(\geq 0)$ and $\beta(\geq 0)$ are chosen differently to lead to different combinations. The integrals (15.6) can be approximated by the rules (15.14) of integration shown later.

The combinations of Ritz-Galerkin and finite difference methods (simply written as RGM-FDMs) are designed to seek a solution $u_h \in V_h$ such that

$$\widehat{a_h}(u_h, v) = f_h(u, v), \quad \forall v \in V_h , \qquad (15.7)$$

where

$$\widehat{a_h}(u,v) = \widehat{\iint_{S-}}\triangledown u \triangledown v ds + \iint_{S+}\triangledown u \triangledown v ds + \widehat{D}(u,v) , \quad (15.8)$$

$$f_h(v) = \widehat{\iint_{S-}} f v ds + \iint_{S+} f v ds ,$$

$$\widehat{\iint_{S-}}\triangledown u \triangledown v ds = \sum_{ij}\left[\widehat{\iint_{\square_{ij}}}\triangledown u \triangledown v ds + \widehat{\iint_{\triangle_{ij}}}\triangledown u \triangledown v ds\right],$$

$$\widehat{\iint_{S-}} f v ds = \sum_{ij}\left[\widehat{\iint_{\square_{ij}}} f v ds + \widehat{\iint_{\triangle_{ij}}} f v ds\right].$$

The approximate integrals are given by

$$\widehat{\iint_{\square_{ij}}}\triangledown u \triangledown v ds = \widehat{\iint_{\square_{ij}}}(u_x v_x + u_y v_y)ds \qquad (15.9)$$

$$= \frac{h_i k_j}{2}\left[u_x(i+\frac{1}{2},j)v_x(i+\frac{1}{2},j) + u_x(i+\frac{1}{2},j+1)v_x(i+\frac{1}{2},j+1)\right]$$

$$+ \frac{h_i k_j}{2}\left[u_y(i,j+\frac{1}{2})v_y(i,j+\frac{1}{2}) + u_y(i+1,j+\frac{1}{2})v_y(i+1,j+\frac{1}{2})\right],$$

and $u_x(i+\frac{1}{2},j) = u_x((x_i + x_{i+1})/2, y_j)$. For the right triangles with the vertices (i,j), $(i+1,j)$ and $(i,j+1)$, we choose

$$\widehat{\iint_{\triangle_{ij}}}\triangledown u \triangledown v ds = \widehat{\iint_{\triangle_{ij}}}(u_x v_x + u_y v_y)ds$$

$$= \frac{h_i k_j}{2} \left[u_x(i + \frac{1}{2}, j) v_x(i + \frac{1}{2}, j) + u_y(i, j + \frac{1}{2}) v_y(i, j + \frac{1}{2}) \right],$$

$$\widehat{\iint_{\square_{ij}} fvds} = \frac{h_i k_j}{4} \sum_{\ell,p=0}^{1} f(i + \ell, j + p) v(i + \ell, j + p),$$

$$\widehat{\iint_{\triangle_{ij}} fvds} = \frac{h_i k_j}{8} \left[2f(i, j) v(i, j) + f(i + 1, j) v(i + 1, j) \right.$$

$$\left. + f(i, j + 1) v(i, j + 1) \right]. \tag{15.10}$$

Two kinds of combinations are obtained from (15.7).

I. Penalty Combination with

$$\alpha = \beta = 0. \tag{15.11}$$

II. The generalized combinations with

$$\alpha + \beta = 1. \tag{15.12}$$

In (15.12), the following three typical combinations are most important (see Chapter 11).

1. Combination I: $\alpha = 0$ and $\beta = 1$,
2. Combination II: $\alpha = 1$ and $\beta = 0$,
3. Symmetric Combination: $\alpha = \beta = \frac{1}{2}$.

In computation, the approximate integrals $\widehat{D}(u, v)$ on Γ_0 are evaluated by integration rules. Denote by Z_k the difference nodes on Γ_0, and let

$$\Gamma_0 = \cup_{k=1}^{N_1} \Gamma_0^{(k)}, \quad \Gamma_0^{(k)} = Z_{k-1} Z_k . \tag{15.13}$$

We choose the integration rules

$$\int_{\Gamma_0} \xi \eta d\ell \approx \widehat{\int_{\Gamma_0} \xi \eta d\ell} = \int_{\Gamma_0} \widehat{\xi \eta} d\ell$$

$$= \sum_{k=1}^{N_1} \frac{\overline{Z_{k-1} Z_k}}{6} \Big\{ 2\xi(Z_{k-1}) \eta(Z_{k-1}) + \xi(Z_{k-1}) \eta(Z_k) +$$

$$+ \xi(Z_k) \eta(Z_{k-1}) + 2\xi(Z_k) \eta(Z_k) \Big\},$$

$$\tag{15.14}$$

where $\overline{Z_{k-1}Z_k}$ denotes the length of $Z_{k-1}Z_k$, and $\hat{\xi}$ and $\hat{\eta}$ are the piecewise linear interpolatory functions of ξ and η on the division (15.13) along Γ_0. For the interior boundary Γ_0 in a slant up direction we may choose the integration approximation

$$
\begin{aligned}
\int_{\Gamma_0} \frac{\partial u^-}{\partial n}(v^+ - v^-)d\ell &= \int_{\Gamma_0} \frac{\partial u^-}{\partial x}(v^+ - v^-)dy + \int_{\Gamma_0} \frac{\partial u^-}{\partial y}(v^+ - v^-)dx \\
&\approx \widehat{\int_{\Gamma_0} \frac{\Delta u^-}{\Delta x}}(v^+ - v^-)dy + \widehat{\int_{\Gamma_0} \frac{\Delta u^-}{\Delta y}}(v^+ - v^-)dx \\
&= \widehat{\int_{\Gamma_0} \frac{\Delta u^-}{\Delta n}}(v^+ - v^-)d\ell, \quad\quad\quad (15.15)
\end{aligned}
$$

where

$$
\begin{aligned}
\frac{\Delta u^-}{\Delta x}\big|_{(x_i,y_j)\in\Gamma_0} &= \frac{u^-(x_i + h_i, y_j) - u^-(x_i, y_j)}{h_i}, \\
\frac{\Delta u^-}{\Delta y}\big|_{(x_i,y_j)\in\Gamma_0} &= \frac{u^-(x_i, y_j + k_j) - u^-(x_i, y_j)}{k_j}, \quad\quad (15.16)
\end{aligned}
$$

$u^-(x_i + h_i, y_j) \in S^-$ and $u^-(x_i, y_j + k_j) \in S^-$.

In Chapter 11, superconvergence can be reached by the new norms:

$$
\overline{\|v\|}_h = \left(\overline{\|v\|}_{1,S-}^2 + \|v\|_{1,S+}^2 + \frac{P_c}{h^\sigma} \overline{\|v^+ - v^-\|}_{0,\Gamma_0}^2 \right)^{\frac{1}{2}}, \quad\quad (15.17)
$$

$$
\overline{\|v\|}_1 = \left(\overline{\|v\|}_{1,S-}^2 + \|v\|_{1,S+}^2 \right)^{1/2},
$$

where the norms with discrete summation are

$$
\overline{\|v\|}_{1,S-}^2 = \overline{|v|}_{1,S-}^2 + \overline{\|v\|}_{0,S-}^2,
$$

$$
\overline{|v|}_{1,S-}^2 = \sum_{ij} \left[\widehat{\iint_{\Box_{ij}}} (\nabla v)^2 ds + \widehat{\iint_{\triangle_{ij}}} (\nabla v)^2 ds \right],
$$

$$
\overline{\|v\|}_{0,S-}^2 = \sum_{ij} \left[\widehat{\iint_{\Box_{ij}}} v^2 ds + \widehat{\iint_{\triangle_{ij}}} v^2 ds \right].
$$

The discrete formulas, $\widehat{\iint_{\Box_{ij}}} (\nabla v)^2 ds$, $\widehat{\iint_{\triangle_{ij}}} (\nabla v)^2 ds$, $\widehat{\iint_{\Box_{ij}}} v^2 ds$ and $\widehat{\iint_{\triangle_{ij}}} v^2 ds$, are given in (15.9) − (15.10). Also, the error norm on Γ_0 is

$$
\overline{\|v^+ - v^-\|}_{0,\Gamma_0}^2 = \widehat{\int_{\Gamma_0}} (v^+ - v^-)^2 d\ell,
$$

where the integration $\widehat{\int_{\Gamma_0}} v^2 d\ell$ is given in (15.14).

Under a matching formula between L and h as (15.1), the superconvergence rates of

$$\overline{\overline{\|\epsilon\|}}_h = O(h^{2-\delta}), \quad \text{and} \quad \overline{\overline{\|\epsilon\|}}_h = O(h^{3/2}) \tag{15.18}$$

can be achieved in Chapter 11 by all combinations for the quasiuniform partitions:

$$S^- = \cup_{ij}\Box_{ij}, \quad \text{and} \quad S^- = (\cup_{ij}\Box_{ij}) \cup (\cup_{ij}\triangle_{ij})$$

respectively, where $\delta(>0)$ is an arbitrary small number.

It is well known that the generalized solution derivatives using linear finite element have the errors with $O(h)$. However, Eq. (15.18) implies that the average solution derivatives at the specific nodes have the errors with $O(h^{2-\delta})$ or $O(h^{3/2})$. Note that the superconvergences in (15.18) are scored just by means of the new norms (15.17) involving discrete summation.

15.3 DOMAIN DECOMPOSITION METHODS TO COMBINATIONS

Although the advanced domain decomposition methods have been recently developed by Yserentant (1986), Bramble, Pasciak and Xu (1990) and Smith and Widlund (1990) into parallel multilevel preconditioners, we will still follow the typical approaches of Bjorstad and Widlund (1986), Bramble, Pasciak and Schatz (1986), and Manteuffel and Parter (1990).

Besides the above we here mention some references on DDMs: Schwarz (1869), Borgers (1990), Borgers and Widlund (1990), Bernardi, Maday and Patera (1993), Bramble, Pasciak and Schatz (1988, 1989), Bramble, Pasciak and Xu (1990), Cayco, Foster and Swann (1995), Greenstadt (1982, 1993), Kioustelidis (1989), Kondrat'ev (1967), Mckerrell, Delves and Phillips (1981), Swan (1993), Whiteman and Goodsell (1991), Whiteman (1991) and Ying (1991).

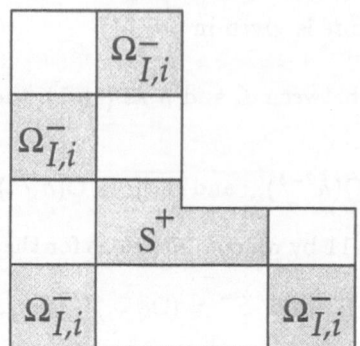

Figure 15.1   The partition of the DDMs where $S^+ \subset \Omega_I$ in Embedding Technique I.

15.3.1   Embedding Techniques

Let S be also split by Γ into Ω_I and Ω_{II} in DDMs:

$$\bar{S} = \partial S \cup \Omega_I \cup \Omega_{II} \cup \Gamma, \tag{15.19}$$

where Ω_I, Ω_{II} and Γ are two subregions and their interface, respectively. To distinguish partitions between DDMs and combinations, we call *subregions* Ω_I, Ω_{II} and *interface* Γ in DDMs, but *subdomains* S^+, S^- and *common boundary* Γ_0 in combinations, where

$$\bar{S} = \partial S \cup S^+ \cup S^- \cup \Gamma_0. \tag{15.20}$$

The key treatments of the DDMs applied to the combinations are how to embed S^+, S^- and Γ_0 into Ω_I, Ω_{II} and Γ. In this chapter we will propose two kinds of embedding techniques.

1. **Embedding Technique I.** Let the regular subdomain S^- be divided into subregions Ω_I and Ω_{II} as the traditional DDMs, and let S^+ belong to one subregion only, i.e.,

$$S^+ \subset \Omega_I. \tag{15.21}$$

   Hence the common boundary Γ_0 must belong to the interface Γ (see Figure 15.1):

$$\Gamma_0 \subset \Gamma.$$

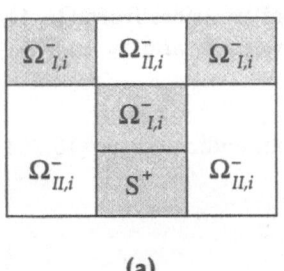 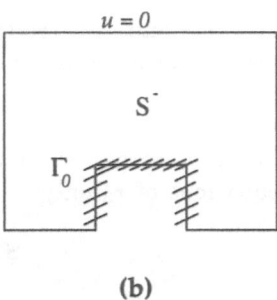

(a)   (b)

Figure 15.2 The partition of the DDM when $\Omega^+ \cup \Gamma^*$, and the subdomain Ω^- in Embedding Technique II.

2. **Embedding Technique II.** Note from Eq. (15.1) that the number of the coefficients D_l is much less than that of the variables $v_{ij} \in \{\Gamma \cap \partial S^+\}$. Hence we may simply attain the singular domain S^+ to interface Γ, and call the interface *zone* Γ^* instead,

$$\Gamma^* = \Gamma \cup \Gamma_0 \cup S^+.$$

So we have (see Figure 15.2)

$$S^+ \subset \Gamma^* \ and \ \Gamma \subset \Gamma^*. \tag{15.22}$$

The subdomain S^- is also divided into Ω_I and Ω_{II} as in Embedding Technique I.

Two embedding techniques I and II may be employed to all four combinations in (15.11) and (15.12), but only Embedding Technique II to the nonconforming combination.

Denote by x_1, x_2 and x_3 the variables in Ω_I, Ω_{II} and their interface Γ (or Γ^*), respectively. Then the unknown expansion coefficients $\{D_i\}$ may be included in x_1 or x_3 depending upon Embedding Techniques I or II. Denote

$$\mathbf{x} = (x_1, x_2, x_3)^T.$$

The equations (15.7) are then reduced to a system of linear algebraic equations:

$$\mathbf{Ax} = \mathbf{b}, \tag{15.23}$$

where $\mathbf{b} = (b_1, b_2, b_3)^T$, and

$$\mathbf{A} = \begin{bmatrix} \mathbf{A}_{11} & \mathbf{0} & \mathbf{A}_{13} \\ \mathbf{0} & \mathbf{A}_{22} & \mathbf{A}_{23} \\ \mathbf{A}_{13}^T & \mathbf{A}_{23}^T & \mathbf{A}_{33} \end{bmatrix}. \tag{15.24}$$

In (15.24), A_{11}, A_{22} and A_{33} are the coefficient matrices from Ω_I, Ω_{II} and Γ (or Γ^*), respectively, A_{13} the relation matrices between x_1 and x_3, and A_{23} the matrix between x_2 and x_3.

By using the block Gaussian elimination, we can again reduce (15.23) to a system of algebraic equations of x_3 only:

$$S x_3 = \widehat{b}_3, \tag{15.25}$$

where the Schur matrix

$$S = A_{33} - A_{13}^T A_{11}^{-1} A_{13} - A_{23}^T A_{22}^{-1} A_{23}, \tag{15.26}$$

and the vector

$$\widehat{b}_3 = b_3 - A_{13}^T A_{11}^{-1} b_1 - A_{23}^T A_{22}^{-1} b_2.$$

We may solve x_3 from (15.25) by the preconditioned conjugate gradient method (PCGM) (see Golub and van Loan (1989)). Then x_1 in Ω_I and x_2 in Ω_{II} can be sought in parallel.

For all four combinations, we choose the same preconditioner matrix

$$S_1 = A_{33}^{(1)} - A_{13}^T A_{11}^{-1} A_{13}, \tag{15.27}$$

where $A_{33}^{(1)}$ excludes the matrix components of the contribution of Ω_{II} to A_{33}, but includes that of S^+ if using (15.22) in Embedding Technique II. Note that for the generalized combinations, their preconditioner matrices S_1 are chosen simply as the same as those for Penalty Combination.

Remark 15.1 *It is natural to use the Neumann-Dirichlet domain decomposition method with partitioning $\Omega_I = S^- \cup S^+$, where the corresponding matrix A_{11} is given in (15.28) below. Then Embedding Technique I may lead to Embedding Technique II. However, only Technique I allows us to choose*

$$\Omega_I = S^+, \quad \Omega_{II} = S^-,$$

where $S^+ = \cup_i S_i^+$, $S^- = \cup_i S_i^-$. The significance of Technique I lies in that the size of S_i^+ should be chosen as large as possible to save the CPU time. Note that S^+ may be used even for the subdomains without singularities.

15.3.2 DDMs to Penalty Combination

The penalty integral

$$D_P(v,v) = \frac{P_c}{h^\sigma} \int_{\Gamma_0} (v^+ - v^-)^2 d\ell,$$

plays an important role to match the admissible functions v^+ and v^-. The variables v_{ij} on Γ_0 are also included in x_3, but $\{D_\ell\}$ may be included in x_1 and x_3 based on (15.21) and (15.22) respectively. We then have from Bjorstad and Widlund (1986),

$$\begin{bmatrix} \mathbf{A}_{11} & 0 & \mathbf{A}_{13} \\ 0 & \mathbf{A}_{22} & \mathbf{A}_{23} \\ \mathbf{A}_{13}^T & \mathbf{A}_{23}^T & \mathbf{A}_{33} \end{bmatrix} \begin{bmatrix} \mathbf{A}_{11} & 0 & \mathbf{A}_{13} \\ 0 & \mathbf{A}_{22} & \mathbf{A}_{23} \\ \mathbf{A}_{13}^T & 0 & \mathbf{A}_{33}^{(1)} \end{bmatrix}^{-1} \begin{bmatrix} 0 \\ 0 \\ y \end{bmatrix} = \begin{bmatrix} 0 \\ 0 \\ SS_1^{-1}y \end{bmatrix},$$

where y are the sequential solutions of x_3 on the interface boundary (or *zone*). The operation involving the inverse matrix can be carried out by two following steps.

Step I. Solve the Dirichlet problem on $\Omega_{II} \cup \Gamma$

$$\begin{bmatrix} \mathbf{A}_{22} & \mathbf{A}_{23} \\ 0 & \mathbf{I} \end{bmatrix} \begin{bmatrix} x_2 \\ y \end{bmatrix} = \begin{bmatrix} 0 \\ y \end{bmatrix},$$

where $\mathbf{I}$ is the unity matrix.

Step II. Solve the Neumann-Dirichlet problem on $\Omega_I \cup \Gamma$

$$\begin{bmatrix} \mathbf{A}_{11} & \mathbf{A}_{13} \\ \mathbf{A}_{13}^T & \mathbf{A}_{33}^{(1)} \end{bmatrix} \begin{bmatrix} x_1 \\ x_3 \end{bmatrix} = \begin{bmatrix} 0 \\ y \end{bmatrix}. \tag{15.28}$$

In fact, the preconditioner matrix S_1 in (15.27) is the Schur matrix of the above equation.

15.3.3 DDMs to Generalized Combinations

Finally, let us describe the renovated DDMs to the generalized combinations. Denote the linear equations obtained from (15.7)

$$\mathbf{A}_{\alpha,\beta}\mathbf{x} = \mathbf{b}, \tag{15.29}$$

where the vector $\mathbf{b}$ is the known vector, and the associated matrix

$$\mathbf{A}_{\alpha,\beta} = \begin{bmatrix} \mathbf{A}_{11}[\alpha,\beta] & 0 & \mathbf{A}_{13}[\alpha,\beta] \\ 0 & \mathbf{A}_{22}[\alpha,\beta] & \mathbf{A}_{23}[\alpha,\beta] \\ \mathbf{A}_{13}^T[\alpha,\beta] & \mathbf{A}_{23}^T[\alpha,\beta] & \mathbf{A}_{33}[\alpha,\beta] \end{bmatrix}.$$

Denote the preconditioner matrix of Penalty Combination with $\alpha = \beta = 0$

$$\mathbf{B} = \mathbf{B}_P = \begin{bmatrix} A_{11}^P & 0 & A_{13}^P \\ 0 & A_{22}^P & A_{23}^P \\ (A^P)_{13}^T & 0 & (A^P)_{33}^{(1)} \end{bmatrix} = \begin{bmatrix} A_{11}[0,0] & 0 & A_{13}[0,0] \\ 0 & A_{22}[0,0] & A_{23}[0,0] \\ A_{13}^T[0,0] & 0 & A_{33}^{(1)}[0,0] \end{bmatrix}. \quad (15.30)$$

We may perform the computation

$$\mathbf{A}_{\alpha,\beta}\mathbf{B}_P^{-1}\mathbf{y} = \mathbf{b}, \quad \mathbf{x} = \mathbf{B}_P^{-1}\mathbf{y},$$

where $\mathbf{y} = (y_1, y_2, y_3)^T$. The operation of the inverse operations $\mathbf{B}_P^{-1}\mathbf{y}$ are similar to Steps I and II in Section 15.3.2.

Although integral $\widehat{D}(u,v)$ in (15.6) when $\beta \neq 0$ is involved with the variables $v_{ij}^- \in \Gamma_0$, $v_{i+1,j}^- \in S^-$ and $v_{i,j+1}^- \in S^-$, the matrix $A_{33}^{(1)}[0,0]$ in the preconditioner matrix $\mathbf{B}_P$ is involved only with $v_{ij}^- \in \Gamma_0$. This simplification leads to much ease in practical application of DDMs to the generalized combinations. Moreover, it is due to a few terms of singular functions needed in S^+ that the parallel algorithms described above are simple to carry out, and need a little extra computation.

15.4   PRECONDITIONING CONDITION NUMBER

We shall show that the new Embedding Technique II will not cause deterioration of preconditioning condition numbers in the renovated DDMs. Analysis on Embedding Technique I appears elsewhere.

Let the matrices $\mathbf{A}$ and $\mathbf{B}$ be positive definitive and symmetric. Then the preconditioning condition number is defined by the following ratios of the maximal and minimal eigenvalues of $\mathbf{B}^{-1}\mathbf{A}$ (see Courant and Hilbert (1953)):

$$Cond. \, (\mathbf{B}^{-1}\mathbf{A}) = Cond. \, (\mathbf{A}\mathbf{B}^{-1}) = \lambda_{\max}(\mathbf{B}^{-1}\mathbf{A})/\lambda_{\min}(\mathbf{B}^{-1}\mathbf{A}), \quad (15.31)$$

where

$$\lambda_{\max}(\mathbf{B}^{-1}\mathbf{A}) = \max_{\|\mathbf{x}\| \neq 0} \frac{\mathbf{A}(\mathbf{x},\mathbf{x})}{\mathbf{B}(\mathbf{x},\mathbf{x})}, \quad \lambda_{\min}(\mathbf{B}^{-1}\mathbf{A}) = \min_{\|\mathbf{x}\| \neq 0} \frac{\mathbf{A}(\mathbf{x},\mathbf{x})}{\mathbf{B}(\mathbf{x},\mathbf{x})},$$

$\mathbf{A}(\mathbf{x},\mathbf{x}) = \mathbf{x}^T\mathbf{A}\mathbf{x}$, and the Euclidean norm $\|\mathbf{x}\| = (\sum_{i=1}^n \mathbf{x}_i^2)^{1/2}$. We obtain the following lemma from Bjorstad and Widlund (1986).

Lemma 15.1 *Let* **A** *and* **B** *be the matrices in (15.24) and (15.30), and* **S** *and* **S**$_1$ *be the matrices (15.26) and (15.27). Then the preconditioning condition number has the bounds*

$$Cond.\ (\mathbf{S}_1^{-1}\mathbf{S}) \leq Cond.\ (\mathbf{B}^{-1}\mathbf{A}).$$

Consider the Poisson equation on S^-

$$-\triangle u = f \text{ in } S^-, \tag{15.32}$$

with the mixed Neumann-Dirichlet condition (see Manteuffel and Parter (1990), and Mroz (1989)):

$$u = 0 \text{ on } \partial S \setminus \Gamma_0, \tag{15.33}$$

$$\frac{\partial u}{\partial n} = 0 \text{ on } \Gamma_0, \tag{15.34}$$

where Γ_0 is the common boundary of S^+ and S^-. Suppose that the finite difference method (or the finite element method) is used in S^-. Let A_a and B_a be the associated matrix and the preconditioner matrix in the traditional DDMs. Many reports on DDMs, such as Bjorstad and Widlund (1986), Bramble, Pasciak and Schatz (1986), Chan (1989), Dryja (1982) and Mroz (1989), have given the bounds

$$\alpha_0 \mathbf{B}_a(\mathbf{x}_a, \mathbf{x}_a) \leq \mathbf{A}_a(\mathbf{x}_a, \mathbf{x}_a) \leq \alpha_1 \mathbf{B}_a(\mathbf{x}_a, \mathbf{x}_a), \tag{15.35}$$

where $\mathbf{A}(\mathbf{x}, \mathbf{x}) = \mathbf{x}^T \mathbf{A} \mathbf{x}$,

$$\alpha_0 = 1, \alpha_1 = C(1 + \ell n^2 h), \tag{15.36}$$

and C is a bounded constant independent of h. Moreover, the study of Manteuffel and Parter (1990) discovers that the matrices A_a and B_a have resulted from the homogeneous Dirichlet boundary conditions on the same portion of the boundary ∂S^-. Below, the analysis shows that the bounds of preconditioning condition numbers have no deterioration against those from the existing bounds (15.35) and (15.36) in DDMs.

Let us first consider Penalty Combination. Suppose that Figure 15.2b is the partition in the DDMs, where the shaded subregions are chosen as Ω_I in the capacity matrix method, and the singular subdomain S^+ is regarded as in (15.22). We have the following theorem.

Theorem 15.1 *Let (15.35) hold. For the DDMs of Penalty Combination by Embedding Technique II, there exist the bounds of the preconditioning condition number.*

$$Cond.(\mathbf{S}_{1P}^{-1}\mathbf{S}_P) \leq \alpha_1/\alpha_0, \tag{15.37}$$

where S_P and S_{1P} are the corresponding Schur and preconditioner matrices.

Proof. We will modify the formulation of (15.32)–(15.34), to obtain the parallel Penalty Combination. In fact, Eqs. (15.32)–(15.34) can be rewritten in a weak form

$$\widehat{\iint_{S-}} \nabla u \, \nabla v ds = \widehat{\iint_{S-}} f v ds, \tag{15.38}$$

where $v(= v^-)$ is the piecewise bilinear and linear function in (15.4), satisfying the Dirichlet condition (15.33). On the other hand, Penalty Combination (15.7) with $\alpha = \beta = 0$ is written as

$$\widehat{\iint_{S-}} \nabla u \, \nabla v ds + \widehat{\iint_{S+}} \nabla u \, \nabla v ds + \frac{P_c}{h^\sigma} \widehat{\int_{\Gamma_0}} (u^+ - u^-)(v^+ - v^-) d\ell = f_h(v). \tag{15.39}$$

Therefore, to obtain (15.39), we should add two more integrals in the left side in (15.38). Based on (15.22) the subvector x_3 will include $v_{ij} (\in \Gamma_0)$ and D_l (see Figure 15.2b). Denote

$$\mathbf{E}(D_l, D_l) = \iint_{S+} (\nabla v)^2 ds.$$

Let the vector $\mathbf{y}_d$ denote $v_{ij} \in \Gamma_0$ and $\{D_\ell\}$, and

$$\mathbf{A}^*(\mathbf{y}_d, \mathbf{y}_d) = \frac{P_c}{h^\sigma} \widehat{\int_{\Gamma_0}} (v^+ - v^-)^2 d\ell \geq 0.$$

Both $\mathbf{A}^*(\mathbf{y}_d, \mathbf{y}_d)$ and $\mathbf{E}(D_\ell, D_\ell)$ are included in $\mathbf{A}_{33}^{(1)}$ in (15.27). Also denote $\mathbf{x}_d = \mathbf{x} \vee \mathbf{y}_d \vee \{D_l\}$, then

$$\mathbf{A}_P(\mathbf{x}_d, \mathbf{x}_d) = \mathbf{A}_a(\mathbf{x}, \mathbf{x}) + \mathbf{E}(D_\ell, D_\ell) + \mathbf{A}^*(\mathbf{y}_d, \mathbf{y}_d),$$
$$\mathbf{B}_P(\mathbf{x}_d, \mathbf{x}_d) = \mathbf{B}_a(\mathbf{x}, \mathbf{x}) + \mathbf{E}(D_\ell, D_\ell) + \mathbf{A}^*(\mathbf{y}_d, \mathbf{y}_d).$$

Consequently, from assumption (15.35), we have the following bounds.

$$\lambda_{\min}(\mathbf{B}_P^{-1}\mathbf{A}_P) = \min_{\|\mathbf{x}_d\| \neq 0} \frac{\mathbf{A}(\mathbf{x}_d, \mathbf{x}_d)}{\mathbf{B}(\mathbf{x}_d, \mathbf{x}_d)}$$

$$= \min_{\|\mathbf{x}_d\| \neq 0} \frac{\mathbf{A}_a(\mathbf{x}, \mathbf{x}) + \mathbf{A}^*(\mathbf{y}_d, \mathbf{y}_d) + \mathbf{E}(D_l, D_l)}{\mathbf{B}_a(\mathbf{x}, \mathbf{x}) + \mathbf{A}^*(\mathbf{y}_d, \mathbf{y}_d) + \mathbf{E}(D_l, D_l)} \geq \min_{\forall \mathbf{x} \neq 0} \frac{\mathbf{A}_a(\mathbf{x}, \mathbf{x})}{\mathbf{B}_a(\mathbf{x}, \mathbf{x})} \geq \alpha_0.$$

Similarly,

$$\lambda_{\max}(\mathbf{B}_P^{-1}\mathbf{A}_P) = \max_{\|\mathbf{x}_d\| \neq 0} \frac{\mathbf{A}(\mathbf{x}_d, \mathbf{x}_d)}{\mathbf{B}(\mathbf{x}_d, \mathbf{x}_d)}$$

$$= \max_{\|\mathbf{x}_d\| \neq 0} \frac{\mathbf{A}_a(\mathbf{x}, \mathbf{x}) + \mathbf{A}^*(\mathbf{y}_d, \mathbf{y}_d) + \mathbf{E}(D_l, D_l)}{\mathbf{B}_a(\mathbf{x}, \mathbf{x}) + \mathbf{A}^*(\mathbf{y}_d, \mathbf{y}_d) + \mathbf{E}(D_l, D_l)} \leq \max_{\forall \mathbf{x} \neq 0} \frac{\mathbf{A}_a(\mathbf{x}, \mathbf{x})}{\mathbf{B}_a(\mathbf{x}, \mathbf{x})} \leq \alpha_1,$$

to obtain from (15.31)

$$Cond.(\mathbf{B}_P^{-1}\mathbf{A}_P) \leq \alpha_1/\alpha_0.$$

The desired results (15.37) are obtained from Lemma 15.1. ∎

We now study the DDMs of Combinations I and II and Symmetric Combination, by choosing the same preconditioner matrix B_P, (15.30). We need the following lemma.

Lemma 15.2 *Suppose that* $\exists \mu > 0$ *such that*

$$\left\|\frac{\partial v^+}{\partial n}\right\|_{0,\Gamma_0} \leq CL^\mu \, \|v^+\|_{0,\Gamma_0}, \ \forall v \in V_h, \tag{15.40}$$

where C is a bounded constant independent of u, v, h *and* L. *Also assume*

$$\sigma \geq 1 \text{ when } \beta > 0 \tag{15.41}$$

and

$$\sigma > 0 \text{ and } h^\sigma L^{2\mu} \leq C \text{ when } \alpha > 0. \tag{15.42}$$

Then when $P_c(> 0)$ *is chosen suitably large but independent of* L, h *and* v, *the following* V_h-*elliptic and bilinear inequalities hold.*

$$C_0 \, \overline{\overline{\|v\|}}_h^2 \ \leq \ \widehat{a_h}(v,v), \ \forall v \in V_h , \tag{15.43}$$

$$|\widehat{a_h}(u,v)| \ \leq \ C \, \overline{\overline{\|u\|}}_h \, \overline{\overline{\|v\|}}_h, \ \forall u, v \in V_h, \tag{15.44}$$

where $\widehat{a_h}(u,v)$ *is defined in (15.7), and the norm* $\overline{\overline{\|v\|}}_h$ *is given in (15.17).*

Proof. We prove only (15.43); the proof of (15.44) is similar. We have from (15.7)

$$\widehat{a_h}(v,v) \ = \ \overline{|v|}_{1,S-}^2 + |v|_{1,S+}^2 + \frac{P_c}{h^\sigma} \, \overline{\overline{\|v^+ - v^-\|}}_{1,\Gamma_0}^2$$

$$- 2 \int_{\Gamma_0} (\alpha \frac{\partial v^+}{\partial n} + \beta \frac{\Delta v^-}{\Delta n})(v^+ - v^-) d\ell. \tag{15.45}$$

For simplicity, we assume $\partial S^- \cap \Gamma \neq \emptyset$ and $\partial S^+ \cap \Gamma \neq \emptyset$. Applying the Poincare-Friedrichs inequality we conclude that there exists a positive constant $C_0 > 0$ independent of $v \ (\in H^1(S^-))$ and $v \ (\in H^1(S^+))$ satisfying (15.3) such that

$$|v|_{1,S-} \geq C_0 \|v\|_{1,S-}, \ |v|_{1,S+} \geq C_0 \|v\|_{1,S+}.$$

For the piecewise bilinear and linear function $v^- \in V_h$, the following norms are equivalent to each other (see Lemma 11.1 of Chapter 11):

$$\overline{|v|}_{1,S-} \sim |v|_{1,S-}, \quad \overline{\|v\|}_{1,S-} \sim \|v\|_{1,S-}, \quad \forall v \in V_h.$$

Therefore, we can see that

$$\overline{|v|}_{1,S-}^2 + |v|_{1,S+}^2 \geq a_0(\overline{\|v\|}_{1,S-}^2 + \|v\|_{1,S+}^2), \tag{15.46}$$

where $a_0(> 0)$ is also a bounded constant independent of L, h and v.

Next, based on assumption (15.40) and the Sobolev imbedding theorem, we have by noting the integration rules in (15.14)

$$\left| \widehat{\int_{\Gamma_0}} \frac{\partial v^+}{\partial n}(v^+ - v^-)d\ell \right| \leq C\|\frac{\partial v^+}{\partial n}\|_{0,\Gamma_0}\overline{\|v^+ - v^-\|}_{0,\Gamma_0}$$

$$\leq CL^\mu\|v^+\|_{0,\Gamma}\overline{\|v^+ - v^-\|}_{0,\Gamma_0} \leq CL^\mu\|v^+\|_{1,S+}\overline{\|v^+ - v^-\|}_{0,\Gamma_0}.$$

It follows from (15.15)—(15.16) that

$$\left| \widehat{\int_{\Gamma_0}} \frac{\Delta v^-}{\Delta n}(v^+ - v^-)d\ell \right| \leq C\left(\left\|\frac{\Delta v^-}{\Delta x}\right\|_{0,\Gamma_0} + \left\|\frac{\Delta v^-}{\Delta y}\right\|_{0,\Gamma_0} \right)\overline{\|v^+ - v^-\|}_{0,\Gamma_0} $$

$$\leq Ch^{-\frac{1}{2}}\overline{\|v^-\|}_{1,S-}\overline{\|v^+ - v^-\|}_{0,\Gamma_0}. \tag{15.47}$$

Combining (15.45)–(15.47) yields

$$\widehat{a_h}(u,v) \geq \left[a_0 - \frac{C}{P_c}(\alpha L^\mu h^{\frac{\sigma}{2}} + \beta h^{\frac{\sigma-1}{2}}) \right] \overline{\|v\|}_h^2.$$

The uniformly V_h- elliptic inequality (15.43) holds under assumptions (15.41) and (15.42) and a suitably large P_c but independent of L, h and v. ∎

Remark 15.2 *Lemma 15.2 is, in fact, Lemma 11.5 of Chapter 11 without proof, which is also useful in the analysis for combinations of RGM-FDM. It is interesting to note that the assumptions (15.40)–(15.42) are exactly the same as those in Lemma 9.2 of Chapter 9 for combinations of RGM-FEM.*

Theorem 15.2 *Let (15.35) and all the conditions in Lemma 15.2 hold. By using the preconditioner matrix B_P as (15.30) in the DDMs, there exist the bounds for the generalized combinations,*

$$Cond.(\mathbf{S}_{1P}^{-1}\mathbf{S}_{\alpha,\beta}) \leq \frac{\beta_1}{\beta_0}\frac{\alpha_1}{\alpha_0}, \tag{15.48}$$

where $\mathbf{S}_{\alpha,\beta}$ is the Schur matrix with parameters α and β.

Proof. By noting the definition (15.17), we have

$$\overline{\overline{\|\mathbf{v}\|}}_h^2 = \mathbf{A}_P(\mathbf{x}_d, \mathbf{x}_d), \quad \widehat{a_h}(v, v) = \mathbf{A}_{\alpha,\beta}(\mathbf{x}_d, \mathbf{x}_d), \quad \forall v \in V_h. \tag{15.49}$$

From on Lemma 15.2, when $P_c(> 0)$ is chosen suitably large but independent of L, h, σ and $\mathbf{x}_d$, there exist the bounds

$$\beta_0 \mathbf{A}_P(\mathbf{x}_d, \mathbf{x}_d) \leq \mathbf{A}_{\alpha,\beta}(\mathbf{x}_d, \mathbf{x}_d) \leq \beta_1 \mathbf{A}_P(\mathbf{x}_d, \mathbf{x}_d), \quad \beta_0 \leq \beta_1,$$

where $\mathbf{A}_P = \mathbf{A}_{0,0}$, and β_0 and β_1 are two bounded constants independent of L, h, σ and $\mathbf{x}_d$. The matrices $\mathbf{A}_P$ and $\mathbf{A}_{\alpha,\beta}$ are the associated matrices obtained from Penalty Combinations and the generalized combinations in (15.7) respectively. Also from Theorem 15.1, we have

$$\alpha_0 \mathbf{B}_P(\mathbf{x}_d, \mathbf{x}_d) \leq \mathbf{A}_P(\mathbf{x}_d, \mathbf{x}_d) \leq \alpha_1 \mathbf{B}_P(\mathbf{x}_d, \mathbf{x}_d),$$

to obtain

$$\alpha_0 \beta_0 \mathbf{B}_P(\mathbf{x}_d, \mathbf{x}_d) \leq \mathbf{A}_{\alpha,\beta}(\mathbf{x}_d, \mathbf{x}_d) \leq \alpha_1 \beta_1 \mathbf{B}_P(\mathbf{x}_d, \mathbf{x}_d).$$

This leads to

$$Cond.(\mathbf{B}_P^{-1} \mathbf{A}_{\alpha,\beta}) \leq \frac{\beta_1}{\beta_0} \frac{\alpha_1}{\alpha_0}.$$

Consequently, the desired results (15.48) are obtained from Lemma 15.1. ∎

15.5 NUMERICAL EXPERIMENTS FOR MOTZ PROBLEM

Consider Motz's problem which solves the Laplace equation on a rectangle S $(-1 < x < 1, 0 < y < 1)$

$$\triangle u = \frac{\partial^2 u}{\partial x^2} + \frac{\partial^2 u}{\partial y^2} = 0 \text{ in } S,$$

with the mixed Neumann-Dirichlet boundary conditions

$$u\big|_{x<0 \cap y=0} = 0, \qquad u\big|_{x=1} = 500,$$

$$\frac{\partial u}{\partial y}\big|_{y=1} = \frac{\partial u}{\partial y}\big|_{y=0 \cap x>0} = \frac{\partial u}{\partial x}\big|_{x=-1} = 0.$$

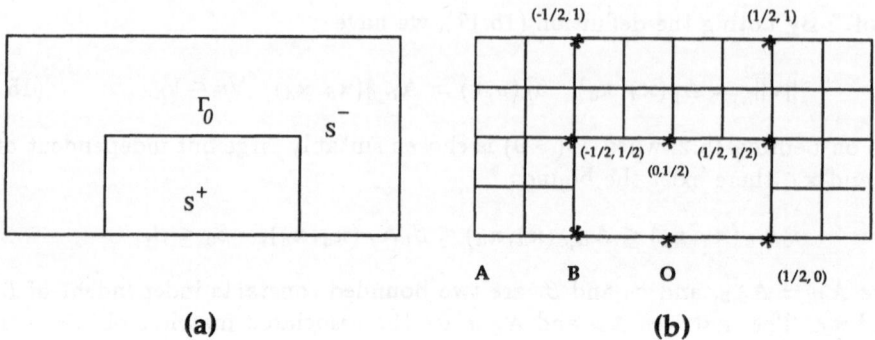

Figure 15.3   Partition of Motz's problem in Combinations.

We split S by Γ_0 into two subdomains S^+ and S^-, where the singular subdomain S^+ is chosen as the rectangle $(-\frac{1}{2} < x < \frac{1}{2}, 0 < y < \frac{1}{2})$, and S^- as the rest of S (see Figure 15.3a). The Ritz-Galerkin method and the finite difference method are used in S^+ and S^- respectively, with the admissible functions in Chapter 11:

$$
v = \begin{cases} v^- & \text{in } S^- \\ v^+ = \sum_{\ell=0}^{L} D_\ell r^{\ell+\frac{1}{2}} \cos(\ell + \frac{1}{2})\theta \text{ in } S^+, \end{cases} \tag{15.50}
$$

where (r, θ) are the polar coordinates, and D_ℓ are the coefficients to be sought. The subdomain S^- is again divided into small squares as shown in Figure 15.3b.

Three typical combinations of the generalized combinations are chosen with the following parameters:

(I) Combination I ($\alpha = 0$ and $\beta = 1$), $P_c = 10$ and $\sigma = 2$,

(II) Combination II ($\alpha = 1$ and $\beta = 0$), $P_c = 10$ and $\sigma = 2$,

(III) Symmetric Combination ($\alpha = \beta = \frac{1}{2}$), $P_c = 10$ and $\sigma = 2$.

The values of $\sigma = 2$ satisfy the condition $\sigma > 1$ in Lemma 15.2. The number $(L+1)$ of singular functions in v^+ of (15.50) is chosen by

$$
\begin{aligned}
M_S &= 2 \text{ and } L + 1 = 4; \quad M_S = 4 \text{ and } L + 1 = 5; \\
M_S &= 8 \text{ and } L + 1 = 6,
\end{aligned} \tag{15.51}
$$

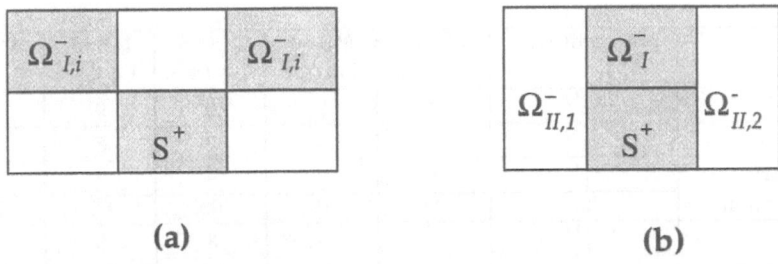

Figure 15.4 Partitions of Motz's problem in the DDMs of combinations: (a) $S^+ \subset \Omega_I$, (b) $S^+ \subset \Gamma^*$.

where M_S is the division number along the boundary $\overline{AB}$, e.g., $M_S = 2$ for Figure 15.3b. Note that Eqs. (15.51) satisfying (15.1) indicate that the number $(L+1)$ grows very slowly with $h \longrightarrow 0$. Table 11.1 in Chapter 11 has provided already the computational results of Symmetric Combination. It is easy to discover the superconvergence rate $O(h^{2-\delta})$ of the generalized derivatives of solutions, where $0 < \delta << 1$. The computational results of Combinations I and II are similar.

Next, for the DDMs of combinations, there are two partitions in Figure 15.4, corresponding to two Embedding Techniques (15.21) and (15.22). Since (15.22) is new to the traditional partitions in the DDMs, we will carry out numerical experiments to (15.22) only. Choose Figure 4b as the computational model. S^- is divided into three subregions

$$S^- = \Omega_I^- \cup \Omega_{II,1}^- \cup \Omega_{II,2}^-,$$

with

$$\Omega_I = \Omega_I^-, \quad \Omega_{II} = \Omega_{II,1}^- \cup \Omega_{II,2}^-$$

where the interface *zone*

$$\Gamma^* = \Gamma \cup \Gamma_0 \cup S^+,$$

and Γ is the common boundary Ω_I and Ω_{II}.

For the preconditioned conjugate gradient method, choose the zero initial values:

$$(u_h)_{ij}^{(0)} = 0 \text{ on } \Gamma \cup \Gamma_0; \quad D_\ell^{(0)} = 0 \quad ,\ell = 0, 1, \cdots, L.$$

Methods	preconditioner	MS=2 L+1=4	MS=4 L+1=5	MS=6 L+1=5	MS=8 L+1=6	MS=10 L+1=6
Comb. I	$S_p^{(1)}$	10	9	9	9	9
Comb. II	$S_p^{(1)}$	7	8	8	8	8
Symmetric Combination	$S_p^{(1)}$	8	9	9	9	9
	no	23	45	62	88	109
Penalty Combination	$S_p^{(1)}$	7	8	8	8	8
	no	23	45	62	88	109
Nonconforming Combination	$S_p^{(1)}$	7	8	8	8	8

Table 15.1   The iteration number of the DDM for five combinations.

Then the sequences $(u_h)_{ij}^{(k)}$ on $\Gamma \cup \Gamma_0$ and $D_\ell^{(k)}$ can be obtained from DDMs of the combinations. To measure the iterative errors we have computed the following sequential errors,

$$\triangle_{\max}^{(k)} = \max\left\{ \max_{(i,j)\in\Gamma\cup\Gamma_0} |(u_h^-)_{ij}^{(k)} - (u_h^-)_{i,j}^{(k-1)}|, \max_\ell |D_\ell^{(k)} - D_\ell^{(k-1)}|\right\},$$

(15.52)

$$\|\delta^{(k)}(u_h^+ - u_h^-)\|_{0,\Gamma_0} = \left\{ \int_{\Gamma_0} [((u_h^+)^{(k)} - (u_h^-)^{(k)}) - ((u_h^+)^{(k-1)} - (u_h^-)^{(k-1)})]^2 d\ell \right\}^{1/2}.$$

(15.53)

Table 15.1 presents the iteration number k in the DDMs of combinations until the reduced ratios satisfy

$$\triangle_{\max}^{(k)}/\triangle_{\max}^{(1)} \leq \frac{1}{2} \times 10^{-6}.$$

(15.54)

It can be seen from Table 15.1 that the minimal iteration number k will increase very slowly as $M_S \longrightarrow \infty$, if the preconditioner matrix B_P, i.e., S_{1P} is employed. This fact coincides with the analysis in Section 15.4. The iteration number of Combination I and Symmetric Combination involving $\frac{\partial u^-}{\partial n}$ as $\beta > 0$ seems to be a little larger; this can be explained by a factor $\beta_1/\beta_0 \geq 1$ in Theorem 15.2.

For simplicity, we mainly list the results of Symmetric Combination. For numerical results from the parallel Penalty Combinations, the reader may refer to Li (1996b, c). Tables 15.2 provides the iteration process of D_ℓ by Symmetric Combination at

Iteration Number	$\tilde{D}_0$	$\tilde{D}_1$	$\tilde{D}_2$	$\tilde{D}_3$	$\tilde{D}_4$	$\tilde{D}_5$
1	403.4410	9.9719	35.3193	−45.1905	1.3526	−15.9483
2	403.8949	90.9584	22.8809	−2.2814	5.4176	4.9447
3	400.1616	87.6210	15.1350	−9.7039	0.8605	0.2952
4	401.1080	87.5939	17.0915	−8.9698	1.8895	0.6640
5	401.0889	87.6469	16.9628	−8.7955	1.7431	0.7521
6	401.0888	87.6455	16.9641	−8.7950	1.7440	0.7333
7	401.0888	87.6455	16.9647	−8.7936	1.7430	0.7340
8	401.0887	87.6455	16.9646	−8.7937	1.7425	0.7340
9	401.0887	87.6455	16.9646	−8.7937	1.7425	0.7340
Combination Coeffs.	401.0887	87.6455	16.9646	-8.7937	1.7425	0.7340
True Coeffs.	401.1625	87.6559	17.2379	-8.0712	1.4403	0.3311

Table 15.2 Coeffieicents by Symmetric Combination with by $M_S = 8$ and $L+1 = 6$.

$M_S = 8$ and $L + 1 = 6$. The limit solutions are exactly the same as the solutions in Table 11.1 of Chapter 11. The sequential errors, $\delta^{(k)}$ and $\triangle_{max}^{(k)}$, are given in Tables 15.3 and 15.4, as well as drawn in Figures 15.5 and 15.6 by the DDMs for Symmetric Combination; only 7–9 iterations are needed to achieve a deduction of $\frac{1}{2} \times 10^{-6}$ for sequential errors. Note that the deduction of sequential errors does not increase much when M_s increase (i.e., h decreases), see Figure 15.6.

Concluding Remarks. Let us summarize the main novelties in this chapter.

1. In order to implement the combined methods into parallel computing, we apply the substructuring iteration method by new embedding techniques (15.22), in which the entire singular subdomain S^+ is regarded as the interface *zone* in DDMs. Because of a few singular functions needed, the inverse computation for the novel preconditioner matrices will consume a little more CPU time. The significance of this chapter lies in that the different combinations can be easily implemented into parallel by the existing DDMs.

2. To solve singularity problems, the finite difference method and the Ritz-Galerkin method are combined by using the hybrid techniques, which are more flexible in coupling various numerical methods. Although the hybrid integrals are

Iteration	Symmetric Combination				
Number	$Ms = 2$	4	6	8	10
1	3.372	1.022	0.5780	0.4009	0.3079
2	1.511	0.5238	0.3151	0.2205	0.1721
3	0.3124	0.1622	0.1185	$0.9205 * 10^{-1}$	$0.7555 * 10^{-1}$
4	0.2055	0.1082	$0.7021 * 10^{-1}$	$0.5273 * 10^{-1}$	$0.4193 * 10^{-1}$
5	$0.7666 * 10^{-2}$	$0.1365 * 10^{-2}$	$0.7535 * 10^{-3}$	$0.4748 * 10^{-3}$	$0.3440 * 10^{-3}$
6	$0.3595 * 10^{-3}$	$0.2755 * 10^{-3}$	$0.9769 * 10^{-4}$	$0.2920 * 10^{-4}$	$0.1847 * 10^{-4}$
7	$0.1100 * 10^{-3}$	$0.1005 * 10^{-4}$	$0.1412 * 10^{-4}$	$0.6741 * 10^{-5}$	$0.4583 * 10^{-5}$
8	$0.7631 * 10^{-6}$	$0.5593 * 10^{-5}$	$0.2949 * 10^{-5}$	$0.8937 * 10^{-6}$	$0.7318 * 10^{-6}$
9	/	$0.2606 * 10^{-7}$	$0.1307 * 10^{-6}$	$0.6726 * 10^{-7}$	$0.4707 * 10^{-7}$

Table 15.3   Error norms $\|\delta^{(k)}(u_h^+ - u_h^-)\|_{0,\Gamma_0}$ by Symmetric Combination.

Iteration	Symmetric Combination				
Number	$Ms = 2$	4	6	8	10
1	400.4	402.8	403.3	403.4	403.5
2	81.72	79.58	79.12	80.99	80.91
3	6.703	7.621	7.547	7.746	7.738
4	1.339	2.071	1.981	1.956	1.937
5	0.2899	0.1571	0.1755	0.1742	0.1883
6	$0.1398 * 10^{-1}$	$0.3225 * 10^{-1}$	$0.2821 * 10^{-1}$	$0.1876 * 10^{-1}$	$0.1639 * 10^{-1}$
7	$0.1412 * 10^{-2}$	$0.1051 * 10^{-2}$	$0.3323 * 10^{-2}$	$0.1396 * 10^{-2}$	$0.1167 * 10^{-2}$
8	$0.4824 * 10^{-4}$	$0.1427 * 10^{-3}$	$0.1927 * 10^{-3}$	$0.4926 * 10^{-3}$	$0.4686 * 10^{-3}$
9	/	$0.3463 * 10^{-5}$	$0.2063 * 10^{-4}$	$0.3275 * 10^{-5}$	$0.3422 * 10^{-5}$

Table 15.4   Error norms Δ_{Max} of the iteration solutions by Symmetric Combination.

complicated since they involve solution derivatives along Γ_0, we may choose the simple penalty preconditioner matrix instead. An analysis and numerical experiments have shown that such a substitution will not cause a deterioration on condition numbers, compared to that of the parallel Penalty Combination and the traditional DDMs as well.

3. In this book. the FDM is regarded as a kind of FEMs, in order to integrate FDM into the combining family. As a consequence, the FDM may *share* the

rich treasure of FEMs in both theory and application. This Chapter and Chapter 11 also display significance of such a linkage.

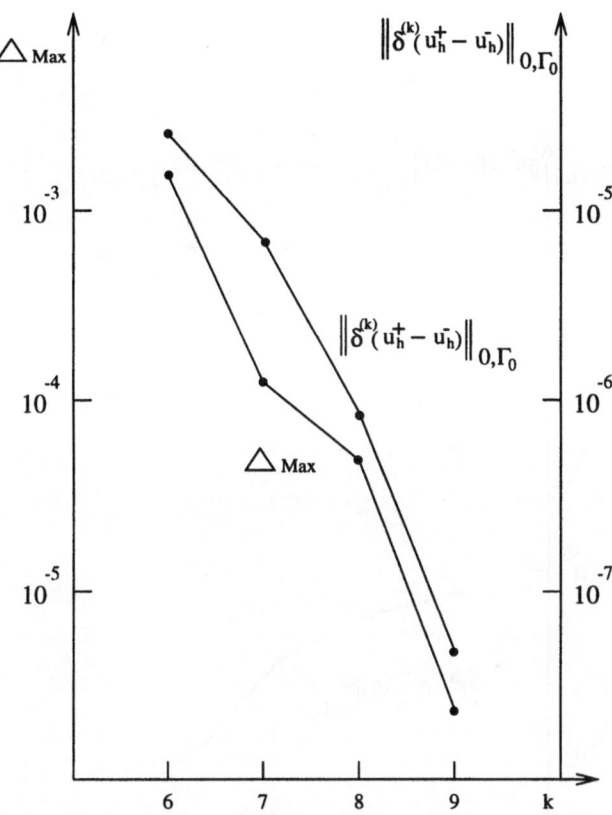

Figure 15.5   The curves of $\triangle_{Max}$ and $\|\delta^{(k)}(u_h^+ - u_h^-)\|_{0,\Gamma_0}$ versus k by Symmetric Combination as $M_S = 8$ and $L + 1 = 8$ with preconditioner S_{1p}.

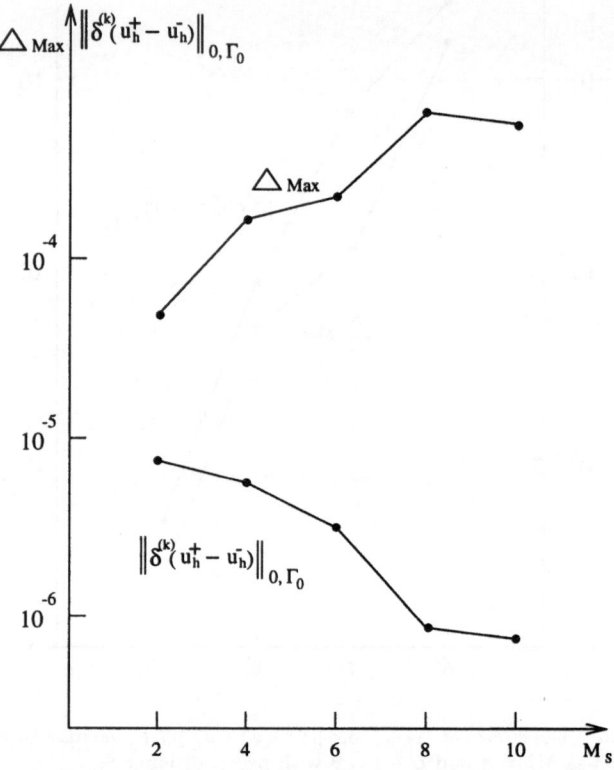

Figure 15.6   The curves of $\triangle_{Max}$ and $\|\delta^{(k)}(u_h^+ - u_h^-)\|_{0,\Gamma_0}$ versus M_S by Symmetric Combination as $k = 8$ with preconditioner S_{1p}.

16

SCHWARZ ALTERNATING METHOD

A discrete technique of the Schwarz alternating method is presented in this last chapter, to combine the Ritz-Galerkin and finite element methods. This technique is well suited for solving singularity problems in parallel, and requires a little more computation for large overlap of subdomains. The convergence rate of the iterative procedure, which depends upon overlap of subdomains, will be studied. Also a balance strategy will be proposed to couple the iteration number with the element size used in the FEM. For the crack-infinity problem of singularity the total CPU time by the technique in this chapter is much less than that by the nonconforming combination in Chapter 12.

16.1 INTRODUCTION

Domain decomposition methods (DDMs) have become very attractive owing to parallel performance. Of a number of DDM reports in Bjorstad and Widlund (1986), Bramble, Pasciak and Schatz (1986), Chan (1989), Dryja (1982), Lions (1988, 1989, 1990), and Schwarz (1869), only a few such as Anderson (1989) and Yanik (1989) are concerned with solution singularity. In this chapter, we will study the parallel algorithms for singularity problems by considering the crack-infinity problem of singularities in Chapter 12

$$-\triangle u + u = 0, \quad \text{in } S^*, \tag{16.1}$$

$$u = 1, \quad \text{on } y = 0, \tag{16.2}$$

$$u = 1, \quad \text{on } x = 0 \cap 0 < y < 1, \tag{16.3}$$

where $\triangle = (\partial^2/\partial x^2) + (\partial^2/\partial y^2)$, and S^* is the upper semi-plane but excluding the crack section

$$x = 0 \cap 0 < y < 1. \tag{16.4}$$

Since there exist two kinds of rather strong singularities of the problem (16.1)–(16.4): a crack singular point $(0, 1)$ and the infinity, the existing domain decomposition methods should be modified for applications to solving Eqs. (16.1)–(16.4). We have developed some variation and innovation of the iterative substructing methods

423

in Chapter 15. Below, we will report a numerical technique of the Schwarz alternating method of Schwarz (1869) and Lions (1988) to implement the combination of RGM-FEM into parallel.

In the nonconforming combination, the RGM and the FEM are used in the subdomains with and without solution singularity, respectively. In the Schwarz alternating method, the subdomains used have to be overlapped with each other, and the RGM and the FEM can be carried out independently and alternatively. In fact, Miller (1965) presented numerical analogy using the RGM-FDM to the Schwarz alternating procedure, and provided error bounds of the solutions. Chan, Hou and Lions (1991) reported some convergence results related for L-, T- and C-shaped domains in the domain decomposition methods. This chapter, however, attempts to study the following questions:

1. How to evaluate convergence rates of the procedure for the typical problem (16.1)–(16.4), whose domains are not rectangles as in Chan, Hou and Lions (1991), but circles or sectors?

2. How to choose the iteration number in the Schwarz alternating method?

We will first describe in the next section a discrete algorithm of the iterative procedures, and then provided in Sections 16.3 and 16.4 analysis and numerical experiments . As a consequence, we can find convergence rates of iteration, and derive a matching strategy of the iteration number with the boundary length of finite elements used.

16.2 DISCRETE TECHNIQUE

Based on the symmetry of geometry of (16.1)–(16.4) we can solve the following problem on only half the region: $S\{(x, y), x > 0 \ \text{and} \ y > 0\}$,

$$- \Delta u + u = 0, \quad \text{in} \ S, \tag{16.5}$$

$$u = 1, \quad \text{on} \ y = 0, \tag{16.6}$$

$$u = 1, \quad \text{on} \ x = 0 \cap 0 < y < 1, \tag{16.7}$$

$$\frac{\partial u}{\partial x} = 0, \quad \text{on} \ x = 0 \cap y > 1. \tag{16.8}$$

For separation of singularities, we split the solution domain S into three subdomains S_1, S_2 and S_3, where S_1 is the region $\{(r, \theta), r \leq r_1, 0 \leq \theta \leq \pi\}$ including the crack

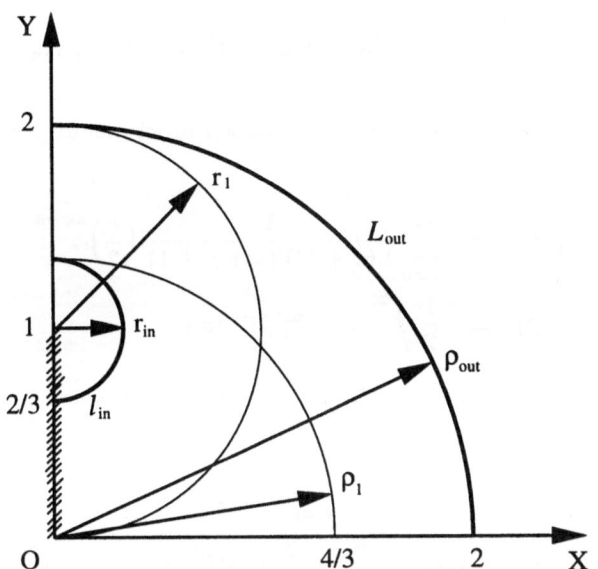

Figure 16.1 Partition of the infinite domain S into S_1, S_2 and S_3: Partition I.

singular point $(0,1)$, $S_2 = \{(\rho, \phi), \rho \geq \rho_1, 0 \leq \phi \leq \pi/2\}$ with the infinity, and S_3 a bounded subdomain with the interior and exterior boundaries (see Figure 16.1):

$$l_{in} = \{(r,\theta),\ r = r_{in},\ 0 \leq \theta \leq \pi\},$$
$$L_{out} = \{(\rho,\phi),\ \rho = \rho_{out},\ 0 \leq \phi \leq \pi/2\},$$

where the polar coordinates (r, θ) and (ρ, ϕ) have the origins $(0, 1)$ and $(0, 0)$ respectively. The partition of Figure 16.1 has the overlaps:

$$S_i \cap S_j \neq \emptyset,\ i, j = 1, 2, 3.$$

Since the solution in S_3 is smooth such that $u \in H^2(S_3)$, the FEM with piecewise linear approximation can be employed, to obtain an optimal convergence rate of approximate solutions. Assume that the true solution on l_{in} and L_{out} is known, we may solve the Dirichlet problem in S_3 by the FEM.

On the other hand, the true solution in S_1 and S_2 can be expanded as in Chapter 12,

$$u = \cosh(r\sin\theta) + \sum_{n=0}^{\infty} a_n I_{n+\frac{1}{2}}(r)\sin\left(\left(n + \frac{1}{2}\right)\theta\right),\ r \leq \min(1, \rho_{out} - 1),$$

$$(16.9)$$

$$u = e^{-\rho \sin \phi} + \sum_{n=0}^{\infty} c_n K_{2n+1}(\rho) \sin((2n+1)\phi), \quad \rho \geq 1 + r_{in}, \qquad (16.10)$$

where $I_\mu(r)$ and $K_n(\rho)$ are the Bessel and Hankel functions for a purely imaginary argument defined by

$$I_\mu(r) = \sum_{k=0}^{\infty} \frac{1}{\Gamma(k+1)\Gamma(k+\mu+1)} \left(\frac{r}{2}\right)^{2k+\mu}, \qquad (16.11)$$

$$K_n(\rho) = \frac{1}{2} \int_{-\infty}^{\infty} e^{-\rho \cosh \eta - n\eta} d\eta. \qquad (16.12)$$

Hence, the following expansions with finite terms are chosen:

$$u_I = u_I(a_n) = \cosh(r \sin \theta) + \sum_{n=0}^{L} a_n \frac{I_{n+\frac{1}{2}}(r)}{I_{n+\frac{1}{2}}(1)} \sin\left(\left(n + \frac{1}{2}\right)\theta\right), \qquad (16.13)$$

$$u_{II} = u_{II}(c_n) = e^{-\rho \sin \phi} + \sum_{n=0}^{LL} c_n \frac{K_{2n+1}(\rho)}{K_{2n+1}(1)} \sin((2n+1)\phi), \qquad (16.14)$$

where $I_{n+\frac{1}{2}}(1)$ and $K_{2n+1}(1)$ in denominators are used as factors. Approximate values of the coefficients a_n and c_n can be obtained from the RGM.

Define the norms on the arcs $l_r \{(r,\theta), r = const \text{ and } 0 \leq \theta \leq \pi)\}$ and $L_\rho \{(\rho,\phi), \rho = const \text{ and } 0 \leq \phi \leq \pi/2\}$ such that

$$|u|_{I,r} = \left(\int_{l_r} u^2 d\ell\right)^{\frac{1}{2}} = \left[\int_0^\pi u^2(r,\theta) r d\theta\right]^{\frac{1}{2}},$$

$$|u|_{II,\rho} = \left(\int_{L_\rho} u^2 d\ell\right)^{\frac{1}{2}} = \left[\int_0^{\pi/2} u^2(\rho,\phi) \rho d\phi\right]^{\frac{1}{2}}.$$

In order to use the RGM and the FEM, the approximate solutions $\tilde{u}_I$ and $\tilde{u}_{II}$ can be found by choosing the coefficients a_n and c_n such that

$$|\tilde{u}_I(\tilde{a}_n) - u_h|_{I,r_1} = \min_{\forall a_n} |u_I(a_n) - u_h|_{I,r_1},$$

$$|\tilde{u}_{II}(\tilde{c}_n) - u_h|_{II,\rho_1} = \min_{\forall c_n} |u_{II}(c_n) - u_h|_{II,\rho_1},$$

where u_h is the finite element approximation obtained.

Below, let us summarize the numerical technique of the Schwarz alternating method by the following four steps.

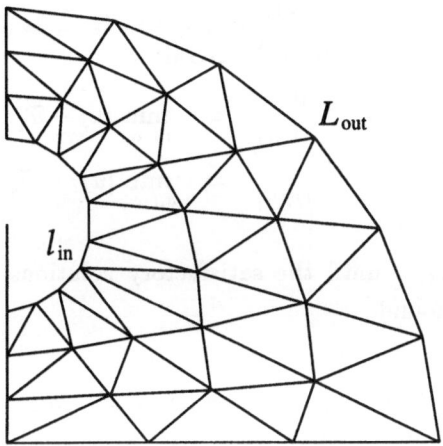

Figure 16.2 Triangulation of S_3 into finite elements.

Step 1. Let the initial approximation be

$$\tilde{u}_I^{(0)} = 1 \text{ on } l_{in} \text{ and } \tilde{u}_{II}^{(0)} = 1 \text{ on } l_{out} \ (\ i.e., \ \tilde{u}_h^{(0)} = 1 \text{ in } S_1 \cup S_2).$$

Step 2. The finite element approximation $u_h^{(2k+1)} \in V^{(k)}, k = 0, 1, 2, \ldots$, are obtained in odd iterations by

$$I(\tilde{u}_h^{(2k+1)}) = \min_{u_h \in V^{(k)}} I(u_h),$$

with the energy

$$I(v) = \iint_{\widehat{S}_3^h} \left(\left(\frac{\partial v}{\partial x}\right)^2 + \left(\frac{\partial v}{\partial y}\right)^2 + v^2 \right) ds,$$

where $\widehat{S}_3^h$ is the union of triangular elements of S_3 (see Figure 16.2), and $V^{(k)}$ denotes the space of the functions satisfying

$$u_h^{(2k+1)} = \tilde{u}_I^{(2k)}, \qquad \text{on } l_{in}, \tag{16.15}$$

$$u_h^{(2k+1)} = \tilde{u}_{II}^{(2k)}, \qquad \text{on } L_{out}, \tag{16.16}$$

$$u_h^{(2k+1)} = 1, \qquad \text{on } y = 0 \cap 0 \le x \le \rho_{out}, \tag{16.17}$$

$$u_h^{(2k+1)} = 1, \qquad \text{on } x = 0 \cap 0 \le y \le 1 - r_{in}. \tag{16.18}$$

Step 3. The solutions $\widetilde{u}_I^{(2k+2)}$ and $\widetilde{u}_{II}^{(2k+2)}$, $k = 0, 1,\ldots$, as (16.13) and (16.14) are obtained in even iterations from the RGM:

$$\left|\widetilde{u}_I^{(2k+2)} - \widetilde{u}_h^{(2k+1)}\right|_{I,r_1} = \min_{\text{all } a_n} \left|u_I - \widetilde{u}_h^{(2k+1)}\right|_{I,r_1},$$

$$\left|\widetilde{u}_{II}^{(2k+2)} - \widetilde{u}_h^{(2k+1)}\right|_{II,\rho_1} = \min_{\text{all } c_n} \left|u_{II} - \widetilde{u}_h^{(2k+1)}\right|_{II,\rho_1}.$$

Step 4. Return to Step 2 until the satisfactory solutions $\widetilde{u}_h^{(2k+1)}$, $\widetilde{u}_I^{(2k+2)}$ and $\widetilde{u}_{II}^{(2k+2)}$ have been found.

Step 3, which is indeed of the least squares approach, can easily used for different subdomains S_1 and S_2, and this technique requires only a little more computation even for large overlap of subdomains (cf., Chan, Hou, and Lions (1991), Lions (1988, 1990), and Tang (1989)).

The method of Wigley (1987, 1988) may be interpreted as a variation of the Schwarz alternating method. For instance, when the discrete approximation with an analytic solution on the entire solution domain is sought, the expansion coefficients $\widetilde{a}_h$ and $\widetilde{c}_h$ can be evaluated directly from

$$\widetilde{a}_n = \frac{2}{\pi} \frac{I_{n+1/2}(1)}{I_{n+1/2}(r)} \int_0^\pi [u_h(r,\theta) - \cosh(r\sin\theta)] \sin\left(n + \frac{1}{2}\right)\theta d\theta, \qquad (16.19)$$

$$\widetilde{c}_n = \frac{4}{\pi} \frac{K_{2n+1}(1)}{K_{2n+1}(\rho)} \int_0^{\pi/2} [u_h(\rho,\phi) - e^{-\rho\sin\phi}] \sin(2n+1)\phi d\phi. \qquad (16.20)$$

Eqs. (16.19) and (16.20) are a little simpler than Step 3, but the latter from the least squares sense yields smaller errors of numerical solutions.

It is worthwhile to point out that Steps 1-4 can also be applied to problems without singularities. Analytical expansions of u may replace the singular expansions (16.13) and (16.14). Since the Ritz-Galerkin method using particular solutions (or expansions) requires much less CPU time (see Section 16.4 below), the techniques in this chapter can also be used for solving the problems with nonuniform smoothness of solutions in parallel.

16.3 CONVERGENCE RATES OF ITERATIVE PROCEDURE

It is well known from Lions (1988, 1989, 1990) that the Schwarz alternating procedure is convergent in the maximal norm $|u|_{\infty,S}$ and the energy norm

$$E = \sum_{i=1}^{3} \iint_{S_i} \left[\left(\frac{\partial u}{\partial x}\right)^2 + \left(\frac{\partial u}{\partial y}\right)^2 + u^2 \right] ds, \tag{16.21}$$

provided that there are some uniform overlaps:

$$S_1 \cap S_3 \neq \emptyset, \quad S_2 \cap S_3 \neq \emptyset,$$

which can be guaranteed by

$$r_1 > r_{in}, \quad \rho_1 < \rho_{out}. \tag{16.22}$$

Combining (16.22), (16.9) and (16.10) yields the bounds of radii:

$$r_{in} < \quad r_1 \quad \leq \min(1, \ \rho_{out} - 1), \tag{16.23}$$

$$1 + r_{in} \leq \quad \rho_1 \quad < \rho_{out}. \tag{16.24}$$

In this section, we focus on the convergence rates of the iterative procedure as Steps 1-4, first to derive a prior rate and then to conduct an empirical rate.

Define the norms

$$|u|_{r,\rho} = \left\{ |u|_{I,r}^2 + |u|_{II,\rho}^2 \right\}^{\frac{1}{2}}, \tag{16.25}$$

$$|u|_B = \left\{ |u|_{in}^2 + |u|_{out}^2 \right\}^{\frac{1}{2}}, \tag{16.26}$$

$$|u|_{in} = |u|_{I,r_{in}}, \quad |u|_{out} = |u|_{II,\rho_{out}}. \tag{16.27}$$

The convergence of $|u|_{\infty,S}$ and (16.21) implies the convergence of $|u|_B$, since

$$|u|_B \leq C|u|_{\infty,S} \text{ and } |u|_B \leq CE,$$

where C is a bounded constant independent of k.

The L-norm, which is equivalent to (16.25)–(16.27), is defined by

$$[v]_{r,\rho} = \left([v]_{in,r}^2 + [v]_{out,\rho}^2 \right)^{\frac{1}{2}},$$

where r and ρ are constant, and

$$[v]_{in,r} = \left[\int_0^\pi v^2(r, \ \theta) \ d\theta \right]^{\frac{1}{2}}, \ [v]_{out,\rho} = \left[\int_0^\pi v^2(\rho, \ \phi) \ d\phi \right]^{\frac{1}{2}}.$$

In fact, $[v]_{in,r}$ is an L-norm of the function $u(r, \theta)$ on $[0, \pi]$. Also denote

$$[v]_B = [v]_{r_{in},\rho_{out}} = \left([v]_{in}^2 + [v]_{out}^2\right)^{\frac{1}{2}},$$

where

$$[v]_{in} = [v]_{in,r_{in}}, \quad [v]_{out} = [v]_{out,\rho_{out}}.$$

The limits are

$$\tilde{u}_I = \tilde{u}_I(\tilde{a}_n) = \lim_{k \to \infty} \tilde{u}_I^{(2k)}\left(\tilde{a}_n^{(2k)}\right), \tag{16.28}$$

$$\tilde{u}_{II} = \tilde{u}_{II}(\tilde{c}_n) = \lim_{k \to \infty} \tilde{u}_{II}^{(2k)}\left(\tilde{c}_n^{(2k)}\right), \tag{16.29}$$

as $k \to \infty$, where $u_I(a_n)$ and $u_{II}(c_n)$ are given in (16.13) and (16.14). Let the errors

$$\tilde{\varepsilon}^{(2k)} = \begin{cases} \tilde{u}_I^{(2k)} - \tilde{u}_I & \text{in } S_1, \\ \tilde{u}_{II}^{(2k)} - \tilde{u}_{II} & \text{in } S_2, \end{cases} \tag{16.30}$$

we have

$$\left[\tilde{\varepsilon}^{(2k)}\right]_{in,r} = \left[\frac{\pi}{2}\sum_{n=0}^{L}\left(\tilde{a}_n^{(2k)} - \tilde{a}_n\right)^2 \frac{I_{n+\frac{1}{2}}^2(r)}{I_{n+\frac{1}{2}}^2(1)}\right]^{\frac{1}{2}}, \tag{16.31}$$

$$\left[\tilde{\varepsilon}^{(2k)}\right]_{out,\rho} = \left[\frac{\pi}{4}\sum_{n=0}^{LL}\left(\tilde{c}_n^{(2k)} - \tilde{c}_n\right)^2 \frac{K_{2n+1}^2(\rho)}{K_{2n+1}^2(1)}\right]^{\frac{1}{2}}. \tag{16.32}$$

Suppose that there exist the positive constants $\alpha_1 < 1$ and $\alpha_2 \leq 1$ such that (see Miller (1965))

$$\left[\tilde{u}_I^{(2k)} - \tilde{u}_I\right]_{in} \leq \alpha_1 \left[\tilde{u}_I^{(2k)} - \tilde{u}_I\right]_{in,r_1}, \tag{16.33}$$

$$\left[\tilde{u}_{II}^{(2k)} - \tilde{u}_{II}\right]_{out} \leq \alpha_1 \left[\tilde{u}_{II}^{(2k)} - \tilde{u}_{II}\right]_{out,\rho_1}, \tag{16.34}$$

$$\left[\tilde{u}^{(2k+2)} - \tilde{u}\right]_{r_1,\rho_1} \leq \alpha_2 \left[\tilde{u}^{(2k+1)} - \tilde{u}\right]_B, \tag{16.35}$$

where $\tilde{u} = \tilde{u}_I$ or $\tilde{u}_{II}$. It follows from (16.15) and (16.16) that

$$\left[\tilde{\varepsilon}^{(2k)}\right]_B \leq (\alpha_1\alpha_2)\left[\tilde{\varepsilon}^{(2k-2)}\right]_B \leq (\alpha_1\alpha_2)^k\left[\tilde{\varepsilon}^{(0)}\right]_B \to 0 \text{ as } k \to \infty. \tag{16.36}$$

Note that the estimates (16.33)–(16.35) result from the solution errors in Steps 2 and 3. We shall prove a prior estimate for α_1:

Theorem 16.1 *Let (16.23) and (16.24) hold, there exist the bounds:*

$$\left[\tilde{u}_I^{(2k)} - \tilde{u}_I\right]_{in} \leq \left(\frac{r_{in}}{r_1}\right)^{\frac{1}{2}} \left[\tilde{u}_I^{(2k)} - \tilde{u}_I\right]_{in,r_1}, \qquad (16.37)$$

$$\left[\tilde{u}_{II}^{(2k)} - \tilde{u}_{II}\right]_{out} \leq e^{-(\rho_{out}-\rho_1)} \left[\tilde{u}_{II}^{(2k)} - \tilde{u}_{II}\right]_{out,\rho_1}.$$

Proof. Since the general terms in series (16.11) and the integrand in (16.12) are positive, we have

$$I_\mu(r_{in}) = \sum_{k=0}^{\infty} \frac{\left(\frac{r_{in}}{r_1}\right)^{2k+\mu} \left(\frac{r_1}{2}\right)^{2k+\mu}}{\Gamma(k+1)\Gamma(k+\mu+1)} \leq \left(\frac{r_{in}}{r_1}\right)^{\mu} I_\mu(r_1),$$

$$K_n(\rho_{out}) = \frac{1}{2}\int_{-\infty}^{\infty} e^{-\rho_1 \cosh \eta - n\eta} \times e^{-(\rho_{out}-\rho_1)\cosh \eta} d\eta \leq e^{-(\rho_{out}-\rho_1)} K_n(\rho_1).$$

Then from (16.31) and (16.32) we obtain

$$\left[\tilde{u}_I^{(2k)} - \tilde{u}_I\right]_{in} \leq \left[\frac{r_{in}}{r_1}\frac{\pi}{2}\sum_{n=0}^{L}(\tilde{a}_n^{(2k)} - \tilde{a}_n)^2 \frac{I_{n+\frac{1}{2}}^2(r_1)}{I_{n+\frac{1}{2}}^2(1)}\right]^{\frac{1}{2}}$$

$$\leq \left(\frac{r_{in}}{r_1}\right)^{\frac{1}{2}} \left[\tilde{u}_I^{(2k)} - \tilde{u}_I\right]_{in,r_1},$$

$$\left[\tilde{u}_{II}^{(2k)} - \tilde{u}_{II}\right]_{out} \leq \left[e^{-2(\rho_{out}-\rho)}\frac{\pi}{4}\sum_{n=0}^{LL}(\tilde{c}_n^{(2k)} - \tilde{c}_n)^2 \frac{K_{2n+1}^2(\rho_1)}{K_{2n+1}^2(1)}\right]^{\frac{1}{2}}$$

$$\leq e^{-(\rho_{out}-\rho_1)} \left[\tilde{u}_{II}^{(2k)} - \tilde{u}_{II}\right]_{out,\rho_1}. \qquad \blacksquare$$

Define the ratio

$$R = \lim_{k\to\infty} \frac{\left[\tilde{\varepsilon}^{(2k)}\right]_B}{\left[\tilde{\varepsilon}^{(2k+2)}\right]_B}, \qquad (16.38)$$

which is the reciprocal of the contraction factor $\alpha_1\alpha_2$ in (16.36). The convergence rate is given by Hageman and Young (1981):

$$Convergence\ Rate = \log R.$$

The iterative procedure converges when $R > 1$. The larger R, the faster the convergence of Steps 1-4. Hence we may choose R to measure the convergence of the Schwarz alternative method.

For partitioning shown in Figure 16.1, we choose

$$r_{in} = \frac{1}{3}, \ \rho_{out} = 2, \ r_1 = 1, \ \rho_1 = \frac{4}{3}. \tag{16.39}$$

Note that there exists also an overlap between S_1 and S_2. Then a prior estimate R^* of R for this partition follows from Theorem 16.1 and the property $\alpha_2 \leq 1$:

$$R^* \leq \min\left\{ \left(\frac{r_1}{r_{in}}\right)^{\frac{1}{2}}, \ e^{\rho_{out}-\rho_1} \right\} = 1.73. \tag{16.40}$$

In practice, using the following sequential errors is more convenient since the solution limit (16.28) and (16.29) may be unknown:

$$\left|\Delta\tilde{\varepsilon}^{(2k)}\right|_B = \left[\left|\Delta\tilde{\varepsilon}_I^{(2k)}\right|_{in}^2 + \left|\Delta\tilde{\varepsilon}_{II}^{(2k)}\right|_{out}^2 \right]^{\frac{1}{2}}, \tag{16.41}$$

where

$$\Delta\tilde{\varepsilon}_I^{(2k)} = \tilde{u}_I^{(2k)} - \tilde{u}_I^{(2k-2)}, \quad \Delta\tilde{\varepsilon}_{II}^{(2k)} = \tilde{u}_{II}^{(2k)} - \tilde{u}_{II}^{(2k-2)}. \tag{16.42}$$

The trial computation in Li (1994) exhibits the empirical ratio:

$$\bar{R} = \frac{\left|\Delta\tilde{\varepsilon}^{(2k)}\right|_B}{\left|\Delta\tilde{\varepsilon}^{(2k+2)}\right|_B} = 2.8.$$

From (16.38) and the bounds

$$\left|\tilde{\varepsilon}^{(2k)}\right|_B \leq \frac{1}{\bar{R}^{k-1}(\bar{R}-1)} \left|\Delta\tilde{\varepsilon}^{(2)}\right|_B$$

in Schwarz (1989), we may simply regard

$$R \approx \bar{R} = 2.8. \tag{16.43}$$

The empirical value of $R = 2.8$ is much better than $R^* = 1.73$ in (16.40). An investigation on R versus overlap of subdomains is also undertaken in Li (1994).

16.4 ERROR ESTIMATES OF SOLUTIONS AND MATCHING STRATEGY BETWEEN ITERATION NUMBER AND ELEMENT SIZE

The following true errors are important in error estimates:

$$\left| \varepsilon^{(2k)} \right|_B = \left\{ \left| \widetilde{u}_I^{(2k)} - u_I \right|_{in}^2 + \left| \widetilde{u}_{II}^{(2k)} - u_{II} \right|_{out}^2 \right\}^{\frac{1}{2}},$$

where u_I and u_{II} are given in (16.13) and (16.14) with the true coefficients a_n and c_n given in Section 3.6.3 of Chapter 3.

For saving CPU time the iteration number k should be chosen not too large, but just to balancing errors from the iterative procedure and from the Ritz-Galerkin-FEM methods. In this section, we shall derive error bounds of the obtained solutions in S_3, and then lead to a useful matching formula.

Define a continuous solution u^* in S_3 that satisfies (16.5)–(16.8) and

$$u^* = \widetilde{u}_I \text{ on } l_{in}, \quad u^* = \widetilde{u}_{II} \text{ on } L_{out},$$

where $\widetilde{u}_I$ and $\widetilde{u}_{II}$ are given in (16.28) and (16.29). Also denote the norm

$$\|v\|_{\frac{1}{2},B} = \left(\|v\|_{\frac{1}{2},l_{in}}^2 + \|v\|_{\frac{1}{2},L_{out}}^2 \right)^{\frac{1}{2}}, \tag{16.44}$$

where $\|v\|_{\frac{1}{2},l_{in}}$ is the Sobolev half-norm, then we have the following lemma.

Lemma 16.1 *Let*

$$u \in H^2(S_3), \tag{16.45}$$

then there exists a bounded constant C independent of r_{in}, ρ_{out}, L, LL and h such that

$$\|u^* - u\|_{\frac{1}{2},B} \leq C \left[r_{in}^{L+\frac{3}{2}} + \left(\frac{1}{\rho_{out}} \right)^{2LL+3} + h^{\frac{3}{2}} \right]. \tag{16.46}$$

Proof. Let $\widehat{u}$ in S_3 satisfy (16.5), (16.8) and

$$\widehat{u} = u_I \text{ on } l_{in}, \quad \widehat{u} = u_{II} \text{ on } L_{out},$$

where u_I and u_{II} are given in (16.13) and (16.14) with the true coefficients a_n and c_n. By means of the triangular inequality, we have

$$\|u^* - u\|_{\frac{1}{2},B} \leq \|u^* - \widehat{u}\|_{\frac{1}{2},B} + \|\widehat{u} - u\|_{\frac{1}{2},B}. \qquad (16.47)$$

It is easy to obtain from (16.44), (16.9), (16.10), (16.13) and (16.14)

$$\|\widehat{u} - u\|_{\frac{1}{2},B} \leq C \left\{ \left[\sum_{l=L+1}^{\infty} a_l^2 (l + \frac{1}{2})^2 \frac{I_{l+\frac{1}{2}}^2(r_{in})}{I_{l+\frac{1}{2}}^2(1)} \right]^2 \right.$$

$$\left. + \left[\sum_{l=LL+1}^{\infty} c_l^2 (2l+1)^2 \frac{K_{2l+1}^2(\rho_{out})}{K_{2l+1}^2(1)} \right]^2 \right\}^{\frac{1}{2}}. \qquad (16.48)$$

From (16.45) and Chapter 12,

$$a_l = O\left(\frac{1}{l}\right), \qquad c_l = O\left(\frac{1}{l}\right), \qquad (16.49)$$

$$\frac{I_{l+\frac{1}{2}}(r_{in})}{I_{l+\frac{1}{2}}(1)} \leq C r_{in}^{l+\frac{1}{2}}, \qquad \frac{K_{2l+1}(\rho_{out})}{K_{2l+1}(1)} \leq C \left(\frac{1}{\rho_{out}}\right)^{2l+1}. \qquad (16.50)$$

Therefore, combining (16.48)-(16.50) yields

$$\|\widehat{u} - u\|_{\frac{1}{2},B} \leq C \left\{ \sum_{l=L+1}^{\infty} r_{in}^{2l+1} + \sum_{l=LL+1}^{\infty} \left(\frac{1}{\rho_{out}}\right)^{4l+2} \right\}^{\frac{1}{2}}$$

$$\leq C \left\{ r_{in}^{L+\frac{3}{2}} + \left(\frac{1}{\rho_{out}}\right)^{2LL+3} \right\}. \qquad (16.51)$$

On the other hand, the expansions $\widetilde{u}_I$ and $\widetilde{u}_{II}$ in (16.28) and (16.29) are computed from the finite element approximation of $\widehat{u}$ on l_{in} and L_{out}. Then

$$\|u^* - \widehat{u}\|_{\frac{1}{2},B} \leq C h^{\frac{3}{2}}, \qquad (16.52)$$

where h is the maximal boundary length of quasiuniform elements. The desired result (16.46) is obtained from (16.47), (16.51) and (16.52). ∎

Theorem 16.2 *Let (16.45) hold, and suppose that there exists a constant $R(> 1)$ independent of h, k, L and LL such that*

$$\left| \Delta \widetilde{\varepsilon}^{(2k+2)} \right|_B \leq \frac{1}{R} \left| \Delta \widetilde{\varepsilon}^{(2k)} \right|_B, \qquad (16.53)$$

where h is the maximal length of quasiuniform elements used in S_3. Then the errors of the solutions from Step 2, the FEM, have the bounds:

$$\left\| \tilde{u}_h^{(2k+1)} - u \right\|_{1,\widehat{S}_3^h} \leq C \left\{ h + \frac{1}{R^k} + r_{in}^{L+\frac{3}{2}} + \left(\frac{1}{\rho_{out}} \right)^{2LL+3} \right\}, \tag{16.54}$$

where C is also a bounded constant C independent of r_{in}, ρ_{out}, h, k, L and LL.

Proof. By means of the triangular inequality again we obtain

$$\left\| \tilde{u}_h^{(2k+1)} - u \right\|_{1,\widehat{S}_3^h} \leq \left\| \tilde{u}_h^{(2k+1)} - u_h^* \right\|_{1,\widehat{S}_3^h} + \left\| u_h^* - u^* \right\|_{1,\widehat{S}_3^h} + \left\| u^* - u \right\|_{1,\widehat{S}_3^h}. \tag{16.55}$$

The second term on the right side of (16.55) is bounded by

$$\left\| u_h^* - u^* \right\|_{1,\widehat{S}_3^h} \leq C h \left| u^* \right|_{2,\widehat{S}_3^h}.$$

We also have for the third term

$$\left\| u^* - u \right\|_{1,\widehat{S}_3^h}^2 = \left\| u^* - u \right\|_{1,S_3}^2 + \left\| u^* - u \right\|_{1,S_3 \cap \widehat{S}_3^h}^2. \tag{16.56}$$

Since $S_3 \cap \widehat{S}_3^h = O(h^3)$ and $u^* - u \in H^2(S_3)$ due to (16.45),

$$\left\| u^* - u \right\|_{1,S_3 \cap \widehat{S}_3^h}^2 = O(h^3). \tag{16.57}$$

Moreover, let $w = u^* - u$,

$$\|w\|_{1,S_3}^2 = \oint_{\partial S_3} w \frac{\partial w}{\partial n} d\ell = \int_{\ell_{in}} w \frac{\partial w}{\partial n} d\ell + \int_{\ell_{out}} w \frac{\partial w}{\partial n} d\ell$$

$$\leq \|w\|_{\frac{1}{2},\ell_{in}} \left\| \frac{\partial w}{\partial n} \right\|_{-\frac{1}{2},\ell_{in}} + \|w\|_{\frac{1}{2},\ell_{out}} \left\| \frac{\partial w}{\partial n} \right\|_{-\frac{1}{2},\ell_{out}}.$$

Since w satisfies equation (16.5), then

$$\left\| \frac{\partial w}{\partial n} \right\|_{-\frac{1}{2},\ell_{in}} \leq C\|w\|_{1,S_3}, \qquad \left\| \frac{\partial w}{\partial n} \right\|_{-\frac{1}{2},\ell_{out}} \leq C\|w\|_{1,S_3}.$$

It follows by noting the definition (16.44) that

$$\|w\|_{1,S_3} \leq C \left(\|w\|_{\frac{1}{2},\ell_{in}} + \|w\|_{\frac{1}{2},\ell_{out}} \right) \leq C\|v\|_{\frac{1}{2},B}. \tag{16.58}$$

We can see the bounds of the second term in (16.55) from (16.56)–(16.58) and Lemma 16.1:

$$\|u^* - u\|_{1,\widehat{S}_3^h} \leq C\left(\|u^* - u\|_{\frac{1}{2},B} + h^{\frac{3}{2}}\right)$$

$$\leq C\left\{r_{in}^{L+\frac{3}{2}} + \left(\frac{1}{\rho_{out}}\right)^{LL+3} + h^{3/2}\right\}. \quad (16.59)$$

Since $\widetilde{u}_h^{(2k+1)}$ and u_h^* are the finite element approximations from different values of solutions on l_{in} and L_{out}, we can provide bounds of the first term in the right side of (16.55), based on the extension theory for the FEM in DDM in Bramble, Pasciak and Schatz (1986), and Dryja (1982):

$$\left\|\widetilde{u}_h^{(2k+1)} - u_h^*\right\|_{1,\widehat{S}_3^h} \leq C\left\|\widetilde{u}_h^{(2k+1)} - u_h^*\right\|_{\frac{1}{2},\partial\widehat{S}_3^h}$$

$$\leq C\left\{\left|\widetilde{u}^{(2k+1)} - u_h^*\right|_B + \left\|\widetilde{u}_h^{(2k+1)} - u_k^*\right\|_{\frac{1}{2},S_3\cap\partial\widehat{S}_3^h}\right\}$$

$$\leq C\left\{\left|\widetilde{\varepsilon}^{(2k)}\right|_B + h^{\frac{3}{2}}\right\} \leq C\left\{\frac{1}{R^k}\left|\widetilde{\varepsilon}^{(2)}\right|_B + h^{\frac{3}{2}}\right\}. \quad (16.60)$$

Note that the last step of (16.60) follows from (16.53). Consequently, combining (16.55), (16.59) and (16.60) leads to (16.54). ∎

To grant the optimal convergence rate:

$$\left\|\widetilde{u}_h^{(2k+1)} - u\right\|_{1,\widetilde{S}_3^h} = O(h),$$

the following relations have to be fulfilled

$$\frac{1}{R^k} = O(h), \quad (16.61)$$

$$r_{in}^{L+\frac{3}{2}} = O(h), \quad (16.62)$$

$$\left(\frac{1}{\rho_{out}}\right)^{2LL+3} = O(h). \quad (16.63)$$

From (16.62) and (16.63) an optimal matching for balancing h and $L_S(= L + 1)$ and $LL_S(= LL + 1)$ has been obtained in Chapter 12. Below, an optimal strategy for balancing k and h can be found from (16.61):

Corollary 16.1 *Let all the conditions in Theorem 16.2 hold, and a pair (k_0, h_0) be an optimal balance by trial computation. Then any pair (k_1, h_1) of $k_1(\geq k_0)$*

k				8		14		20	
M_s	N_s	L	LL	$|\Delta\widetilde{\varepsilon}^{(2k)}|_B$	R	$|\Delta\widetilde{\varepsilon}^{(2k)}|_B$	R	$|\Delta\widetilde{\varepsilon}^{(2k)}|_B$	R
2	6	4	2	4.42×10^{-4}	2.86	9.84×10^{-6}	2.78	2.52×10^{-9}	2.77
3	9	5	3	4.37×10^{-4}	2.90	9.81×10^{-6}	2.78	2.52×10^{-9}	2.77
4	12	5	3	4.21×10^{-4}	2.91	9.15×10^{-6}	2.80	2.23×10^{-9}	2.79
6	18	6	4	4.12×10^{-4}	2.93	8.80×10^{-6}	2.81	2.20×10^{-9}	2.80
8	24	6	4	4.05×10^{-4}	2.93	8.52×10^{-6}	2.81	2.05×10^{-9}	2.80
12	36	7	5	4.03×10^{-4}	2.94	8.46×10^{-6}	2.82	2.03×10^{-9}	2.80

Table 16.1 The sequential errors $\|\delta\widetilde{\varepsilon}^{(2k)}\|_b$ and R versus different M_s by Steps 1-4.

and $h_1(\leq h_0)$ satisfying

$$k_1 = k_0 + \frac{\ln\left(\frac{h_0}{h_1}\right)}{\ln R}$$

is also an optimal balance. In particular, when

$$\frac{h_1}{h_0} = \frac{1}{2}, \quad R \approx 2, \tag{16.64}$$

then

$$k_1 \approx k_0 + 1. \tag{16.65}$$

Eqs. (16.64) and (16.65) are significant since only one more iteration is needed for optimal balance while the boundary length of elements used in S_3 decreases to its half!

From $M_S = O(\frac{1}{h})$ and the trial computation given in Table 16.1, the pair

$$k = 8, \quad M_S = 3 \tag{16.66}$$

can be regarded optimal in balance for the partition of Figure 16.1. Since $R(> 2)$ does not decrease as $h \to 0$, we choose from Corollary 16.1 the new good pairs in balance such as:

$$k = 9, \quad M_S = 6; \quad k = 10, \quad M_S = 12. \tag{16.67}$$

M_s	N_s	L	LL	k	Max	$\|\varepsilon\|_{0,S}$	$\|\varepsilon\|_1$
2	6	4	2	8	5.93×10^{-2}	4.22×10^{-2}	0.226
3	9	5	3	8	3.24×10^{-2}	1.91×10^{-2}	0.154
4	12	5	3	9	2.14×10^{-2}	1.09×10^{-2}	1.19×10^{-2}
6	18	6	4	9	1.17×10^{-2}	4.98×10^{-3}	8.05×10^{-2}
8	24	6	4	10	7.22×10^{-3}	2.76×10^{-3}	6.10×10^{-2}
12	36	7	5	10	3.71×10^{-3}	1.31×10^{-3}	4.16×10^{-2}
M_s	N_s	L	LL	k	$|\varepsilon|_{in}$	$|\varepsilon|_{out}$	$|\varepsilon|_B$
2	6	4	2	8	7.75×10^{-3}	1.37×10^{-2}	1.57×10^{-2}
3	9	5	3	8	3.10×10^{-3}	5.68×10^{-3}	6.47×10^{-3}
4	12	5	3	9	1.94×10^{-3}	3.08×10^{-3}	3.58×10^{-3}
6	18	6	4	9	8.40×10^{-4}	1.37×10^{-3}	1.61×10^{-3}
8	24	6	4	10	3.38×10^{-4}	6.88×10^{-4}	7.67×10^{-4}
12	36	7	5	10	2.45×10^{-4}	3.59×10^{-4}	4.34×10^{-4}

Table 16.2 Error norms versus different M_s by Steps 1-4.

Based on (16.66) and (16.67), the computed results are given in Table 16.2, where the norm

$$\|\varepsilon\|_1 = \left(\|\varepsilon\|_{1,S_1}^2 + \|\varepsilon\|_{1,S_2}^2 + \|\varepsilon\|_{1,\widehat{S}_3^h}^2 \right)^{\frac{1}{2}},$$

with $\varepsilon = \widetilde{u}_h^{(2k+1)} - u$. The curves in Figures 16.3 and 16.4 drawn from Table 16.2 exhibit the following asymptotic relations:

$$\|\varepsilon\|_1 = O(h), \qquad \|\varepsilon\|_{0,S} = O(h^2), \tag{16.68}$$

$$\|\varepsilon\|_{\infty,S} = O(h^{\frac{5}{3}}), \qquad |\varepsilon|_B = O(h^2). \tag{16.69}$$

Eqs. (16.68) and (16.69) verify the analysis on the optimal balance, as well as Theorem 16.2 and Corollary 16.1. It is remarkable that only 10 times the iterations in Steps 1-4 are required to match the finest triangulation in S_3 with $M_S = 12$.

Compared Steps 1-4 with the nonconforming combination. Two error norms $\|\varepsilon\|_1$ and $\|\varepsilon\|_{0,S}$ are optimal, but the norm $\|\varepsilon\|_{\infty,S}$ has a lower order $O(h^{\frac{5}{3}})$ than $O(h^{2-\delta})$ obtained in Chapter 12, where $\delta(>0)$ is arbitrarily small.

Finally, let us compare CPU time of Steps 1-4 with that of the nonconforming combination obtained in Chapter 12. Suppose that the Gaussian elimination method is used for solving the algebraic equations obtained. When we order the variables

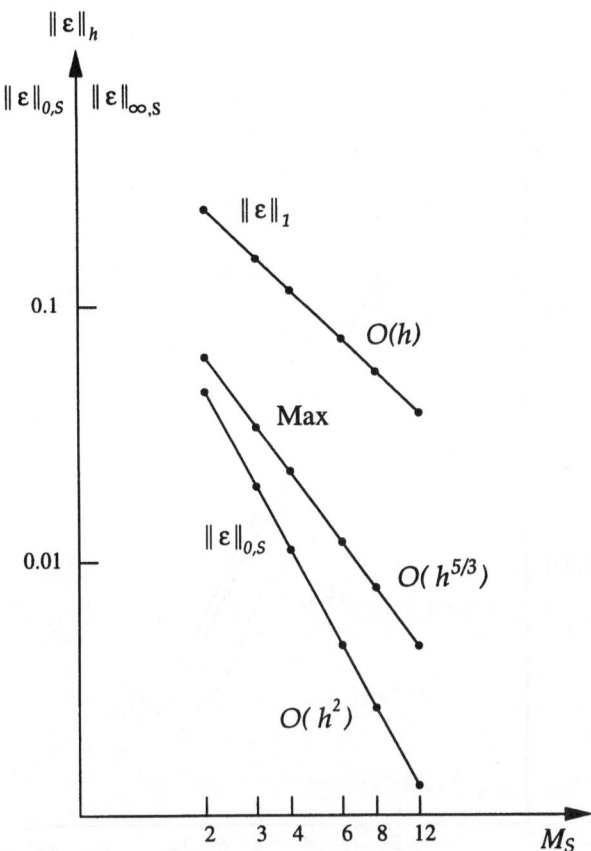

Figure 16.3   The error curves of $||\varepsilon||_h$, $||\varepsilon||_0$ and $||\varepsilon||_{\infty,S}$ versus M_S.

at element nodes along the radius direction, the band width of the stiff matrix in Step 2 is M_S. Since we need the LU decomposition only once, and repeatedly use the forward-backward procedure, the number of arithmetic operations in Schwarz (1989) is

$$\frac{1}{2}M_S(M_S + 3)N_S M_S^* + k\left(2N_S M_S^*(M_S + 1)\right),$$

where $M_S^* = M_S - 1$ and $N_S M_S$ is the total number of interior nodes of finite elements.

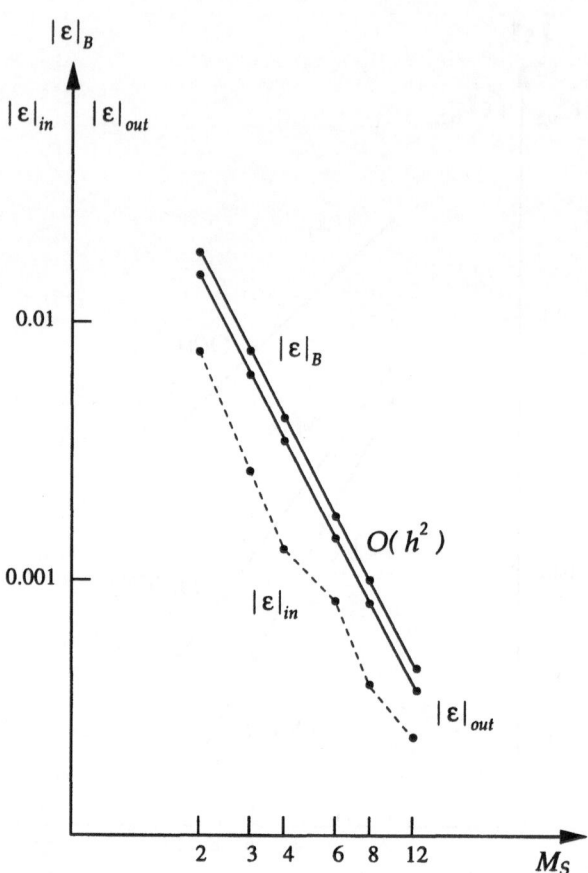

Figure 16.4   The error curves of $|\varepsilon|_{in}$, $|\varepsilon|_{out}$ and $|\varepsilon|_B$ versus M_S.

In Step 3, we need approximately

$$k\left[N_S(L_S + L_{SS}) + \frac{L_S^3}{3} + \frac{LL_S^3}{3}\right]$$

operations (when using (16.19) and (16.20), only $kN_S(L_S + L_{SS})$ operations are required). Therefore, the total number of arithmetic operations in Steps 1-4 is approximately

$$\frac{1}{2}M_S(M_S + 3)N_S M_S^* + k\left[2N_S M_S^*(M_S + 1) + N_S(L_S + L_{SS}) + \frac{L_S^3}{3} + \frac{LL_S^3}{3}\right].(16.70)$$

On the other hand, the order of variables at nodes in the nonconforming combination in Chapter 12 should be arranged along the radian direction, because the continuity constraints of admissible functions are enforced along l_{in} between S_3 and S_1, and along L_{out} between S_3 and S_2. Hence the band width of the associated matrix is $N_S + 1$, and the number of arithmetic operation is

$$\frac{1}{2}(N_S M_S^* + L_S + LL_S)(N_S + 1)(N_S + 4) + 2(N_S + 2)(N_S M_S^* + L_S + L_{SS}), \quad (16.71)$$

by the Gaussian elimination method.

When the optimal matchings (16.61), (16.62) and (16.63) are used, then

$$k = O(|\ln h|), \quad L_S = O(|\ln h|), \quad LL_S = O(|\ln h|).$$

The main parts of (16.70) and (16.71) are reduced to $c_0 M_S^2$ and $c_0 N_S^2$ as $h \to 0$, where $c_0 = \frac{1}{2} N_S M_S^*$. By noting the fact $N_S = 3 M_S$, we have the following corollary.

Corollary 16.2 *Let all conditions in Theorem 16.2 and the optimal matching (16.61)–(16.63) hold. Suppose that the Gaussian elimination method is used for solving the algebraic equations obtained from the FEM, the CPU time of Steps 1-4 in this chapter is about 1/9 of that in the nonconforming combination in Chapter 12 as $h \to 0$.*

In summary, the Schwarz alternating method is adopted to implement the combinations into parallel. In the Schwarz alternating method, the elliptic boundary problem is split into several (even many) smaller subproblems with some overlaps. For solving the subproblems, the numerical methods should be chosen to fix best the subproblems so that they are usually different. This leads to the combined methods that have been developed in this book. Although we take as an example only the parallel combinations for the RGM-FEM in Chapter 12, the Schwarz alternating method can be adopted to other combinations as well.

Remark 16.1 *The Schwarz alternating method originated by Schwarz (1869) is a kind of combinations, the combinations that split an entire problem into (or many) subproblems, solve the subproblems separately, and integrate the partial solutions through iteration to obtain the solution of the original problem. It is due to the Schwarz alternating method that the domain decomposition methods become a rich area of parallel computing. Moreover, the combined methods described in this book split the solution domain into several (or many) subdomains, choose different admissible functions, integrate these functions through coupling techniques, and solve* *the united algebraic equations simultaneously. Compared the combined methods with the domain decomposition methods, there are many things in common: domain partition, interface matching, parallelism, synchronization and domain combination. Chapters 15 and 16 demonstrate with samples how efficiently to apply the existing domain decomposition methods to the combined methods. Furthermore, Corollary 16.2 also exhibits a surprise that the proper matching of the domain decomposition methods and the combinations may greatly save CPU time.*

EPILOGUE

This whole book deals with nothing more than *combining* because the combining is so important:

It is true that without combining, there are no marriage and no family and no true human being;

It is true that with combining, our world becomes so diverse and so sophisticated and so flourish;

It is true that the combining brings us many fantasies and many miracles and many surprises.

A number of surprises have been already reported in this book: collaboration in combinations, extreme accuracy of the leading coefficients of Motz's solutions with near 1000 significant digits, the most accurate solutions for Motz's and the crack-infinity problems, a few terms as $L = O(|\ln h|)$ of particular solutions used as in RGM-FEM and RGM-FDM, high efficiency in solving singularity and unbounded domain problems, facile programming for the algorithms of combinations, significant saving of CPU time and computer storage, etc. We truly believe that more surprises will come by using the combined methods, the combinations that may integrate all merits of the existing methods. (This is also an echo to Strang and Fix (1973, p. 135)).

Finally, let us address again the philosophy guiding the study on the combined methods and in this book.

No method is perfect. Once several numerical methods have appeared for solving elliptic boundary value problems, their comparisons and combinations are inevitable, to seek a better performance, e.g., in saving CPU time, or to solve complicated problems, in particular for those involving singularities.

When combining different numerical methods, the following steps may be taken into account.

443

Step 1: Analyze solution behavior, partition the solution domain S into several subdomains S_i with suitable geometrical shapes, and choose the numerical methods in the subdomains to fit best the solution behavior wherein.

Step 2: Choose coupling techniques to match the different numerical methods along their common boundary Γ_0.

Step 3: Seek the optimal parameters involved in the coupling techniques, e.g., the penalty power σ when using the penalty integrals on Γ_0.

Step 4: Find the optimal matching relations of different methods, for instance, the matching pairs, L and h, of RGM-FEM (or RGM-FDM).

Step 5: Determine, by trial computation on a coarse partition, some empirical parameters to be used in computation, e.g., the penalty constant P_c, and the sample of good matching pairs between h and L.

Step 6: Carry out applicable computations.

Step 7: Examine and analyze the computed results.

Step 8: Change the domain partition, coupling techniques, or coupling parameters, etc., and go back to the above steps for retrial computations if the computed results are not satisfactory.

Note that when combining different methods, users must follow the matching rules described in this book; otherwise disasters may occur. If we speak of combining, coupling and matching as a kind of marriage in Bossavi (1988, 1991), we would like to cite the following quotation as a warning.

Never say that marriage has more of joy than pain.
—— Alcestic (438 B.C.) ——

Euripides (485–406 B.C.)

in Familiar Quotations (Sixteenth Edition), Eds. by Bartlett J. and Kaplan J., Little, Brown and Company, Boston, 1992.

REFERENCES

Abdel-Messieh, Y. S. and Thatcher, R. W. (1990), Estimating the form of some three-dimensional singularities, *Commun. Appl. Numer. Methods*, 6(5):333-341.

Abramowitz, M. and Stegun, I. A. (1972), *Handbook of Mathematical Functions: with Formulas, Graphs and Mathematical Tables*, Dover Publications, Inc., New York.

Adams, R. A. (1975), *Sobolev Spaces*, Academic Press, New York.

Ahlfords, L. V. (1979), *Complex Analysis(Third Ed.)*, McGrow Hill Book Company.

Akin, J. E. (1976), The generation of elements with singularities, *Inter. J. Numer. Methods Engrg.*, 10(6):1249–1259.

Akin, J. E. (1979a), Finite elements, singularities and fracture, in Whiteman, J. R., editor, *MAFELAP 1978*, 35–54, Academic Press, London.

Akin, J. E. (1979b), Elements for the analysis line singularities, in Whiteman, J. R., editor, *MAFELAP 1978*, 65–75, Academic Press, London.

Akin, J. E. (1982), *Application and Implementation of Finite Element Methods*, Academic Press, London.

Ames, W. F. (1977), *Numerical Methods for Partial Differential Equations*, Academic Press, New York.

Amini, S. and Kirkup, S. M. (1995), Solution of Helmholtz in the exterior domain by elementary boundary integral methods, *J. Comp. Phys.*, 118:208–221.

Anderson, C. R. (1989), Domain decomposition techniques and the solution of Poisson's in infinite domains, in Chan, T. F. et. al., editors, *Domain Decomposition Methods*, 129–139, SIAM, Philadelphia.

Arnold, D. (1982), An interior penalty finite element method with discontinuous elements, *SIAM J. Numer. Anal.*, 19:742–760.

Ascher, U. M., Mattheij, R. M. M. and Russell, R. D. (1988), *Numerical Solution of Boundary Value Problems for Ordinary Differential Equations*, Prentice Hall.

Atkinson, K. E. (1985), The numerical evaluation of particular solutions for Poisson's equation, *IMA J. Numer. Anal.*, 5:319–338.

Atkinson, K. E. (1989), *An Introduction to Numerical Analysis (2nd Edition)*, John Wiley & Sons, New York.

Atkinson, K. and de Hoog, F. (1984), The numerical solution of Laplace's equation on a wedge, *IMA J. Numer. Anal.*, 4:19–41.

Axelsson, O. and Barker, V. A. (1984), *Finite Element Solution of Boundary Value Problems: Theory and Computation*, Academic Press.

Aziz, A. K., Dorr, M. R. and Kellogg, R. B. (1982), A new approximation method for the Helmholtz equation in an exterior domain, *SIAM J. Numer. Anal.*, 19(5):899–908.

Aziz, A. K. and Kellogg, R. B. (1981), Finite element analysis of a scattering problem, *Math. Comp.*, 37:261–272.

Aziz, A. K., Kellogg, R. B. and Stephen, A. B. (1985), Least squares methods for elliptic systems, *Math. Comp.*, 44:53–70.

Babuska, I. (1970), Finite element method for domains with corners, *Computing*, 6:264–273.

Babuska, I. (1972), The finite element method for infinite domains I, *Math. Comp.*, 26(117):1–11.

Babuska, I. (1973a), The finite element method with Lagrangian multipliers, *Numer. Math.*, 20:179–192.

Babuska, I. (1973b), The finite element method with penalty, *Math. Comp.*, 27:221–228.

Babuska, I. (1974), Solution of problem with interface and singularities, in Boor, C., editor, *Mathematical Aspects of Finite Elements in Partial Differential Equations*, 213–277, Academic Press, New York.

Babuska, I. (1976a), Solution of interface problems by homogenization I, *SIAM J. Math. Anal.*, 7(5):603–634.

Babuska, I. (1976b), Solution of interface problems by homogenization II, *SIAM J. Math. Anal*, 7(5):635–645.

Babuska, I. (1977), Solution of interface problems by homogenization III, *SIAM J. Math. Anal*, 8(6):923–937.

Babuska, I. and Aziz, A. Z. (1972), Survey lectures on the mathematical foundations of the finite element method, in Aziz, A. K., editor, *The Mathematical Foundations of the Finite Element Method with the Application to Partial Differential Equations*, 3–359, Academic Press, New York.

Babuska, I. and Dorr, M. R. (1981), Error estimates for the combined h and p versions of the finite element method, *Numer. Math.*, 37:257–277.

Babuska, I. and Guo, B. Q. (1988), The $h-p$ version of the finite element method for domains with curved boundaries, *SIAM J. Numer. Anal.*, 25(4):837–861.

Babuska, I. and Guo, B. Q. (1989), The $h-p$ version of the finite element method for problems with nonhomogeneous essential boundary condition, *Comp. Meth. Appl. Mech. Engrg.*, 74:1–28.

Babuska, I., Guo, B. Q. and Osborn, J. E. (1989), Regularity and numerical solution of eigenvalue problems with piecewise analytic data, *SIAM J. Numer. Anal.*, 26(6):1534–1560.

Babuska, I., Guo, B. Q. and Stephan, E. P. (1990), The $h-p$ version of the boundary element method with geometric mesh on polygonal domains, *Comput. Methods Appl. Mech. Engrg*, 80:319–325.

Babuska, I. and Kellogg, R. B. (1975), Nonuniform error estimates for the finite element method, *SIAM J. Numer. Anal.*, 12(6):868–875.

Babuska, I., Kellogg, R. B. and Pitkaranta, J. (1979), Direct and inverse error estimates for finite elements with mesh refinements, *Numer. Math.*, 33:447–471.

Babuska, I. and Miller, A. (1984), The post-processing approach in the finite element method I. Calculation of displacements, stresses and other higher derivatives of the displacements, II. The calculation of stress intensity factors, *Inter. J. Numer. Methods Engrg.*, 20:1085–1109, 1111–1129.

Babuska, I. and Oh, H. S. (1990), The p-version of the finite element method for domains with corners and for infinite domains, *Numer. for Meth. PDE*, 6(4):371–392.

Babuska, I. and Osborn, J. (1980), Analysis of finite element methods for second order boundary value problems using mesh dependent norms, *Numer. Math.*, 34:41–62.

Babuska, I. and Osborn, J. (1991), Eigenvalue problems in P. G. Ciarlet and J. L. Lions, editors, *Finite Element Methods(Part I)*, 641–788, North-Holland.

Babuska, I. and Rheinboldt, W. C. (1978), Error estimates for adaptive finite element computations, *SIAM J. Numer. Anal.*, 15:736–754.

Babuska, I. and Rheinboldt, W. C. (1984), Adaptive finite element processes in structural mechanics, in Birkhoff, G. and Schoenstat, A., editors, *Elliptic Problem Solvers II*, Academic Press.

Babuska, I. and Rosenzweig, M. B. (1972), A finite element scheme for domains with corners, *Numer. Math.*, 20:1–21.

Babuska, I., Stroulis, T., Gangaraj, S. K. and Upadhyay, C. S. (1997), Pollution error in the h-version of the finite element method and the local quality of the recovered derivatives,

Comput. Methods Appl. Mech. Engrg., 140:1–37.

Babuska, I. and Suri, M. (1987), The optimal convergence rate of the p-version of finite element method, *SIAM J. Numer. Anal.*, 24:750–776.

Babuska, I. and Suri, M. (1990), The p- and h − p versions of the finite element method an overview, *Comput. Methods Appl. Mech. Engrg.*, 80:5–26.

Babuska, I., Szabo, B. A. and Katz, I. N. (1981), The p-version of the finite element method, *SIAM J. Numer. Anal.*, 18(3):515–545.

Babuska, I. and Zlamal, M. (1973), Nonconforming elements in the finite element method with penalty, *SIAM J. Numer. Anal*, 10(5):863–875.

Baker, G. A. (1977), Finite element methods for elliptic equations using nonconforming elements, *Math. Comp.*, 31:45–59.

Bank, R. E. and Rose, D. J. (1987), Some error estimates for the box method, *SIAM J. Numer. Anal.*, 24:777–787.

Bank, R. E. and Scott, L. R. (1989), On the conditioning of finite element equations with highly refined meshes, *SIAM J. Numer. Anal.*, 26(6):1383–1394.

Barbosa, H. J. C. and Hughes, T. J. R. (1992), Boundary Lagrange multipliers in finite element methods: error analysis in natural norms, *Numer. Math.*, 62:1–15.

Barnhill, R. E. and Whiteman, J. R. (1974), Singularities due to re-entrant boundaries in elliptic problems, in *Numerische Methoden bei Differentialgleichungen und Mit Funktional Analytischen*, 29–45, Internat. Schriftenreihe Numer. Math., Band 19, Birkhauser, Basel.

Barnhill, R. E. and Whiteman, J. R. (1975), Error analysis of Galerkin methods for Dirichlet problems containing boundary singularities, *J. Inst. Math. Appl.*, 15:121–125.

Barrett, J. W. and Elliott, C. M. (1986), Finite element approximation of the Dirichlet problem using the boundary penalty method, *Numer. Math.*, 49:343–366.

Barrett, J. W. and Elliott, C. M. (1987), Fitted and unfitted finite-element methods for elliptic equations with smooth interfaces, *IMA J. Numer. Anal.*, 7:283–300.

Barsoum, R. S. (1976), On the use of isoparametric finite elements in linear fracture mechanics, *Inter. J. Numer. Meth. Engrg.*, 10:25–27.

Barsoum, R. S. (1977), Triangular quarter-point elements as elastic and perfectly-plastic crack tip elements, *Inter. J. Numer. Meth. Engrg.*, 11:85–98.

Bath, K. J. and Wilson, E. L. (1976), *Numerical Methods in Finite Element Analysis*, Prentice-Hall Inc., Englewood Cliffs, New Jersey.

Bayliss, A., Gunzburger, M. and Turkel, E. (1982), Boundary conditions for the numerical solution of elliptic equations in exterior regions, *SIAM J. Appl. Math.*, 42(2):430–451.

Beagles, A. E. and Whiteman, J. R. (1986), Finite element treatment of boundary singularities by augmentation with nonexact singular functions., *Numer. Meth. for PDE*, 2(2):113–121.

Begue, C., Bernardi, C., Debit, N., Maday, Y., Kariadakis, G. E., Mavriplis, C. and Patera, A. T. (1989), Non-conforming spectral element-finite element approximations for partial differential equations, *Comput. Methods Appl. Mech. Engrg.*, 75:109–125.

Bellman, S. and Adomian, G. (1985), *Partial Differential Equations: New Methods for their Treatment and Solution*, D. Reidel Publishing Company.

Belytschko, T. and Lu, Y. Y. (1994), A variationally coupled FE-BE method for transient problems, *Inter. J. Numer. Methods Engrg.*, 37:91–105.

Berger, H., Warnecke, G. G. and Wendland, W. L. (1994), Coupling of FEM and BEM for transonic flows, in Whiteman, J. R., editor, *The Mathematics of Finite Elements and Applications*, 323–350, Wiley-Interscience Publication, Chichester.

Bergman, S. (1969), *Integral Operations in the Theory of Partial Differential Equations*, Springer-Verlag, New York.

Bergman, S. and John, G. H. (1961), Application of the method of the kernel function for solving boundary value problems, *Numer. Math.*, 3:209–225.

Bergman, S. and John, G. H. (1965), Numerical solution of boundary value problems by the method of integral operator, *Numer. Math.*, 7:42–65.

Bermejo, R. (1991), Analysis of an algorithm for the Galerkin-characteristic method, *Numer. Math.*, 60:163–194.

Bernardi, C., Debit, N. and Maday, Y. (1990), Coupling finite element and spectral methods: first results, *Math. Comp.*, 54(189):21–39.

Bernardi, C. and Karageorghis, A. (1996), Methodes spectrales dans une portion de disque, *Numer. Math.*, 73:265–289.

Bernardi, C. and Maday, Y. (1997), Spectral methods in Ciarlet P. G. and Lions L. G. editors, *Techniques of Scientific Computing (Part II)*, 209-485, Elsevier, Amsterdam.

Bernardi, C., Maday, Y. and Landriani, G. S. (1989/90), Nonconforming matching conditions for coupling spectral and finite element methods, *Appl. Numer. Math.*, 6(1-2):65–84.

Bernardi, C., Maday, Y. and Patera, A. T. (1993), Domain decomposition by the Mortar element method, in Kaper, H. G. and Garbey, M., editors, *Asymptotic and Numerical Methods for Partial Differential Equations with Critical Parameters*, 269–286, Kluwer Academic Publishers, Dordrecht, NATO Adv. Sci. Inst. Ser. C Math. Phys. Sci., 384.

Bettess, P. (1977), Infinite elements, *Inter. J. Numer. Methods Engrg.*, 11:53–64.

Bettess, P., Emson, C. and Chiam, T. C. (1984), A new mapped infinite element for exterior wave problem, in Lewis, R. W., et. al. editors, *Numerical Methods in Coupled Systems*, 489–504, Wiley - Interscience Publication.

Bielak, J. and MacCamy, R. C. (1991), Symmetric finite element and boundary integral coupling methods for fluid-solid interaction, *Quar. Appl. Math.*, 49(1):107–119.

Birkhoff, G. (1972), Angular singularities of elliptic problem, *J. Appl. Th.*, 6:215–230.

Birkhoff, G. and Lynch, R. E. (1984), *Numerical Solution of Elliptic Problems*, SIAM, Philadelphia.

Bjorck, A. (1990), Least squares methods, in Ciarlet, P. G. and Lions, J. L., editors, *Handbook of Numerical Analysis, Vol. 1*, 465–652, North–Holland.

Bjorstad, P. E. and Widlund, O. B. (1986), Iterative methods for the solution of elliptic equations on the regions partitioned into substructure, *SIAM J. Numer. Anal.*, 23:1097–1120.

Blum, H. (1982), A simple and accurate method for the determination of stress intensity factors and solutions for problems on domains with corners, in Whiteman, J. R., editor, *MAFELAP 1981*, 57–64, Academic Press, London.

Blum, H. (1987), On Richardson extrapolation for linear finite elements on domains with reentrant corners, *ZAMM*, 67(5):T351–T353.

Blum, H. (1988a), Extrapolation techniques in the numerical treatment of corner and crack singularities, in Whiteman, J. R., editor, *The Mathematics of Finite Elements and Applications VI*, 555–562, Academic Press, London.

Blum, H. (1988b), Numerical treatment of corner and crack singularities, in *Finite Element and Boundary Element Techniques from Mathematical and Engineering Point of View*, 171–212, Springer, Vienna.

Blum, H. and Dobrowolski, M. (1982), On finite element methods for elliptic equations on domains with corners, *Computing.*, 28:53–63.

Blum, H. and Rannacher, R. (1988), Extrapolation techniques for reducing the pollution effect of reentrant corners in the finite element method, *Numer. Math.*, 52(5):539–564.

Blum, H. and Rannacher, R. (1990), Finite element eigenvalue computation on domains with reentrant corners using Richardson extrapolation, *J. Comp. Math.*, 8(4):321–332.

Bochev, P. B. and Gunzburger, M. D. (1994), Analysis of least squares finite element methods for the Stokes equations, *Math. Comp.*, 63(208):479–506.

Bohmer, K. and Locker, J. (1988), Asymptotic expansions for the discretization error of least squares solutions of linear boundary value problems, *Math. Comp.*, 51(183):75–91.

Boisvert, R. F. (1981), Families of high order accurate discretizations of some elliptic problems, *SIAM J. Sci. Stat. Comput.*, 2(3):268–284.

Borgers, C. (1990), A triangulation algorithm for fast elliptic solvers based on domain imbedding, *SIAM J. Numer. Anal.*, 27(5):1187–1196.

Borgers, C. and Widlund, O. B. (1990), On finite element domain imbedding methods, *SIAM J. Numer. Anal.*, 27(4):963–978.

Bossavit, A. (1988), Mixed methods and the marriage between finite elements and boundary elements, in Brebbia, C. A., editor, *Boundary Elements X, Vol. 1*, 447–458, Comput.

Mech., Southampton.

Bossavit, A. (1991), Mixed methods and the marriage between "mixed" finite elements and boundary elements, *Numer. Meth. for PDE*, 7:347–362.

Boulbrachene, M., ph. Cortey-Dumont and Miellou, J. C. (1988), Mixing finite element and finite differences in a subdomain method, in Glowinski, R. et. al., editors, *First Inter. Symposium on Domain Decomposition Methods for PDEs*, 198–216, SIAM, Philadelphia.

Bourlard, M., Dauge, M., Lubuma, M. S. and Nicaise, S. (1992), Coefficients of the singularities for elliptic boundary value problems on domains with conical points III. finite element methods on polygonal domains, *SIAM J. Numer. Anal.*, 29(1):136–155.

Bourlard, M., Dauge, M. and Nicaise, S. (1989), Error estimates on the coefficients obtained by the singular function method, *Numer. Func. Anal. Optim.*, 10(11-12):1077–1113.

Bourlard, M., Nicaise, S. and Paquet, L. (1990), An adapted Galerkin method for the resolution of Dirichlet and Neumann problems in a polygonal domain, *Math. Methods Appl. Sci.*, 12(3):251–265.

Bowers, K. L. and Lund, J. (1987), Numerical solution of singular Poisson problems via the Sinc-Galerkin method, *SIAM J. Numer. Anal.*, 24(1):36–51.

Boyd J. F. (1989), *Chebyshev and Fourier Spectral Methods, Lectures Notes in Engineering*, 49, Springer, Berlin.

Bramble, J. H. (1981), The Lagrange multiplier methods for Dirichlet's problem, *Math. Comp.*, 37(155):1–12.

Bramble, J. H., Ewing, R. E., Parashkevov, R. R. and Pasciak, J. E. (1992), Domian decomposition methods for problems with partial refinements, *SIAM J. Sci, Stat. Comp.*, 13:397–410.

Bramble, J. H., Hubbard, B. E. and Zlamal, M. (1968), Discrete analogous of the Dirichlet problem with isolated singularities, *SIAM J. Numer. Anal.*, 5:1–25.

Bramble, J. H. and King, J. T. (1997), A finite element method for interface problems in domains with smooth boundaries and interfaces, *Comp. Math.*, 6:109–138.

Bramble, J. H. and Nitsche, J. A. (1973), A generalized Ritz-least-squares method for Dirichlet problems, *SIAM J. Numer. Anal.*, 10:81–93.

Bramble, J. H., Pasciak, J. E. and Schatz, A. H. (1986), The construction of preconditioners for elliptic problems by substructuring I, *Math. Comp.*, 47:103–134.

Bramble, J. H., Pasciak, J. E. and Schatz, A. H. (1988), The construction of preconditioners for elliptic problems by substructuring III, *Math. Comp.*, 51(184):415–430.

Bramble, J. H., Pasciak, J. E. and Schatz, A. H. (1989), The construction of preconditioners for elliptic problems by substructuring IV, *Math. Comp.*, 53(187):1–24.

Bramble, J. H., Pasciak, J. E. and Xu, J. (1990), Parallel multilevel preconditioner, in Chan, T., Glowinski, E., Perieux, J. and Wildlund, O., editors, *Third Int. Symposium on Domain Decompositions for PDEs*, 341–357, SIAM, Philadelphia.

Bramble, J. H. and Schatz, A. H. (1970), Rayleigh-Ritz-Galerkin methods for Dirichlet's problem using subspaces without boundary conditions, *Comm. Pure Appl. Math.*, 23:653–675.

Bramble, J. H. and Schatz, A. H. (1971), Least squares methods for 2mth order elliptic boundary-value problems, *Math. Comp.*, 25(113):1–32.

Bramble, J. H. and Schatz, A. H. (1977), Higher order local accuracy by averaging in the finite element method, *Math. Comp.*, 31:94–111.

Brebbia, C. A. (1978), *The Boundary Element Method for Engineering*, Pentch Press, London.

Brebbia, C. A. and Dominguez, J. (1989), *Boundary Elements, An Introductory Course*, McGraw-Hill Book Company, New York.

Bremi, P. (1977), Combination of analytical and numerical calculation methods, in Descloux, J. and Marti, J., editors, *Numerical Analysis*, 195–208, Internat. Ser. Numer. Math., Birkhauser and Basel.

Brezzi, F. and Fortin, M. (1991), *Mixed and Hybrid Finite Element Methods*, Springer-Verlag, New York.

Brezzi, F. and Johnson, C. (1979), On the coupling of boundary integral and finite element methods, *Calcolo*, 16(2):189–201.

Brezzi, F., Johnson, C. and Nedelec, J. C. (1978), On the coupling of boundary integral and finite element methods, in *Proceedings of the Fourth Symposium on Basic Problems of Numerical Mathematics*, 103–114. Prague.

Brichau, F. and Deconinck, J. (1992), A numerical model coupling galvanic corrosion and ohmic voltage drop in buried pipelines, in Brebbia, C. A., Dominguez, J. and Paris, F., editors, *Boundary Element Technology, VII*, 389–403, Comput. Mech., Southampton.

Brink, U. and Stephan, E. P. (1996), Convergence rates for the coupling of FEM and collocation BEM, *IMA J. Numer. Anal.*, 16:93–110.

Browder, F. E. (1962), Approximation by solutions of partial differential equations, *Amer. J. Math.*, 84:134–160.

Bui, T. D. and Oppenheim, A. K. (1987), Evaluation of wind effects on model buildings by the random vortex method, *Appl. Numer. Math.*, 13:195–207.

Burda, P. (1979), The conforming finite element method for elliptic problems with interfaces, *Acta Polytechnica Prace CVUT*, 4(1):51–75.

Burda, P. (1993), A finite element algorithm for elliptic problems with interfaces and singularities, *ZAMM*, 73(7-8):T666–T670, Bericht uber die Wissenschaftliche Jahrestagung der GAMM.

Cai, Z. (1991), On the finite volume method, *Numer. Math.*, 58:713–735.

Cai, W., Lee, H. C. and Oh, H. S. (1993), Coupling of spectral methods and the p version of the finite element method for elliptic boundary value problems containing singularities, *J. Comp. Phys.*, 108(2):314–326.

Canuto, C. and Funaro, D. (1988), The Schwarz algorithm for spectral methods, *SIAM J. Numer. Anal.*, 25(1):24–40.

Canuto, C., Hariharan, S. I. and Lustman, L. (1985), Spectral methods for exterior elliptic problem, *Numer. Math.*, 46:505–520.

Canuto, C., Hussaini, M. Y., Quarteroni, A. and Zhang, T. A. (1988), *Spectrsl Methods in Fluid Dynamics*, Springer-Verlay, Berlin.

Canuto, C., Maday, Y. and Quarteroni, A. (1982), Analysis of the combined finite element and Fourier interpolation, *Numer. Math.*, 39:205–220.

Canuto, C. and Pietra, P. (1990), Boundary and interface conditions within a finite element preconditioner for spectral methods, *J. Comp. Phys.*, 91:310–343.

Cao, Z. Y. and Cheung, Y. K. (1992), *Semi-Analytic Numerical Methods (in Chinese)*, Defence Industry Publishing, Beijing.

Carey, G. E. and Oden, T. T. (1986), *Finite Elements Fluid Mechanics VI*, Prentice Hall, Englewood Cliffs.

Carey, G. F. and Shen, Y. (1989), Convergence studies of least-squares finite elements for first order systems, *Comm. Appl. Numer. Meth.*, 5:427–434.

Carstensen, C. (1996), Coupling of FEM and BEM for interface problems in viscoplasticity and plasticity with hardening, *SIAM J. Numer. Anal.*, 33(1):171–207.

Carstensen, C. and Stephan, E. P. (1995), Coupling of FEM and BEM for a nonlinear interface problem: the $h - p$ version, *Numer. Meth. for PDE*, 11:539–554.

Cayco, M., Foster, L. and Swann, H. (1995), On the convergence rate of the cell discretization algorithm for solving elliptic problems, *Math. Comp.*, 64(212):1397–1419.

Cecot, W. and Orkisz, J. (1987), On coupling of the boundary element and τ function methods, in *Boundary Elements IX, Vol. 1*, 435–445, Comput. Mech., Southampton.

Chan, T. F. (1989), Analysis of domain decomposition preconditioners, *SIAM J. Numer. Anal.*, 24:382–390.

Chan, T. F., Hou, T. Y. and Lions, P. L. (1991), Geometry related convergence results for domain decomposition algorithms. *SIAM J. Numer. Anal.*, 28:378–391.

Chen, C. and Huang, Y. (1995), *High Accuracy Theory of Finite Element Methods (in Chinese)*, Human Science and Technology Publishing.

Chen, G. and Zhou, J. (1992), *Boundary Element Methods*, Academic Press, New York.

Cheney, E. W. (1966), *Introduction to Approximation Theory*, McGraw-Hill, New York.

Cheung, Y. K., Guo, D., Tham, L. G. and Cao, Z. (1986), Infinite ring elements, *Inter. J.*

Numer. Methods Engrg., 23:385–396.

Cheung, Y. K., Jin, W. G. and Zienkiewicz, O. C. (1989), Direct solution procedure for solution of harmonic problems using complete, nonsingular, Trefftz functions, *Comm. Appl. Numer. Meth.*, 5(3):159–169.

Cheung, Y. K., Jin, W. G. and Zienkiewicz, O. C. (1991), Solution of Helmholtz equation by Trefftz method, *Inter. J. Numer. Methods Engrg.*, 32:63–78.

Chorin, A. J. (1973), Numerical studies of slightly viscous flow, *J. Fluid Mech.*, 57:785–796.

Chow, S. N., Duninger, D. R. and Miklavcic, M. (1990), Galerkin approximations for singular linear elliptic and semilinear parabolic problem, *Applicable Analysis*, 40:41–52.

Ciarlet, P. G. (1978), *The Finite Element Method for Elliptic Problem*, North–Holland, New York.

Ciarlet, P. G. (1991), Basic error estimates for elliptic problems, in Ciarlet P. G. and Lions L. G. editors, *Finite Element Methods (Part I)*, 17–352, North-Holland, Amsterdam.

Ciric, I. R. and Wong, S. H. (1986), Inversion transformation for the finite-element solution of three-dimensional exterior-field problems, *Inter. J. Comp. Math. EEE*, 5(2):109–119.

Colin, F. and Floquet, A. (1986), Combination of finite element and integral transform techniques in a heat conduction quasi-static problem, *Inter. J. Numer. Methods Engrg.*, 23:13–23.

Collatz, L. (1981), Some numerical test problems with singularities, in Whiteman, J. R., editor, *MAFELAP IV*, 65–76, Academic Press, London, New York.

Cordero, C. A., Dodson, C. T. J. and Leon, M. D. (1989), *Differential Geometry of Frame Bundles*, Klumer Academic Publishers.

Costabel, M. (1987), Symmetric methods for the coupling of finite elements and boundary elements, in *Boundary Elements IX, Vol. 1*, 411–420, Comput. Mech., Southampton.

Costabel, M. (1989), On the coupling of finite and boundary element methods, in *Proceedings of the Inter. Symposium on Numerical Analysis*, 1–14, Middle East Tech. Univ., Ankara.

Costabel, M., Ervin, V. J. and Stephan, E. P. (1990), Symmetric coupling of finite elements and boundary elements for a parabolic-elliptic interface problem, *Quar. Appl. Math.*, 48(2):265–279.

Costabel, M. and Stephan, E. P. (1990), Coupling of finite and boundary element methods for an elastoplastic interface problem, *SIAM J. Numer. Anal.*, 27(5):1212–1226.

Coulaud, O., Funaro, D. and Kavian, O. (1990), Laguerre spectral approximation of elliptic problems in exterior domains, *Comp. Meth. Appl. Mech. Engrg.*, 80:451–458.

Courant, R. (1943), Variational methods for the solution of problems of equilibrium and vibrations, *Bulletin of AMS*, 49:2165–2187.

Courant, R. and Hilbert, D. (1953), *Methods of Mathematical Physics, Vol. 1*, Wiley-Interscience Publishers, New York.

Cox, C. L. and Fix, G. J. (1984), On the accuracy of least squares methods in the presence of corner singularities, *Comput. Math. Applic.*, 10(6):463–475.

Crank, J. and Furzeland, R. M. (1977), The treatment of boundary singularities in axially symmetric problems containing discs, *J. Inst. Math. Appl.*, 20(3):355–370.

Cruse, T. A. (1988), *Boundary Element Analysis in Computational Fracture Mechanics*, Kluwer Academic Publishers, Dordrecht.

Dasgupta, G. (1984), Computation of exterior potential fields by infinite substructuring, *Comput. Methods Appl. Mech. Engrg*, 46(3):295–305.

Davis, P. J. and Rabinowitz, P. (1984), *Methods of Numerical Integration*, Academic Press, San Diego.

Delves, L. M. (1979), Global and regional methods, in Gladwell, L. and Wait, R., editors, *A Survey of Numerical Methods for Partial Differential Equations*, Clarendon Press, Oxford.

Delves, L. M. and Freeman, T. L. (1981), *Analysis of Global Expansion Methods: Weakly Asymptotically Diagonal Systems*, Academic Press, London, New York.

Delves, L. M. and Hall, C. A. (1979), An implicit matching principle for global element calculations, *J. Inst. Math. Appl.*, 23(2):223–234.

Demkowicz, L., Devloo, P. and Oden, J. T. (1985), On an h–type mesh-refinement strategy based on minimization of interpolation errors, *Comput. Methods Appl. Mech. Engrg.*, 53:67–89.

Deville, M. O. and Mund, E. H. (1990), Finite-element preconditioning for pseudospectral solutions of elliptic problems, *SIAM J. Sci. Stat. Comput.*, 11(2):311–342.

Dixon, L. C. W., Morgan, D. J. and Harrison, D. (1979), On singular cases arising from Galerkin's method, in Whiteman, J. R., editor, *Mathematics of Finite Elements and Applications III*, 217–225, Academic Press, London.

Dobrowolski, M. (1981), *Numerical approximation of elliptic interface and corner problems*, Habilitationsschrift, Bonn.

Dobrowolski, M. (1985), On finite element methods for nonlinear elliptic problems on domains with corners, in Grisvard, P., Wendland, W. and Whiteman, J. R., editors, *Singularities and Constructive Methods for their Treatment*, 85–103, Springer, Berlin, New York.

Donnelly, J. D. P. (1969), Eigenvalues of membranes with reentrant corners, *SIAM J. Numer. Anal.*, 6:47–61.

Dorny, C. N. (1968), Finite-difference approximation of the exterior problem for Poisson's equation, *J. Comp. Phys.*, 2:363–380.

Dorr, M. R. (1986), The approximation of solutions of elliptic boundary-value problems via the p-version of the finite element method, *SIAM J. Numer. Anal.*, 23:58–77.

Dorr, M. R. (1989), On the discretization of interdomain coupling in elliptic boundary-value problems, in Chan, T. F., et. al., editors, *Domain Decomposition Methods*, 17–37, SIAM, Philadelphia.

Dorr, M. R. (1991), A domain decomposition preconditioner with reduced rank interdomain coupling, *Appl. Numer. Math.*, 8:333–352.

Driscoll, T. A. (1997), Eigenmodes of isospectral drums, *SIAM Review*, 39(1):1-17.

Dryja, M. (1982), A capacity matrix method for Dirichlet problem in polygon region, *Numer. Math.*, 39:51–64.

Duff, G. F. D. and Naylor, D. (1966), *Differential Equations of Applied Mathematics*, John Wiley & Sons, Inc.

Duran, R., Muschietti, M. A. and Rodriguez, R. (1991), On the asymptotic exactness of error estimators for linear triangular finite elements, *Numer. Math.*, 59:107–127.

Dussault, J. P. (1995), Numerical stability and efficiency of penalty algorithms, *SIAM J. Numer. Anal.*, 32(1):296–317.

Dyksen, W. R., Houstis, E. N., Lynch, R. E. and Rice, J. R. (1984), The performance of the collocation and Galerkin methods with Hermite bi-cubics, *SIAM J. Numer. Anal.*, 27:695–715.

E, W. N., Huang, H. C. and Han, W. M. (1987), Error analysis of local refinements of polygonal domains, *J. Comp. Math.*, 5(1):89–94.

Eisenstat, S. C. (1974), On the rate of convergence of the Bergman-Vekua method for the numerical solution of elliptic boundary-value problem, *SIAM J. Numer. Anal.*, 11:654–680.

Eisenstat, S. C. and Schultz, M. H. (1972), Computational aspects of the finite element method, in Aziz, A. K., editor, *The Mathematical Foundations of the Finite element Method*, Academic Press, New York.

El Misiery, A. E. M. and Ortiz, E. L. (1986), Tau-lines: A new hybrid approach to the numerical treatment of crack problems based on Tau method, *Comput. Methods Appl. Mech. Engrg.*, 56:265–282.

El-Namoury, A. R. M. (1992), On the finite element method for a singular problem, *Appl. Math. Comp.*, 49:239–249.

Eriksson, K. (1985a), Improved accuracy by adapted mesh-refinements in the finite element method, *Math. Comp.*, 44(170):321–343.

Eriksson, K. (1985b), Finite element methods of optimal order for problems with singular data, *Math. Comp.*, 44(170):345–360.

Eriksson, K. (1985c), High-order local rate of convergence by mesh-refinement in the element method, *Math. Comp.*, 45(171):109–142.

Eriksson, K. (1986), Some error estimates for the p-version of the finite element method,

SIAM J. Numer. Anal., 23(2):403–411.

Eriksson, K. and Johnson, C. (1988), An adaptive finite element method for linear elliptic problems, *Math. Comp.*, 50(182):361–383.

Eriksson, K. and Thomee, V. (1984), Galerkin methods for singular boundary value problems in one space dimension, *Math. Comp.*, 42(166):345–367.

Ervin, V. J. and Stephan, E. P. (1990), A boundary element Galerkin method for a hypersingular integral equation on open surfaces, *Math. Meth. Appl. Sci.*, 13(4):281–289.

Fairweather, G. (1978), *Finite Element Galerkin Methods for Partial Differential Equations*, Marcel Dekker, Inc., New York.

Falk, R. S. (1976), A Ritz method based on a complementary variational principle, *RAIRO Anal. Numer.*, 10:39–48.

Fares, N. and Li, V. C. (1989), 3-D boundary element method for cracks in a plate, *Inter. J. Numer. Methods Engrg.*, 28:1703–1713.

Feistauer, M., Felcman, J. and Lukacova-Medvidova, M. (1995), Combined finite element-finite volume solution of compressible flow, *J of Comp. and Appl. Math.*, 61:179–199.

Feistauer, M., Felcman, J. and Lukacova-Medvidova, M. (1997), On the convergence of a combined finite volume–finite element for nonlinear convection–diffusion problems, *Numer. Meth. for PDE*, 13:163–190.

Feng, K. et. al. (1978), *Methods of Numerical Computation(in Chinese)*, Defence Industry Publishing, Beijing.

Feng, K. (1979), On the theory of discontinuous finite elements, *Math. Numer. Sincia*, 1:378–385.

Feng, K. and Shi, Z. C. (1981), *Mathematical Theory of Elasticity Structure(in Chinese)*, Academic Publishing, Beijing.

Feng, K. and Yu, D. H. (1994), A theorem for the natural integral operator of harmonic equation (in Chinese), *Math. Num. Sinica*, 16:221–226.

Ferket, P. J. J. and Reusken, A. A. (1996), A finite difference discretization method for elliptic problems on composite grids, *Computing*, 56:343–369.

Fix, G. J. (1973), Singular finite element methods, in Dwoyer D. L., et. al., editors, *Finite Elements, Theory and Application*, 50–66, Springer Verlag, New York.

Fix, G. J., Gulati, S. and Wakoff, G. I. (1973), On the use of singular functions with finite element approximations, *J. Comp. Phys.*, 13:209–238.

Fix, G. J., Liang, G. and Lee, D. N. (1982), Penalty-hybrid finite element method, *Comput. Math. Appl.*, 8(5):393–399.

Fix, G. M. (1976), Hybrid finite element methods, *SIAM Review*, 18(3):460–485.

Fletcher, C. A. J. (1984), *Computational Galerkin Methods*, Springer-Verlay.

Fornberg, B. (1988), Generation of finite difference formulas on arbitrarily spaced grids, *Math. Comp.*, 51(184):699–706.

Forsythe, G. E. (1958), Singularity and near singularity in numerical analysis, *Amer. Math. Monthly*, 65:229–240.

Forsythe, G. E. and Wasow, W. R. (1960), *Finite Difference Methods for Partial Differential Equations*, Wiley, New York.

Fortin, M. (1977), An analysis of the convergence of mixed finite element methods, *RAIRO Anal. Numer.*, 11:341–354.

Fox, L. (1971), Some experiments with singularities in linear elliptic partial differential equations, *Proc. Roy. Roc. A.*, 323:179–190.

Fox, L. (1979), Finite differences and singularities on elliptic problems, in Gladwell, L. and Wait, R., editors, *A Survey of Numerical Methods for Partial Differential Equations*, 42–69, Clarendon Press, Oxford.

Fox, L., Henrici, P. and Moler, C. (1967), Approximations and bounds for eigenvalues of elliptic operators, *SIAM J. Numer. Anal.*, 4:89–102.

Fox, L. and Snakar, R. (1969), Boundary singularities in linear elliptic differential equations, *J. Inst. Math. Appl.*, 5:340–350.

Franke, R. (1978), On the computation of optimal approximations in Sard corner spaces, *SIAM J. Numer. Anal.*, 15(4):791–800.

Fufner, A. (1985), *Weighted Sobolev Spaces*, John Wiley and Sons.

Fujii, H. and Yamaguti, M. (1980), Structure of singularities and its numerical realization in nonlinear elasticity, *J. Math. Kyoto Univ.*, 20(3):489–590.

Funaro, D., Quarteroni, A. and Zanolli, P. (1988), An iterative procedure with interface relaxation for domain decomposition methods, *SIAM J. Numer. Anal.*, 25(6):1213–1236.

Galerkin, B. G. (1915), Series solution of some problems of elastic equilibrium of rods and plates (Russian), *Vestnik. Inzh. Tech.*, 19:897–908.

Gardner, L. R. T., Gardner, G. A. and Dag, I. (1993), Hermite infinite elements and graded quadratic B spline finite elements, *Inter. J. Numer. Methods Engrg.*, 36(19):3317–3332.

Gatica, G. N. and Hsiao, G. C. (1989), The coupling of boundary element and finite element methods for a nonlinear exterior boundary value problem, *Zeitschrift fur Analysis und Ihre Anwendungen*, 8(4):377–387.

Gatica, G. N. and Hsiao, G. C. (1992), On the coupled BEM and FEM for a nonlinear exterior Dirichlet problem in R^2, *Numer. Math.*, 61:171–214.

Gerdes, K. and Demkowicz, L. (1996), Solution of 3D Laplace and Helmholtz equations in exterior domains using hp infinite elements, *Comp. Methods Appl. Mech. Engrg.*, 137:239–273.

Girault, V. (1976), A combined finite element and Marker and cell method for solving Navier-Stokes equations, *Numer. Meth*, 26:39–59.

Givoli, D. (1992), *Numerical Methods for Problems in Infinite Domains*, Elsevier, New York.

Givoli, D., Rivkin, L. and Keller, J. B. (1992), A finite element method for domains with corners, *Inter. J. Numer. Methods Engrg.*, 35:1329–1345.

Godunov, S. K. and Ryabenkii, V. S. (1987), *Difference Schemes*, North–Holland, New York.

Goldlewski, E. and Raviart, P. A. (1996), *Numerical Approximation of Hyperbolic Systems of Conservative Law*, Springer-Verlay.

Goldstein, A. A. (1974), Optimization with corners, in *Nonlinear programming*, 2, 215–230, Academic Press, New York.

Goldstein, C. I. (1981a), The finite element method with nonuniform mesh sizes for unbounded domains, *Math. Comp.*, 36(154):387–404.

Goldstein, C. I. (1981b), The finite element method with non-uniform mesh sizes applied to the exterior Helmholtz problem, *Numer. Math.*, 38:61–82.

Goldstein, C. I. (1981c), A numerical method for solving elliptic boundary problems in unbounded domain, in Schultz, M. H., editor, *Elliptic Problem Solvers*, 307–313, Academic Press, New York.

Goldstein, C. I. (1982), A finite element method for solving Helmholtz type equations in waveguides and other unbounded domains, *Math. Comp.*, 39(160):309–324.

Goldstein, C. I. (1983), The solution of exterior interface problems using a variational method with Lagrange multipliers, *J. Math. Anal. Appl.*, 97(2):480–508.

Golub, G. H. and van Loan, C. F. (1989), *Matrix Computations (2nd Edition)*, The Johns Hopkins, Baltimore and London.

Goodrich, R. F. and Stenger, F. (1970), Movable singularities and quadrature, *Math. Comp.*, 24:283–300.

Gottlieb, D. and Orszag, S. A. (1977), *Numerical Analysis of Spectral Methods:Theory and Applications*, SIAM, Philadelphia.

Gradshteyn, I. S. and Ryzhik, I. M. (1980), *Table of Integrals, Series and Products*, Academic Press, New York.

Grannell, J. J. (1987), On simplified hybrid methods for coupling of finite elements and boundary elements, in *Boundary Elements IX, Vol. 1*, 447–460, Comput. Mech., Southampton.

Grannell, J. J. (1988), BEM-FEM coupling for screen and infinite domain potential problems, in *Boundary Elements X, Vol. 1*, 459–475, Comput. Mech., Southampton.

Gratkowski, S. (1986), More on a simple infinite element, *COMPEL. Int. J. Comp. Math. in EEE.*, 5(4):191–194.

Greenspan, D. and Warten, R. M. (1962), On the approximate solution of Dirichlet-type problems with singularities on the boundary, *J. Franklin Inst.*, 273:187–200.

Greenstadt, J. (1982), The cell discretization algorithm for elliptic partial differential equations, *SIAM J. Sci. Statis. Comp.*, 3(3):261–288.

Greenstadt, J. (1993), Solution of elliptic systems of partial differential equations by cell discretization, *SIAM J. Sci. Comp.*, 14(3):627–653.

Greenstadt, J. (1995), The removal of overshoot in PDE solutions by the use of special basis functions, *Comput. Methods Appl. Mech. Engrg.*, 120:45–64.

Gregory, J. A., Fishelov, D., Schiff, B. and Whiteman, J. R. (1978), Local mesh refinement with finite elements for elliptic problems, *J. Comp. Phys.*, 29(1):133–140.

Grisvard, P. (1985), *Elliptic Problems in Nonsmooth Domains*, Pitman Advanced Publishing Program, Boston.

Grisvard, P. (1992), *Singularities in Boundary Value Problems*, Springer–Verlag, Berlin Heidelberg.

Grisvard, P., Wendland, W. and Whiteman, J. R. (1985), *Singularities and Constructive Methods for their Treatment, Lecture Notes in Mathematics*, Springer–Verlag.

Gropp, W. D. and Keyes, D. E. (1992), Domain decomposition with small mesh refinement, *SIAM J. Sci. Comp.*, 13:967–993.

Grosch, C. E. and Orszag, S. A. (1977), Numerical solution of problems in unbounded regions: coordinate transforms, *J. Comp. Phys.*, 25(3):273–295.

Guenther, R. B. (1975), On the numerical treatment of partial differential equations in the neighborhood of isolated singularities with applications, in *Numerische Behandlung von Differentialgleichungen*, 93–110, Internat. Schriftenreihe Numer. Math., Band.

Gui, W. and Babuska, I. (1986), The h, p and $h - p$ versions of the finite element method in 1 dimension III, the adaptive $h - p$ version, *Numer. Math.*, 49:659–683.

Guirguis, G. H. (1987a), On the coupling boundary integral and finite element methods for the exterior Stokes problem in 3D, *SIAM J. Numer. Anal.*, 24(2):310–322.

Guirguis, G. H. (1987b), A third order boundary condition for the exterior Stokes problem in three dimensions, *Math. Comp.*, 49(180):379–389.

Guo, B. Q. and Babuska, I. (1986a), The $h - p$ version of finite element method I, *Comput. Mech.*, 1:21–42.

Guo, B. Q. and Babuska, I. (1986b), The $h - p$ version of finite element method I, *Comput. Mech.*, 1:203–220.

Guo, B. Q. and Oh, H. S. (1994), The $h - p$ version of the finite element method for problems with interfaces, *Inter. J. Numer. Methods Engrg.*, 37:1741–1762.

Guo, B. Y. (1988), *Difference Methods for Partial Differential Equations* (in Chinese), Academic Publishing.

Gupta, A. K. (1978), A finite element for transition from a fine to a coarse grid, *Inter. J. Numer. Methods Engrg.*, 12:35–45.

Habashi, W. G. (1979), The finite element method in the solution of unbounded potential flows, *Inter. J. Numer. Methods Eengrg.*, 14:1347–1358.

Hackbusch, W. (1989), On first and second order box scheme, *Computing*, 41:277–296.

Hackbusch, W. (1992), *Elliptic Differential Equations: Theory and Numerical Treatment*, Springer–Verlag.

Hageman, L. A. and Young, D. M. (1981), *Applied Iterative Methods*, Academic Press, San Diego.

Hall, C. A. and Porsching, T. A. (1990), *Numerical Analysis of Partial Differential Equations*, Prentice–Hall, Englewood Cliffs, New Jersey.

Han, H. D. (1982a), The error estimates for the infinite element method for eigenvalue problems, *RAIRO Anal. Numer.*, 16(2):113–128.

Han, H. D. (1982b), The numerical solutions of interface problems by infinite element method, *Numer. Math.*, 39:39–50.

Han, H. D. and Wu, X. N. (1985), Approximation of infinite boundary condition and its application to finite element methods, *J. Comp. Math.*, 3(2):179–192.

Han, H. D. and Wu, X. N. (1986), Approximation of boundary condition at infinity and its application to boundary element methods, in *Boundary Elements (Beijing, 1986)*, 127–134, Pergamon, Oxford Elmsford, NY.

Han, H. D. and Ying, L. A. (1979), An iterative method in the infinite element (in Chinese), *Math. Numer. Sinica*, 1:91–99.

Harari, I. and Hughes, T. J. R. (1991), Finite element methods for the Helmholtz equation in an exterior domain: model problems, *Comput. Meth. Appl. Mech. Engrg.*, 87(1):59–96.

Harari, I. and Hughes, T. J. R. (1992), Galerkin/least-squares finite element methods for the reduced wave equation with nonreflecting boundary conditions in unbounded domains, *Comput. Meth. Appl. Mech. Engrg.*, 98(3):411–454.

He, Y. N. and Li, K. T. (1990), The coupling method of finite elements and boundary elements for radiation problem, *ACTA Math. Appl. Sinica*, 6:95–112.

Hebeker, F. K. (1987), On a new boundary element spectral method, in *Large Scale Scientific Computing*, 180–193, Birkhauser, Boston, Mass.

Heinrichs, W. (1989), Improved condition number for spectral methods, *Math. Comp.*, 53(187):103–119.

Hendry, J. A. and Delves, L. M. (1979), The global element method applied to a harmonic mixed boundary value problem, *J. Comp. Phys.*, 33(1):33–44.

Hendry, J. A., Delves, L. M. and Phillips, C. (1979), Numerical experience with the global element method, in Whiteman, J. R., editor, *MAFELAP III*, 341–348, 341–348.

Hennart, J. P., Jaffre, J. and Roberts, J. E. (1988), A constructive method for deriving finite elements of nodal type, *Numer. Math.*, 53:701–738.

Henshell, R. D. and Shaw, K. G. (1975), Crack tip finite elements are unnecessary, *Inter. J. Numer. Methods Engrg.*, 9:495–507.

Heuer, N. and Stephan, E. P. (1991), Coupling of the finite element and boundary element method– the h p version., *ZAMM*, 71(6):T584–T586.

Hoog, F. R. D. and Weiss, R. (1976), Difference methods for boundary value problems with a singularity of the first kind, *SIAM J. Numer. Anal.*, 13(5):775–813.

Houstis, E. N., Vavalis, E. A. and Rice, J. R. (1988), Convergence of $O(h^4)$ cubic spline collocation methods for elliptic partial differential equations, *SIAM J. Numer. Anal.*, 25:54–90.

Hsiao, G. C. (1986), Boundary element methods for a class of exterior singular perturbation problems, in *BAIL IV*, 89–97, Boole, Dun Laoghaire.

Hsiao, G. C. (1988a), Boundary element methods for exterior problems in elasticity and fluid mechanics, in Whiteman, J. R., editor, *The Mathematics of Finite Elements and Applications VI*, 323–341, Academic Press, London.

Hsiao, G. C. (1988b), The coupling of BEM and FEM– a brief review, in *Boundary Elements X, Vol. 1*, 431–445, Comput. Mech., Southampton.

Hsiao, G. C. (1990), The coupling of boundary element and finite element methods, *ZAMM*, 70(6):T493–T503.

Hsiao, G. C. and Porter, J. F. (1988), The coupling of BEM and FEM for the two-dimensional viscous flow problem, *Appl. Anal.*, 27:79–108.

Huffel, S. V. and Vandewalle, J. (1991), *The Total Least Squares Problem: Computational Aspects and Analysis*, SIAM, Philadelphia.

Hughes, T. J. R. and Akin, J. E. (1980), Techniques for developing "special" finite element shape functions with particular reference to singularities, *Inter. J. Numer. Methods Engrg.*, 15(5):733–751.

Huiskamp, G. (1991), Difference formulas for the surface Laplacian on a triangulated surface, *J. Comp. Phys.*, 95:477–496.

Ikebe, Y., Lynn, M. S. and Timlake, W. P. (1969), The numerical solution of the integral equation formulation of the single interface Neumann problem, *SIAM J. Numer. Anal.*, 6:334–346.

Ingraffea, A. R. and Manu, C. (1980), Stress-intensity factor computation in three dimensions with quarter-point elements, *Inter. J. Numer. Methods Engrg.*, 15(10):1427–1445.

Jensen, S. (1991), An H_0^m interpolation result, *SIAM J. Math. Anal.*, 22(3):785–791.

Jensen, S. S. and Suri, M. (1992), On the L_2 error for the p version of the finite element method over polygonal domains, *Comput. Methods Appl. Mech. Engrg.*, 97(2):233–243.

Jespersen, D. (1978), Ritz-Galerkin methods for singular boundary value problems, *SIAM J.*

Numer. Anal., 15(4):813–834.

Jiang, S. and Li, K. T. (1987), The coupling of finite element method and boundary element method for two dimensional Helmholtz equation in an exterior domain, *J. Comp. Math.*, 5(1):21–37.

Jirousek, J. and Guex, L. (1986), The hybrid-Trefftz finite element model and its application to plate bending, *Inter. J. Numer. Methods Engrg.*, 23:651–693.

Johnson, C. and Nedelec, J. C. (1980), On the coupling of boundary integral and finite element methods, *Math. Comp.*, 35(152):1063–1079.

Kantorovich, L. V. and Akilov, G. P. (1964), *Functional Analysis in Normal Spaces*, Pergamon Press, New York.

Kantorovich, L. V. and Krylov, V. I. (1958), *Approximate Methods of Higher Analysis*, Wiley-Interscience Publication, New York.

Karageorghis, A. (1994), Conforming spectral methods for Poisson problems in cuboidal domains, *J. Sci. Comp.*, 9(3):341–350.

Kariadakis, G. E., Maxriplis, C. and Patera, A. T. (1989), Non-conforming spectral element finite element approximation for partial differential equations, *Comp. Meth. Appl. Mech. Engrg.*, 75:109–125.

Karpilovskaja, E. B. (1963), Convergence of the collocation method, Soviet Math, Dokl, 151:766–769.

Kellogg, R. B. (1971), Singularities in interface problems, in *Numerical Solution of Partial Differential Equations, II (SYNSPADE 1970)*, 351–400, Academic Press, New York.

Kellogg, R. B. (1972), Higher order singularities for interface problems, in Aziz, A. K., editor, *The Mathematical Foundations of the Finite Element Method with Applications to Partial Differential Equations*, 589–602, Academic Press, New York.

Kellogg, R. B. (1975), On the Poisson equations with intersecting interfaces, *Appl. Anal.*, 4:101–129.

Kermode, M., McKerrell, A. and Delves, L. M. (1985), The calculation of singular coefficients, *Comput. Meth. Appl. Mech. Engrg.*, 50(3):205–215.

King, J. T. (1974), New error bounds for the penalty method and extrapolation, *Numer. Math.*, 23:153–165.

King, J. T. (1975), A quasioptimal finite element method for elliptic interface problems, *Computing*, 15(2):127–135.

King, J. T. and Serbin, S. M. (1978), Computational experiments and techniques for the penalty method with extrapolation, *Math. Comp.*, 32(141):111–126.

Kioustelidis, J. B. (1989), L_2 error bounds for approximate solutions of elliptic partial differential equations with Dirichlet boundary conditions, *Computing*, 43(2):133–140.

Kirsch, A. and Monk, P. (1990), Convergence analysis of a coupled finite element and spectral method in acoustic scattering, *IMA J. Numer. Anal.*, 9:425–447.

Kober, H. (1957), *Dictionary of Conformal Representations*, Dover, London.

Kondrat'ev, V. A. (1967), Boundary problems for elliptic equations with conical or angular points, *Trans. Moscow Math. Soc.*, 16:227–313.

Kozlov, V., Maz'ya, V. and Rozin, L. (1994), On certain hybrid iterative methods for solving boundary value problems, *SIAM J. Numer. Anal.*, 31(1):101–110.

Kreß, R. and Spassov, W. T. (1983), On the condition number of boundary integral operators for the exterior Dirichlet problem for the Helmholtz equation, *Numer. Math.*, 42:77–95.

Krizek, M. and Neittaanmaki, P. (1984), Superconvergence phenomenon in the finite element method arising from averaging gradients, *Numer. Math.*, 45:105–116.

Krizek, M. and Neittaanmaki, P. (1987), On a global-superconvergence of the gradient of linear triangular elements, *J. Comput. Appl. Math.*, 18:221–223.

Kufner, A. (1980), *Weighted Sobolev Spaces*, John Wiley & Sons.

Laasonen, P. (1958), On the truncation error of discrete approximations into the solutions of Dirichlet problems in a domain with corners, *J. Assoc. Comput. Mach.*, 5:32–38.

Ladyzhenskaya, O. A. (1985), *The Boundary Value Problems of Mathematical Physics*, Springer–Verlag, New York.

Lakshmikantham, V. and Trigiante, D. (1988), *Theory of Difference Equations: Numerical*

Methods and Applications, Academic Press, Boston.

Lamp, U., Schleicher, T., Stephan, E. and Wendland, W. L. (1984), Galerkin collocation for an improved boundary element method for a plane mixed boundary value problem, *Computing*, 33:269–296.

Lascaux, P. and Lesaint, P. (1975), Some nonconforming finite element for the plate bending problem, *Rev. Francaise Automat Informat. Recherche Operalionnelle Sér Rouge Anal. Numer.*, 12(1):9–53.

Lee, P. (1993), A Lagrange multiplier method for the interface equations from electromagnetic applications, *SIAM J. Numer. Anal.*, 30(2):478–506.

Lefeber, D. (1989), *Solving Problems with Singularities Using Boundary Elements*, Comp. Mech. Publ., Southampton.

Lehman, R. S. (1959), Developments at an analytic corner of solution of elliptic differential equations, *J. Math. Meth.*, 8:727–760.

Lentini, M. A. and Keller, H. B. (1980), Boundary value problems on semi-infinite intervals and their numerical solution, *SIAM J. Numer. Anal.*, 17(4):577–604.

Li, K. T. and He, Y. N. (1989), The coupling of boundary integral and finite element methods for the Navier-Stokes equations in an exterior domain, *J. Comp. Math.*, 7(2):157–173.

Li, Z. and Reed, M. B. (1995), Convergence analysis for an element-by-element finite element method, *Comput. Methods Appl. Mech. Engrg.*, 123:33–42.

Li, Z. C. (1980), The combined method between original energy method and finite element method for Laplace's boundary value problems with singularities (in Chinese), *Math. Numer. Sinica*, 2:319–328.

Li, Z. C. (1983a), An approach for combining the Ritz-Galerkin and finite element methods, *J. Appr. Th.*, 39(2):132–152.

Li, Z. C. (1983b), On the reduced rate of convergence for a nonconforming combined method, *SIAM J. Numer. Anal.*, 20(1):86–93.

Li, Z. C. (1986), A nonconforming combined method for solving Laplace's boundary value problems with singularities, *Numer. Math.*, 49:475–497.

Li, Z. C. (1987), Numerical methods for elliptic boundary value problems with singularities, *Appl. Math. Notes*, 12(1 & 2):14–22.

Li, Z. C. (1988), A note on Kellogg's eigenfuctions of periodic Sturm-Liouville system, *Appl. Math. Letters*, 1:123–126.

Li, Z. C. (1989a), An approach combining the Ritz-Galerkin and finite difference methods, *Numer. Meth. for PDE*, 5:279–295.

Li, Z. C. (1989b), A combined method for solving elliptic problems on unbounded domains, *Comput. Meth. Appl. Mech. Engrg.*, 73:191–208.

Li, Z. C. (1989c), A nonconforming combination for solving elliptic problems with singularities, *J. Comp. Phys.*, 80:288–313.

Li, Z. C. (1990), *Numerical Methods for Elliptic Problems with Singularities: Boundary Methods and Nonconforming Combinations*, World Scientific, Singapore.

Li, Z. C. (1991), On nonconforming combinations of various finite element methods for solving elliptic boundary value problems, *SIAM J. Numer. Anal.*, 28:446–475.

Li, Z. C. (1992a), Optimal convergence rates for the combined methods of different finite element methods, *Numer. Meth. for PDE*, 8(3):203–220.

Li, Z. C. (1992b), Penalty-combined approaches to the Ritz-Galerkin and finite element methods for singularity problems of elliptic equations, *Numer. Methods for PDE*, 8:33–57.

Li, Z. C. (1994), The Schwarz alternating method for singularity problems, *SIAM J. Sci. Comput.*, 15(5):1064–1082.

Li, Z. C. (1996a), Combinations of Ritz-Galerkin and finite difference method, *Inter. J. Numer. Methods Engrg.*, 39:1839–1857.

Li, Z. C. (1996b), Parallel penalty combinations for singularity problems, *Engineering Analysis with Boundary Elements*, 18:119–130.

Li, Z. C. (1996c), Domain decomposition methods to penalty combinations for singularity problems, in Glowinski, R., Periaux, J., Shi, Z. C. and Wildlund, O., editors, *Domain*

Decomposition Methods in Sciences and Engineering, 73–81, Joh Wiley & Sons, England.

Li, Z. C., (1997a), Penalty combinations of the Ritz-Galerkin and finite difference methods for singularity problems, *J. Comp. and Appl. Math.*, 81:1–17

Li, Z. C. (1997b), Superconvergence of coupling techniques in combined methods for elliptic equation with singularities, Technical report, Dept. Appl. Math. National Sun Yat-sen University, Kaohsiung, presented at the 17th Biennial Conf. on Numerical Analysis, Univ. of Dundee, Scotland, UK, 24–27 June 1997.

Li, Z. C. (1997c), Domain decomposition methods for Poisson's equation with singular points by combined methods, Technical report, Dept. Appl. Math. National Sun Yat-sen University, Kaohsiung.

Li, Z. C. and Bui, T. D. (1988a), Generalized hybrid-combined methods for singularity problems of homogeneous equations, *Inter. J. Numer. Methods Engrg.*, 26:785–803.

Li, Z. C. and Bui, T. D. (1988b), A new kind of combinations between the Ritz-Galerkin and finite element methods for singularity problems, *Computing*, 40:29–50.

Li, Z. C. and Bui, T. D. (1988c), Six combinations of the Ritz-Galerkin and finite element methods for elliptic boundary value problems, *Numer. Meth. for PDE*, 4:197–218.

Li, Z. C. and Bui, T. D. (1990a), Coupling strategy for matching different methods in solving singularity problems, *Computing*, 45:311–319.

Li, Z. C. and Bui, T. D. (1990b), The simplified hybrid-combined methods for Laplace equation with singularities, *J. Comp. and Appl. Math.*, 29:171–193.

Li, Z. C. and Bui, T. D. (1992a), Coupling techniques in boundary-combined methods, *Engineering Analysis with Boundary Elements*, 10:75–85.

Li, Z. C. and Bui, T. D. (1992b), Penalty-combined method and applications in solving elliptic problems with singularities, *Comput. Meth. Appl. Mech. Engrg.*, 97:291–316.

Li, Z. C. and Chou, K. C. (1977), Studies in the combination rates of liquid phase fast reaction system, *Scientia Sinica*, 20:197–220.

Li, Z. C. and Liang, G. P. (1980), On the combined methods of the boundary value problems of elliptic equations (in Chinese), *Math. Numer. Sinica*, 2:192–194.

Li, Z. C. and Liang, G. P. (1981), On Ritz-Galerkin-FEM combined method of solving the boundary value problem of elliptic equations, *Scientia Sinica*, 24:1497–1508.

Li, Z. C. and Liang, G. P. (1983), On the simplified hybrid-combined method, *Math. Comp.*, 41(163):13–25.

Li, Z. C., Lin, Q. and Yan, N. N. (1997), Global superconvergence in combinations of Ritz-Galerkin-FEM for singularity problems, Technical report, Dept. Appl. Math. National Sun Yat-sen University, Kaohsiung.

Li, Z. C., Lu, T. T., Chen, L. C. and Chen, J. H. (1997), Very high accurate solutions of Motz's problem, Technical report, Dept. of Appl. Math., National Sun Yat-sen University, Kaohsiung.

Li, Z. C. and Mathon, R. (1990a), Boundary methods for elliptic problems on unbounded domains, *J. Comp. Phy*, 89:414–431.

Li, Z. C. and Mathon, R. (1990b), Error and stability analysis of boundary methods for elliptic problems with interfaces, *Math. Comp.*, 54:41–61.

Li, Z. C., Mathon, R. and Sermer, P. (1987), Boundary methods for solving elliptic problem with singularities and interfaces, *SIAM J. Numer. Anal.*, 24:487–498.

Li, Z. C., Shi, Z. W., Zhou, G. Q. and Chou, K. C. (1981), A semianalytic method for computing the concentration distribution in enzyme-substrate fast reaction systems, *J. Comp. Chem.*, 2:273–277.

Li, Z. C. and Wang, S. (1997), The finite volume method and application in combinations, Technical report, Dept. of Appl. Math. National Sun Yat-sen University, Kaohsiung.

Liang, G. P. (1979), 'mixed element' in finite element method(in Chinese), *Math. Numer. Sinica*, 1:342–346.

Liang, G. P. (1980a), Semi-analysis method and its error estimation(in Chinese), *Math. Numer. Sinica*, 2:217–228.

Liang, G. P. (1980b), General finite strip method and error estimation (in Chinese), *Math. Numer. Sinica*, 2:336–344.

Liang, G. P. and Gu, Y. G. (1978), Semi-analytic method for the plane plastic problem with crack (in Chinese), *Math. Numer. Sinica*, 1:65–78.

Liang, G. P. and He, J. H. (1992), The non-conforming domain decomposition method for elliptic problems with Lagrangian multipliers (in Chinese), *Math. Numer. Sinica*, 14(2):207–215.

Liang, G. P., Kong, L. and He, J. H. (1996), A new finite element space for singularity problems(in Chinese), *Math. Numer. Sinica*, 18(1):1–7.

Liang, G. P. and Liang, P. (1990), Non-conforming domain decomposition with hybrid method, *J. Comp. Math.*, 8(4):363–370.

Lin, J. L. and Rheinboldt, W. C. (1994), A posteriori finite element error estimators for indefinite elliptic boundary value problems, *Numer. Func. Anal. Optim.*, 15(3–4):335–356.

Lin, Q. (1991a), Fourth order eigenvalue approximation by extrapolation on domains with reentrant corners, *Numer. Math.*, 58(6):631–640.

Lin, Q. (1991b), Superconvergence of FEM for singular solution, *J. Comp. Math.*, 9(2):111–114.

Lin, Q. and Whiteman, J. R. (1990), Superconvergence of recovered gradients of finite element approximations on nonuniform rectangular and quadrilateral meshes, in Whiteman, J. R., editor, *The Mathematics of Finite Elements and Applications VII*, 563–571, Academic Press, London.

Lin, Q. and Yan, N. N. (1996), *The Construction and Analysis of High Efficient FEM (in Chinese)*, Hobei University Publishing.

Lin, Q. and Zhu, Q. (1994), *The Preprocessing and Postprocessing for the Finite Element Method (in Chinese)*, Shanghai Scientific & Technical Publishers.

Lin, T. C. (1985), The numerical solution of Helmholtz's equation for the exterior Dirichlet problem in three dimensions, *SIAM J. Numer. Anal.*, 22:670–686.

Linde, H. J. V. (1974), High-order finite-difference methods for Poisson's equation, *Math. Comp.*, 28(126):369–391.

Lions, P. L. (1988), On the Schwarz alternating method I, in Glowinski, R. et. al., editors, The First Inter, *Symposium on Domain Decomposition Method for PDE*, 1–42, SIAM, Philadelphia.

Lions, P. L. (1989), On the Schwarz alternating method II, Stochastic interpretation and order properties, in Chan, T. F. et. al., editors, *Domain Decomposition Methods*, 47–70, SIAM, Philadelphia.

Lions, P. L. (1990), On the Schwarz alternating method III, A variant for nonoverlaping subdomains, in Chan, T. F. et. al., editors, *Domain Decomposition Methods*, 202–223, SIAM, Philadelphia.

Lions, J. L. and Magenes, E. (1968), *Problems aux Limites non Gomogenes et Applications*, Vol. 1, Travaux et Recherches Mathematiques, no. 17, Dunod, Paris.

Lu, T. T. and Horng, M. Y. (1996), Boundary approximation method for Motz problem, Technical Report, Dept. of Appl. Math. National Sun Yat-sen University, Kaohsiung.

Lu, T., Shih, T. M. and Liem, C. H. (1992), *Domain Decomposition Methods:Techniques of Numerical Solution for PDE (in Chinese)*, Academic Publishing.

Lucas, T. R. and Reddien, G. W. (1972), Some collocation methods for nonlinear boundary value problems, *SIAM J. Numer. Anal.*, 9:341–356.

MacCamy, R. C. and Marin, S. P. (1980), A finite element method for exterior interface problems, *Int. J. Math. Math. Sci.*, 3(2):311–350.

MacKinnon, R. J. and Carey, G. F. (1988), Analysis of material interface discontinuities and superconvergent fluxes in finite difference theory, *J. Comp. Phys.*, 75:151–167.

MacKinnon, R. J. and Carey, G. F. (1990), Nodal superconvergence and solution enhancement for a class of finite-element and finite difference methods, *SIAM. J. Sci. Stat. Comput.*, 11:343–353.

Maita, S. K. (1992), A multicorner variable order singularity triangle to model neighbouring singularities, *Inter. J. Numer. Methods Engrg.*, 35(2):391-408.

Mandel, J. and Tezaur, R. (1996), Convergence of a substructuring method with Lagrange multipliers, *Numer. Math.*, 73:473–487.

Manteuffel, T. A. and Parter, S. C. (1990), Preconditioning and boundary conditions, *SIAM J. Numer. Anal.*, 27:656–694.

Marchuk, G. I. (1982), *Methods of Numerical Mathematics*, Springer–Verlag, New York.

Margulies, M. (1981), Combination of boundary element and finite element methods, in Brebbia, C. A., editor, *Progress in Boundary Element Methods*, 258–288, Pentch Press, London.

Marti, J. T. (1986), *Introduction to Sobolev Spaces and Finite Element Solution of Elliptic Boundary Value Problems*, Academic Press, London.

Masmoudi, M. (1987), Numerical solution for exterior problems, *Numer. Math.*, 51(1):87–101.

Mathon, R. and Johnston, R. L. (1977), The approximate solution of elliptic boundary-value problems by fundamental solutions, *SIAM J. Numer. Anal.*, 14:638–658.

Mathon, R. and Sermer, P. (1982), Numerical solution of the Helmholtz equation, *Congressus Numerantium*, 34:313–330.

Matveyenko, V. P. and Borzenkov, S. M. (1996), Semianalytical singular element and its application to stress calculation and optimization, *Inter. J. Numer. Methods Engrg.*, 39:1659–1680.

McArthur, K. M., Bowers, K. L. and Lund, J. (1990), The Sinc method in multiple space dimensions: model problems, *Numer. Math.*, 56(8):789–816.

McCormick, S. F. (1989), *Multilevel Adaptive Methods for Partial Differential Equations*, SIAM, Philadelphia.

McCormick, S. F. (1992), *Multilevel Projection Methods for Partial Differential Equations*, SIAM Publication, Philadelphia.

McKerrell, A., Delves, L. M. and Phillips, C. (1981), A note on Chebyshev expansion methods for the solution of elliptic partial differential equations, *J. Comp. Phys.*, 41:444–452.

McLean, W. (1984), Corner singularities and boundary integral equations, in *Contributions of Mathematical Analysis to the Numerical Solution of Partial Differential Equations*, 197–213, Austria. Nat. Univ., Canberra.

McLean, W. (1986), A special Galerkin method for a boundary integral equation, *Math. Comp.*, 47:597–607.

Meddahi, S., Valdes, J., Menedez, O. and Perez, P. (1996), On the coupling of boundary integral and mixed finite element methods, *J. Comp. Appl. Math.*, 69:113–124.

Medina, F. and Taylor, R. L. (1983), Finite element techniques for problems of unbounded domains, *Inter. J. Numer. Methods Engrg.*, 19:1209–1226.

Mercier, B. (1989), *An Introduction to the Numerical Analysis of Spectral Methods*, Springer.

Mergelies, M. (1961), Combination of the boundary element and finite element methods, in Brebbia, C. A., editor, *Progress in Boundary Element Methods*, 258–288, Pentch Press, London.

Michavila, F., Gavete, L. and Diez, F. (1988), Two different approaches for the treatment of boundary singularities, *Numer. Meth. for PDE*, 4:255–282.

Mickens, R. E. (1994), *Nonstandard Finite Difference Models of Differential Equations*, World Scientific, Singapore.

Mikhlin, S. G. (1964), *Variational Methods in Mathematical Physics*, MacMillan.

Mikhlin, S. G. (1971), *The Numerical Performance of Variational Methods*, Wolters-Noordhoff.

Miller, K. (1965), Numerical analogs to the Schwarz alternating procedure, *Numer. Math.*, 7:91–103.

Miller, J. J. H. and Wang, S. (1994), An exponentially fitted finite volume method for the numerical solution of 2D unsteady incompressible flow problems, J. Comp. Phys., 115:56–64.

Mitchell, A. R. and Griffiths, D. G. (1980), *The Difference Method in Partial Differential Equations*, John Wiley & Sons.

Mitchell, A. R. and Wait, R. (1977), *The Finite Element Method in Partial Differential Equation*, Wiley, New York.

Moriya, K. (1985), Infinite elements for the analysis of unbounded domain problems, in Whiteman, J. R., editor, *The Mathematics of Finite Elements and Applications V*, 283–290, Academic Press, London, New York.

Morley, L. S. D. (1973/74), Finite element solution of boundary-value problems with nonremovable singularities, *Philos. Trans. Roy. Soc. London Ser. A*, 275:463–488.

Morris, J. L. and Wait, R. (1979), Crack-tip elements with curved boundaries and variable nodes, *Appl. Math. Model.*, 3(4):259–262.

Morton, K. W. (1996), *Numerical Solution of Convection Diffusion Problems*, Chapman & Hall, London.

Motz, H. (1947), The treatment of singularities of partial differential equations by relaxation methods, *Quart. App. Math.*, 4:371–377.

Mroz, M. (1989), Domain decomposition method for elliptic mixed boundary methods, *Computing*, 42:45–59.

Nakao, M. T. (1987), Superconvergence of the gradient of Galerkin approximations for elliptic problems, *J. Comput. Appl. Math.*, 20:341–348.

Nazarov, S. A. and Plamenevsky, D. A. (1994), *Elliptic Problems in Domains with Piecewise Smooth Boundaries*, Walter de Gruyter, Berlin.

Nitsche, J. A. (1971), Uber ein variationsprinzip zur losung von Dirichlet problemem bei verwendung von teilraumen, die keinen randbekingungen unterwofen sind, *Adhes. Math. Sem. Univ.*, 36:9–15, Hamburg.

Oden, J. T. (1991), Finite elements: An Introduction, in Ciarlet, P. G. and Lions, J. L., editors, *Finite Element Methods(Part I)*, 3–15, North-Holland.

Oden, J. T. and Carey, G. F.(1983), *Finite Elements, Vol. IV. Mathematical Aspects*, Prentice-Hall, Inc., Englewood Cliffs, N.J.

Oden, J. T. and Reddy, J. N. (1976), *An Introduction to the Mathematical Theory of Finite Elements*, John Wiley & Sons, New York.

Oganesyan, L. A. (1990), Finite element method (FEM) for a singular elliptic equation, *Soviet J. Numer. Anal. Math. Model.*, 5(4-5):299–326.

Ogen, G. and Schiff, B. (1983), Constrained finite elements for singular boundary value problems, *J. Comp. Phys.*, 51(1):65–82.

Oh, H. S. and Babuska, I. (1992), The p−version of the finite element method for the elliptic boundary value problems with interfaces, *Comput. Meth. Appl. Mech. Engrg.*, 97:211–231.

Oliveriral, F. A. (1980), Collocation and residual correction, *Numer. Math.*, 36:27–31.

Olson, L. G., Georgiou, G. C. and Schultz, W. W. (1991), An efficient finite element method for treating singularities in Laplace's equation, *J. Comp. Phys.*, 96(2):391–410.

Orikasa, T., Honma, T., Fukai, I. and Washisu, S. (1983), Finite element methods for unbounded field problems and application to two-dimensional taper, *Inter. J. Numer. Methods Engrg.*, 19:157–168.

Papamichael, N. (1985), The treatment of singularities in orthonormalization methods for numerical conformal mapping, in Grisvard, P., Wendland, W. and Whiteman, J. R., editors, *Singularities and Constructive Methods for their Treatment*, 207–227, Springer, Berlin.

Papamichael, N. (1990), Numerical conformal mapping onto a rectangle with applications to a solution of Laplacian problems, *J. Comput. Appl. Math.*, 28:63–83.

Papamichael, N. and Whiteman, J. R. (1973), A numerical conformal transformation method for harmonic mixed boundary value problems in polygonal domains, *Z. Angew. Math. Phys.*, 24:304–316.

Pathria, D. and Karniadakis, G. E. (1995), Spectral element methods for elliptic problems in nonsmooth domains, *J. Comp. Phys.*, 122:83–95.

Pekhlivanov, A. I., Lazarov, R. D., Carey, G. F. and Chow, S. S. (1992), Superconvergence analysis of approximate boundary-flux calculations, *Numer. Math.*, 63:483–501.

Percell, P. and Wheeler, M. F. (1978), A local residual finite element procedure for elliptic equations, *SIAM J. Numer. Anal.*, 15:705–714.

Percell, P. and Wheeler, M. F. (1980), A C^1 finite element collocation method for elliptic equations, *SIAM J. Numer. Anal.*, 17:605–622.

Pereyra, V., Proskurowski, W. and Widlund, O. (1977), High order fast Laplace Solvers for the Dirichlet problem on general regions, *Math. Comp.*, 31(137):1–16.

Petersdorff, T. V. and Stephan, E. P. (1990), Decompositions in edge and corner singularities for the solution of the Dirichlet problem of the Laplacian in a polyhedron, *Math. Nachr.*, 149:71–104.

Phillips, C. and Delves, L. M. (1980), A fast implementation of the global element calculations, *J. Inst. Math. Appl.*, 25:177–197.

Phillips, T. N. (1986), On the numerical treatment of boundary singularities in elliptic problems, *J. Comp. Phys.*, 64(2):459–472.

Phillips, T. N. and Davies, A. R. (1988), On semi-infinite spectral elements for Poisson problems with re-entrant boundary singularities, *J. Comp. Appl. Math.*, 21(2):173–188.

Piltner, R. (1985), Special finite elements with holes and internal cracks, *Inter. J. Numer. Methods Engrg.*, 21(8):1471–1485.

Pitkaranta, J. (1979), Boundary subspaces for the finite element method with Lagrange multipliers, *Numer. Math.*, 33:273–289.

Pitkaranta, J. (1981), The finite element method with Lagrange multipliers for domains with corners, *Math. Comp.*, 37(155):13–30.

Quarteroni, A. and Valli, A. (1994), *Numerical Approximation of Partial Differential Equations*, Springer-Verlag, Berlin.

Quarteroni, A. and Zampieri, E. (1992), Finite element preconditioning for Legendre spectral collocation applications to elliptic equations and systems, *SIAM J. Numer. Anal.*, 29:917–936.

Rangogni, R. (1986), Numerical solution of the generalized Laplace equation by coupling the boundary element method and the perturbation method, *Appl. Math. Model.*, 10(4):266–270.

Rank, E. (1989), Adaptive h and hp versions for boundary integral element methods, *Inter. J. Numer. Methods. Engrg.*, 28(6):1335–1349.

Rapley, C. W. (1985), Turbulent flow in a duct with cusped corners, *Inter. J. Numer. Methods Fluids*, 5:155–167.

Raviart, P. A. and Thomas, J. M. (1977), Primal hybrid finite element methods for 2nd order elliptic equations, *Math. Comp.*, 31(138):391–413.

Reitich, F. (1991), Singular solutions of a transmission problem in plane linear elasticity for wedge-shaped region, *Numer. Math.*, 59:179–216.

Rektoys, K. (1975), *Variational Methods in Mathematical Sciences and Engineering*, D. Reidel Publishing Company.

Richtmyer, R. D. and Morton, K. W. (1967), *Difference Methods for Initial-Value Problems*, Interscience Publishers, New York.

Ritz, W. (1909), Über eine nene Methode zur Lösung gewissen Variations-Probleme der mathematischen Physik, *J. Reine Angew. Math.*, 135:1–61.

Roberts, J. E. and Thomas, J. M. (1991), Mixed and hybrid methods, in Ciarlet, P. G. and Lions, J. L., editors, *Finite Element Methods(Part I.)*, 523–640, North-Hollaud.

Roos, H. G., Stynes, M. and Tobiska, L. (1996), *Numerical Methods for Singularly Perturbed Differential Equations: Convection-Diffusion and Flow Problems*, Springer–Verlag, Berlin.

Rosser, J. B. and Papamichael, N. (1975), A power series solution of a harmonic mixed boundary value problem, MRC, Technical report, University of Wisconsin.

Rude, U. (1989), Local corrections for eliminating the pollution effect of reentrant corners, in *Proceedings of the Fourth Copper Mountain Conference on Multigrid Methods*, 365–382, SIAM, Philadelphia.

Ruotsalainen, K. (1987), On the boundary element method with mesh refinement on curves with corners, *J. Comp. Appl. Math.*, 20:373–378.

Russell, R. D. (1977), A comparison of collocation and finite differences for two point boundary value problems, *SIAM J. Numer. Anal.*, 14:19–39.

Russell, R. D. and Shampine, L. F. (1972), A collocation method for boundary value problems, *Numer. Math.*, 19:1–28.

Samarskii, A. A. and Andblev, B. B. (1976), *Difference Methods for Elliptic Equations*, Science Publication, Moscow.

Schatz, A. H. and Wahlbin, L. B. (1976), Maximum norm error estimates in the finite element method for Poisson equation on plane domains with corners, in *Approximation Theory II*, 541–547, Academic Press, New York.

Schatz, A. H. and Wahlbin, L. B. (1977), Interior maximum norm estimates for finite element methods, *Math. Comp.*, 31:414–442.

Schatz, A. H. and Wahlbin, L. B. (1978), Maximum norm estimates in the finite element method on plane polygonal domains, Part I, *Math. Comp.*, 32(141):73–109.

Schatz, A. H. and Wahlbin, L. B. (1979), Maximum norm estimates in the finite element method on plane polygonal domain, Part II, *Math. Comp.*, 33(146):465–492.

Schatz, A. H. and Wahlbin, L. B. (1995), Interior maximum-norm estimates for finite element methods, Part II, *Math. Comp.*, 211:907–928.

Schatz, A. H. and Wang, J. (1996), Some new error estimates for Ritz-Galerkin methods with minimal regularity assumptions, *Math. Comp.*, 65:19–27.

Schiff, B. (1988), Finite element eigenvalues for the Laplacian over an L shaped domain, *J. Comp. Phys.*, 76(2):233–242.

Schiff, B., Fishelov, D. and Whiteman, J. R. (1979), Determination of a stress intensity factor using local mesh refinement, in Whiteman, J. R., editor, *MAFELAP 1978*, 55–64, Academic Press, London.

Schmitz, H., Volk, K. and Wendland, W. (1993), Three-dimensional singularities of elastic fields near vertices, *Numer. Meth. for PDE*, 9:323–337.

Schwarz, H. A. (1869), Uber einige Abbildungsaufgaben. *Ges. Math. Abh.*, 11:65–83.

Schwarz, H. R. (1989), *Numerical Analysis, A Comprehensive Introduction*, John Wiley, Chichester.

Scott, R. (1976), Optimal L sup-inf estimates for the finite element method on irregular meshes, *Math. Comp.*, 30:681–697.

Seager, M. K. and Carey, G. F. (1990), Adaptive domain extension and adaptive grids for unbounded spherical elliptic PDEs, *SIAM J. Sci. Statis. Comp.*, 11(1):92–111.

Sewell, G. (1988), *The Numerical Solution of Ordinary and Partial Differential Equations*, Academic Press.

Sgallari, F. (1985), Primal-dual variational problems by boundary and finite elements, *Appl. Math. Model.*, 9(4):246–252.

Sheldon, J. W. (1958), Algebraic approximations for Laplace's equation in the neighborhood of interfaces, *Math. Tables Aids Comput.*, 12:174–186.

Shi, Z. C. (1984), On the convergence rate of the boundary penalty method, *Inter. J. Numer. Methods Engrg.*, 20:2027–2032.

Shi, Z. C. (1985), Convergence properties of two nonconforming finite elements, *Comp. Meth. Appl. Engrg.*, 48:123–137.

Silvester, P. and Hsieh, M. S. (1971), Finite element solution of 2-dimensional exterior field problems, *Proc. IEE*, 118:1743–1747.

Smith, B., Bjorstad, P. and Gropp, W. (1996), *Domain Decomposition, Parallel Multilevel Methods for Elliptic Partial Differential Equations*, Cambridge University Press.

Smith, B. F. and Widlund, O. B. (1990), A domain decomposition algorithm using a hierarchical basis, *SIAM J. Sci, Stat. Comp.*, 11:1212–1220.

Sobolev, S. L. (1963), *Application of Functional Analysis in Mathematical Physics*, AMS, Providence.

Solecki, J. S. and Swedlow, J. L. (1984), On quadrature and singular finite elements, *Inter. J. Numer. Methods Engrg.*, 20:395–408.

Steger, K. (1990), Corner singularities of solutions of the potential equation in three dimensions, in Bank, R. E., Bulirsch, R. and Merten, K., editors, *Mathematical Modelling and Simulation of Electrical Circuits and Semiconductor Devices*, 283–297, Birkhauser, Basel.

Stenger, F. (1979), A Sinc-Galerkin method of solution of boundary value problems, *Math. Comp.*, 33(145):85–109.

Stenger, F. (1993), *Numerical Methods Based on Sinc and Analytic Functions*, Springer-

Verlag, New York.

Stephan, E. (1979), Conform and mixed finite element schemes for the Dirichlet problem for the bi-Laplacian in plane domains with corners, *Math. Meth. Appl. Sci.*, 1(3):354–382.

Stephan, E. (1988), Singularities of the Laplacian at corners and edges of three-dimensional domains and their treatment with finite element methods, *Math. Meth. Appl. Sci.*, 10:339–350.

Stephan, E. and Wendland, W. L. (1983), Boundary element method for membrane and torsion crack problems, *Comput. Methods Appl. Mech. Engrg.*, 36(3):331–358.

Stephan, E. P. (1986), A boundary integral equation method for three dimensional crack problems in elasticity, *Math. Meth. Appl. Sci.*, 8(4):609–623.

Stephan, E. P. (1992), Coupling of finite elements and boundary elements for some nonlinear interface problems, *Comput. Methods Appl. Mech. Engrg.*, 101:61–72.

Stephan, E. P. and Costabel, M. (1986), A boundary element method for three dimensional crack problems, in *Innovative Numerical Methods in Engineering*, 351–360, Comput. Mech., Southampton.

Stephan, E. P. and Wendland, W. L. (1984a), An augmented Galerkin procedure for the boundary integral method applied to two-dimensional screen and crack problems, *Applicable Analysis*, 18:183–219.

Stephan, E. P. and Wendland, W. L. (1984b), The boundary integral method for two-dimensional screen and crack problems, 3–18, Comput. Mech. Centre, Southampton.

Stephan, E. and Whiteman, J. R. (1988), Singularities of the Laplacian at corners and edges of three dimensional domains and their treatment with finite element methods, *Math. Meth. Appl. Sci.*, 10(3):339–350.

Stephen, E. P. and Suri, M. (1989), On the convergence of the p-version of the boundary element Galerkin method, *Math. Comp.*, 52:31–48.

Stevenson, R. (1993), Robustness of multi grid applied to anisotropic equations on convex domains and on domains with re-entrant corners, *Numer. Math.*, 66:373–398.

Strang, G. and Fix, G. J. (1973), *An Analysis of the Finite Element Method*, Prentice–Hall, Englewood Cliffs, New Jersey.

Straughan, B. (1992), *The Energy Method, Stability and Nonlinear Convection*, Springer–Verlag, New York.

Strikwerda, J. C. (1989), *Finite Difference Schemes and Partial Differential Equations*, Wadsworlh & Brook.

Summers, D. M., Hanson, T. and Wilson, C. B. (1985), A random vortex simulation of wind-flow over a building, *Inter. J. Numer. Meth. in Fluids*, 5:849–871.

Suli, E., Jovanovic, B. and Ivanovic, L. (1985), Finite difference approximations of generalized solutions, *Math. Comp.*, 45(172):319–327.

Swann, H. (1993), On the use of Lagrange multiplies in domain decomposition for solutioning elliptic problem, *Math. Comp.*, 60:49–78.

Swartz, B. and Wendroff, B. (1974), The relation between the Galerkin and collocation methods using smooth splines, *SIAM J. Numer. Anal.*, 11:994–996.

Symm, G. T. (1967), Numerical mappings of exterior domains, *Numer. Math.*, 10:437–445.

Tang, W. P. (1989), Relief from the pain of overlap, Generalized Schwarz Splitting. Tech. Report, Dept. of Computer Science. Waterloo University.

Thatcher, R. W. (1975), Singularities in the solution of Laplace's equation in two dimensions, *J. Inst. Math. Appl.*, 16(3):303–319.

Thatcher, R. W. (1976), The use of infinite grid refinement at singularities in the solution of Laplace's equation, *Numer. Math.*, 25:163–178.

Thatcher, R. W. (1978), On the finite element method for unbounded regions, *SIAM J. Numer. Anal.*, 15(3):466–477.

Theocaris, P. S., Tsamasphyros, G. and Theotokoglou, E. E. (1984), A combination of the finite element and singular-integral equation methods for the solution of the generally cracked body, *Inter. J. Numer. Methods Engrg.*, 20:2065–2075.

Thomas, J. M. (1992), Finite element matching methods, in *Fifth Inter. Symposium on Domain Decomposition Methods for PDEs*, 99–105, SIAM, Philadelphia.

Thomas, J. W. (1995), *Numerical Partial Differential Equations: Finite Difference Methods*, Springer–Verlag, New York.

Thomée, V. (1990), Finite difference methods for linear parabolic equations, in Ciarlet, P. G. and Lions, J. L., editors, *Handbook of Numerical Analysis, Vol. 1*, 5–196, North–Holland.

Thompson, G. M. and Whiteman, J. R. (1985), Analysis of strain representation in linear elasticity by both singular and nonsingular finite elements, *Numer. Meth. for PDE*, 2:85–104.

Tikhonov, A. N. and Samarskii, A. A. (1973), *Equations of Mathematical Physics*, The MacMillan Company, New York.

Tong, P., Pian, T. H. H. and Lasry, S. J. (1973), A hybrid element approach to crack problems in plane elasticity, *Inter. J. Numer. Methods Engrg.* 7:297–308.

Trefftz, E. (1926), Ein gegenstuck zum Ritz'schen Verfahren, *Proc. 2nd Int. Congr. Appl. Mech., Zürch.*

Tsamasphyros, G. and Giannakopoulos, A. E. (1985), The mapped elements for solution of cracked bodies, *Comput. Methods Appl. Mech. Engrg.*, 49:331–342.

Utku, M. and Carey, G. F. (1982), Boundary penalty techniques, *Comp. Meth. Appl. Mech. Engrg.*, 30:103–118.

Varga, R. S. (1962), *Matrix Iterative Analysis*, Prentice-Hall Inc., Englewood Cliffs, New York.

Vasilopoulos, D. (1988), On the determination of higher order terms of singular elastic stress fields near corners, *Numer. Math.*, 53(1-2):51–95.

Veidinger, L. (1968), On the order of convergence of finite-difference approximations to the solution of the Dirichlet problem in a domain with corners, *Studia-Sci. Math. Hungar.*, 3:337–343.

Veidinger, L. (1978), On the order of convergence of finite element methods for the Neumann problem, *Stud. Sci. Math. H.*, 13:411–422.

Vekua, I. N. (1967), *New Methods for Solving Elliptic Equation*, North–Holland, Amsterdam.

Vorozhtsov, E. V. and Yanenko, N. N. (1990), *Methods for the Localization of Singularities in Numerical Solutions of Gas Dynamics Problems*, Springer-Verlag, New York.

Wahlbin, L. B. (1984), On the sharpness of certain local estimates for H^1 projections into finite element spaces:influence of a reentrant corner, *Math. Comp.*, 42(165):1–8.

Wahlbin, L. B. (1991), Local behavior in finite element method in Ciarlet, P. G. and Lions, J. L., editors, *Finite Element Methods I*, 353–522, North–Holland.

Wahlbin, L. B. (1995), *Superconvergence in Galerkin Finite Elements*, Springer, Berlin.

Wait, R. (1977), Singular isoparametric finite elements, *J. Inst. Math. Appl.*, 20:133–141.

Wait, R. (1978), Finite element methods for elliptic problems with singularities, *Comput. Methods Appl. Mech. Engrg.*, 13:141–150.

Wait, R. (1979), A curved macro-element for cracks and corners, *J. Inst. Math. Appl.*, 24:471–480.

Wait, R. and Mitchell, A. R. (1971), Corner singularities in elliptic by finite element methods, *J. Comp. Phys.*, 8:45–52.

Wathen, A. J. and Baines, M. J. (1985), On the structure of the moving finite-element equations, *IMA J. Numer. Anal.*, 5(2):161–182.

Watson, G. N. (1944), *A Treatise on the Theory of Bessel Functions (2nd Edition)*, Cambridge University Press, Cambridge.

Wendland, W. L. (1983), Boundary element methods and their asymptotic convergence, in *Theoretical Acoustics and Numerical Techniques*, 135–216, Springer, Vienna.

Wendland, W. L. (1988), On asymptotic error estimates for combined BEM and FEM, in *Finite Element and Boundary Element Techniques from Mathematical and Engineering Point of View*, 273–333, Springer, Vienna.

Wendland, W. L. (1990), On the coupling of finite elements and boundary elements, in *Discretization Methods in Structural Mechanics (Vienna, 1989)*, 405–414, Springer, Berlin.

Wendland, W. L., Schmitz, H. and Bumb, H. (1988), A boundary element method for three dimensional elastic fields near reentrant corners, in Whiteman, J. R., editor, *The Mathematics of Finite Elements and Applications VI*, 313–322, Academic Press, London-New

York.

Wendland, W. L. and Stephan, E. P. (1990), A hypersingular boundary integral method for two dimensional screen and crack problems, *Arch. Rational Mech. Anal.*, 112:363–390.

Wheeler, M. F. (1977), A C^0-collocation-finite element method for two-point boundary value problems and one space dimensional parabolic problems, *SIAM J. Numer. Anal.*, 14:71–90.

Wheeler, M. F. and Whiteman, J. R. (1987), Superconvergent recovery of gradients on subdomains from piecewise linear finite element approximations, *Numer. Meth. for PDE*, 3:357–374.

Whiteman, J. R. (1968), Treatment of singularities in a harmonic mixed boundary value problem by dual series methods, *Quart. J. Mech. Appl. Math.*, 21:41–50.

Whiteman, J. R. (1970), Numerical solution of a harmonic mixed boundary-value problem by the extension of a dual series method, *Quart. J. Mech. Appl. Math.*, 23:449–455.

Whiteman, J. R. (1974), Galerkin methods for two dimensional elliptic problems containing boundary singularities, in *Proceedings of the C. Caratheodory Inter. Symposium*, 629–634, Greek Math. Soc., Athens.

Whiteman, J. R. (1978), Numerical methods for elliptic problems containing boundary singularities, in *Proceedings of the Fourth Symposium on Basic Problems of Numerical Mathematics*, 173–181, Charles Univ., Prague.

Whiteman, J. R. (1981), Finite element methods for elliptic problems containing boundary singularities, in *Numerical Treatment of Differential Equations, Vol. 3*, 199–206, Birkhauser, Basel.

Whiteman, J. R. (1982), Finite elements for singularities in two- and three-dimensions, in Whiteman, J. R., editor, *The Mathematics of Finite Elements and Applications IV*, 37–55, Academic Press, London.

Whiteman, J. R. (1984), Singularities in three-dimensional elliptic problems and their treatment with finite element methods, in Griffiths, D. F. and Watson, G. A., editors, *Numerical Analysis (Dundee, 1983)*, 264–275, Springer, Berlin-New York.

Whiteman, J. R. (1986), Singularities in two and three dimensional elliptic problems and finite element methods for their treatment, in Vosmansky, J. and Zlamal, M., editors, *EQUADIFF 6*, 345–352, Springer, Berlin-New York.

Whiteman, J. R. (1987), Finite element methods for treating problems involving singularities, with applications to linear elastic fracture, in Ortiz, E. L., editor, *Numerical Approximation of Partial Differential Equations*, 109–120, North-Holland, Amsterdam.

Whiteman, J. R. (1991), Calculation of the forms of singularities in elliptic boundary value problems, in Albrecht, J., et. al, editors, *Numerical Treatment of Eigenvalue Problems, Vol. 5*, 237–243, Birkhauser, Basel.

Whiteman, J. R. and Akin, J. E. (1980), Finite elements, singularities and fracture, in Whiteman, J. R., editor, *The Mathematics of Finite Elements and Applications III*, 35–54, Academic Press, London.

Whiteman, J. R. and Barnhill, R. E. (1973), Finite element methods for elliptic mixed boundary value problems containing singularities, in *Proceedings of EQUADIFF 3*, 261–267, Folia Fac. Sci. Natur. Univ. Purkynianae Brunensis, Ser. Monograph., Tomus 1, Purkyne Univ. and Brno.

Whiteman, J. R. and Goodsell, G. (1988), On some finite element error estimates for stress intensity factors in mode RmI linear elastic fracture problems, in *Transactions of the Fifth Army Conference on Applied Mathematics and Computing*, 541–548, U.S. Army Res. Office, Research Triangle Park, NC.

Whiteman, J. R. and Goodsell, G. (1991), A survey of gradient superconvergence for finite element approximations to second order elliptic problems on triangular and tetrahedral meshes, in Whiteman, J. R., editor, *MAFELAP 1990*, 55–74, Academic Press, London.

Whiteman, J. R. and Papamichael, N. (1972), Treatment of harmonic mixed boundary problems by conformal transformation methods, *ZAMP*, 23:655–664.

Whiteman, J. R. and Schleicher, K. T. (1984), Introduction to the treatment of singularities in elliptic boundary value problems using finite element techniques, in *The Mathematical*

    Basis of Finite Element Methods, 169–183, Oxford Univ. Press, New York.

Widlund, O. B. (1997), Preconditioners for spectral and Mortar finite element methods, in Glowinski, R., Periaux, J., Shi, Z. C. and Wildlund, O., editors, *Domain Decomposition Methods in Sciences and Engineering*, pp. 19–32, John Wiley & Sons, Chichester.

Wigley, N. M. (1964), Asymptotic expansions at a corner of solutions of mixed boundary value problems, *J. Math. Mech.*, 4:549–576.

Wigley, N. M. (1968), On a method to subtract off a singularity at a corner for the Dirichlet or Neumann problem, *Math. Comp.*, 23:395–401.

Wigley, N. M. (1987), Stress intensity factors and improved convergence estimates at a corner, *SIAM J. Numer. Anal.*, 24:350–354.

Wigley, N. M. (1988), An efficient method for subtracting off singularities at corners for Laplace's equation, *J. Comp. Phys.*, 78:369–377.

Wolf, J. P. and Song, C. (1996), *Finite-Element Modelling of Unbounded Media*, John Wiley and Sons.

Wu, X. and Han, H. (1997), A finite–element method for Laplace- and Helmholtz boundary value problems with singularities, *SIAM J. Numer. Anal.*, 34:1037–1050.

Yanik, E. G. (1989), A Schwarz alternating procedure using spline collection methods, *Inter. J. Numer. Methods Engrg.*, 28:621–627.

Ying, L. A. (1978), The infinite similar element method for calculating stress intensity factors, *Scientia Sinica*, 21:19–43.

Ying, L. A. (1982), A note on the singularity and the strain energy of singular elements, *Inter. J. Numer. Methods Engrg.*, 18(1):31–39.

Ying, L. A. (1986), Infinite element approximation to axial symmetric Stokes flow, *J. Comp. Math.*, 4(2):111–120.

Ying, L. A. (1991), Infinite element method for elliptic problems, *Science in China (Scientia Sinica). Series A. Mathematics, Physics, Astronomy and Technological Sciences*, 34(12):1438–1447.

Yoseph, P. B. and Israeli, M. (1986), Asymptotic finite element method for boundary value problems, *Inter. J. Numer. Methods Fluids*, 6:21–34.

Yserentant, H. (1986a), The convergence of multilevel methods for solving finite element equations in the presence of singularities, *Math. Comp.*, 47(176):399–409.

Yserentant, H. (1986b), On the multi-level splitting of finite element spaces, *Numer. Math.*, 49:379–412.

Yu, D. H. (1985), Approximation of boundary conditions at infinity for a harmonic equation, *J. Comp. Math.*, 3(3):219–227.

Yu, D. H. (1987), A posteriori error estimates and adaptive approaches for some boundary element methods, in *Boundary Elements IX, Vol. 1*, 241–256, Comput. Mech., Southampton.

Yu, D. H. (1993), *Mathematical Theory of National Boundary Element Methods(in Chinese)*, Academic Publishing, Beijing.

Yu, D. H. (1996), A domain decomposition method based on natural boundary reduction over unbounded domain(in Chinese), *Math. Numer. Sinica*, 16:448–459.

Zang, T. A., Wong, Y. S. and Hussaini, M. Y. (1982), Spectral multigrid methods for elliptic equations, *J. Comp. Phys.*, 48:485–501.

Zenger, C. and Gietl, H. (1978), Improved difference schemes for the Dirichlet problem of Poisson's equation in the neighborhood of corners, *Numer. Math.*, 30:315–332.

Zhang, S. (1990), Optimal order nonnested multigrid methods for solving finite element equations II, on nonquasiuniform meshes, *Math. Comp.*, 55(192):439–450.

Zhou, G. H. and Rannacher, R. (1996), Pointwise superconvergence of the streamline diffusion finite element method, *Numer. Meth. PDE*, 12:123–145.

Zhou, T. X. and Feng, M. F. (1993), A least squares Petrov-Galerkin finite element method for stationary Navier-Stokes equations, *Math. Comp.*, 60:331–543.

Zhu, C. D. and Lin, Q. (1989), *The Hyperconvergence Theory of Finite Elements* (in Chinese), Human Science and Technology Publishing.

Zielinski, A. P. (1988), Trefftz method: elastic and elastoplastic problems, *Comput. Methods*

Appl. Mech. Engrg., 69:185–204.

Zielinski, A. P. and Zienkiewicz, O. C. (1985), Generalized finite element analysis with T-complete boundary solution functions, *Inter. J. Numer. Methods Engrg.*, 21:509–528.

Zienkiewicz, O. C. (1977), *The Finite Element Method*, McGraw-Hill, Third edition, London.

Zienkiewicz, O. C. (1984), Coupling problems and their numerical solution, in Lewis, R. W., editor, *Numerical Methods in Coupled Systems*, A Wiley-Interscience Publication.

Zienkiewicz, O. C. (1990), Where now finite elements, in Whiteman, J. R., editor, *MAFEALP*, Academic Press, London.

Zienkiewicz, O. C., Bando, K., Bettess, P., Emson, C. and Chiam, T. C. (1985), Mapping infinite elements for exterior wave problems, *Inter. J. Numer. Meth. Engrg.*, 21:1229–1251.

Zienkiewicz, O. C. and Cheung, Y. K. (1967), *The Finite Element Method in Structural and Continuum Mechanics*, McGraw-Hill Publishing.

Zienkiewicz, O. C., Emson, C. and Bettess, P. (1983), A novel boundary infinite element, *Inter. J. Numer. Meth. Engrg.*, 19:393–404.

Zienkiewicz, O. C., Gallagher, R. H. and Lewis, R. W. (1994), International Journal for numerical methods in engineering: The first 25 years and the future, *Inter. J. Numer. Methods Engrg.*, 37:2151–2158.

Zienkiewicz, O. C., Kelley, D. W. and Bettess, P. (1977), The coupling of the finite element method and boundary solution procedures, *Inter. J. Numer. Methods Engrg.*, 11(2):355–375.

Zienkiewicz, O. C. and Morgan, K. (1983), *Finite Elements and Approximation*, John Wiley & Sons, New York.

GLOSSARY OF SYMBOLS

BAM : *the boundary approximateion method.*

BEM : *the boundary element method.*

CM : *the collocation method.*

DDM : *the domain decomposition method.*

FDM : *the finite difference method.*

FEM : *the finite element method.*

FVM : *the finite volume method.*

LSM : *the least squares method.*

RGM : *the Ritz − Galerkin method.*

SAM : *the Schwarz alternating method.*

C : *a positive bounded constant.*

D_l : *the expansion coefficients.*

$\widetilde{D_l}$: *the approximate coefficient of D_l.*

P_c : *the penalty constant.*

S : *the solution domain.*

S^+ : *the subdomain of S possibly involving solution singularities.*

S^- : *another subdomain of S with smooth solutions.*

$S_h = (\cup_{ij} \square_{ij}) \cup (\cup_{ij} \triangle_{ij})$: *the partition of S.*

$\widehat{S_1^h}$: *the triagulation region of S_1 so that $\widehat{S_1^h} \approx S_1$.*

u : *the true solution.*

u_I : *the interpolant of u.*

$u_h, \widetilde{u_h}$: *the approximate solutions.*

u_h^N, u_h^P, u_h^C : *the numerical solutions by combinations.*

V_h : *a finite dimensional subspace, $V_h \subset H^1(S)$.*

V_h^0 : *a finite dimensional subspace, $V_h^0 \subset H_1^0(S)$.*

471

w : *the weight costant.*

Π_p : *the interpolation operator by combinations.*

Γ : *the exterior boundary of the solution domain.*

Γ_D : *the exterior boundary with the Dirichlet boundary condition.*

Γ_M : *the exterior boundary with the Robin boundary condition.*

Γ_N : *the exterior boundary with the Neumann boundary condition.*

Γ_0 : *the interface boundary, artificial or material.*

Ω_I, Ω_{II} : *the subregions in the domain decomposition method.*

α, β : *coupling parameters.*

σ : *penalty power.*

λ : *Lagrange multiplier.*

$\lambda_{\max}$: *the maximal eigenvalue.*

$\lambda_{\min}$: *the minmal eigenvalue.*

$\hat{\Box}, \hat{\triangle}$: *the reference rectangle and triangle.*

$\widehat{\int}, \widehat{\iint}$: *approximation of the integrals* $\int$, $\iint$.

$u = f_L + R_L$: *the solution expansion.*

$f_L = \displaystyle\sum_{i=1}^{L} a_i \phi_i$: *the leading terms.*

$R_L = \displaystyle\sum_{i=L+1}^{\infty} a_i \phi_i$: *the remainder.*

$Cond.(\mathbf{B}) = \left\{ \dfrac{\lambda_{Max}(\mathbf{B})}{\lambda_{Min}(\mathbf{B})} \right\}$: *the condition number of the associated matrix* $\mathbf{B}$.

$I_\mu(r) = \displaystyle\sum_{k=0}^{\infty} \dfrac{1}{\Gamma(k+1)\Gamma(k+\mu+1)} \left(\dfrac{r}{2}\right)^{2k+\mu}$: *the Bessel functions for a purely imaginary argument.*

$K_n(\rho) = \dfrac{1}{2} \displaystyle\int_{-\infty}^{\infty} e^{-\rho\cosh\eta - n\eta} d\eta$: *the Hankel functions for a purely imaginary argument.*

$L + 1 = O(|\ln h|)$: *the coupling relation in combinations.*

$span\{\psi_i\}$: *spanned by the basis functions* ψ_i.

$u_\nu = \dfrac{\partial u}{\partial \nu}$: *the normal derivative of u.*

$\Delta = \dfrac{\partial^2}{\partial x^2} + \dfrac{\partial^2}{\partial y^2}$: *the Laplace operator.*

$\Delta_\epsilon = \epsilon^+ - \epsilon^-$: *the difference of ϵ on the interface from both sides.*

$\overline{|u|}_{1,\square_{ij}}, \overline{|u|}_{1,\triangle_{ij}}$: *the norms involing discrete summation.*

$\overline{|u|}_{0,\square_{ij}}, \overline{|u|}_{0,\triangle_{ij}}$: *the norms involing discrete summation.*

$\max\limits_S |\epsilon| = \|\epsilon\|_{\infty,S}.$

$H^1(S) = \{v, \ v_x, \ v_y \in L^2(S)\}.$

$H_0^1(S) = \{v, \ v_x, \ v_y \in L^2(S), \ and \ v|_{\Gamma_D} = 0\}.$

$H_*^1(S) = \{v, \ v_x, \ v_y \in L^2(S), \ and \ v|_{\Gamma_D} = g\}.$

$u^+ = u|_{S+} \ or \ u^+ = u|_{S_2}.$

$u^- = u|_{S-} \ or \ u^- = u|_{S_1}.$

$$\|v\|_{l,\partial S} = \left[\sum_{|\alpha|\leq l} \int_{\partial S} \left(\frac{\partial^\alpha v}{\partial s^\alpha} \right)^2 d\ell \right]^{\frac{1}{2}}.$$

$$|v|_{l,\partial S} = \left[\sum_{|\alpha|= l} \int_{\partial S} \left(\frac{\partial^\alpha v}{\partial s^\alpha} \right)^2 d\ell \right]^{\frac{1}{2}}.$$

$$\|v\|_{m,S} = \left\{ \sum_{|\alpha|\leq m} \int_S |D^\alpha v|^2 dx \right\}^{\frac{1}{2}}.$$

$$|v|_{m,S} = \left\{ \sum_{|\alpha|= m} \int_S |D^\alpha v|^2 dx \right\}^{\frac{1}{2}}.$$

$\|v\|_1 = \left\{ \|v\|_{1,S+}^2 + \|v\|_{1,S-}^2 \right\}^{\frac{1}{2}}.$

$|v|_1 = \left\{ |v|_{1,S+}^2 + |v|_{1,S-}^2 \right\}^{\frac{1}{2}}.$

$|v|_B = [v, v]^{\frac{1}{2}}.$

$\|u\|_1 = \left(\|u\|_{1,S_1}^2 + \|u\|_{1,S_2}^2 \right)^{\frac{1}{2}} \ or$

$\|u\|_1 = \left(\|u\|_{1,\widehat{S}_1^h}^2 + \|u\|_{1,S_2}^2 \right)^{\frac{1}{2}}.$

$$\|u\|_h = \left(\|u\|_{1,S_1}^2 + \|u\|_{1,S_2}^2 + \frac{1}{h^\sigma} \|u^+ - u^-\|_{0,\Gamma_0}^2 \right)^{\frac{1}{2}} \ \text{or}$$

$$\|u\|_h = \left(\|u\|_{1,S_1}^2 + \|u\|_{1,S_2}^2 + \frac{P_c}{h^\sigma} \|u^+ - u^-\|_{0,\Gamma_0}^2 \right)^{\frac{1}{2}}.$$

$$\|u\|_H = \left(\|u\|_{1,S_1}^2 + \|u\|_{1,S_2}^2 + \|\frac{\partial u}{\partial n}\|_{-\frac{1}{2},\Gamma_0}^2 \right)^{\frac{1}{2}}.$$

$$\|u\|_{h,P_c} = \left(\|u\|_{1,S_1}^2 + \|u\|_{1,S_2}^2 + \frac{P_c}{h^\sigma} \|u^+ - u^-\|_{0,\Gamma_0}^2 \right)^{\frac{1}{2}}.$$

$$\|u\|_{h,1} = \left(\|u\|_{1,S_1}^2 + \|u\|_{1,S_2}^2 + \frac{1}{h^\sigma} \|u^+ - u^-\|_{0,\Gamma_0}^2 \right)^{\frac{1}{2}}.$$

$$\overline{\|v^+ - v^-\|}_{0,\Gamma_0} = \left(\widehat{\int_{\Gamma_0}} (v^+ - v^-)^2 d\ell \right)^{\frac{1}{2}}.$$

$$\|v\|_{\frac{1}{2},\Gamma_0} = \left(\|v\|_{0,\Gamma_0}^2 + \int_{\Gamma_0} \int_{\Gamma_0} \frac{(v(P) - v(Q))^2}{(P-Q)^2} d\ell(Q) d\ell(P) \right)^{\frac{1}{2}}.$$

$$\|u\|_{-\frac{1}{2},\Gamma_0} = \sup_v \frac{\left| \int_{\Gamma_0} uv \, d\ell \right|}{\|v\|_{\frac{1}{2},\Gamma_0}}.$$

$$\overline{\|u\|}_{1,S_1} = \left(\sum_{ij} \overline{\|u\|}_{1,\square_{ij}}^2 + \overline{\|u\|}_{1,\triangle_{ij}}^2 \right)^{\frac{1}{2}}.$$

$$\overline{|u|}_{1,S_1} = \left(\sum_{ij} \overline{|u|}_{1,\square_{ij}}^2 + \overline{|u|}_{1,\triangle_{ij}}^2 \right)^{\frac{1}{2}}.$$

$$\overline{\|u\|}_1 = \left(\overline{\|u\|}_{1,S_1}^2 + \|u\|_{1,S_2}^2 \right)^{1/2}.$$

$$\overline{|u|}_1 = \left(\overline{|u|}_{1,S_1}^2 + |u|_{1,S_2}^2 \right)^{1/2}.$$

$$\overline{\|u\|}_h = \left(\overline{\|u\|}_{1,S_1}^2 + \|u\|_{1,S_2}^2 + \frac{P_c}{h^\sigma} \overline{\|u^+ - u^-\|}_{0,\Gamma_0}^2 \right)^{1/2}.$$

$$\|u\|_h^* = \left(\|u\|_{1,S_1}^2 + \|u\|_{1,S_2}^2 + \frac{P_c}{h^\sigma} \overline{\|u^+ - u^-\|}_{0,\Gamma_0}^2 \right)^{1/2}.$$

INDEX

Other *Mathematics and Its Applications* titles of interest:

W.L. Miranker: *Numerical Methods for Stiff Equations and Singular Perturbation Problems*. 1980, 220 pp. ISBN 90-277-1107-0

K. Rektorys: *The Method of Discretization in Time and Partial Differential Equations*. 1982, 470 pp. ISBN 90-277-1342-1

L. Ixary: *Numerical Methods and Differential Equations and Applications*. 1984, 360 pp. ISBN 90-277-1597-1

B.S. Razumikhin: *Physical Models and Equilibrium Methods in Programming and Economics*. 1984, 368 pp. ISBN 90-277-1644-7

A. Marciniak: *Numerical Solutions of the N-Body Problem*. 1985, 256 pp.
ISBN 90-277-2058-4

Y. Cherruault: *Mathematical Modelling in Biomedicine*. 1986, 276 pp.
ISBN 90-277-2149-1

C. Cuvelier, A. Segal and A.A. van Steenhoven: *Finite Element Methods and Navier-Stokes Equations*. 1986, 500 pp. ISBN 90-277-2148-3

A. Cuyt (ed.): *Nonlinear Numerical Methods and Rational Approximation*. 1988, 480 pp. ISBN 90-277-2669-8

L. Keviczky, M. Hilger and J. Kolostori: *Mathematics and Control Engineering of Grinding Technology. Ball Mill Grinding*. 1989, 188 pp. ISBN 0-7923-0051-3

N. Bakhvalov and G. Panasenko: *Homogenisation: Averaging Processes in Periodic Media. Mathematical Problems in the Mechanics of Composite Materials*. 1989, 404 pp. ISBN 0-7923-0049-1

R. Spigler (ed.): *Applied and Industrial Mathematics*. Venice-1, 1989. 1991, 388 pp. ISBN 0-7923-0521-3

C.A. Marinov and P. Neittaanmaki: *Mathematical Models in Electrical Circuits. Theory and Applications*. 1991, 160 pp. ISBN 0-7923-1155-8

Z. Zlatev: *Iterative Improvement of Direct Solutions of Large and Sparse Problems*. 1991, 328 pp. ISBN 0-7923-1154-X

M. Vajtersic: *Algorithms for Elliptic Problems*. 1992, 352 pp.
ISBN 0-7923-1918-4

V. Kolmanovskii and A. Myshkis: *Applied Theory of Functional Differential Equations*. 1992, 223 pp. ISBN 0-7923-2013-1

A.D. Egorov, P.I. Sobolevsky and L.A. Yanovich: *Functional Integrals: Approximate Evaluation and Applications*. 1993, 426 pp. ISBN 0-7923-2193-6

G.I. Marchuk: *Adjoint Equations and Analysis of Complex Systems*. 1994
ISBN 0-7923-3013-7

Other *Mathematics and Its Applications* titles of interest:

A. Bakushinsky and A. Goncharsky: *Ill-Posed Problems: Theory and Applications.*
1994, 256 pp.   ISBN 0-7923-3073-0

G.N. Milstein: *Numerical Integration of Stochastic Differential Equations.* 1995,
170 pp.   ISBN 0-7923-3213-X

V.V. Shaidurov: *Multigrid Methods for Finite Elements.* 1995, 331 pp.
                                                                 ISBN 0-7923-3290-3

A.N. Tikhonov, A.V. Goncharsky, V.V. Stepanov and A.G. Yagola: *Numerical
Methods for the Solution of Ill-Posed Problems.* 1995, 264 pp.
                                                                 ISBN 0-7923-3583-X

H.W. Engl, M. Hanke and A. Neubauer: *Regularization of Inverse Problems.* 1996,
328 pp.   ISBN 0-7923-4157-0

G.I. Marchuk: *Mathematical Modelling of Immune Response in Infectious Diseases.* 1997, 358 pp.   ISBN 0-7923-4528-2

V. Feinberg, A. Levin and E. Rabinovich: *VLSI Planarization.* 1997, 190 pp.
                                                                 ISBN 0-7923-4510-X

S.L. Sobolev[†] and V.L. Vaskevich: *The Theory of Cubature Formulas.* 1997,
438 pp.   ISBN 0-7923-4631-9

Zi Cai Li: *Combined Methods for Elliptic Equations with Singularities, Interfaces
and Infinities.* 1998, 480 pp.   ISBN 0-7923-5084-7